Triple Canopy

Triple Canopy

A Warrior's Journey from Grenada to Iraq

Patrick O'Kelley

Triple Canopy: A Warrior's Journey from Grenada to Iraq
by Patrick O'Kelley

ISBN 978-1-956904-03-1

Printed in the United States of America

Published by Blacksmith LLC
Fayetteville, North Carolina

www.BlacksmithPublishing.com

Direct inquiries and/or orders to the above web address.

iv

To those who ran to the sound of the guns and didn't return:

1LT Robert Breitmayer, A Company, 2nd Battalion, 325th Infantry, 82nd Airborne Division, killed in Operation Gallant Eagle, 1982

CPT Michael Ritz, B Company, 2nd Battalion, 325th Infantry, 82nd Airborne Division, killed in action, Grenada, 1983

SSG Gary Epps, B Company, 2nd Battalion 325th Infantry, 82nd Airborne Division, killed in action, Grenada, 1983

PVT Wesley McDavid, 3rd Ranger Battalion, killed in training accident, 1984

PFC Russell Hobgood, 3rd Ranger Battalion, killed in training accident, 1985

SGT Sean O'Kelley, 3rd Ranger Battalion, killed 1987

SSG Gregory Fronius, 7th Special Forces Group, killed in action, El Salvador, 1987

SGT Franklin Dennis Winters, 1st Ranger Battalion, killed in training accident, 1987

SSG Christopher Cummings, 10th Special Forces Group, killed in training accident, 1990

SGT Leonard Russ, Group Support Company, 5th Special Forces Group, killed, Saudi Arabia, 1991

SGM Patrick Hurley, SFOD-D, killed in action, Iraq, 1991

SFC Robert Deeks, 5th Special Forces Group, killed by a land mine, Somalia, 1993

SFC Randy Shughart, SFOD-D, killed in action, Somalia, 3 October 1994

SFC Daniel Petithory, 5th Special Forces Group, killed in action, Afghanistan, 5 December 2001

vi

Contents

Preface

I have written several books over the years, all about the Revolutionary War in the Carolinas. Friends would ask me when I was going to write a book about what I did during my 20 years of service in the Army. I would tell them that I wasn't ready to sit down and write it just yet. What they didn't know was that I had already written it in raw form, as it happened. Six months before I joined the Army, I started to keep a diary. I did not know what I would see in my time in the service and I thought that it might be interesting enough to put in a book someday. At first, I did not know if I was going to stay in the Army for just the first four-year enlistment, but I kept writing in my diary, at least once a week. The diary turned into a second book, then a third, then a dozen. It went everywhere I went. I knew at times that I was breaking the rules… OPSEC is the nickname for it… operational security… but I continued to write anyway. During the first Gulf War I ran out of pages, but I found an Iraqi ledger in a bunker, so I was able to continue. Towards the end of my time in service, I would write out entries on a computer, print it out, and then glue it into the last book of the collection, book number 16. So much of what you will read here is not a 30-year-old memory of a 53-year-old man, but you will be reading the thoughts of a 21-year-old, or a 30-year-old, as they are happening.

I did not want to write down every single thing that happened in my military career, mainly because it would be too many pages and would bore the reader. So, I decided to condense it to three events and three units: the invasion of Grenada with the 82nd Airborne Division; a friendly fire death while in the 3rd Ranger Battalion; and Desert Storm with the 5th Special Forces Group. I wrote this as a historian, so I have included recollections and accounts of others who were there with me. There are many stories in this book that have never been told before. I also did not want to sugar coat what happened during each chapter. In the Army there are heroes and legends, but there are also scoundrels and cowards. Some of the latter went on to do great things, so to mention them in the book by their name would hurt their reputations. This book was not written to do that, but to tell what happened. To keep their identity concealed I decided to use a style that I have seen in 18th century manuscripts. I would only write the first letter of their last name, but keep the rest blank, such as SSG Smith now becoming SSG S____.

You, the reader, will also notice two distinct writing styles in this book. When I wrote of Grenada and Desert Storm, I wrote it as a historical narrative. However, when I wrote the story of the death of PFC Hobgood in the 3rd Ranger Battalion, the words have dialogue and seem to be written like a novel. This is due to my own version of therapy. I had a hard time dealing with Hobgood's death, so in the weeks after his funeral I wrote everything down as it happened, but I changed all the names and wrote it like a novel. Thirty years later I had to try to remember all the names and change them back again, but the words spoken by the people in the story were the ones actually spoken during that time.

Throughout the book you will also see drawings that I made while I was still in the Army. At one time I was called the "King of Locker Room Art" in Fort Bragg and my pictures were taped up all around the 82nd Airborne and Special Forces. Every now and then I would see a plagiarized mural, plaque or t-shirt with my art on it, but I didn't mind much.

There are many more stories I could tell you, the reader, of my time in the Army, such as the deadliest peacetime jump, Operation Gallant Eagle in 1982, where we had more men killed and wounded than we did when we went to war in Grenada a year later. Or the operation into Somalia a year before there was a "Blackhawk Down". Perhaps the anti-drug missions on the California and Mexican border, in the mountains of San Bernardino or Death Valley. There was the border surveillance of East Germany and Czechoslovakia borders during the Cold War, or the time I was part of a climbing team in the Himalayan Mountains in Pakistan, all the while keeping an ear out for Arabic speakers and the location of a new threat to America, a rich Saudi named Bin Laden. There was also a rescue mission to go into Sierra Leone and find a lost A-Team, and the training missions in Mali, all the while keeping an eye out for mercenary armies that may want to do to them what they had done in other countries, such as the war over the Blood Diamonds. There was much that I saw and did in 20 years, but this will be my only book. My love is history, and this is merely a detour from my passion. The rest are in those faded diaries, for my children to read if they want.

Patrick O'Kelley
Barbecue Township, NC
September 2021

Triple Canopy

noun

1.   Thickest jungle with vegetation growing at three levels, often reaching up more than 50 feet.

From "Urban Dictionary"

Triple Canopy

1.   Term referring to a soldier in the U.S. Army who wears the Special Forces, Ranger, and Airborne tabs on their left shoulder.
     a.   Only the biggest studs in the army bear the triple canopy.

None of us had ever heard of the island of Grenada or knew where it was. I remember "60 Minutes" doing a story about the Cubans building an airfield there, but we just made jokes about it. I remember one of the guys in the barracks joking about how we should grenade Grenada. Like most news stories, I didn't pay much attention.

Chapter 1: AIRBORNE - The Invasion of Grenada

The week before I went to war for the first time, I was incredibly busy. Prior to the Global War on Terror in 2001 there were different alert levels in the 82nd Airborne Division. These different alert levels were from DRF-1 to DRF-9, since there were nine battalions of infantry in the three brigades.[1] The unit that was on DRF-1 had to be able to recall all their troops within two hours and had to be able to deploy anywhere in the world within 18 hours. Every other level of readiness had a different amount of time to recall, such as six- or 12-hour recall. Only DRF-9 had the same recall time as the DRF-1 unit. This was because the DRF-9 unit was the support element for the DRF-1 unit, and they would help out load the primary unit. I always got a chuckle when the media would state that the 82nd Airborne was on alert. What they failed to realize is that there was always someone on alert at all times.

I was in A Company, 2nd Battalion, 325th Infantry Regiment and we assumed DRF-1 status on October 15th. Prior to that, the Company was busy preparing to assume that role. This entailed preparing Powers of Attorney and updating Wills and records. All of the soldiers living in the barracks, of which I was one, had to fill out personal property inventory sheets. We also had to pack up our

---

[1] "DRF" stood for Division Readiness Force.

duffel bags for DRF-1 and then inventory all the items in those duffle bags. This was to ensure that each soldier had exactly what he needed in case we went to war. I had done this many times in the four years that I had been at Fort Bragg, and we had never gone anywhere yet.

I had just finished reading a novel about Vietnam, titled *The 13th Valley*. In that book the Vietnam soldiers had small ammo cans with all their personal items in it that they carried in their rucksacks. I decided to do the same thing, and I made up what I called my "alert box". Inside that box was paper, pens, envelopes, stamps, shoelaces, shaving cream, toothpaste, toothbrush, razors, sharpening stone and a mirror. While we were getting ready to go on alert status there was a rumor floating around that we would be going to Seneca, New York. According to the rumor there was a storage facility for nuclear weapons there, and there were going to be some "no nuke" demonstrations on October 29th. I had been told that if we did go, I would be assuming my sniper role, and be overwatching the demonstrators.[2]

On Saturday I had planned to go over to Rudy W____'s house near the Fayetteville airport and shoot my M1911 .45 pistol. Rudy was the Commander of the 2nd North Carolina Regiment, the Revolutionary War reenactment unit that I belonged to, and I hung around his home a lot. I dated his daughter, Andrea, for about a year, but our relationship had ended. I was hoping to get back together with her, but it didn't seem like it would happen since her mom, Inge, wasn't too thrilled about it. Just as I was heading to the parking lot to leave, my unit received an alert notification. Not just any alert, but we were told it was N+30! What this meant was that 30 minutes had already gone by, and so now the whole Battalion had to form in an hour and a half. Anyone who missed that alert would receive an Article 15 punishment under the UCMJ.[3]

The Company was turned on its ear as everyone scrambled to meet this short deadline. I was a Sergeant, the A Team Leader in the 3rd Squad. I had to make sure that my Team was squared away in time. I had a couple of dependable soldiers, such as SP/4 Meirs, and Kelly Harris, but I also had two members of the Squad who continued to give me grief. One was Sergeant Bu____, the B

---

[2] In military terms, to "overwatch" a target means to keep an eye on it from a distance, or basically watching over it. The reason why the military would overwatch a demonstration was in case the demonstration became violent. If the demonstrators attempted to seize any military material, they would need to be subdued.

[3] Uniform Code of Military Justice.

Team leader, who thought he should be in charge whenever the Squad Leader, SSG Sengbusch, was not around.[4]

The other was SP/4 Sy___, who was on my Team.[5] I had been Squad Leader of the Weapons Squad, but when I was promoted to Sergeant, I could no longer could be a machine gunner, so I was appointed as the Team Leader of A Team.[6] Unfortunately whenever SSG Sengbusch was not around, I inherited SGT Bu___'s petty power games and childish attitude.

Another thing I did not care about Bu___ was that he was easily scared by the unknown. That summer the Squad had been put on a detail to support the Special Forces school for the "Robin Sage" exercise. During "Robin Sage" the Special Forces candidates conducted missions against us as the OPFOR.[7] My Squad was set up to guard a microwave tower near Camp Mackall. However, each night Bu would go out on the dirt road that led to the microwave tower, yelling at the shadows, "I'm here SF! Come and get us SF!" and he would fire his blank rounds into the woods.

All of these antics led me to try to counsel Bu___ in my room behind closed doors; however, once I closed the door, he thought it was time to fight. He punched me in the eye, and I pinned him down and landed a few punches myself. When it was all over our Platoon sergeant, SFC Howard thought it was all a racial issue since Bu___ was black.[8] It had nothing to do with race, and everything to do with being an effective soldier.

SP/4 Sy___ was just as immature and would disobey my orders just to see if he could get away with it. On one mission, when I was acting Squad leader,

---

[4] SSG stands for Staff Sergeant. Sengbusch eventually went to Delta Force. He served in combat with them in Panama during Operation Just Cause and rose to the rank of CSM. After he retired, he was called out of retirement by a request from General John Nicholson, my old platoon leader. Sengbusch was the brigade CSM for the 1/23rd Infantry, Stryker Battalion when they went into Iraq. After that combat tour, he retired again. Brian Sengbusch died of a massive heart attack, at the age of 60, in June of 2019.
[5] SP/4 stands for Specialist Fourth Class. At that time there was SP/4 to SP/7. Today this has evolved to one rank of "Specialist" which has a pay grade of E-4.
[6] Each platoon had three squads, and a weapons Squad. The Weapons Squad consisted of two M-60 machineguns. The Weapons Platoon in the company was the Mortar Platoon and consisted of three 81mm mortars.
[7] OPFOR = Opposing Forces aka the "bad guys".
[8] SFC stands for Sergeant First Class. SFC Howard had originally been the 3rd Platoon Sergeant. While he was there, he was one of the better Platoon Sergeants. He had served with the 101st Air Assault in Vietnam and was one of the more experienced NCOs we had. He left the 82nd Airborne to do a few years in Germany and when he returned, he became the 1st Platoon Sergeant.

I was tired of his attitude and relieved him. He was the acting Team Leader and he didn't think I had the authority to relieve him, but I did it anyway, since it was the only way I could figure out to accomplish the mission. It worked, though it led to hard feelings between us.

Once an alert is called there is a mad rush to get certain things done quickly. There were final items to be checked out from supply and packed into the DRF-1 bags. We had to draw our M-17A1 protective masks[9] from the NBC room,[10] but we were also issued live filters and actual M-258 decontamination kits.[11] Normally the protective masks would have a set of filters in them that were "good enough" for training. They would stop the non-lethal chemicals, such as CS gas, but they deteriorated after a period of hours.[12] When we were called out on alert the live filters would be issued to replace those inside the mask. These filters were in packages and would not be opened until we were absolutely sure we were going to war. Until then they were just packed away in the rucksacks. The soldiers drew their weapons but did not draw any ammunition yet. This would not be given out until we were down at the airfield ready to board the aircraft.

---

[9] These are gas masks, but no one in the military calls it that, the same as no one ever calling an M-16 rifle a "gun".

[10] Nuclear, Biological, Chemical.

[11] The decontamination kit had a series of chemicals, that when mixed together would supposedly decontaminate skin, or equipment. Hardly any soldier believed it would be effective, since we had been told horror stories of what nerve agent or blister agent would do to a human body. We never did find out whether it was effective though.

[12] CS stands for 2-chlorobenzalmalononitrile. It was used by the military to flush out the Viet Cong in their tunnels underground. CS gas is the nastiest thing that the military can throw at you in a training environment. It makes pepper spray look like a day at the beach. CS is not supposed to be used by any law enforcement since it could possibly kill someone in an enclosed space. This is not the tear gas that the police would throw at rioters. The reaction to CS is burning of the skin, tearing of the eyes, snot dripping out of your nose, and choking until you vomit. Even after the CS gas is gone you can feel it burning your skin for hours afterwards.

Our Company marched down to Deglopper Field and we were manifested several times. This process put us into chalks for each of the aircraft that was to depart and go to "war".[13] Our platoon only had one person missing, SP/4 DeJesus. After the final manifest the alert was called off. It had just been a practice alert like the hundreds of others I had been through. All of the sensitive items and weapons were turned back in to the arms, supply and NBC rooms. Once this was completed, we were finally released at 6:00 on Saturday night.

---

[13] The term for a group of paratroopers that would be assigned to a specific airplane was known as a "chalk". This term came from WWII when the aircraft would have the numbers of men and vehicles written on the side, in chalk. A "stick" of paratroopers would be the exact number who could jump out at any one time in a single pass.

The first part of the weekend was shot so I watched a VCR movie that had been set up at the CQ desk for most of the night.[14] I decided to go over to Rudy's on Sunday, shoot my pistol, make some dice out of lead balls for reenactments, and take a big bucket of Kentucky Fried Chicken with me.[15] Even though the weekend started with an alert, while I was at Rudy's there was a second "telephonic" alert. This was pretty unusual.

In each company there is an alert roster of soldiers who live outside of the barracks because they are married or have higher rank. The way an alert roster works is that the person at the top of the list, the Company Commander, is called. He then calls the next people on the list, and they call the next name, and it goes on until the last name on the list is called. If anyone is not home or does not answer the phone, they are skipped and the next person is called. The last person will then call the CQ desk and tell them if anyone was skipped. A telephonic alert is not one where anyone has to come in. It is merely a check of the telephonic alert roster.

Those in the barracks are not included on this telephonic alert. I had once leased a pager, a fairly new idea in the 1980s. Unfortunately, the First Sergeant of the Company did not want anyone on the alert roster who was not officially living off post. So even though I had a pager, I could not be contacted if there was an alert. Having no reason to own a pager anymore, I got rid of it.

At 3:00 on Monday morning the Company had another call out, but this time the "alert" was for the Company to clean the weapons that might have gotten dirty on Saturday's call out. This was because a Division Maintenance Team would be inspecting the Company, and there could not be any dirt or dust on the weapons.[16] For us in A Company, Monday would be a long workday.

In 1979 Maurice Bishop led a bloodless coup against Sir Eric Gairy on the island of Grenada. Bishop created the Provisional Revolutionary Government (PRG). Bishop was a Marxist who looked towards Fidel Castro's Cuba as a role

---

[14] The VCR was still new to us, and only a few soldiers had one. CQ stood for "Charge of Quarters". In every military barracks there is one or two soldiers who awake, all night. They answer the phones, make sure the place is secure and nothing is catching on fire. The soldier who had CQ would normally get the next day off to recover.
[15] The 2nd North Carolina Regiment is one of the largest Revolutionary War reenactment units. The members portray the Continental Army of 1780 and do living history and battle reenactments from Canada to Florida. In the 1980s had its own Sutler store and would sell these dice to spectators.
[16] Division Maintenance Team, also known as D-MET.

model for Grenada's political and economic problems. Bishop invited assistance from Cuba and other Communist countries, which conflicted with America's foreign policy. Bishop didn't help his cause when he refused to hold free elections. Tensions escalated when Bishop and the PRG began to construct an international airport at Point Salines. This would be built with mostly Cuban workers and half the cost was to be paid by Cuba.

Bishop stated that the purpose for the airport was to improve the decreasing tourist trade. The main airport on Grenada, Pearls Airport, could only handle smaller twin engine aircraft. Though this was the reason Bishop gave, U.S. observers noted that no hotels were being built for this influx of thousands of tourists. After the invasion, documents proved that the Cubans planned to use the airport as a staging base for their soldiers in Angola, Africa. The airport could also refuel Soviet aircraft on the way to Nicaragua.

At the very end of the airport was the True Blue Medical campus. This was a medical school that had over 600 American citizens as students. Most of the staff was also from the United States. Prior to the invasion intelligence showed that all the students lived on the campus at the end of the Point Salines runway. Though the Marxist leaders did not like being dependent on an American dominated school, they could not resist the $2.5 million that the students spent locally each year.

By 1983 the Grenadian army, known as the People's Revolutionary Armed Forces (PRA), had grown larger than that of any other Eastern Caribbean Island nations. When Ronald Reagan visited Barbados in 1982 other Caribbean leaders expressed their fear that Grenada would attempt to expand communism in the Caribbean. On the island of Grenada there were two opposing viewpoints on the future of the Revolutionary government. Prime Minister Maurice Bishop thought that economic progress was slow and wanted closer ties to Cuba. Deputy Prime Minister Bernard Coard wanted not just to continue the course, but to speed up the progression to a Marxist state and have closer ties with the Soviet Union.

Two days before our Battalion assumed DRF-1 status Coard ordered Bishop to step down from his position as Prime Minister. Coard was able to do this because he had the backing of the military under General Hudson Austin. Bishop was then placed under house arrest. When several of his Ministers resigned in protest, they were arrested too. Over the next few days there were angry meetings amongst the new leadership, demanding that Bishop be court martialed for being a traitor to the revolution. Bishop was also accused of trying to kill Coard and is wife. Bishop tried to defend himself, but it was no use.

17

Triple Canopy

On the day that the 2/325 assumed primary alert status, in Grenada the news media was denied entry at Pearls Airport. Journalists who had slipped in were subject to body searches, but they were able to smuggle out film. Stories filtered out that there were demonstrations in the streets of Grenada by pro-Bishop sympathizers. Shops and businesses were closed, and Pearls Airport shut down for several days. No one was coming in or going out of the island.

Bernard Coard met with others and suggested moving Bishop to Cuba. As for the protesters, Coard said "They can stay in the streets for weeks, after a while they are bound to get tired and hungry and want peace." Some of the men at the meeting said Bishop should be reinstated as Prime Minister. These men, along with others, were arrested on October 17th and placed in the Richmond Hill prison. On October 18th Maurice Bishop was told that if he did not accept the decision of the Central Committee martial law would be imposed on the tiny island nation.

On Tuesday, October 18th the paratroopers of my Battalion knew they would be doing another call out, but this time it was to do a real mission. It wasn't "real world", but a training mission known as an EDRE. The 2nd Battalion was alerted at 10:00 that night, but the whole Battalion wasn't needed. The Battalion stood in formation with all of their gear until it was determined which unit would do the mission. B Company was chosen to do the main mission, which was to perform an airmobile assault to rescue eleven American civilians from 120 enemy.

A Company, 1st Platoon, along with the Weapons Squad and one Engineer was chosen to secure and destroy three bridges and one dam. My Squad, the 3rd Squad, was sent with the engineer to destroy the dam. SSG Sengbusch had been detached to do another mission, so I was the acting Squad leader for this mission. Since I was the only qualified sniper in the Company, I took along my M-21 rifle, though I figured I would not need it.

The Battalion was released, and we returned to the barracks to get whatever sleep we could until 0430 in the morning, when breakfast would be served. I decided to skip breakfast, and go with more sleep, so I stayed in bed until 0800. We had decided to travel light, no rucksacks. I wore a set of nylon Rhodesian chest pouches to carry the magazines for my M-21.

18

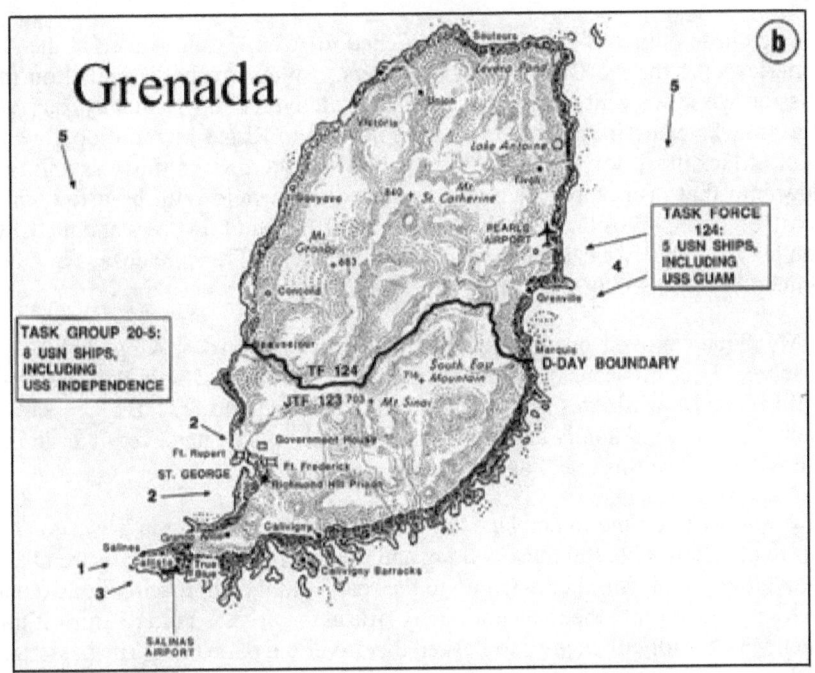

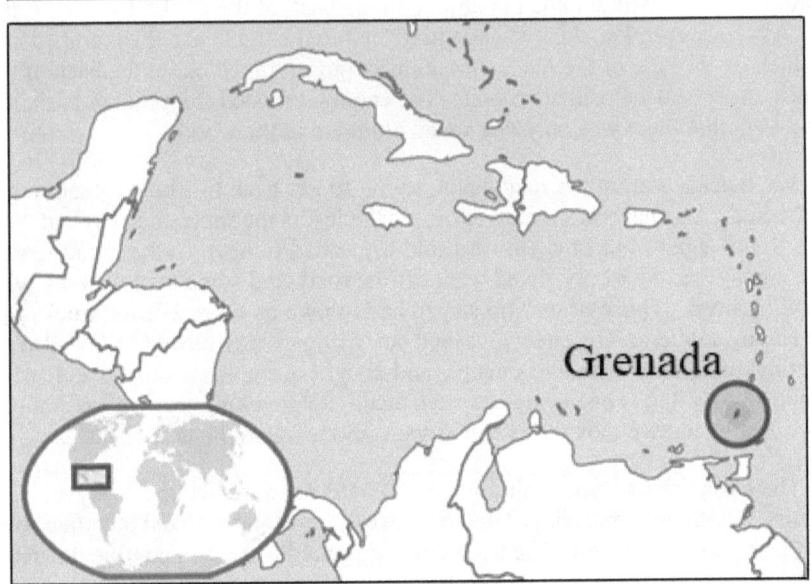

The chosen units of A Company marched to Area J and waited at the All American PZ for the Blackhawk helicopters. We were back-briefed on the mission while we waited, and then lunch was delivered to us. Finally, our two Blackhawks came in, picked us up, and flew us to Rhine-Luzon drop zone at Camp Mackall. I don't think we realized how big a mission this was until we flew into that drop zone. We thought that there would only be a few units involved, but C-130s had parachuted in several 105mm howitzers and the CSC gun jeeps. When we came in with our two choppers there were dozens of other helicopters off-loading their paratroopers.

My Squad moved out to the dam, with SFC Howard moving with us to observe. I set up security, while the engineer moved to "blow up" the dam. Unfortunately, while this was going on SFC Howard and SGT Bu____ sat on their rucks, talking loudly and making noise. I always treated every mission as if it were real, so this really annoyed me.

When it came time to detonate the fake explosives, we all ran back, down a dirt road. I saw a 2½ ton truck coming and told everyone to get off the road and form a hasty ambush. I crawled up to the road, and the approaching truck, and then stepped out into the road aiming my rifle at the driver. I didn't know if this truck was friendly or enemy, so I asked the driver the password. The passenger in the truck jumped out and ran around to the back of the truck. I was carrying a LAW rocket, so I unslung it, and "fired" it into the truck, and then jumped into a ditch on the side of the road. About thirty guys jumped out of the back of the truck and a heavy firefight erupted. The enemy retreated from our ambush, not knowing that there was only one squad out there in the woods.

We quickly ran up the road again, trying to get back to Rhine Luzon to get extracted. Another truck appeared, or maybe it was the same one. I yelled "Off the Road" again, but SFC Howard told my squad to keep walking and ignore the enemy. Some of my squad went off the road, and some kept walking with SFC Howard. Those of us who stayed had to take on three 2½ ton trucks full of enemy soldiers. The enemy jumped out, firing as they moved. We ended up getting mixed up with the enemy, and fought them from an old defensive fighting position. The enemy retreated again, not knowing that we were only a Fire Team, and we took off as fast as we could towards the drop zone.

The trucks came back a third time and I had the squad set up a hasty ambush again. The enemy must have known we were there, and realized that they were not going to get through. The trucks turned around instead of coming down the road, leaving us alone.

20

When we got to Rhine-Luzon I discovered that SFC Howard had already taken our helicopter back to Fort Bragg, leaving us on the drop zone to fend for ourselves. I was able to get a ride for my squad with an artillery unit in their Blackhawks. We finally returned to the barracks at 11:45 that night, extremely tired from the four days of call outs, practice alerts and the EDRE. The mission was successful and all of the American citizens, portrayed by civilian employees and wives of the paratroopers, were rescued unharmed.

At the same time that I was marching to Area J in Fort Bragg to rendezvous with the Blackhawks, crowds of students in Grenada were marching through the streets, shouting slogans for Bishop and against Coard. The police and soldiers along the march watched impassively as the crowd yelled "We want Maurice! We want Maurice!" Foreign Minister Union Whiteman had returned from addressing the United Nations in New York and was shouting to a crowd of thousands. Whiteman told them that they should march to Bishop's house and rescue him.

Inside his house Bernard Coard felt threatened. This was a realistic threat since there were no soldiers in between the approaching crowd and himself. He

RECON

telephoned Fort Frederick and ordered three BTR-60PB armored personnel carriers to come to his assistance at once.[17] Only two of the BTR-60s were able to get to Coard's home, which was right beside Bishop's house. The crowd was too packed with people to allow the third BTR-60 to approach. The situation was not critical though, and the crews of the BTR-60s opened the hatches and relaxed on top of the armored vehicles. They were seen talking and giving the milling crowd "high fives".

At Bishop's house next door, the guards on the gate were feeling the pressure. The officer of the guard, Lieutenant Iman Abdullah, fired his submachinegun in the air and then ordered the APC crews to also fire their heavy machineguns in the air.[18] The crowd was silent for a few seconds, and then began yelling "Shoot us! Kill us!" and surged over the gate. By 10:00 that morning Bishop and some of his men were freed.

Bishop decided not to go after Coard in the house next door, but instead got into a car and drove to Fort Rupert, the PRA headquarters.[19] Fort Rupert had a narrow approach, a radio for communication, and plenty of arms, ammunition, food and water. At 11:00 Bishop and several hundreds of supporters occupied the fort. The crowd moved past the PRA guards shouting "We get we leader, fuck Coard!"

Bishop sent a messenger to the Cuban embassy asking for assistance from the hundreds of armed Cuban workers at the airfield. Fidel Castro realized the diplomatic mess that Grenada was creating, and ordered that no Cubans become involved in the internal problems on the island.

Once inside the security of Fort Rupert Bishop ordered that the PRA lay down its arms. He also ordered that the telephone lines from Coard's house and to Fort Frederick be cut so that he could not organize any resistance.

---

[17] The BTR-60PB was a six wheeled armored personnel carrier. It carried a 7.62mm machinegun and a 14.5mm machinegun in the turret (.57 caliber).
[18] APC stands for Armored Personnel Carrier. Iman Abdullah was also known as Callistus Bernard.
[19] Fort Rupert was built on a high hill on the north side of St. George's harbor. St. George was the capital of the island. Fort Rupert originally had been named Fort Royal, when the French built it in 1705. After the British had gained the island with the Treaty of Paris 1763, they renamed it Fort George. In July of 1779 Admiral Comte D'Estaing attacked Grenada, focusing on Fort George and Hospital Hill. After four days of fighting the British surrendered. In the Treaty of Paris of 1783 Grenada was returned to the British. In 1979 the fort was renamed Fort Rupert, in honor of Rupert Bishop, the father of the Prime Minister.

What Bishop did not know was that Coard and Hudson Austin, along with others, had ridden to Fort Frederick on the east side of St. Georges.[20] Fort Frederick was the home of three platoons of PRA and six BTR-60PBs. Coard and his Central Committee devised a plan to take back Fort Rupert and kill Bishop in the process. Bishop's death could be explained to the Grenadian people and the world much easier if he died resisting the attack.

Three BTR-60s, under the command of Officer Cadet "Connie" Mayers, moved to Fort Rupert. Mayers had been trained by the Cubans, and he had also been in the United States Army's Berlin Brigade. Two of the BTR-60s continued up the steep road to the Fort, while the third remained outside the Fort to form a military cordon. Mayers did not hesitate, and immediately began firing when he neared the Fort's gates. Supporters of Bishop crowded together inside the fort. Some of them had weapons, but most were unarmed men, women and children. They were not expecting an assault upon the colonial fortification.

The attack was started by a soldier firing an RPG rocket launcher at the headquarters building. The rocket roared in, flying low, striking a vehicle in front of the building that erupted in a ball of flame. Mayers' large 14.5 mm machinegun on the BTR-60s turret raked the crowds from left to right.[21] Behind Mayers, Iman Abdullah's BTR-60 also began firing on the crowd. The PRA soldiers dismounted and began firing RPGs and small arms. While they moved forward, they threw grenades towards the supporters of Bishop.

There was no place to run, since the BTR-60s blocked the only exit. Many of the people jumped over the walls of the Fort, falling twenty feet onto the rocks below. Some of the reporters who had remained on Grenada were able to film the horror as men, women and children were cut down, or tried to leap to safety on the rocks below.

The defenders were not helpless, and after the initial shock they returned fire with AK-47 rifles and Makarov 9mm pistols. Mayers was hit four times in the groin by AK-47 fire and would die within the hour. Three other attackers were killed in the assault, but it was one sided. Approximately forty of Bishop's supporters were killed, and over a hundred wounded in a day that became known as "Bloody Wednesday".

---

[20] Fort Frederick had been built by the British upon Richmond Hill in 1791. A mental hospital shared the hillside, and would later be bombed accidentally by US aircraft.
[21] The 14.5 mm is a .57 caliber.

Bishop and seven of his supporters were captured and lined up against the wall of the upper fort. They remained there for almost an hour, while Coard and the Central Committee decided what to do. Finally, Abdullah was given a piece of paper and he told the eight "you shall be executed by fire." One of the eight was a woman, Jacqueline Creft, who pleaded that she was pregnant. Andy Mitchell, one of Coard's men, yelled out "No fucking Comrade at this time!" Abdullah gave the command to fire. One of the executioners, Fabian Gabriel, described it, "Bodies tripe open. Fitzroy Bain tripe was out, Maurice Bishop back, belly and neck was cut, Jacqueline Creft hand fly off. The bodies were lying there, virtually in bits of blood. Pieces of flesh of the bodies were also stuck on the wall."[22] After a minute the firing stopped. As the bodies were being carried to the bottom square and loaded onto trucks a single white flare was fired, signifying to Coard and the Central Committee that Bishop was dead. Unfortunately, the United States did not know about Bishop's execution until after the invasion was well under way.

On Friday we had a short day, since the Battalion had such a busy week. In the morning there was a battalion run. The Battalion Commander, Lieutenant Colonel Jack Hamilton, told the formation that we would only be doing a short run, a mere two miles.[23] We knew that it had to be a joke. No one ran only two miles for physical training unless it was in a protective mask.[24] In the end the Battalion ran six miles. What made it a good day was that everyone was released by noon.

That weekend I decided to take a risk and go outside of the two-hour alert radius. I figured that no alerts would be called on a Friday when we had just been released early. When I first came to the 82nd Airborne in 1979 the Battalion had been on DRF-1. I went out at that time and watched *Apocalypse Now*, a really long movie. While I was at the movie the company had a callout and I missed the two-hour recall by ten minutes. I received a "desk drawer" Article 15 punishment.

An Article 15 is non-judicial punishment. An offense can merit a Court Martial, or a case that is determined by an officer judge, however a soldier can choose to accept Article 15 punishment. This means the soldier waves his right to a Court Martial and accepts a lesser punishment by pleading guilty. Normally

---

[22] Tripe means intestines.
[23] He is also known as "Mad Jack" Hamilton.
[24] Units running in full chemical suit with masks could occasionally be seen on Ardennes Street during morning PT (physical training).

this would go in a soldier's permanent record, but there was a type of unofficial Article 15 known as a "desk drawer" punishment. The Article 15 is not submitted to the soldier's permanent record, and is instead kept in the desk file of the Commander. When the soldier leaves the unit that file is destroyed or is given to the soldier.

I decided to risk going outside the two-hour alert radius, to drive to the Charlotte Coliseum gun show that was about 2 ½ hours away. My main intent was to look for a large sized flintlock, so that I could get Rudy Weimann to build what was known as a "wall gun". I didn't really think I would find such a thing, but I liked going to gun shows.[25] I did buy a WWII British Enfield rifle for $90, and took it to Rudy's the next day to shoot it.

I hung out at Rudy's quite a bit because I just didn't feel like sitting around in the barracks doing nothing. On Saturday, October 22nd, I spent the day helping Rudy dig up some stumps in his front yard. While I was helping him move the stumps, a puppy ran up and wanted to play. I goofed around with the dog, and it ended up splitting open my lip with its claw. Little did I know that this split lip would hurt over the next few days when I was on the island of Grenada.

On Sunday I hung out in the barracks, and that was the day America heard of the bombing of the Marine barracks in Beirut. I didn't understand it, since Islamic suicide bombing was not typical at that time. I wrote in my journal "In Beirut some Kamikaze drove up to the Marine HQ and blew up 2,000 lbs of explosives. 146 of them are dead now." My brother Sean was in the Marines at the time, but he was in Okinawa, so I knew he was safe. The day before, I had received a black silk kimono from him in the mail, so I was not worried about him. I didn't think we would go over to Beirut, since the mission was a peacekeeping one intending to keep the Israelis and Syrians from killing each other. The 82nd Airborne typically was not a unit to do peacekeeping missions.

Towards the end of the day, I learned that a friend of mine, Brian Jeznach was returning to the Army and would be coming into Fort Bragg on Monday. Jeznach had been in the A Company Commo section when I first arrived there in 1979. He had gotten out of the Army in 1981 after his first enlistment with the rank of SP/4, but he missed the Army life and decided to return. He would be assigned to the Special Forces as a PFC.[26] He wouldn't be on an operational

---

[25] I finally did get a wall gun, but not until 2003. It was one of a kind, made by Ray Tingle of Kentucky.
[26] PFC stands for Private First Class, and would be an E-3 pay grade.

team in Special Forces, but would be assigned to a "candy stripe" unit.[27] I was going to find him on Monday, and take him out on the town to catch up on things.

After Bishop and his supporters were gunned down at Fort Rupert, the Revolutionary Military Council (RMC) put a curfew in effect. Initially no Grenadians were allowed outside of their house, and those who did venture out would be shot on sight. However, on Friday, while our Battalion was running six miles down the roads of Fort Bragg, the curfew was lifted between the hours of 10:00 A.M. and 2:00 P.M.

The RMC did not want the United States to intervene militarily, so General Hudson Austin personally assured Dr. Geoffrey Bourne, the Vice Chancellor of St. George's University School of Medicine that the medical students were in no danger. Dr. Bourne was allowed to travel anywhere on the island, with police escort, to get any supplies needed for the students during the curfew. Fidel Castro in Havana also urged the RMC to leave the American students alone and to not stop any from leaving or being evacuated.

Bishop had been friends with Castro, visiting him in Cuba just prior to Bishop's death. Fidel Castro stated that the killers of Bishop "deserve exemplary punishment". Castro also wrote that the United States "will now try

---

[27] In the 1980s Special Forces went through various changes with their headgear. In the 1960s and 1970s only Special Forces qualified soldiers could wear the Green Beret. For some reason this changed, and those assigned to Special Forces, even though they weren't Special Forces qualified, were allowed to wear the Green Beret. The only way to tell who was an operator and truly Special Forces qualified was to look at the patch on the beret. If it was the full patch, then the person wearing it was Special Forces qualified. However, if the patch was only a small rectangle, with the unit colors in it, then they were a support soldier. These small, rectangular colored patches became known as "candy stripes". This continued until the end of the 1980s, when the "Special Forces" tab was created. After that, all personnel wore Green Berets, and the mark of the Special Forces qualified operator was the tab. This changed around 1992, when the support personnel either wore a maroon beret, if they were airborne qualified, or a BDU baseball cap. There are many people who believe that there were female Special Forces soldiers due to the appearance of women with Green Berets on their heads in the 1980s. However, there are no women on a Special Forces Operational Detachment, nor has there ever been when this book was written in 2013 (2020 edit: As of this year, due to some serious political band-standing and lowering of the standards before the female soldier ever showed up, the first female passed through Special Forces Qualification).

to profit from this tragedy". Two days after Bishop's execution Coard and his wife withdrew from the events happening on the island and went into seclusion.

Once word was received that the PRA had fired into crowds of people, the United States Atlantic Command began planning for the evacuation of American citizens on the island of Grenada. They planned for evacuation in both a friendly environment, and a hostile one. They soon discovered that there was very little known about the small island. Vice President George Bush called in the Special Situation Group to determine a course of action. It was decided that the 22nd Marine Amphibious Unit (MAU) would be diverted to Grenada to assist in any evacuation. These 1,900 Marines had been on their way to relieve the Marines in Beirut before the deadly Islamic attack on October 23rd.

President Reagan authorized the course of action and gave it to Lieutenant Colonel Oliver North to plan the details. The Joint Chiefs of Staff (JCS) prepared for a full-scale operation to seize the island as an alternative. Late Friday night the Marines in the 22nd MAU, the Rangers in the 1st and 2nd Ranger Battalions and the 82nd Airborne Division were alerted to the possible mission.

On Saturday, October 22nd there was an emergency meeting of the Organization of East Caribbean States (OECS) in Barbados. They agreed to cut all links to Grenada until further notice. No new issues of currency to Grenada would be shipped, and telecommunications on Trinidad would be severed. The OECS contemplated invading Grenada to rescue Bishop, but several member nations did not agree to using military force. The United States would not intervene until there was a formal written request from the OECS. This finally arrived late on the night of Sunday, October 23rd. Once this request was received the alerted military began to plan for the invasion of Grenada as a primary option. President Reagan had been golfing in Augusta, Georgia and was briefed about the possible US military intervention.

On Sunday President Reagan flew back to Washington and met with his advisors in the White House situation room. He was briefed on the situation during one of the worst weekends of his administration. Reagan had to make a decision on what to do about both the Beirut bombings that had killed 241 Marines, and the situation on the island of Grenada.

Though the students seemed to be in no danger on Grenada, Reagan was worried that the situation might change. He had been briefed on how the PRA had fired into crowds that contained women and children. President Reagan also did not want another hostage situation like the one that haunted his predecessor, Jimmy Carter, when Iran took hostages at the American Embassy in Tehran. In addition, Reagan wanted to end the spread of Communism in his own hemisphere. The Chief of Staff of the Soviet Armed Forces bragged about how Communism was spreading, "Over two decades ago there was only Cuba in Latin America, today there are Nicaragua, Grenada, and a serious battle is going on in El Salvador." This threat to the American students would be the excuse Reagan needed to eliminate the most militarized island in the region. Finally, Reagan knew that a military victory would raise the public's spirit and morale after the devastating attack in Beirut.

Knowing that this would be a full-scale military invasion Reagan commented, "Well, if we've got to go there, we might as well do all that needs to be done." Though there was a need for secrecy, the American media soon learned of the diversion of the Marines to Grenada. Word of an impending invasion by the OECS was also divulged by Guyana. Once the media learned, so did the RMC on Grenada.

After work on Monday, I went out to the 19th Replacement Depot trying to find Jeznach.[28] I soon found out that he was looking for me too. We eventually ran across each other driving on opposite sides of the street. We went out to Fayetteville and goofed around, ending up at a theater to watch a movie, *The Right Stuff*. This was a really long movie. Halfway through the movie I dutifully called up the company to see if there had been an alert. I didn't want a repeat of what happened when I missed that callout after I had first arrived at Fort Bragg in 1979.

Due to that earlier punishment, I always made sure that if I was gone for a long period of time, I would call the CQ desk to make sure that there were no callouts. If I had my pager it would have solved the problem, but since I lived in the barracks no one would call it anyway. The CQ told me that there was nothing happening and I stayed to watch the rest of the movie. After the movie I drove Jeznach back to the Replacement Depot and we noticed convoys of trucks on the road. I figured someone was heading out to the field, but it sure was a heck of a lot of trucks.

I dropped off Jeznach and returned to the A Company barracks. It was about midnight and when I came into the building the CQ desk was empty. I yelled down the hallway "Is anyone here?" There was no answer. I then jokingly yelled "I'm going to steal your CQ desk!" Still no answer. I started to head up the stairs and I noticed that the backdoor was chained shut. This was unusual.

---

[28] Replacement Depot also known as "Reppel Deppel".

I looked out the window of the back door and saw the whole Battalion formed on the field.

Crap!

There had been a callout.

Missed another one!

The plan of the invasion was for the 1st and 2nd Ranger Battalions to secure Point Salines airfield by conducting an operation known as airfield seizure. This involved a combination of jumping onto the airfield by parachute, and landing aircraft on the airfield with support gun jeeps. The Rangers were also to secure the True Blue medical campus and the PRA barracks at Calvigny. The SEALs and Delta Force would attack special targets in St. Georges. The Marines who had been diverted to the island were to capture Pearls airfield in the north. Initially H-Hour was to be 0200, with all objectives seized by sunrise. The 82nd Airborne was to land on Point Salines airfield at H+4, or around 0600 on October 25th.[29] By D+1 all the Special Operations forces and the Marines would withdraw off the island, leaving the 82nd Airborne there to maintain law and order and to mop up any resistance. The 82nd Airborne would eventually turn over its duties to the Caribbean military allies.

Facing this large military force were the defenders of Grenada. Once the Grenadian government learned of the upcoming invasion the Grenadian People's Revolutionary Militia (PRM) was ordered to mobilize, but the order was largely ignored. Estimates of the enemy's strength vary depending on the source. The RMC may have had as many as 1,200 PRA soldiers. Mark Adkin in his book "Urgent Fury, The Battle for Grenada" estimated that there were a little less than 500 regulars of the PRA and less than 250 of the PRM.

These men would crew two anti-aircraft (AA) batteries of six ZU-23s each and three platoons of two 12.7mm AA each.[30] There were more AA batteries

---

[29] When an invasion happened, the day that it began was known as D-Day. The exact hour is known as H-Hour. These are used for planning purposes. For example, if a bridge must be blown the day before the invasion it would happen on D-1. If a building had to be captured within three hours of the invasion, it had to happen by H+3. So, D+1 would be one day after the invasion date. Of course, the most famous D-Day is the one in Normandy in 1944.

[30] The ZU-23 is a twin barreled 23mm cannon, and is one of the best anti-aircraft weapons in the world.

on the island, but not enough soldiers to man them. There was also a motorized company of sixteen BTR-60PBs and eight GAZ trucks. With them were a small anti-tank squad of two 75mm recoilless rifles and a mortar platoon of three 82mm mortars. Their Cuban allies consisted of 43 military advisors and 635 armed construction workers. Though critics of the invasion downplayed the abilities of the Cuban engineers, it was later learned that many of these men were combat veterans of the wars in Angola and Ethiopia.

The enemy troops were dispersed to locations where invasion seemed imminent, mainly the Salines and Pearls airfields and locations throughout St. Georges. The defense of the Salines airfield was mainly left to the Cuban military and construction workers. General Hudson Austin and the RMC leadership moved into the tunnels underneath Fort Frederick where they would be safe from airstrikes, but by doing so they no longer were able to communicate with their units.

On the night of October 23rd twelve men from SEAL Team 6 and four Air Force Combat Control Teams (CCT) were supposed to jump near the USS *Clifton Sprague*, located 30 kilometers from the southwest tip of Grenada.[31] These men were to land in the water near the ship, and land on the island. Once there they would observe Salines airfield prior to the Ranger airborne operations. The SEALs parachuted in with two Boston Whaler boats to help them conduct the mission.[32] Many historians have stated that the SEALs did a HALO jump (High Altitude Low Opening), or a military free fall, but this is not true. The SEALs jumped at an altitude between 800 and 1000 feet, using static line parachutes from the ramp of the aircraft. The boats were also dropped by parachute from that altitude.[33]

Unfortunately, tragedy struck when four SEALs came up missing after a lengthy recovery operation. They were Kenneth Butcher, Kevin Lundberg, Stephen Morris, and Robert Schanberger. These men were never found and were the first casualties of the invasion. They also lost one of the Boston Whaler boats. The surviving SEALs boarded the remaining boat and headed towards the island. As they neared the shore, they spotted a Grenadian patrol boat. The SEALs cut the engines and let the boat drift, remaining undetected. Unfortunately, the wake of the patrol boat, and the choppy sea had swamped the boats motors and they could not be restarted. With dawn approaching fast the

---

[31] HALO is a parachute insertion similar to skydiving. HALO stands for High Altitude, Low Opening. HAHO would be High Altitude, High Opening. The gag within the HALO teams is that you don't want a HANO, or High Altitude, No Opening.
[32] Extracted from *Grenada, Operation Urgent Fury*, located at http://www.warboats.org/grenada.htm on 18 November 2013
[33] Facebook post from Joe Muccia on 03 December 2013

SEALs paddled their boat back out into the ocean and rendezvoused with the destroyer USS *Caron.*[34]

It was decided to conduct a second attempt to insert the SEALs and CCT on the night of October 24th. However, to give them time to have eyes on the Salines airfield the H-Hour had to be delayed until 0500 of October 25th. On Monday a second SEAL Team successfully parachuted near the USS *Clifton Sprague* and boarded two Zodiac boats. As they approached the beach under the cover of darkness the rough sea swamped the boats again and the CCT lost most of their equipment. The mission was aborted for a second time and the boats limped back to the Navy destroyer. Due to this the Rangers did not have any intelligence of what was on the airfield and due to the delayed H-Hour all the special operations forces would have to take their objectives without the cover of darkness.[35]

As I was scrambling to change into my uniform my Platoon Leader, 1st Lieutenant Nicholson entered my barracks room and asked where the hell I had been.[36] He wasn't mad, but quickly let me know that the callout had already progressed to N+3. I was an hour late for the alert. LT Nicholson then told me it wasn't an alert, it was real. We were going to WAR! He asked if I had any 9mm ammunition. I only owned a .45 caliber pistol, so I told him I didn't have any.[37] The lieutenant decided if we were going to war, he was going to carry his personal 9mm. It was not authorized, but it was a war, so we figured the

---

[34] In ship sizes, the biggest warship was a battleship, and then a cruiser, and then the smaller destroyers. In modern warfare we no longer need large ships, such as battleships, and now we have enough firepower on the smaller destroyers that we no longer need the larger ships, though there are a few cruisers left in the Navy's inventory. However, the Navy has planned to get rid of them and do not have any plans for new ones being built as of 2020.

[35] There are many parts of this story that I know that I have gotten wrong. This is due to the classified nature of the SEAL missions, even to this day (2021). A fellow historian and Marine Joe Muccia is writing the most comprehensive history of Operation Urgent Fury, but until he can get clearance to write about the special operations missions, that book will remain unpublished.

[36] In 2013 Lieutenant John Nicholson was Major General "Mick" Nicholson, Commanding General of the 82nd Airborne Division (2020 edit: He became General Nicholson in charge of all of all the US and NATO forces in Afghanistan until 2018 and retired that year, after serving 35 years in military).

[37] The US Army had not progressed to the Beretta M9 9mm yet, and all the pistols in the 82nd Airborne were M1911A1 .45 calibers, such as the one I owned. Company commanders and those paratroopers on crew served weapons were the only ones who carried pistols, but I sometimes carried one as the Company sniper.

rules were considered less strict. I asked him where we were going, and he said we were going somewhere south, maybe South America, but he told me to hurry up, get dressed and get outside.

Luckily all my bags had been packed by me earlier, and all I had to do was don my uniform and get to DeGlopper field. I grabbed my personal .45 caliber out of my car, along with a box of hollow points. I put the pistol in a leather Army issued shoulder holster. I also grabbed 40 rounds of hollow point ammunition for my sniper rifle out of my car.[38]

Both my M16A1 rifle and my M-21 Sniper rifle and scope had been drawn from the arms room by my squad and were waiting in formation for me. I did not want to carry two rifles, and I ended up getting in an argument with LT Nicholson over which one I needed more. Normally my sniper rifle would be carried in the company trains, but we were told that we may not link up with our vehicles for a long time. I wanted to carry the M-16, since we would be jumping into a hot drop zone, but I was over-ruled by LT Nicholson. I ended up carrying both weapons, strapping my M-21 sniper rifle on the side of my rucksack. I packed my .45 pistol in my rucksack, along with the ammunition from my car, not wanting to draw attention to it.

Our Company was the Initial Readiness Company (IRC) and would be going in first, but our Battalion's C Company would not be going. Our normal Charlie Company was COHORT, and it was considered that they were not trained or competent enough to go to war.[39] What took their place was B Company, 2nd

---

[38] Soldiers are not allowed to keep ammunition or weapons in their personal automobiles, but many do anyway. They are supposed to put them in the unit arms room, but it is such a bureaucratic pain in the butt to get the weapon out if you want to practice with it.

[39] Around this time the Army was testing an idea known as COHORT. COHORT stood for Cohesion, Operational Readiness and Training. In this concept a group of soldiers would complete basic training together, and then be assigned to a unit. The entire unit would be "created" at the same time. The soldiers would spend their entire tour within this unit. The theory of this was that there would be more esprit de corps and higher morale due to the camaraderie within the unit. The problems with this would be that everyone under the rank of SGT would be inexperienced and would be required to do additional training to become combat ready. By the time of the Operation Urgent Fury the 2/325's COHORT company was not ready to be deployed. In the end the COHORT system failed due to a variety of reasons, but mainly because it was almost impossible to keep an entire COHORT company or a COHORT battalion's soldiers in place for the life of the COHORT unit. This idea had already been done, with disastrous consequences, by the British Army in WWI. They created the "Pals Battalions", that kept men from the same area all together, however after the blood bath of the Battle of Somme, and entire towns losing most of their young men, the Pals Battalions were discontinued. An

Battalion, 505th Infantry under Captain Mark Rocke. This unit was notified that they would be going with us and only had a short time to get ready. They were re-designated "C Company" so that there wouldn't be any confusion with having two B Companies. Also, within our Brigade the 1st Battalion, 325th Infantry would not be going, since they were scheduled for Sinai peacekeeping duty in January.[40] Our 1st Battalion would be replaced with 2nd Battalion of the 508th Infantry.

As I moved through the formation on DeGlopper Field I saw one paratrooper in our company who had recently broken his arm and I saw him cutting off his cast in the formation with the help of another soldier, so he would be able to go to war. This paratrooper was going to go to war, no matter what. However, I also saw soldiers trying to come up with reasons for not going. There was one soldier in our platoon who said he couldn't go to war because he had recently converted to a religion known as "Holiness" and he just couldn't kill anyone now. I had never heard of it and thought he was just being a pussy. SFC Howard wasn't going to put up with any of it and told him to grab his gear and get his ass on the vehicles that were pulling up to the parade field.

I made it just in time for the accountability formation, strapped on all my gear and then boarded the 80 Packs with the Battalion.[41] We moved out to CLACC in the early morning darkness.[42] The weather was not too cold, but there was a chill in the air. The Battalion moved into some barracks in a fenced-in area near Green Ramp at Pope Air Force Base.[43] I took the time to rearrange my rucksack the way I wanted it. I took off my BDUs and put on a pair of "Cammies" which were lighter.[44] I took out all the cold weather gear from

---

example of the losses was The Sheffield City Battalion that had lost 495 dead and wounded in one day on the Somme.

[40] Due to the peace treaty between Israel and Egypt, the Sinai Peninsula was patrolled by multinational force observers (MFO) of the United Nations. Their job is to ensure that neither country violates the treaty. The 1/325 would deploy there for six months. Many of the 1st Battalion troopers were angry that they could not be part of the largest combat operation since Vietnam.

[41] The 80 Pack was a tractor-trailer "cattle car" that held 80 soldiers.

[42] CLACC stood for Central Loading Area Control Center. This was located beside Pope Air Force Base.

[43] Green Ramp was the area where paratroopers loaded onto the aircraft on Pope Air Force base. Each section of the airfield is color coded. Green Ramp in the early 1980s was entirely out in the open and there were no buildings to wait in, like there are now. Only the "heavy drop" area had a building for the officer in charge. In 2011 Pope Air Force Base was absorbed into Fort Bragg and was redesignated Pope Field.

[44] BDUs or Battle Dress Uniform was fairly new and we had only been wearing them for about a year. They were heavy cotton fabric and took a long time to dry. Prior to this type of camouflage uniform the 82nd Airborne would wear a solid green uniform,

inside my rucksack and put it in my duffle bag. I figured I wouldn't need it if we were going someplace in South America. Finally, I laid down, using the rucksack for a pillow, and got about an hour of sleep. I did not now that I would not sleep again for the next 24 hours.

Around 0300 I woke when I heard the Battalion moving around. We gathered outside and marched into another fenced-in area where we were issued equipment for the unknown mission facing us. We filed by cardboard containers and metal ammo cans on steel palates and were given a mosquito net, a Kevlar flak jacket,[45] two 2-quart canteens, two poncho liners,[46] three C-Ration meals, and all the ammunition they thought we needed for the mission.

My ammunition issue consisted of two smoke grenades, two LAW rockets, 140 rounds of 5.56mm ball ammunition, and 90 rounds of 5.56mm tracer ammunition.[47] I was supposed to receive 240 rounds of 7.62mm MATCH ammunition for my sniper rifle but my company didn't get any, so I jumped into B Company's line to get some.[48] I also had three magazines of ammunition for my .45 pistol.

Since I had two weapons, I was issued both the basic load for a Team leader, and the basic load for a sniper. The other two sniper rifles in the company were carried by paratroopers from different platoons. They were not snipers and just carried the weapons broken down in their rucksacks during the entire

---

nicknamed the "pickle suit". It was totally unsuitable for field use, so we also were issued a set of ERDL camouflage fatigues made up of lighter cotton "rip stop" material that we just called "cammies". The paratroopers were only allowed to wear these camouflage fatigues when they went to the field or they were on DRF-1.

[45] Kevlar was fairly new, and we had only recently been issued the new Kevlar helmets the year before. We had never worn the flak jacket, and only knew of them from pictures from the Vietnam War. These flak jackets were made of Kevlar, but they would not stop a rifle bullet. They were only meant to stop shrapnel from grenades or mortars. This was the days before anyone ever thought of wearing body armor.

[46] The poncho liner was much coveted by the 82nd Airborne paratroopers. It looked like a camouflage quilt. Normally we would have to buy these at Clothing Sales or in surplus stores, but this time we were issued it from the war stocks. Sometime around the new millennia the soldiers began calling it the "woobie", but no one ever called it that in 1983.

[47] The M72A2 LAW rocket was a light anti-armor weapon, capable of penetrating one foot of steel plate. It had an effective range of about 200 meters. The 66mm rocket was replaced in the 1990s by the AT-4 84mm rocket. Both the LAW and the AT-4 are one shot weapons, and could not be reloaded.

[48] MATCH ammunition is supposed to be better than the standard ammunition 7.62mm machinegun ammunition given to the M-60 gunners, and therefore more accurate.

operation.[49] We were issued the LAW rockets because there were no hand grenades available. Each man carried multiple LAWs to use in place of a grenade. We did not get any grenades until D+4.[50] We also didn't get issued any Claymores or anti-personnel mines. The M-203 grenadiers weren't issued any M-203 vests or pouches and had to carry all their grenades in their rucksacks, tied to their LCEs or in their cargo pockets.[51]

We moved back to the fenced in holding area where our rucksacks and duffle bags were, and then sorted through all of the gear that we didn't want to take on the mission. We still had no idea of where we were going, or what we would do, but with the issue of the mosquito net, and the rumor we were going to South America, we figured that we weren't going to be fighting in Siberia. The consensus was that it would be a tropical location. Everyone began to get rid of any "hawk" gear.[52] Nothing but the basics would go with us.

I took out everything except a poncho, rubberized wet weather suit, field jacket liner, three pair of socks, two T-shirts, and toilet articles such as a razor,

---

[49] One of these two soldiers, SP/4 Schmidt, made the *The Big Picture* photograph in the January 1984 *LIFE* magazine. They were pictured shaking each other's hands after they returned from Grenada, and the barrels for the M-21 rifles are visible poking up out of their rucksacks.

[50] According to *The Rucksack War, US Army Operational Logistics in Grenada 1983* "General Trobaugh had decided not to issue grenades at the individual issue ammunition point because of safety concerns during the flight to Grenada. Instead, the company commanders received boxes of grenades that they stored in their follow-on vehicles. Distribution to the troops would occur after they reached the island. The sole exception was Captain Ritz. The supply of grenades was exhausted at the ammunition point when Ritz's Company B passed through, so he and his men received none".

[51] The M-203 Grenade launcher was a pump action weapon attached to the bottom of an M-16 rifle. It replaced the Vietnam era M-79 grenade launcher. The M-203 fired a 40mm grenade that had about the same killing radius of a regular hand grenade but it was able to fire out to 400 meters. Besides an explosive grenade it could also fire smoke, flare and buckshot for close range. Each squad had two grenadiers, one in each Fire Team. The grenadiers would carry the various rounds in a grenadier vest, but during Grenada the 82nd Airborne grenadiers did not get the vests. One of the few photographs of the 82nd Airborne in action appeared on the cover of the book *Urgent Fury, the Battle for Grenada*. In that photograph two troopers from B Company can be seen going into one of the Cuban Warehouses in the Calliste Area. One of the men was a grenadier, and his 40mm rounds in their plastic packaging, can be seen tied to his LCE and flopping around.

[52] Hawk gear was the nickname of cold weather gear, such as the poncho, long underwear, gloves or field jackets. "The Hawk" was the nickname for the piercing cold that would bite into you when you were wet or the wind was blowing hard. The gear was also nicknamed "snivel" gear.

toothbrush and soap. I also still carried the "alert box" that I made earlier that was full of odds and ends. I packed it all in a large ALICE pack that I had bought the year before.[53] It could hold more equipment than the medium ALICE pack that was normally carried and the soldiers using that issued rucksack could barely fit their equipment inside it.

I threw out the three C-Ration meals, and instead packed just one MRE meal that I had been issued during Operation Bright Star in August.[54] The rest of my rucksack was full of ammunition, food, water, a .45 pistol, M-21 sniper rifle, ART-1 scope and M49 spotting scope. The weight of my rucksack was about 150 pounds and I couldn't stand without another person's assistance.

We left the duffle bags inside that fenced in area and marched back to the barracks. A lot of the soldiers left some of their gear behind because they weren't able to fit it all in their duffle bags. Clothing, C-rations and all sorts of items lay on the ground. Many of the men knew that if they lost something, they would have to pay for it, but they didn't want the extra weight on this real-world mission. We were issued maps with the contingency items, but the maps were grainy Xeroxed sheets that we couldn't comprehend.

When we went back into the barracks, we were told we would be able to get some sleep. I took the time to exchange my practice filters in my M17A1 protective mask with the real combat filters. I had just laid my head down to get some sleep when we were told to get outside and get on the 80-Pack trucks. When we marched outside to get on the cattle cars, they weren't there. We didn't have time to wait, so we marched to Green Ramp with all our heavy combat gear.

We arrived at Green Ramp at 0500 and got into rows behind C-141 aircraft to begin rigging up our gear for a parachute drop. A few men in each squad

---

[53] The ALICE pack was the rucksack. ALICE stood for All-Purpose Lightweight Individual Carrying Equipment and consisted of the LCE and the rucksack. LCE stood for Load-Carrying Equipment and was the pistol belt, suspenders and ammunition pouch. The ALICE pack was replaced after the attacks on September 11th by the MOLLE system (pronounced "Molly").

[54] MRE or "meals ready to eat" were new and we did not get issued them except on rare occasions. We had last been issued them when we were in Egypt during the annual Bright Star mission. We were issued the MREs for the "show of force" jump in Sudan. Libya had been making threatening gestures towards Sudan, so our battalion jumped near Khartoum to make Mohammar Qaddafi back down. A "show of force" was a threat... saber rattling, that usually worked. The normal field food was C-Rations, which was the same canned food that had not changed since Vietnam. The canned food was bulkier and heavier than MREs, but it was a lot tastier.

carried the M-47 DRAGON missile and they rigged up their DRAGON missile jump packs.[55] We all began loading our ammunition with no real guidance. Some men loaded an entire magazine with tracers, to set something on fire, or to mark a target. I loaded the last four rounds of my M-16 magazines with tracers. I did this because when the magazine was about to empty, I would notice the tracers going down range.

I still had the M-21 sniper rifle and I decided to give it to someone who had no weapon at all. I handed it to one of the assistant gunners of the M-60 machineguns, PFC Dodson. All he had was a .45 caliber pistol. I gave him a quick class on how to use the rifle, and then gave him three loaded magazines.[56] I did not give him the ART-1 scope, but figured he could use the rifle like a standard M-14. I gave my ART-1 scope to another paratrooper in my squad, SP/4 Meirs, who would take better care of it, and told him to wrap it up in something so it would not get damaged.

While we were waiting on the airfield an officer ran over to our platoon and told us our destination. We were going to a place called Grenada. None of us had ever heard of it, or knew where it was. I didn't even know it was an island until after I had been there for a half a day. The officer gathered us around him and he kneeled down on the concrete and rolled out a giant aerial photograph of an airstrip. He told us that there were an unknown number of hostages being held by the enemy. He said that the 2nd Ranger Battalion had already jumped in and they had suffered one killed and five wounded in a firefight. He said they were being surrounded and we needed to rescue them. We were told that we would jump onto the airfield in the photograph. It was surrounded by water on three sides and it looked like a peninsula. The officer pointed to an intersection of roads on the photograph and said that our platoon would assemble on that intersection.[57] From there we were supposed to find some American civilians

---

[55] The M-47 Dragon was a medium anti-tank missile. It was wire guided, which meant that after it was fired the gunner could move the sight in the scope, and the missile would fly in that direction. The gunner had to remain kneeling or sitting when he fired, and could not take the crosshairs of the scope off the target until the missile struck. If he did move, the missile would fly off and not hit the target. The missile had a range of 1,000 meters and it could penetrate 18 inches of armor. The missile could only be fired once, and could not be reloaded. After it had been fired the gunner had to remove the sight, which was reusable. The M-47 Dragon was replaced in the 1990s with the FGM-148 Javelin missile.

[56] PFC Dodson was the assistant gunner to the M60 machine gun. Assistant Gunners did not carry rifles, so I was going to give him mine. Before the operation was over PFC Dodson would end up with a captured AK-47.

[57] This was the intersection of roads at the eastern end of the airfield, that would become the site of an armored attack on the airfield a few hours after we arrived.

who lived on the island and get them out of there. The entire operations order for my first combat mission took less than five minutes and didn't really tell us what we would do once we assembled on that road intersection.

The Battalion lined up in chalk order behind the C-141 aircraft. I would be on the first aircraft along with the Company Commander, Captain Charles Jacoby, the Battalion Commander, LTC "Mad Jack" Hamilton and the 82nd Airborne Division Commander, Major General Edward Trobaugh.[58] I had a small 35mm camera that I always carried in a separate ammo pouch and I had SP/4 Meirs take a picture of me. No one knew what would happen when we got to Grenada and I figured if I was killed this final picture of me would be found on my body. After we boarded the aircraft, I tried to get some sleep. This ended up being impossible because we were all fighting the upcoming battle in our minds.[59] Even if some could sleep, they were startled awake from time to time as the aircraft loudspeaker would tell us of what was happening on the island. As we flew southward the loudspeaker told us that the Rangers were still fighting on the airfield, and now the SEALs were surrounded in a fort at St. Georges and being slaughtered. We all wondered what we were about to get into.[60]

---

[58] In 2011 Captain Jacoby was Lieutenant General "Chuck" Jacoby, commander of Northern Command (NORTHCOM) in Colorado Springs, CO.

[59] According to *The Rucksack War, US Army Operational Logistics in Grenada 1983* the first aircraft carrying the 82nd Airborne Division took off at 10:07.

[60] Over the years I have heard numerous units who stated that they were on the first aircraft to land in Grenada, however according to *The Rucksack War, US Army Operational Logistics in Grenada 1983*, "General Trobaugh decided to take most of the key members of the division command group on the first aircraft with him. To make room for them, all the members of Companies B and C shifted to other craft. One platoon from Company A and the company command group, including its commander, Captain Jacoby, accompanied Trobaugh. Colonel Hamilton had initially intended to ride in the same aircraft; however, when Colonel Silvasy learned of Trobaugh's decision, he directed otherwise to mitigate the chance of losing his division commander and the commander of his lead battalion at the same time. Hamilton and the men forming his battalion tactical operations center thus boarded the second aircraft" (2020 edit: Since I wrote that I have been in contact with other paratroopers, who are in my photographs of us waiting to board the aircraft, and they were from Captain Rocke's "C Company". Unfortunately, the actual records of who was on each aircraft and when they took off was destroyed by some 82nd Airborne Division soldier who was told to clean up all the old papers in the museum and get rid of them. So, the lead aircraft did hold more than just A Company, but exactly who was on it, we will never know).

Chapter 1: AIRBORNE - The Invasion of Grenada

In 1983 there were only two Ranger Battalions in the United States Army. The 1st Ranger Battalion was commanded by Lieutenant Colonel Wesley Taylor and was stationed in Hunter Army Airfield near Fort Stewart, Georgia. The 2nd Ranger Battalion was commanded by Colonel Ralph Hagler, and was stationed at Fort Lewis, Washington. The Rangers are the finest light infantry in the world and all the men in the battalions are airborne qualified.[61]

There is a school to earn a Ranger Tab, the US Army Ranger School, but only a very small percentage of graduates from the Ranger school ever get to serve in a Ranger Battalion. Ranger School in the 1980s only lasted 75 days; however, serving in a Ranger Battalion lasted as long as you could continue being a Ranger or until the Army transferred you elsewhere.

Graduates of the Ranger School wore a black and gold "Ranger Tab" on their left shoulder, above their unit patch. The unit patch of the Ranger Battalion in 1983 was an unauthorized black, red and white scroll.[62] The saying in the Battalion is that the tab is just a badge, but the scroll is a way of life.

Each Ranger in the battalion must pass the Ranger Indoctrination Course (RIP) prior to entering the ranks of the Army's "Spartans".[63] Once the Ranger graduated from the RIP course, he was awarded the Black Beret.[64] One of the tasks in RIP is memorizing the Ranger Creed. This is not merely an empty set of words that some may memorize and then forget. Each Ranger would say the Ranger Creed, in formation, prior to physical training each day. Each Ranger would live his life by the Ranger Creed, and if any Ranger violated the Ranger Creed he would be removed from the Battalion.[65]

---

[61] Airborne qualified = paratrooper.

[62] Prior to Grenada the Ranger Battalion scroll was unauthorized as a unit patch. The scrolls were created in WWII, but they were unauthorized then too, and the official patch of the Rangers in WWII was a yellow and blue diamond with the word "Ranger" in it. It proved to be very unpopular with the Rangers because it looked like the "Sunoco" emblem. Appropriately the first scrolls that were authorized in US Army regulations were the two Ranger Battalion combat patches for Grenada.

[63] RIP was replaced with RASP (Ranger Assessment and Selection Course) in 2010.

[64] In 2001 the headgear for the entire United States Army became the Black Beret. The reason for giving the entire Army the same headgear as the Rangers was never explained satisfactorily to the Rangers. The Rangers were simply told to find a new color, or wear Black like everyone else. The Rangers switched to the Tan Beret, but the Rangers still felt betrayed by Chief of Staff General Eric Shinseki, who pushed for the Army-wide wearing of the Black Beret. Old Rangers to this day can be seen wearing T-shirts, depicting a Black Beret, and the words "earned, not issued" on them.

[65] The uniform of the Ranger Battalion during Grenada was solid green cotton, ripstop, "jungle fatigues". If a Ranger was kicked out of the Battalion, he would put on a pair of

When the Joint Special Operations Command (JSOC) was alerted, the commander, Major General Dick Scholtes, thought that the Grenada mission would only need one battalion to pull off the rescue of the students. Scholtes decided that it should be the 1st Ranger Battalion, under LTC Wesley Taylor. The reason for choosing the 1st Rangers was due to them just finishing the yearly REFORGER exercise in Europe and the fact that they were just coming off the Ranger Ready Force (RRF) status.

After telling Taylor to head to Fort Bragg for the initial planning meeting, Scholtes was contacted by the Army Chief of Staff, General John Wickham. Wickham told Scholtes to alert the 2nd Ranger Battalion as well. Scholtes argued that they didn't need another Ranger Battalion and the 2nd Rangers would remain on RRF in case something else popped up in the world. Wickham said no, and ordered Scholtes to "Take 2nd along with 1st. Second Battalion is the RRF. They should go. It's all or none. You either take both or they sit on the sidelines."[66]

Angry at being told how to run the JSOC, Scholtes slammed down the phone and then told his J3 to alert the 2nd Ranger Battalion. LTC Ralph Hagler, 2nd Ranger Battalion commander, soon began his flight to Fort Bragg from his home in Fort Lewis, Washington. On Saturday, October 22nd, both Ranger Battalion Commanders both arrived at Fort Bragg to be briefed on the Ranger's objectives on the island of Grenada. The 1st Ranger Battalion would jump into Point Salines, while Hagler's 2nd Ranger Battalion would seize Pearl's airfield. There was no hard intelligence about either objective; no maps except a photocopy of an old British tourist map, and the officers learned that they could only take half of their battalions due to the shortage in flight crews trained for nighttime

---

BDUs on his way out. The BDU uniform became known as the "Battalion Departure Uniform".

[66] Facebook message from Joe Muccia on October 25, 2020

operations.[67] While Hagler was at Fort Bragg his Battalion flew from Fort Lewis to Hunter Army Airfield, arriving on Sunday.[68]

Taylor chose to just take two companies, A Company under Captain John Abizaid and B Company under Captain Clyde Newman.[69] The reason he only took two companies was due to C Company of 1st Ranger Battalion being assigned to Delta, so Taylor could bring the entire A and B Companies. Meanwhile though Hagler would take all three of his companies, he had to reduce their strength to about 50 to 80 handpicked Rangers. Late in the afternoon on Sunday Hagler learned that he would not be taking Pearl's Airfield, but instead would be following Taylor's 1st Ranger Battalion onto Point Salines and then secure the PRA base at Camp Calvigny, seven and a half miles from the airfield.

On Monday, as the Rangers zeroed their weapons, the two Commanders received the news about the change in H-Hour. Now the entire operation would have to be performed in daylight. With the older plan, jumping in before daylight, Ranger Pathfinders would HALO in ahead of the main element to provide intelligence of the airfield. These pathfinders would freefall from an AC-130 "Spectre" gunship.[70]

Each Ranger Battalion would have five C-130s for the mission. On board these C-130s would be the Rangers who would parachute onto the runway to

---

[67] One of the reasons there was no real time intelligence on Grenada is due to the only CIA officer on Grenada leaving the island right before the operation began because the officer had felt the Grenadian forces were closing in on them. Another issue about the lack of intelligence was due to the relationship between JSOC commander General Dick Scholtes and the commander of the Intelligence Support Activity, Colonel Jerry King. King's command, which had provided some support to the Eagle Claw operation, was left out of the Grenada mission because of the personality conflict between the two commanders. As a result, there was no DoD unit tasked with providing intelligence support to JSOC and the broader US military force deploying for Urgent Fury. As usual, it would be the troops that would pay the price for this failure. (Facebook post from Joe Muccia on October 23, 2021)

[68] Hunter Army Airfield is near Savannah.

[69] Captain John Abizaid would later be the Commanding General in charge of Iraq and Central Command (CENTCOM).

[70] AC-130s are the primary support aircraft of special operations. In 1983 the Spectre gunship had two 20mm Vulcan cannons that could fire 2,500 rounds a minute; two 7.62mm Gatling guns, capable of firing 6,000 rounds a minute; a 40mm cannon; and an automatic 105mm cannon. All of the weapons protruded from the left side of the aircraft. The AC-130 is the most accurate aircraft platform, since all it has to do is circle the target, aiming the left side straight down. The motto of the Spectre gunships is "you can run, but you'll only die tired".

clear the airfield, special gun jeeps armed to the teeth, motorcycles to recon the airfield and to ensure that it was cleared, and disassembled AH-6 or MH-6 black-painted "Little Bird" helicopters to provide air support.[71]

The pilot of the MC-130 that carried LTC Taylor told him that the airfield was blocked with heavy vehicles and they would not be able to land any of the aircraft. The Ranger gun jeeps and the Little Birds would not be able to be used until they cleared the airfield.[72]

Inside the C-130s carrying the 1st Ranger Battalion there was much confusion. Taylor was only able to communicate with the other aircraft through the flight crews' radios because someone had left the hatch radios behind. The Rangers had first been told that everyone except the jeep crews would jump. Then they were told that they would be airlanding. With much cursing the Rangers in one aircraft derigged their parachutes and prepared to land the C-130s onto the airfield. However, the loadmasters then told the Rangers that they would be jumping after all. In just a few minutes the Rangers had to put their main parachutes back on, and rig up their rucksacks and equipments. In the cramped confines of that aircraft, it was chaos.

The 2nd Battalion Rangers decided to leave the reserve parachutes behind since the jump would be too low at 500 feet for the reserve chute to properly open. Normally a parachute drop would happen at 1,100 feet, but the C-130s went lower than that to get below the anti-aircraft fire. As the first C-130 came in low over the runway at 0534, a single PRA searchlight locked onto the aircraft. Everything had gone wrong and for the Rangers it seemed that they were doomed to failure.

Defending Point Salines were several PRA manned AA guns on the steep hills beside the airfield. On the other side of the hills were the old Cuban barracks, and the new barracks 1,000 yards away. The commander of the Cuban defenses was Lieutenant Colonel Orlando Matamoros Lopez. He had been ordered not to engage the Americans unless he was attacked. His "army" consisted of 635 engineers, and 43 military advisors.[73] Matamoros ordered his

---

[71] Rangers were trained to assemble these helicopters quickly once they have rolled off the ramp of the C-130.
[72] Mark Adkins wrote in his book that "Ranger Pathfinders" HALOed into the airfield and two Rangers died when their parachutes did not open. This did not happen and there is no such unit as "Ranger Pathfinders".
[73] One other source claim that besides the Cuban advisors, there were an additional 49 Soviets, 24 North Koreans, 16 East Germans, 14 Bulgarians, and three or four Libyans. These may have been advisors, or communists on vacation to one of the few places that their country allowed them to visit.

men to dig trenches, string telephone wire, and block the runway with whatever engineer equipment they had. Metal stakes were also driven into the airfield, to stop any aircraft from landing. When they heard the firing of the anti-aircraft guns and the roars of the C-130 engines passing overhead the Cubans ran to their trenches in both of their camps.

Hank Collins lived near the airfield and had been hearing aircraft flying overhead since 0300. He had been in the bathroom of his house when he heard "the sound of the circling plane changing and it was coming in close and low. Without pulling up my pants, I ran to the end of my driveway and stood there, buck naked, with my jaw dropped as a C-130 flew right in front of me, with parachutes popping out beneath it. My brain was trying to grasp what was dropping when the dark blob under a nearby chute started kicking its legs. Holy shit!"

Hank ran back into his house and "got under my desk and listened as nothing at first happened. Then the AA crew across the street opened up, and Spectre spoke back in reply. At the time I thought that Spectre were munitions firing from the ground up to the planes. Had I known what the reality was, with Spectre ripping things up across the street, my heart rate would have been elevated, shall we say."[74]

The first C-130 with LTC Taylor ignored the searchlight lighting up his plane and dropped a platoon of Rangers onto the runway. Every PRA soldier with a weapon fired into the air, but none hit anything as the C-130 finished its pass, dove down to 100 feet and then left the area. The following aircraft did not want to fly through all of the anti-aircraft fire directed at the first C-130 and aborted the drop. This left about fifty 1st Ranger Battalion Rangers on the ground to fend for themselves for the next 20 minutes. Two Spectre gunships slowly rotated overhead trying to destroy any opposition, but almost every weapons system on one of the Spectres had malfunctioned. Everything that could go wrong did. The Rangers on the ground were on their own against hundreds of heavily armed enemy soldiers.

A second MC-130 carrying 1st Battalion Rangers was able to drop 20 minutes later, giving Taylor 120 Rangers to continue the fight until the final aircraft was able to drop its Rangers an hour later at 0705.[75] Though there were

---

[74] Letter from Hank Collins to Patrick O'Kelley in May 2014.
[75] The 1st Ranger Battalion had seven C-130s jump in their Rangers. The 2nd Battalion had five C-130s. The 2nd Battalion's aircraft all dropped in sequence after the 1st Battalion, but the 1st Battalion's C-130s dropped out of sequence... aircraft 3, 4, 1, 2, 5, 6, 7. In the middle of the 1st Battalion's jump sequence the Task Force 160th helicopters

only a few Rangers in the southwest corner of the airfield, the PRA was too busy concentrating their fire on the C-130s flying low overhead to be concerned with the enemy on the ground. Taylor and B Company of the 1st Ranger Battalion went to work on the airfield. Construction equipment was hotwired and moved off the airfield.[76] A Cuban bulldozer was hotwired and then used to flatten the iron stakes and push off vehicles blocking the airfield that couldn't be started.

Michael Broland, with the 1st Ranger Battalion, wrote "I landed under canopy watching the hail of green and red tracers just off the south edge of the runway by the NW corner of Bagdad Bay, then quickly jogged 300m east under my heavy RTO ruck to get to the NE corner of the bay ... we cleared towards the heaviest resistance hiding in buildings and positions on the hillside defending this stronghold. It was about halfway across the tarmac ... I looked up just in time to see SGT Tony Davis get up, take a couple steps and fold in half, forward at the waste. This just became very real as the human body doesn't bend like that. I started to get up to run to him but before I even got to my feet SGT Paul Bell was dragging Tony north, so I flopped back down to cover ... under still heavy fire, Paul still dragging Tony and grunting like a power lifter maxing out to get through the hail of rounds to a small area of cover and concealment, so Tony could receive life sustaining aid from a medic... SSG Manuous Boles had the bulldozer fired up and we headed up the hill towards the ADA site..."[77]

"SGT Norm Dietrich riding shotgun squatting on the blade ... firing around the blade ... it all began happening quickly as we advanced although I couldn't tell you actually how long it took to get up to ADA site... some buildings needing clearing, some buildings emptying themselves of combatants voluntarily as the sights and sounds of us Rangers rumbling up the road behind the dozer. M16's and M60's firing and voices shouting commands for pairs and teams to peel off and clear buildings along the way."[78]

Two 82nd Airborne Paratroopers, SGT Charles Spain and SP/4 William Richardson, jumped in with the Rangers. These paratroopers were engineers

---

carrying shot up DELTA troopers landed on the airfield (from a Facebook post from Joe Muccia on 25 October 2013).

[76] Due to Grenada, Rangers were later taught how to hotwire automobiles and equipment. They were also taught how to operate heavy machinery. In 1985 the 3rd Ranger Battalion also included smaller heavy machinery, such as a small bulldozer, in their airborne operations. One idea proposed was to take a small tank, such as the Stuart tank from WWII and rig it up with a bulldozer blade, but the idea of having an armored Ranger unit went against all concepts of being light infantry and the plan was shelved.

[77] Tony Davis was able to get the medical attention and he survived the battle.

[78] Facebook post from Michael Broland on 29 October 2013

44

and were tasked to move any heavy equipment from the runway. Spain told the Fayetteville Observer "When I jumped, they were shooting at us. I could hear the bullets going by me, but the only thing I could think of was the rushing asphalt. I know what hard ground feels like."

When Spain landed it bent the barrel of his M-16 so badly it was useless. He said "I could see a man shooting at me from a house, and he was coming within three or four feet of me. I was shooting back, and wasn't even hitting the house he was in." Spain was able to get an AK-47 off of one of the dead and use it.[79] Ranger Bryan Staggs wrote of another bent barrel caused by landing on the hard tarmac of the airfield, "Erickson bent his 16 on the jump and used his POW (Python) to engage Cubans."[80]

Once the last of the 1st Battalion Rangers had dropped COL Hagler was able to bring in his five aircraft and all his 2nd Battalion Rangers were out in 30 seconds. Miraculously not a single Ranger was killed by enemy fire during the drop. This was mainly due to the Cubans being ordered to not oppose the Americans. The Cubans tried to remain neutral in the fight between Americans and Grenadians, but soon found themselves in the fight. After the invasion Matamoros stated "We thought the U.S. troops were going to evacuate the students and withdraw, but, after they landed, they began attacking our positions and taking prisoners."

Point Salinas

Not every Ranger jumped, and some, especially those with the motorcycles and jeeps, needed to airland. SP/5 James Keen, of the 1st Rangers, had his C-130 "hit by fire and rocked so hard that the man about to jump fell down. By the time the plane circled the airport so we could jump again, the pilot felt the airport was secure enough to land us right on the runway."[81]

As the aircraft slowed to a stop Keen got on his motorcycle and headed off the ramp of the C-130. However, as he did automatic weapons fire poured out of Cuban and PRA positions at the north end of the runway. Keen was only able to get 100 yards when an armor piercing AK-47 bullet punched through the

---

[79] Juan Santos. *I could hear the bullets, Army paratrooper says*, Fayetteville Observer, October 31, 1983
[80] POW stands for Personally Owned Weapon, in this case a .357 Magnum Colt Python. Facebook post from Bryan Staggs on 24 July 2013.
[81] Thomas O'Toole, *In Grenada, an AK47 Round in the Chest,* The Washington Post, November 6, 1983

M1911 pistol he was wearing, and the flak jacket, before it entered his chest. The bullet traveled down his chest to his abdomen, where it lodged in his back.

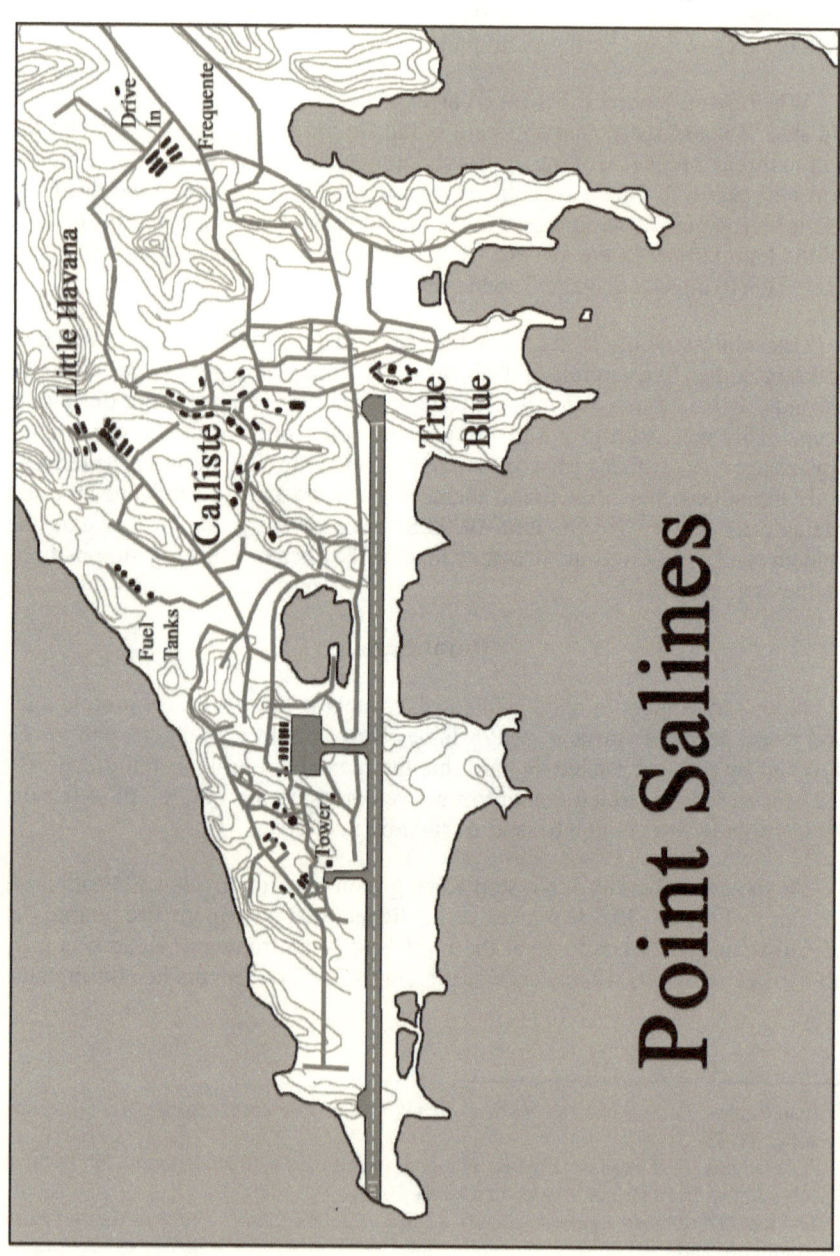

Keen said "The bullet went through my chest but it felt like I'd been hit right in the belly with a baseball bat." He also told a reporter later, "I know I'm lucky to be alive because when I was hit, I was on a motorcycle doing 60 miles an hour and never fell off. I just kept driving because I knew if I didn't get to some cover I'd probably get shot again and really be in trouble."[82]

Keen was able to get to a ditch before he fell off, but it was too shallow. A Ranger yelled at him to run to a deeper ditch that was 50 yards away, but he couldn't run, all he could do was crawl to the other ditch, as the bullets tore up the ground around him.

The Ranger sergeant that had yelled at him to get to cover, ran over to help him, and Keen said, "I kept telling the sergeant to get his knee out of my back and he kept telling me he didn't have his knee in my back. Of course, it wasn't his knee, it was the bullet."[83]

Keen tried to find the bullet hole in his stomach, but couldn't find anything. The Ranger sergeant ripped off Keen's shirt and then discovered the hole in his chest. Keen began throwing up blood, and heard all the enemy gunfire erupting around him. He heard a Ranger yell out, "They're in that house, they're in that house", pointing to a house near the runway. Spectre gunships above the Rangers saw the enemy fire, and fired an artillery round that tore the house apart. Afterwards the enemy gunfire stopped. Keen would later be evacuated by helicopter out to the USS Guam for surgery.[84]

Captain John Abizaid had to get to the True Blue medical campus at the eastern end of the runway, so he ordered two of his platoons to assault across the runway while being supported by M-60 machineguns and LAW rockets. It was at this time that Private Mark Yamane was hit in the neck and became the first Ranger to die by enemy fire.

Ranger Mark Broland who was in Captain Abizaid's company later wrote about how Yamane was killed. "I was … busy in a separate firefight a couple hundred meters east of (the end of the runway) preparing to assault up "goat hill" towards the ADA site. Mark Yamane and Blair Donaldson, two AMAZING HERO Ranger buds, were having their own private war… two vs. many, driving a truck which they had captured from three PRA soldiers and used to drive around inflicting fire upon the enemy, until they themselves in the

---

[82] Ibid.
[83] Ibid.
[84] Ibid.

truck became a sought-after target...they dismounted and used the vehicle as cover it was there that Mark Yamane died in Blair Donaldson's arms."[85]

Ranger Roland Crawford fired his 90mm recoilless rifle at the Cubans in the shack that had fired the weapons that killed Mark Yamane. Smoke rose from the shack when another Ranger, Robert Scott, called in fire from the AC-130 circling in the sky. The Cubans in the shack never had a chance. The burning shack was one of the first things seen, and remembered by paratroopers that arrived later that day.[86]

Sergeant Manouse Boles jumped into a bulldozer's cab, and under heavy fire he drove towards the anti-aircraft positions while his squad crouched behind the vehicle for cover. By 7:30 the Rangers had fought their way past the PRA guards in front of the True Blue medical school and were yelling for the Americans inside to identify themselves. The PRA soldiers wisely retreated northwards into the hills by the runway.

B Company of the 1st Rangers fought their way north to the Cuban's old camp, killing one of the enemy and capturing 22 Cubans. By 10:00 they reached the fuel tanks on a ridge that were 600 meters from the airfield. From this high ground they were able to see down into the new Cuban barracks, known as Little Havana.

Mark Adkins wrote in his book that the Rangers saw the Cubans setting up some mortars, but quickly scattered the enemy crews by firing a captured 12.7 mm anti-aircraft gun into the compound.[87] However Joe Muccia, who interviewed the veterans of that day, wrote "When 1st Platoon, A-1/75 got to the top of the hill, they tried to get the gun working on Little Havana. One of the Rangers hopped behind the gun. But he didn't know that the foot pedals are the triggers. He thought they rotated the gun. Well, he blasted a few rounds off and one of them passed through the sleeve of SGT Buddy Bradshaw's jacket. Once that happened, they un-assed the gun and never tried to bring it to bear on the Cubans. 3rd Platoon, B-1/75 Ranger sniper Bob "Spike" Ollari had a similar incident in their AO as well. But no, the Rangers never purposely put a captured AA gun in action against the Cubans."[88]

A Company had secured the high ground known as the Calliste area at the northeastern end of the runway. The Rangers held their positions on these high

---

[85] Facebook post from Michael Broland on 30 October 2013
[86] Facebook post from Joe Muccia on 22 July 2014
[87] The mortar that the Cubans were setting up also turned out to be a 57mm recoilless rifle and not a mortar. Adkins got many things wrong in his book.
[88] Facebook post from Joe Muccia on 30 October 2013

hills, with their snipers firing down at targets of opportunity, until a Cuban recoilless rifle forced the Rangers to retreat to the reverse side of the hill. The Rangers called in a Marine Cobra attack helicopter to eliminate the enemy recoilless rifle.

Initially the Rangers and Marine pilots could not identify the targets, since they were using two different maps, but soon the Rangers' Forward Air Controller came up with the idea to shine a reflection from a signal mirror onto the Cuban target. The Cobra fired a TOW missile into the building where the 90mm recoilless rifle was located, which collapsed the building.[89] Three Cubans ran for a truck to escape, but another TOW missile destroyed the vehicle. Ranger sniper Eric Foltz wrote "We couldn't talk to them so we just pointed down the valley. On their first pass, one of the Cubans fired an AK47 at the lead Helo. They came back around and lit the building up with a TOW."[90]

Two Rangers on motorcycles, Ron Johnson and Gary Genovese, rode straight towards the Cuban compound, unaware that there were enemy in the buildings. Heavy machinegun fired knocked them from their black-painted Kawasaki 250s. Wounded, they remained in the exposed road, slowly crawling to nearby bushes, and remained there until there was a cease fire later that afternoon.[91] The wounded Rangers were not left on their own since Ranger snipers kept the Cubans from rushing out and capturing them.[92]

At 10:45 Colonel Hagler was amazed to see a C-130 land on the airfield and what appeared to be PRA soldiers offloading. After a few tense moments the 2nd Battalion Rangers determined that these were Caribbean allies from Jamaica, Barbados, Antigua, Dominica, St. Lucia and St. Kitts. No one had told the Rangers that there were any Caribbean allies. These West Indian soldiers and police would be used to guard the hundreds of prisoners that would be captured in the next few days, but for a few dangerous minutes, when they first arrived, they were almost wiped out by the Rangers.

Once the Ranger's C-130s were able to land they could use their gun jeeps. These were standard M151 Jeeps used by the Army, but they were modified to

---

[89] The BGM-71 TOW missile works in much the same way as the M-47 Dragon missile, but it has a longer range. TOW stands for Tube launched, Optically tracked, Wire command link guided. The missile has a range of 3,750 meters and can penetrate all known tank armor. The TOW missile is still in use today, and TOW missiles were used during Operation Iraqi Freedom to kill Uday and Qusay Hussein.
[90] Facebook post from Eric Foltz on 23 July 2014.
[91] Facebook post from Joe Muccia on 23 July 2014.
[92] Some of the Ranger snipers were 'Spike' Ollari, Dave Manges, Eric Foltz, Brian Duffy from B Company, and Preacher Keith from C Co, 1/75.

hold a huge amount of ammunition. During an airfield seizure these jeeps would be used to secure any avenues of approach by the enemy. SGT Randy Cline commanded Gun Jeep 5 and was ordered by CPT Abizaid to secure a road intersection 200 meters north of A Company's position. In the gun jeep with Cline were four Ranger privates. Unfortunately, Cline drove more than 2,000 meters before he realized he was lost, and turned the jeep around. When he drove back down the same dirt road the PRA had noticed them, and now was waiting in an ambush. The Ranger gun jeep was torn apart by RPG rockets and machinegun fire. Though Cline and his Rangers returned fire, the element of surprise was on the side of the Grenadian Army. Cline, Marlin Maynard, Mark Rademacher and Russell Robinson were all killed.[93]

Joe Muccia, the Marine historian who interviewed many Grenada veterans, wrote "Rademacher was still alive after the RPG hit the jeep. He told Romick to head for the airfield to get reinforcements, fired his M203, killing one of the Grenadians and was killed by return fire. He lived the Ranger Creed and didn't falter even though the situation was basically hopeless."[94]

PVT Timothy Romick was wounded, but was able to crawl away from the enemy and get back to his unit on foot. Another Ranger, Rand Miller, wrote that he was "badly wounded and delirious. He was all bloody; shot in five places and was carrying a Soviet AK-47 rifle…He didn't recall how he got the Soviet rifle or didn't even remember how and when he got shot 5 times."

The flight to Grenada took about four hours. When the C-141s were two hours out we were told over the loudspeaker that the Rangers had secured the airfield and we would now be airlanding. I felt two emotions. I was really pissed off that I wasn't going to be jumping into a hot drop zone, like the paratroopers at Normandy in World War II. However, I was also relieved that I wasn't going to be jumping into a hot drop zone, like the paratroopers in Normandy that suffered almost 50% casualties.

At 30 minutes out we were told to de-rig our chutes. We took off our parachutes and our ALICE packs, and the men with the Dragon missile jump packs took those off. While we were de-rigging our Company Commander,

---

[93] Maynard was one of two Rangers who were towed jumpers on the jump into Salines that morning. A towed jumper is when the jumper's parachute does not open and he dangles at the end of his static line until the jumpmaster can either cut him loose or pull them back inside the plane. Maynard and Bill Fedak were both towed and dragged back into the plane, where they landed at Salines later.

[94] Facebook post from Joe Muccia on 22 November 2013.

CPT Jacoby, told me that I would be assigned to his headquarters section as a sniper. I found PFC Dodson, the assistant machine gunner with my M-21 sniper rifle, and took it back. I also took back my ART-1 scope from Specialist Meirs. I gave my LAW rocket to one of the soldiers on my Team that didn't have one. I was going to give my M-16 rifle to PFC Dodson, but CPT Jacoby stopped me and said he wanted it. Company Commanders were only issued a .45 caliber pistol, so I gave him my M-16 and the seven magazines full of ammunition.[95]

Our C-141 came down hard on Point Salines airfield and touched down with a jolt. As soon as it made contact everyone stood up, facing towards the back. There were parachutes and de-rigged equipment everywhere inside the aircraft and in the way of those who were trying to get ready to run off the plane. As our aircraft shot down the runway the tailgate opened up. I was the fifth person in line, right behind CPT Jacoby. In front of him was the Division Commander.

The jumpmaster came down the line and yelled at us that when we got off the plane we were to run to the right, towards the high ground. As the aircraft slowed, we could see the runway speeding past the ramp. Before the airplane slowed to a halt, we were told to get out.[96]

I saw the Division Commander, MG Trobaugh, run out the back, lose his balance, and tumble along the runway. All I could think was "that had to hurt". I then saw the Division SGM run out and tumble along the runway due to the speed of the aircraft. We were not slowing down at all! CPT Jacoby ran off, amazingly keeping his balance and not falling, then it was my turn. I ran out, thinking I would have lightning speed due to all the adrenaline, but it felt like I was barely moving due to all the weight in my rucksack. I stepped onto the airfield, and the forward momentum of the plane threw me backwards. I tumbled along the airfield. I was correct when I saw General Trobaugh

---

[95] Throughout history officers have been lightly armed. The theory is that they should be leading, and not fighting. If they get too wrapped up in fighting, they won't know what is going on.

[96] I have always wondered when did we arrive? In *Urgent Fury, the Battle for Grenada,* Mark Adkin wrote that the 82nd arrived at 2:05 in the afternoon. I remember it being much earlier, and I had always thought that we arrived around 10:00 in the morning. An article written in *Army* magazine matches the time I thought we had arrived at 10:00 (Cragg, Dan. *The U.S. Army in Grenada. Army,* vol. 33, no. 12, Dec. 1983, pp. 29-31). The time of arrival determines who is allowed to wear the Bronze Arrowhead on their Armed Forces Expeditionary Force Medal. Every soldier who was on Grenada was awarded the medal, but the Bronze Arrowhead designates those who were in the first phases of an operation, or the assault landing phase. According to AR 672-3, those soldiers who were on Grenada between the hours of 0635 and 1300 on October 25th are allowed to wear the arrowhead.

bouncing along the ground thinking that it hurt… it did hurt. My first impression of the island was the heat. It was incredibly hot, humid and oppressive. We had left Fort Bragg on a chilly fall morning and now we were only a few degrees above the Equator. I later wrote in my diary that "the heat hit me like a sledgehammer."

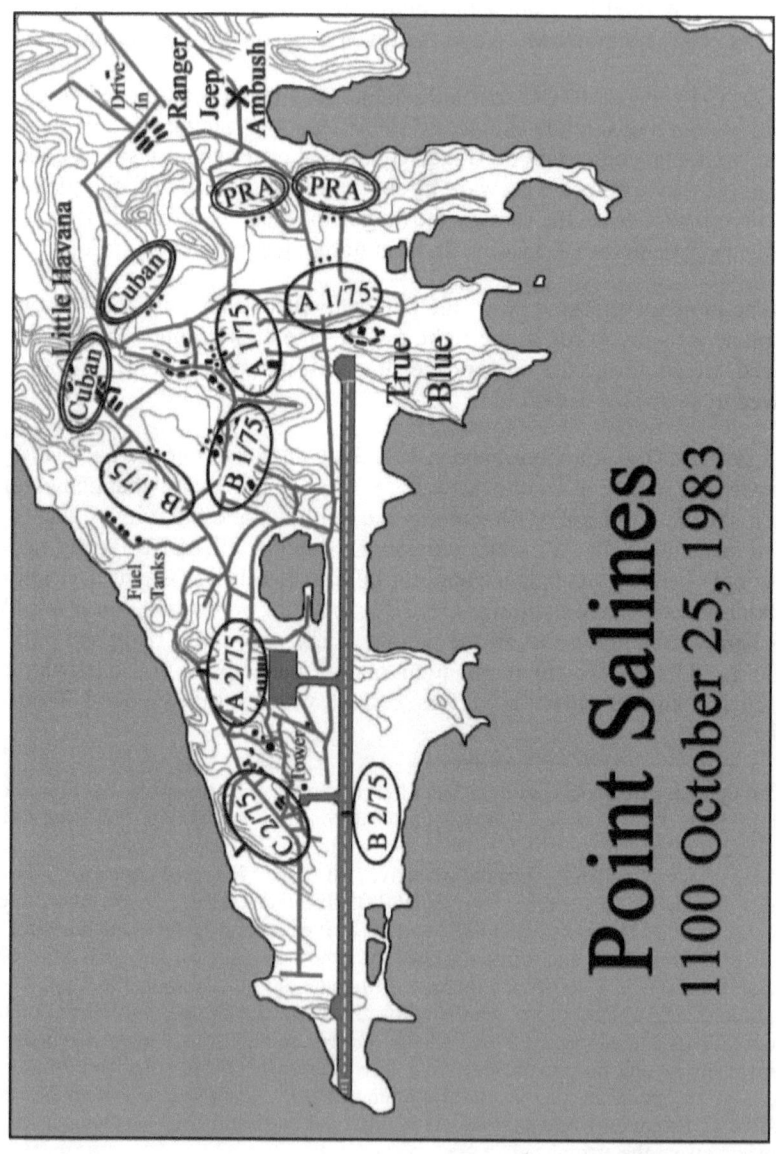

Point Salines
1100 October 25, 1983

My next thought was that someone was shooting at us, since I could hear the rounds snapping above my head. It felt like the bullets were inches away, but since this was my first time under fire they may have much farther off. Though everyone was told to go to the right and run to the hill, everyone ran to the left and took cover on the low ground beside the runway, leading down to the water. Whoever was shooting at us was towards the hill, and we went the opposite way.

I jumped down into that low ground and crouched lower with each shot. I was almost sure it was meant for me. I have always thought that a bullet passing over your head was louder than the rifle that fired it. It sounds like someone clapping their hands right beside your ears and is due to the bullet breaking the speed of sound as it passes by.

As our C-141 turned around at the end of the runway to make way for the next aircraft, I heard the bullets smacking on the aircraft's skin. Beside me a Sergeant Major said "whoever planned this was pretty fucked up!" I agreed with him, and ducked down as the C-141 took off into the air, blowing sand, dust and even more heat all over us. Someone later told me that the CSM who told me it was "pretty fucked up" was the Division Sergeant Major.

Only one airplane could land at a time due to limitations of the runway, so we had to get off of the airfield so the next one could land and unload. After the C-141 banked away into the sky the sniper quit shooting at us. Even though no more bullets were coming our way, no one wanted to move, but we had to. The company rose from the ground and ran to the right side of the runway, where we were told to go to initially.

SP/4 Ray Meier, from my squad, later wrote "I remember when we first exited the C-141, and we were laying on this dirt mound on the side of the runway, and Gen. Trobaugh said "Let's get the hell out of the open, Jack, move out into those hills". Mad Jack then said; Let's go, A Co. then B Co. Then Capt. Jacoby ordered us to move out ...then LT Nicholson said, let's go... That's when I found myself as point man going across that runway towards the tree line. Scared as shit, half way across, I heard someone behind me say, "we should be running". I agreed, and gave the double-time hand signal and we were all up there in no time. Once we got into our hasty positions, we started getting rounds coming our way. I turned to SFC Howard because he was right behind me, and he was curled up in a ball behind a tree, hugging his rifle. I thought, damn he's no good to us right now. Soon after, he shook it off and was back with us." [97]

---

[97] Email from Raymond Meier on September 11th, 2010

Triple Canopy

Mark MacNamara, a young paratrooper with the 2/325th Infantry wrote "As we exited the aircraft, I remember LT Stogner coming out the ramp beside me and I said 'Should we lock and load? "Yes!" Was the answer as we ran towards the hill and path that led up to the top of Goat Hill. There was the sound of firefights in the distance. We saw several shacks along the trail that were all shot up and the first dead Cuban soldier. We moved, leap frog, up the trail and then there was the dead goat, I guess that's why they named it Goat Hill. We saw two or three other dead Cubans and there was a small dog running around...I remember hearing rounds whiz by, but didn't know where they were coming from so, I stayed low and tried to find some cover, but it was bare up there. Looking down the other side of the hill from the airfield was the compound and I saw people running around definitely up to no good. The distance was about 600 meters which is out of M-16 max effective range, and the officers on the hill didn't give the order to fire, and no one else was shooting so we didn't."

"Me and someone else were down just a bit below the jeep trail when three Cubans started firing. Green Tracers were flying up to my right side and red tracers flying down ... it was quite a show, until 2nd Platoon's M-60's started shooting from behind and above my position. 'Damn why did they do that?' I said, 'All that's gonna do is draw fire on us:' And sure enough I saw the green tracers racing up towards us, but thankfully they were high, but I still wasn't safe. For every one tracer there were about eight real machine gun bullets flying over me and the other guy and there wasn't shit to get behind except a few scrub bushes that wouldn't stop piss."

"I yelled up to SGT Hudson that we need to go back because we had no cover. He called the other soldier back and I was left out there with no cover and tracers flying all around. A couple of minutes later SGT Hudson called me back also. I got back in the hole that I had been in earlier looking towards the airport."

As we moved up the hill more of our Battalion came in with the next C-141 and the Cuban snipers off to our right began firing again. I could hear return fire as I neared the top of the hill. As one of our platoons approached the top, they came under fire from Rangers nearby who did not know who we were. Luckily no one was hurt and the Rangers quickly realized their mistake.[98] We continued up the steep hill, following a narrow dirt trail, and struggling under the oppressive heat and the weight of our rucksacks. I passed burned out buildings and enemy corpses left where they had fallen.

Fellow Alpha Company paratrooper Brad Gallardo wrote "David Parker and I cleared the first house we came to on Goat Hill. Once we cleared it, we both went straight to the fridge...no food... But two ice cold Carib beers. We smiled at each other simultaneously grabbed and popped the tops. Drained & healthy belch and back to business."[99]

At the top of the hill was a four barreled 12.7mm anti-aircraft gun beside a house. The gun had either been knocked out or had quit working. Around the gun position were several Rangers observing the enemy. I dropped my heavy rucksack on the ground with theirs. It felt great to get that damn thing off my back. I crouched down with the Rangers and asked them what was out to their front. They were watching a group of buildings and warehouses that were located in a valley about 800 meters from our position.[100] The Rangers told me

---

[98] The platoon that was fired upon told me that the Rangers had done the shooting, but it may have been the PRA or Cubans still around the airfield.
[99] Facebook post from Brad Gallardo on 27 October 2013
[100] This was the new Cuban barracks known as "Little Havana".

Triple Canopy

that inside those buildings were a bunch of Cuban holdouts who wanted to die for their country. The Rangers said that a lot of the Cubans had surrendered earlier and had let hostages go, but these Cubans were not going to give up without a fight. They flew a Cuban flag from one of the buildings.[101]

Our B Company moved to relieve Bravo Company of the 1st Ranger Battalion to the left of our ridgeline. 1LT Jim Bowen of Bravo Company, 2/325, wrote "we were moving forward to relieve B 1/75 in place, 2nd Platoon was moving across a marshy area spread out in fire team wedges in traveling overwatch, with me and the usual gaggle of antennas in the middle. A sniper opened up (the same one who had been harassing the company CP earlier) and rounds started splashing all around us. Every time one splashed, Skule [his RTO] had a comment. POW, splash! "Thank you, sir, that was a good one!" POW, splash! "Thank you, sir, I enjoyed that very much!" POW, splash! "Thank you, sir! Can I have another?!" It was funny."[102]

Before the first aircraft of the 82nd Airborne arrived at Point Salines, CPT Clyde Newman, the commander of B Company, 1st Rangers, decided to attempt to rescue the two wounded Rangers who had been shot off their motorcycles. Newman gathered up some Cuban prisoners and placing them in the front of his men, moved towards the Cuban barracks. He took a Spanish speaking Ranger with him. As they approached the barracks, he told the Cubans to surrender. Two Cubans quickly rose up with their rifles above their heads and gave up. Soon they discovered 150 more Cubans in the compound. There were also 23 wounded Cubans and two dead ones.

A Cuban 120mm mortar crew that had been pinned down all morning by Ranger snipers surrendered. This one mortar could have destroyed any aircraft landing on the airfield. 1LT Jim Bowen, wrote that the mortar probably did not fire since "we were inside of the minimum range... I also believe, based on stuff we found up the hill behind the larger compound, that the battery commander was up that hill. Remember the Soviet doctrine was that battery commanders called their own fires."[103] Though the mortar did not provide accurate fire on the airfield, the Cuban crew had taken the risk and fired rounds that landed randomly around the airfield.

---

[101] These were 1st Battalion Rangers and I most likely spoke to Captain John Abizaid, their commander.
[102] Facebook post from Jim Bowen on 28 July 2014.
[103] Ibid.

Ranger sniper Eric Foltz wrote that "We took the mortar crew out of the game pretty early. They were trying to move ammo from the compound up the hill, but from our position we had a couple of spots where they had to break cover. I had their route figured out by the time Crawford brought his M60 up and we kept them moving after that."[104] All this was happening to the Rangers as I was trudging up that steep hill. In the end eighty of the Cubans decided to continue the fight.

When MG Trobaugh linked up with MG Richard Scholtes, commander of JSOC, he learned that very few of the objectives had been reached.[105] However, there had been some successes such as the Marines to the north securing Pearls airport with little to no opposition.[106]

Though the Marines did not see much action their support aircraft did. Marine AH-1 Cobra attack helicopters fired rockets and machineguns at Fort Frederick. This drew the fire of every weapon around St. Georges harbor. One of the Cobra gunships, piloted by CPT Timothy Howard, was hit. Howard told Camp Lejeune's newspaper "We had begun firing some rockets when we were hit with so much antiaircraft fire, I thought they must have been missiles. I think a round went through both engines." Howard was also hit in the arm and leg. "Next thing I knew, my hand and forearm were lying on the floor in front of me."[107] Amazingly Howard was able to land his helicopter hard on Tanteen Field, a sports field, using his leg to guide the control stick.

---

[104] Ibid.

[105] JSOC stands for Joint Special Operations Command and had only been created a few years before. It was supposed to coordinate all of the different armed services special operations units into one cohesive team.

[106] Kevin O'Hara, flew a CH53 onto Pearls Airport, told me in a Facebook message on 18 October 2020, "I brought 2 jeeps in a shitter (CH-53) on the initial assault! I had one jeep slam into a tree just behind the plane, and the 2nd jeep had nowhere to go so he drove it through the side of our aircraft! It got stuck there and we lifted into a hover and tried to kick it out the back while the pilots tried bouncing the plane! It finally broke free and dropped to the ground from about 50ft. Day 1, initial assault!" I didn't understand what he meant, so he explained it, "The first one shot off the ramp and slammed into a tree directly behind the airplane. The second one moving just as fast as the first one had no place to go, since the first one was just off the ramp and blocking the area behind the plane. The 2nd one hit the brakes and tried to clear the first one, but the airframe was in his way, so he slammed into it! Trying to push the 2nd jeep off the ramp, the tow operator (they were tow jeeps), tried to kick the tow tubes off the stands while we were picking up into a hover and we broke his leg when we picked up! Came back to the ship with the ramp hanging down.(we broke it during the whole bouncing hover maneuver) and the left side ramp fairing tore off! Everyone thought we took an rpg!"

[107] Fred Dodd, *A pilot's story of survival*, The Globe, November 1983.

Howard's co-pilot, CPT Jeb Seagle, was able to pull Howard out before the Cobra was consumed in fire. Seagle moved away to find some assistance leaving Howard in the grass. After he left Howard could see PRA soldiers moving towards the burning helicopter. Howard later told a reporter "I had rounds hit between my legs and I would hear them pop next to my ears. I was laying in an open field; it was a miracle they didn't hit me. They were coming down the hill shooting at me."

Howard felt the earth shake around him and thought it was naval gunfire. What he heard was the approach of a Cobra gunship piloted by CPT John Giguere and LT Jeffrey Scharver. Giguere roared over Howard, blasting the PRA soldiers with 2.75-inch rockets, as their Cobra came under anti-aircraft fire from Fort Rupert and D'Arbeau. While Giguere drew fire from the forts a CH-46 helicopter landed near Howard. AK-47 fire hit the CH-46, but the pilot, Major Melvin DeMars, would not be deterred. Gunnery SGT Kelley Neidigh, a Vietnam veteran, jumped down from the CH-46, threw Howard over his shoulder and ran back into the helicopter. Giguere ran out of ammunition and continued to do dry runs on the anti-aircraft guns to draw their fire. The Cobra was hit repeatedly and then plunged into the bay, killing both Giguere and Scharver.[108]

John Batista witnessed this fight and later wrote "It's embedded in my mind seeing the rocket come up from the ground with the red trail and hit the Cobra square. I remember seeing it drop like a rock."[109] DeMars waited for any sign of CPT Seagle, but realized that he was jeopardizing everyone on board the helicopter. DeMars took off, leaving Seagle, and flew out to sea. Seagle was later found dead, on the beach. Seagle had been cut down by enemy fire when he tried to signal friendly helicopters.[110]

In retaliation for the Cobras being shot down A-7 Corsairs attacked Fort Frederick with rockets, cannon fire and bombs.[111] The RMC leaders took refuge in a tunnel in the fort, which was impervious to anything the A-7s could throw

---

[108] Giguere, Scharver and Seagle posthumously received the Silver Star medal for their actions. Howard and Demars also received the Silver Star.

[109] From a Facebook post from John Batista on 01 November 2013

[110] *Newsweek* magazine decided to publish photos of Seagle lying dead on the beach, which angered many of the veterans of Urgent Fury. Many of the Marines believed that Seagle was murdered by the PRA when he tried to surrender.

[111] The A-7 was a ground attack aircraft that was used effectively in Viet Nam. It replaced the Navy Skyhawks and the Air Force Skyraiders. It was last used in Desert Storm, when it was replaced in the Air Force by the A-10 and the F-16, and in the Navy with the F/A-18. It had a 20mm Vulcan cannon and could carry up to 15,000 pounds of bombs.

at them. Next door to the fort was a mental hospital that flew a Grenadian flag and beside that hospital were anti-aircraft guns firing at the A-7s and Cobras. This building now became a legitimate target, though none of the American pilots knew it was a mental hospital. On most of the maps that the pilots were using it showed this building as Fort Mathews. Seventeen of the patients were killed when the A-7s bombed the anti-aircraft guns. After the explosions knocked down the walls, other patients escaped and roamed the streets. This was a public relations nightmare, which many of the reporters focused on once they were finally allowed on the island.[112]

MG Trobaugh also learned that the Salines airfield was relatively secured, and the medical students at the True Blue campus were safe. However, the Rangers were not at full strength and they could not totally secure the airfield and simultaneously attack the barracks at Calvigny. Trobaugh also learned that a second group of medical students were discovered north of the airfield at Grand Anse.

So far, most of the special ops missions around St. George's had gone horribly wrong. At 0615 SEAL Team 6 flew over St Georges harbor in their seven Blackhawk helicopters and was met with a wall of bullets. Four anti-aircraft guns at Fort Rupert, two at Fort Frederick, all the BTR-60s around the harbor and any PRA soldier with a weapon fired upwards at them. The Blackhawks were hit repeatedly and the SEALs on board were wounded. The helicopter turned back to the USS *Guam*, to offload the wounded and then continue the mission.

On their second attempt two Blackhawks were able to find their objective, the home of Governor-General Sir Paul Scoon, who represented the Queen of England on the island. In 1983 Grenada was still a commonwealth of Britain. Twenty-two SEALs fast-roped into the grounds, while PRA fired upon them from a distance.[113] The SEALs identified Scoon inside the house and placed him in a closet for his protection.

---

[112] Unlike present day conflicts, during Grenada the reporters were kept away from the invasion. Many tried to sneak onto the island but were kept away by the U.S. Navy. Reporters cried that it was censorship and that it was violating the freedom of speech, but President Reagan thought that the reporters would cause more damage than good. The reporters were later allowed on the island, in limited supervised groups, but after almost all of the fighting had ended. Due to this there is very little footage or photographs of the invasion. The primary video that does exist was filmed by civilians on Grenada, or by foreign film crews who were able to get onto the island during the invasion.
[113] At the time of the invasion the tactic of fast roping was considered secret, however today it can be witnessed in many action movies, such as *Blackhawk Down*. Fast roping

According to the Navy SEAL Museum, unfortunately "After setting up at the Governor's mansion, the SEALs realized that their satellite communications equipment was still on their insertion helicopter. As Grenadian and Cuban troops began surrounding the men, the SEALs' only radio ran out of battery power. They were forced to improvise, and used the mansion's land line telephone to call their headquarters to direct AC-130 aircraft fire support on the approaching enemy."[114]

After a short time, a BTR-60 attempted to force its way onto the grounds by smashing through a gate. The SEALs were able to call in an air strike by a Spectre gunship that disabled the APC, but the AC-130 soon had to return to Barbados to get more ammunition. Once the gunship left, the SEALs were besieged in the house, along with the governor-general.[115] One of the SEALs set up a sniper position, armed with his G3 SG-1 rifle. He single handedly kept the PRA at a distance, killing 21 enemy soldiers. The SEALs would remain there, under siege for over 24 hours, until the morning of October 26th.

Richmond Hill

The mission of DELTA and their Ranger support from C Company, 1st Ranger Battalion, was to fly from Barbados into Richmond Hill prison and rescue any political prisoners there.[116] What they found was one of the most

---

is similar to rappelling, but the soldier is not hooked into the rope. Instead he slides down the rope like a fire pole. This is quicker and allows a whole unit to be on the ground in a matter of seconds.

[114] Extracted from *SEAL History: Navy SEALs in Grenada Operation URGENT FURY* at https://navysealmuseum.com/about-navy-seals/seal-history-the-naval-special-warfare-story/seal-history-navy-seals-in-grenada/ on 19 November 2013

[115] Supposedly the SEALs had problems talking in on the radio, so they utilized the telephone inside the mansion to place a call back to Fort Bragg, NC, asking the Watch Officer to contact the *USS Guam* and request fire support. This is the story about someone using a credit card to make a phone call to bring in an air strike that so many veterans have heard about. It has taken on mythological status, and every unit that was there swears that the "credit card air strike" was done by their unit. This was also used in the inaccurate movie "Heartbreak Ridge" starring Clint Eastwood, that showed a Marine Force Recon unit doing everything that the Rangers and SEALS actually did.

[116] The nickname Delta Force stuck after the movie of the same name. Prior to that movie not many knew of the organization. In Special Forces there are different teams. The main operational team is called the A Team and consists of 12 men. The B Team is the company level support asset, and the C Team is the Battalion level asset. Delta Force's actual title was SFOD-D, or Special Forces Operational Detachment – Delta. This was a battalion sized element that was used for special missions. Over the years DELTA has changed its name and moved its training location in an attempt to have some sort of

heavily defended places on the island. The prison itself was built like a fort, with twenty-foot-high walls, and watchtowers around the perimeter. The men of DELTA would have to hover above the prison and fast rope into the yard, while coming under fire from any PRA nearby. Three hundred meters away from the prison and 150 feet higher than the prison compound, was Fort Frederick with all its anti-aircraft guns and its garrison of soldiers.

The pilots of the Blackhawk helicopters expected little to no resistance, but they met a hail of fire coming at them from the fort. The helicopters flew on the same level as the AA guns, firing at them at point blank range. As enemy fire peppered the Blackhawks, their M-60 door gunners returned feeble fire against the enemy. The Delta operators on the inside were being torn apart by the enemy bullets punching through the fuselage.

The pilot of one of the Blackhawks, CPT Keith Lucas, was hit five times in the chest and head and died instantly.[117] His co-pilot, LT Paul Price, was wounded in the head but he maintained control of the helicopter. The Blackhawk flew back towards the coast, but at 0640 it was hit again, locking the controls and driving the helicopter into Amber Belair Hill. Price and some others were able to crawl out of the wreckage before it burst into flames.[118]

As the helicopters hovered over the prison, the DELTA operators saw that the place was abandoned. DELTA operator Eric Haney wrote in his memoir, "The damn place was abandoned! The main gate to the prison was wide open and so were all the doors we could see." As he saw his men get ready to throw out the fast rope, Haney yelled "DO not! Do not! Don't! Don't! Let's get out of here!" All of the leaders agreed and the Blackhawks flew off.[119]

Both Blackhawks and Little Birds of Task Force 160th flew the wounded Delta to the *USS Guam* for medical service. One of the Blackhawks, flown by Major Bob Johnson would not shut down, and was ready to catch on fire on the landing deck. Crewmen rushed forward with their hoses and flooded the intake engines on the Blackhawk before it exploded. Johnson had also been injured and had to be taken to the hospital on board the ship. Overall, 17 of the 39 Delta

---

secrecy. In the late 1990s they were known as CAG or Combat Actions Group. Today they are known by a different title.

[117] Lucas was posthumously awarded one of three Distinguished Flying Cross for his actions during Grenada. The DFC is fourth highest award for combat service. Only the Silver Star, the Distinguished Service Cross and the Medal of Honor are higher.

[118] There is a video of Lucas's helicopter crashing on the peninsula of Bel Air and of the Rangers fast roping onto the site to rescue Price.

[119] Haney, Eric. *Inside Delta Force*, Delacorte Press 2006.

operators and crew of the Task Force 160th helicopters had been wounded, over 40% casualties.[120]

The helicopters carrying the uninjured DELTA soldiers flew to Salines airfield and landed off to the side of the airfield.[121] The DELTA operators pulled security from behind a hill near the terminal as the last of the Rangers jumped onto the airfield. Many Rangers were worried when they looked down and saw helicopters beneath them, with their spinning rotor blades.[122] The DELTA soldiers had two other missions to conduct that morning, but due to the massive casualties during the Richmond Hill Prison mission, the other two missions, and the follow-on mission of the "Hard Rock Charlie" Company Rangers, were aborted.[123]

Around 0700 SEAL Team 6 helicoptered into the Beausejour transmitting station and captured it.[124] They were soon attacked by soldiers of the PRA with a BTR-60 and an 82mm mortar.[125] The SEALs fought back for about an hour, but after suffering casualties and realizing that there was no way they could hold off the PRA with their 9mm MP5 submachine guns, they retreated from the radio station and returned to the beach. There they hid until nightfall, then fought their way to the water and swam the long distance to the USS *Caron*. The tower had to be destroyed but after numerous air strikes and shelling from the 5-inch guns of the *Caron* the radio tower still did not fall.[126]

After hearing of the heavy resistance and of the losses to the Special Forces troops in St. Georges, MG Trobaugh fired off a message to Washington, "Keep sending battalions until I tell you to stop!"[127]

---

[120] Facebook post from Joe Muccia on 30 October 2021.
[121] Some historians have stated that twenty to thirty Americans were killed and wounded in this attack, but they were not included in the official records to ensure secrecy.
[122] Facebook post from Joe Muccia on 29 October 2013.
[123] Facebook post from Joe Muccia on 23 July 2014.
[124] This is also known by some historians as the mission to destroy "Radio Free Grenada", though that station was never taken out.
[125] Adkins wrote that this was SEAL Team 4, but SEAL Team 4 operated on the north coast, and did reconnaissance on Pearl's Airport. Adkins also wrote that the mission happened at 0900, but it happened closer to 0700. The SEALs Blackhawk took off from Barbados 0545 and there was a 45-minute flight.
[126] The holes from the shelling could still be seen in 2013, peppering the concrete parade ground at Fort George.
[127] Critics of the Grenada invasion stated that the United States sent in an overwhelming force to subdue only a handful of enemy. Ironically the opposite criticism was used by pundits about the war in Iraq. They state that not enough troops were sent. Grenada is

CPT Jacoby came over to the anti-aircraft gun where I had taken position and told me to follow him to 1st Platoon's area. Over the next few days, I became CPT Jacoby's "bodyguard" and followed him wherever he would go. The 1st Platoon was located along a ridgeline and they had Rangers intermingled in with them. The Rangers were carrying suppressed MP-5 submachine guns and they wore plain green jungle fatigues and the old "steel pot" helmet. They also were well equipped with M-203 vests and hand grenades that we never got issued. The Rangers in the 1st Platoon area had some PRA prisoners and one of them had been shot through the foot. As a medic took care of the wounded prisoner another Ranger with a 90mm recoilless rifle sighted in on the Cuban barracks below.[128]

I was running out of water and I asked the 1st Platoon if anyone had any extra water, or knew where to get any. They didn't know and were also looking for any place to refill their canteens, since they had the same problem. The lack of water caused many heat casualties in the first day. A few of our paratroopers were trying to question the PRA prisoners, but they couldn't understand their answers. When one of our Spanish speaking soldiers was brought forward to question them, the PRA prisoner replied "I doan speak Spanish mon." It wasn't until that moment that we discovered the language of Grenada was English, but it had such a heavy accent that we thought it was a foreign language.

CPT Jacoby and I returned to the house with the anti-aircraft gun by moving along a narrow dirt trail. I stopped at the house and asked some Ranger snipers what targets were out there, and what the ranges were to the targets. I had no idea what the rules of engagement were, but they told me that they were only allowed to shoot anyone carrying a weapon. There were Grenadian civilians

---

ridiculed for happening too fast and there weren't many casualties. Iraq is ridiculed because the war took too long and there were too many casualties. In all wars there are armchair generals that analyze what should have happened and there are the men on the ground that actually fight them. The axiom of never underestimating your enemy is what should always be used. Hit the enemy with the largest force possible. There is no such thing as a fair fight when men's lives are in danger.

[128] This was the M67 Recoilless rifle, and was a weapon used almost exclusively by the Rangers at this time. Most of the Army had switched to the M47 Dragon missile by 1983, but the Rangers continued to carry the recoilless rifle due to its versatility. It fired both a high explosive round and the anti-personnel flechette round. Many Rangers would carry one of the steel flechette darts in their patrol caps for luck. Though the round could fire 2,300 meters, the effective range was 300 meters. This was replaced in the 1990s by the M3 "Carl Gustav" 84mm recoilless rifle. From talking to other Rangers since then, the Ranger on that 90mm in Grenada was most likely SSG Pickering.

around and they didn't want to shoot them by accident. We were never given any rules of engagement, and were simply told to shoot at anyone who looked like the enemy.

The CO wanted to check on 2nd Platoon's position so I followed him into a house on the ridgeline.[129] I talked to the M-60 team in the house and asked if they had any water, or knew where to get any. SP/4 Edmonds, the gunner, gave me one of his canteens and I gave him one of my empties. That water tasted excellent, but I would need to find a water source soon.

Inside one of the houses someone had found a restaurant placemat with a map of the island on it. It showed St. Georges, the capital, and some cartoon drawings of what the island was famous for. At the top of the placemat, it said "Grenada, Isle of Spices". This was the first time I knew we were on an island, and what the island looked like. I folded it up and carried this placemat around in my pocket for the next ten days as my map.[130]

---

[129] CO stands for Commanding Officer, or in this case CPT Jacoby the company commander.

[130] Carl Wells was with the 319th Military Intelligence Battalion of XVIII Airborne Corps. He knew firsthand about the map situation. He said that he was "an Intel guy on the Corps Planning Cell. ... I knew the map process like the back of my hand. I confirmed this when I got from the island. The Defense Mapping Agency had a Contingency Map Plan that could provide Brigade sized flat stock map packages and aerial photos on 24 - 48 hour notice. A decision was made at the fucking JCS level NOT TO UTILIZE THE DMA CONTINGENCY PLAN. FOR THE FUCKING SAKE OF OPSEC. There are at least a dozen of our brothers who died or were permanently maimed due to that decision. The 2nd night of the fight the Delta S-2 (who was later murdered in Lebanon a few years later) came to the Corps All Source. He let me look at his battle map. I did some quick shuffling and found out there were at least four to five different maps being used on the island at the same time for indirect fire and CAS. None of them had the same grid system except two... 1. Naval Charts used by the Navy for navigation that showed the island, unsure of the scale. 2. Joint Operational Graphics (JOG), Operational Navigation Chart (ONC), and Tactical Pilot Chart (TPC) are three of the products that could have been used by Navy CAS with my probable estimate being the ONC, since the Navy would probably have that on board the carriers somewhere. 3. The Air Force was probably using the TPC since they had enough warning time to quietly procure those. 4. The Rangers/SOF had a locally made map with an arbitrary grid. 5. The 82nd had a map based on one from tourist source or encyclopedia. 6. The Marines had one based on a gas station or tourist map. All the Naval and Joint maps had a UTM grid systems (if you could figure it out from a 1:250000 scale or 1:100000 scale). All of the standard products had the lat/long system, I don't believe that Call For Fire with lat/long was pushed that much at the time. When I took the basic info to my bosses it was the first time I saw a gaggle of field grade officers go pale as a group. The Delta 2 said he

64

The white house with the anti-aircraft gun had become the company CP and the CO, the 2nd Platoon leader and I returned to the house along the narrow dirt trail. I was still wondering where I could get some water, when the air around me crackled with AK-47 rounds. Some Cubans or Grenadians had been waiting right beside the path to ambush us and waited until we were extremely close, then fired a long burst of automatic fire. I jumped over the side of the path so fast that it didn't even register what had happened. Instinct took over. I had not put on my ART-1 scope yet, so I aimed into the bushes by the trail and fired back as fast as I could. When my magazine ran out it became extremely quiet. I called out, asking if anyone was hurt. Amazingly none of us were hit, though we just had a full magazine of AK-47 rounds fired at us, point blank, on full automatic. Suddenly all hell broke loose at the east end of the runway where the C-141s would turn around. It seemed like the fighting had finally caught up to us.

Around 3:00 three Grenadian BTR-60s, under the command of Warrant Officer McEwen, were ordered by Grenadian deputy secretary of defense Ewart Layne to launch a counterattack at the eastern end of the runway. McEwen had twenty-four PRA soldiers in three armored vehicles. Right as the BTRs crested the ridge to the east of the runway A Company of 1st Rangers and A Company of the 2nd Rangers was at the end of the runway, forming a column to march to Grand Anse and rescue the students there.[131]

Abizaid's OP saw the BTR-60s coming fast, one right behind the other.[132] Abizaid had ordered 1st Lieutenant Sydney Farrar to take a patrol out to find out what had happened to the Rangers that were part of the jeep team, Juliet-5. Farrar took SGT Joel Krauss, CPL's Tim Lyle and Tony Nunley, SP4's Kelly Vendon and Max Delo, PFC's John Welton and Jimmy Foxworth, PVT Jose Gordon and medic, SP5 Johnny Bowen.

The Ranger squad cleared a house at the top of a small hill, and captured two prisoners. The POWs were quickly taken to the POW compound at the airfield.

---

now understood why calls for fire/CAS were hitting the wrong place. My boss dragged me to Corps Ops to put a FLASH or CRITIC level message out on it... Some of our guys did not have to die in those days. Facebook post from Carl Wells on 11 September 2021.
[131] Facebook post from Joe Muccia on 23 July 2014. In a later article on his own Facebook page, Joe Muccia wrote that the 2nd Ranger Battalion had begun to stage for a movement to contact with the intent of clearing the PRA training camp at Calivigny
[132] OP stands for Observation Post and would normally consist of a few soldiers in the front of the main lines.

The rest of the squad took up positions around the house, looking for the missing jeep team. They soon had their answer when they saw the burning wreck by the side of the road through their binoculars. As the squad began to climb down the hill, the Rangers spotted the green-painted BTRs rushing down the road towards their position. 1LT Farrar called back on his radio, "I have three Bravo Tango Romeo's moving towards my position!"

The situation was serious, because if the BTRs broke through the Rangers they could have destroyed the C-141 that was taxiing on the airfield, effectively stopping anymore aircraft from coming in, until the wreckage could be cleared. Farrar decided to have his squad attack them first. Farrar yelled over to the squad leader, Joel Krauss and asked him, "SGT Krauss, how big are your balls?"

Krauss's reply was, "Not big enough Sir!" He really didn't want to take on armored vehicles with the few men that were there, but he was also a Ranger. So, when Farrar ordered a few of the Rangers to grab their LAW rockets and follow him, they immediately did what he ordered.

Farrar, Vendon and Lyle all fired their LAWs at the same time, in a tactic known as "volley fire", but it had no effect on the armored vehicles. Ranger CPL Nunley would later tell Joe Muccia, "Two BTR60s came down this same road that we were watching. Someone fired a LAW at the lead vehicle but it bounced off the side from the angle or maybe we were too close. The BTRs keep going and we knew they were headed straight for the A CO perimeter; we were glad they kept going because if they had turned towards us, they would have eaten our lunch."[133]

Whether or not the LAWs were effective didn't matter, the Ranger attack had warned A Company, 1st Rangers that the attack was heading their way, and most of them were prepared and waiting.

The Grenadians drove through the Rangers of the 2nd Platoon and then began firing with their turret mounted cannons. Every Ranger nearby returned fire on the attackers. Ranger Paul Bell would later say that the BTRs had the bad luck of running "into two pissed off Ranger battalions."[134]

---

[133] Facebook post from Joe Muccia on March 10, 2016.
[134] Ibid.

I looked up from where we had been ambushed and saw all the Rangers running towards the sound of the firing. This wasn't AK-47 fire, but was a deep booming, like an automatic cannon. I could hear LAW rockets firing with their loud popping noise. I took off running to the sound of the guns with CPT Jacoby following. We stopped at the end of the ridge and could hear explosions going off. I could see the dirty gray smoke caused by high explosives, but I couldn't make out what the Rangers were shooting at. I saw one of the Rangers down below me rise up and fire his LAW, the backblast peppering me with dirt and rocks. I couldn't see what he fired at since a house blocked my view. I did see the 66mm rocket ricochet off something and fly straight up in the air.

There were figures moving inside a building down below me, but I was not sure if they were friendly or enemy so I held my fire. I needed to see what was going on and it sounded as if every gun on the island had focused on whatever was behind the houses. I thought the Cubans had broken through our perimeter and were counterattacking. I had lost CPT Jacoby as I ran forward, so I was alone as I continued to move forward. I slid down the steep ridge trying to find a target that wasn't concealed by the houses to my front. A Ranger M203 grenadier came up, joined me, and we both moved as a team towards the fighting.

As the three BTR-60s penetrated the Ranger's lines, the drivers must have realized what a fatal mistake this was. The lead BTR stopped quickly and then began backing up just as quick. It slammed into the second BTR that had stopped behind it in the road.

Joe Muccia interviewed several Rangers involved in this fight and he wrote that 2nd Battalion Ranger KK Chinn said "The BTRs arrived and immediately everyone began screaming, "90's FORWARD!" Because of a depression that it sat in; the campus was actually lower than the road the BTRs arrived on. SSG Pohland got the tubes set in a hasty emplacement and began firing at the BTRs." [135]

Ranger Brian Ivers heard the BTR-60s before he saw them. He told Joe Muccia, "I heard engine noise coming from the east and could not believe it when I saw the APCS coming over the far ridge and down into the valley, one behind the other. I was surprised they would try to attack us given the size of our force. I had two LAW rockets with me and was the first to fire at the lead

---

[135] Facebook post from Joe Muccia on 25 November 2013.

APC that was coming toward the airfield.[136] The APC was about 200-300 yards from my position. The rocket hit the very front of the APC and it stopped it for several seconds after impact. Then the vehicle started moving forward again."[137]

Ranger Andy King saw the LAW ricochet off the BTR-60, and told Joe Muccia that it "bounced off like a freakin tennis ball!" Joe Nowak saw some of the Grenadians try to get out of the BTR-60. He told Muccia, "A rear hatch opened and a couple of guys came from the rear of the BTR's, all hell broke loose. I engaged them with a M60".[138]

Ranger Santenalles also opened fire with his M-60 machine gun, and said that he "went through 400 rounds when the 90's showed up and started shooting rounds right over me and my AG.[139] I told Andy to retrieve some ammo from his pack." It took a little while to get the ammo up, and as Santenalles was reloading, a Grenadian soldier appeared about 100 yards in front of his gun. He pulled the trigger and the bullets sawed right through the middle of the enemy soldier, cutting him in half. As he did this Andy King fired a LAW at the BTR-60s.[140]

The Rangers from 1st Battalion opened up on the unlucky enemy armor vehicles. M203 grenadier Ranger Blair fired on one of the BTRs and hit it in the turret with the 40mm grenade round. He quickly reloaded and hit the same vehicle again. Ranger Eddie Payne had a captured RPK machine gun and he fired a full magazine at the enemy vehicles, mixing the green Soviet tracers with the red ones from the M60s. The XO of A Company, Terry Driskill, told Joe Muccia, "You could see silver-green tracers from those commie bullets hitting the three BTR-60s like a Panama rain. Little Fred Swank (another Ranger with a captured machine gun) was right in the way of those bastards. He fired both belts. He fired one magazine after another. He would not burst, just trigger back and empty the thing and grab another. The barrel turned red. The barrel turned white. The bipod turned red. The bipod turned white. Still little Swank ate the magazines up like - well - like his life depended on it, as it did in a way. The little guy burned every damn bullet in that big box, without a breather. Ranger Brent Love said that when the last bullet was fired that Swank got up on a knee and threw the gun forward, at the vehicles, and took up his carbine and started the same thing with it. Not bad for a 17-year-old kid facing three

---

[136] APC stands for "Armored Personnel Carrier".
[137] Facebook post from Joe Muccia on March 10, 2016.
[138] Ibid.
[139] AG is "Assistant Gunner".
[140] Facebook post from Joe Muccia on March 10, 2016.

armored vehicles about 200 yards to his front with nothing between him and them but one line of rocks."[141]

Vietnam veteran and Command Sergeant Major of the 2nd Rangers, James Vovles, yelled out "90's Forward" and told all the other Rangers, that had been filling canteens to prepare for the Calvigny movement, to shoot out the tires of the armored vehicles.

Ranger Jon Krancich witnessed what happened next when Ranger Jimmy Pickering fired his 90mm at the BTRs. Krancich told Joe Muccia, "I heard the fighting and started moving in the direction of the gunfire. As I crested the hill in tall weeds I heard a sound, it turned out to be the last clear sound I ever heard, the sound of a 90-gunner closing the breech of the gun. I looked up from the weeds just in time to see the AG tap the gunner on the head. The gun fired, knocked me back down the hill where I started. My ears were ringing and I couldn't hear a dam thing. I crawled back up the hill to the action, my PLT SGT Donnie Shocklee was yelling for me to join him at the berm, He was a really cool dude, when he saw me and realized what had happened, he used hand and arm signals to bring me in position and then pointed me towards the action. It's all pretty funny now but at the time I saw no humor in it. I started in the campus area and moved up hill to a small berm to my position. Up the hill, we fired small arms at them, the 90 gunners were also firing from our pos prior to Pick moving out and placing effective fire on them. M-16, no effects, I think we were firing our 16's just to make us feel better, all it may have done was to cause them to button up. The 90s making big BOOM's, one from the gun a smaller one from the hit, the BTR's jerked a little and that was pretty much it. Pickering was the 90 Gunner King and in hot demand, morale was real high when we all found out he was up and moving, he took his pos and fired, his shots did a dam dam on the BTRs. He did a hell of a job that day. He was in an exposed position with no regard for his own safety, he just wanted to get the job done and he did. He's a good man."

Delta Force operator Jeff Beatty thought he had a hand in repositioning the 90mm gunners so they could get better fire on the BTRs. He wrote, "I was on my way to the aid station for treatment and I repositioned that 90 crew about 100m further east (out to the next bend in the road). Their original position left the med campus/aid station outside the hasty perimeter. I think my long hair and thick mustache and a little blood/transmission oil on my uniform was all they needed in the way of bona fides. A short while later they took out a btr 60." [142]

---

[141] Ibid.
[142] Facebook post from Jeff Beatty on 13 December 2020.

Ken Bachman wrote "I remember pulling a Law from my jeep and watching the destruction and thought, no need for a law these guys have it under control until I saw a 90 in front of me getting ready to fire. Back blast was not clear, tuck and roll time behind a rock."[143]

Another 90 gunner, Dave Bazemore joined in the attack with Pickering, and Ranger Paul Bell watched as he fired and missed the first BTR with his shot, and but luckily the round slammed into the second one. "Bazemore, positioned about 700 meters from where the vehicles crested a small rise to the east of the airfield, blasted his first round, but the shot was "a bit high over the left-front…on a bit of an angle across the left front corner to the back right side as the vehicle was moving toward us." The 90 round greased the roof of the first one, sliding across and then slamming into the rear of the second BTR-60.[144]

Pickering ran out into the middle of the road, totally exposed to enemy fire, so that he would be able to get a better shot on the BTRs, that just weren't destroyed yet. Pickering outran his AG, so he loaded the round himself and fired. The round hit the hull of the BTR with a "loud clang". Yet another 90mm gunner slid in beside Pickering, Specialist Stephen Long, and aimed at the armored vehicles, surrounded by dust and smoke. His shots did not seem to affect the armored vehicle, but Pickering's next shot hit the left front of the BTR. This also did not do much damage, since the front of any armored vehicles has the most armor. Part of the problem also was due to the both Long and Pickering being 700 meters away, and they were almost out of range of the attacking Grenadian armor.

The Grenadians, not dead, were definitely shocked at the violence being thrown at them, and the driver of the lead BTR-60 put his vehicle in reverse gear and hit the accelerator. However, the second BTR had been stopped by the 90mm round that Bazemore had fired at the rear of the vehicle, and the first vehicle slammed into the second, stopping in the road. As the smoking BTRs began to unload their crews, other Rangers sprinted to the 90 gunners, bringing them more of the 9-pound ammunition.

In *The Rucksack War, US Army Operational Logistics in Grenada 1983* author Edgar Raines wrote "Captain Rocke's Company C was not fully in position at 1530, when three BTR60s attacked. The armored personnel carriers

---

[143] Facebook post from Ken Bachmann on 11 August 2014.
[144] Like most of this account, all of this is from Joe Muccia. One day Muccia will write the definitive history of the battle of Grenada, and all of these stories will be told. I only skimmed what Muccia has offered, not wanting to take too much from his future, and most likely epic, book.

sped down the road that led toward the True Blue Campus directly into the defenses of Captain Abizaid's Company A, 1st Battalion, 75th Infantry. The attackers—members of the Grenadian Army's Motorized Infantry Company—sprayed the landscape with fire, forcing Rocke to redeploy most of his men to a reverse slope. With Grenadian rounds hitting all around them, the infantrymen could not return fire without hitting Abizaid's men. The fire momentarily disrupted both Colonel Hagler's tactical operations center and the 2d Battalion's aid station. Rangers and the few airborne troopers with a field of fire leveled their weapons and responded in kind. A hail of light antitank missiles and 90-mm. recoilless rifle rounds sailed toward the BTRs. Faced with this intense fire, the first two vehicles collided and their occupants fled, leaving two dead behind."[145]

While both vehicles were motionless, they were hit by multiple anti-armor rounds from the 90mm recoilless rifles and the LAW rockets. SGT Kerry Barry, of the 1st Ranger Battalion, wrote "the LAW shot hit the front glacis plate and went under the BTR. At that time the 90-team fired and killed it."[146]

Tim Saint wrote "I did, as well as every other 60 Gunner from my platoon, target the turret [of the BTR60]. The first LAW to hit them was from SPC Love, ACO, 1-75. The impact of the Law hit the back of the first BTR, causing it to stop and the second BTR rammed into it. We kept peppering it with 7.62. It took a couple minutes for the 90s, mortars, and Spectre to get into the fight."[147]

The 2nd Ranger Battalion FO, Bill Spears, had gotten on the radio and told the Ranger mortar crews of A Company to make sure of their settings, since this mission would be "danger close". He told them "Guys, don't mess this up!"[148] The first mortar round was perfect, landing directly on top of the last BTR-60. Spears told the mortars to "fire for effect", which meant that the coordinates were good and quickly fire four rounds for each mortar tube.

One Ranger told *Soldier of Fortune* magazine "Every guy with a 90 or a LAW was yelling, 'Let me at 'em! Let me at 'em! Tubes were firing

---

[145] Raines, Edgar. *The Rucksack War, US Army Operational Logistics in Grenada 1983,* Center Of Military History, United States Army, Washington, D.C., 2010.

[146] Posted by Kevin Barry on the 3-5SFG Yahoo group list on September 4, 2007. The 90 mm that killed one of the BTR-60s was fired by Jim Pickering. He only received an ARCOM with a "V" device for his actions. In any unit, except the Rangers, he would have received a Bronze Star as a minimum.

[147] Facebook post from Joe Muccia on 17 November 2013.

[148] Facebook post from Joe Muccia on March 10, 2016.

everywhere.[149] The rest of us were busy getting out of the way and avoiding the back blast. Then Spectre joined in and there was shit flying everywhere! They blew those suckers to king shit!"[150] SGT James Bradford of the 1st Rangers described the attack by the PRA as a "valiant, heroic, but stupid move."[151]

Seeing that the mortars were firing, the 1st Ranger Battalion mortars of A Company got into the act and also fired. Grenadians that had been in the vehicles got out and ran for the wood line. The 1st Battalion mortars fired white phosphorous rounds into the area that the enemy had run.[152] Rangers watched as the Grenadians were torn to shreds by the effective fire of the mortars. Ranger "Doc" Torres said that he saw one of the PRA soldiers run out of the back of the BTR and "a 60MM mortar round hit him square on the head...dude was there, then an explosion and then he was not there anymore."[153]

The third BTR had gotten off lucky and reversed quickly about 100 meters, turned around and then headed down the road, pretty much unscathed by the horrific firepower of the Rangers. However, the crew did not escape their fate. Above a Spectre gunship was circling. The air liaison for the Rangers, Air Force TSGT Robert Scott said, "I got a call from someone to assist with AC-130's. I moved up to a dirt pile at the east end and called for fire. The AC responded with 105's first and then 20MM."[154]

Ranger sniper Dale Killinger, from A Company, 2nd Rangers, watched what happened through his scope. He told Joe Muccia, the AC-130 "engaged troops as they disembarked the 1st BTR...with the mini-gun. A short while later he used the 105 in the area around the BTR's. I recall a body flying up in the air

---

[149] Hafemeister, Rod. *"The Insertion." Soldier of Fortune*, February 1984, p. 61.
[150] One of the Rangers firing the 90mm Recoilless rifles was SGT Steven Long. Long later went to ROTC in Augusta College and was commissioned an officer in the Quartermaster Corps. Major Steven Long was killed when Islamic terrorists flew one of the hijacked aircraft into the Pentagon on September 11, 2001. He was one of the first military casualties in the ongoing war on Islamic terrorism.
[151] Adkin, Mark. *Urgent Fury, the Battle for Grenada*, Lexington Books, 1989, pg. 225
[152] White phosphorous is a chemical known as a pyrophoric. It ignites as soon as it comes in contact with the air. Nothing puts it out, unless you can get it out of "air". So spraying water or chemicals on it will not stop it burning. It usually burns until it is done, and it will burn through skin and finally stop on maybe a bone. However, it is mainly used as smoke, since though extremely painful, it is not an efficient killer.
[153] Facebook post from Joe Muccia on March 10, 2016.
[154]Ibid.

100 feet or so from the 105. I did see one of the BTR's flip over and a lot of troops were running from them."[155]

As we ran towards the houses, I could see large explosions on the runway and around one of the armored vehicles. I didn't know at that time that this was Spectre firing on the BTR-60s and thought it was artillery fire from one of the 82nd Airborne's 105mm howitzers deployed beside the airfield.[156]

The Ranger grenadier and I ran to the side of a building and I slung my M-21 on my back. Pulling out my .45 pistol I motioned that we needed to clear the building to be able to see out the windows on the other side.[157] He nodded and I ran through the door, crouching low. The house was not large, only a kitchen and a main room, so we were able to clear it quickly and move to the large plate glass window looking out on the fight in the road.[158] The glass from the window was all over the floor of the room and it crunched under our feet. I could hear the explosions from the Spectre gunship going off, and it felt like it was in the street right beside the house we were in, though the explosions were much farther away. The room shook each time there was an explosion.

I reholstered my pistol, unslung the M-21 rifle, and looked over the window sill. I saw two BTR-60s in road, looking to me like it was really close, but was about 500 yards away, not moving, with smoke coming out of the hatches. Behind the vehicles a group of Grenadians in green uniforms moved away from the fight. They were still in the road, and didn't appear to be moving fast, so I did not know if they had surrendered or not.

Later someone told me that there was blood on their faces from the concussion of the explosions and they had been bleeding from the eyes and ears. Suddenly I heard a machinegun opened up on them. The Ranger beside me yelled "get 'em!" and we both began firing. The hot brass ejected from the Ranger's M-16 and went down my shirt. I fired into the middle of the group of PRA soldiers in the road as fast as I could, emptying the twenty-round magazine in a few seconds. After I changed magazines there was no one in the road,

---

[155] Ibid.

[156] This was the 1st Battalion, 320th Field Artillery, but their artillery did not arrive until 0100 on October 26th.

[157] I was worried that that large M-21 would get caught up on something and figured the .45 pistol would be better in close quarters.

[158] I later discovered that this was part of a schoolhouse near the runway. It was still there in 2013.

except a few bodies. From the distance it looked like one of the bodies hung upside down from a BTR-60 by his boot lace. I don't know if I hit any of them or not. I suddenly felt a burning pain in my chest and quickly jumped up. At first, I thought I might have been wounded. I had not felt the brass from the Ranger's rifle until after the fighting had ended, and now it was burning the hell out of me. I yanked my T-shirt out of my pants, allowing the hot spent cartridges to fall to the ground. Since there were no more targets, I looked around the kitchen, and much to my surprise I found various containers filled with water. Finally, I could refill my canteens!

While the Rangers were fighting the BTR-60s, our 1st Platoon had come under fire as they were climbing up Goat Hill. Brad Gallardo wrote "Zamora, SFC Howard and I were sitting there when I snatched Howard's binos from him to watch the Rangers at the end of the airstrip while the BTRs were coming down. I saw a 90 AT team engaging the BTRs and with the first round out I thought they got blown up. I thought, "Oh hell I just seen two Rangers blown the fuck up, till the dust settled and I realized then it was backblast. As I was on my knees watching that, the bushes just started ripping apart all around me. All I heard was zinnggggning and pinging then a sec later "akakakaakakaka!" I got on the radio and called the CO and said we were taking rounds. The lieutenant was nowhere around and I reckoned he was with CPT Jacoby anyhow. The shooters were behind us, but also behind the house on the ridge. They had no idea the rest of the platoon was in front of that house as they were behind it, and behind us. 1st squad moved on the house and captured the guys hiding under the house, with their AKs stashed."

The last thing the Rangers did before they turned over their positions to the 82nd Airborne was to get some of their men back who were trapped by PRA gunfire. A thousand meters east of the runway the PRG mobilization minister, Selwyn Strachan, had set up a CP in a house on top of a hill that overlooked the runway.[159] Right around the time of the armored assault on the airfield a group of Rangers came within range of Strachan's house and became pinned down. Some of the Rangers were wounded, but they weren't able to withdraw due to the intense firepower from the elevated position.

---

[159] PRG stands for People's Revolutionary Government.

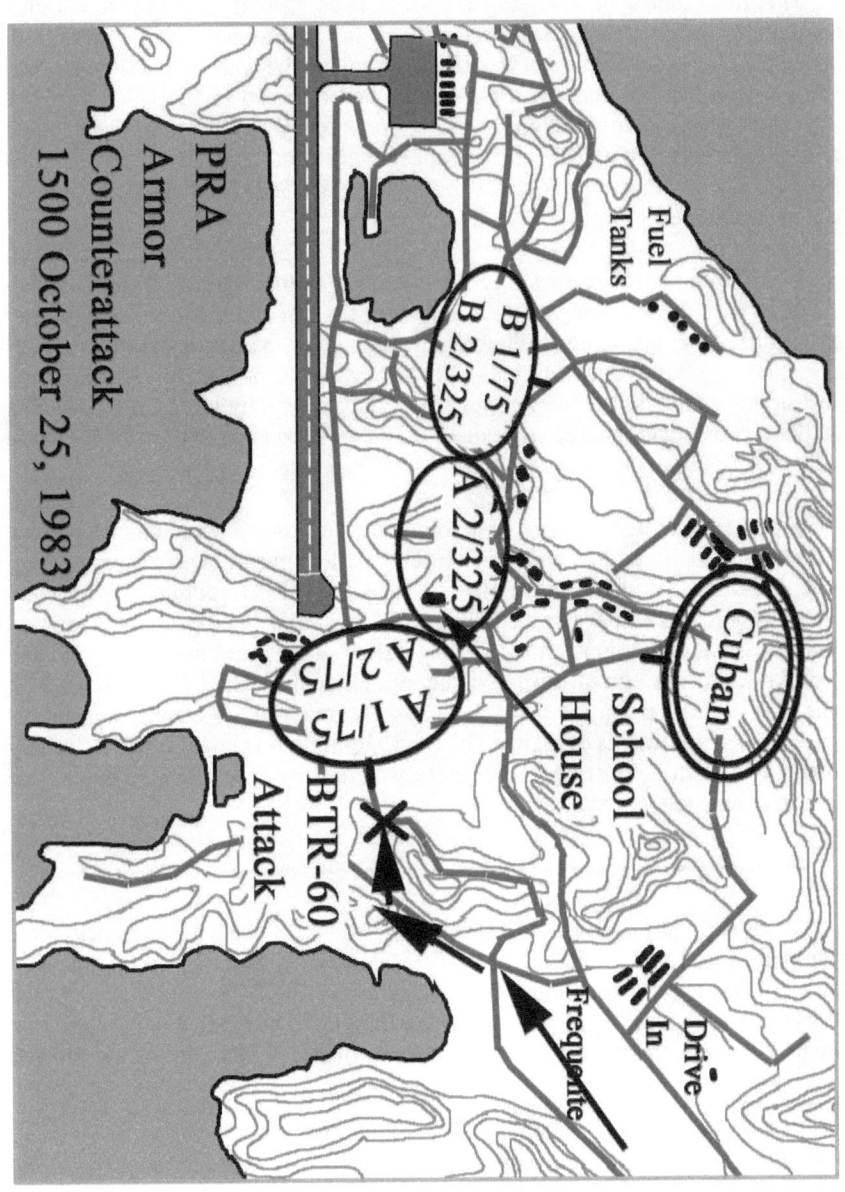

PRA
Armor
Counterattack
1500 October 25, 1983

The fifth stanza of the Ranger Creed states "I will never leave a fallen comrade to fall into the hands of the enemy". The Rangers live by this creed. When CPT Abizaid learned of the trapped Rangers, he wanted to call in an AC-130 gunship. Unfortunately, the gunships had returned to Barbados to refuel and rearm. Abizaid then called in an airstrike of Navy A-7 Corsairs. They made three passes and all the bombs they dropped were duds. For the trapped Rangers this was a good thing, since one fell extremely close and would have killed them if it had exploded.

In Adkin's book *Urgent Fury,* the author states that Abizaid wanted to stop the airstrikes, but the Rangers put out marker panels, and the pilots did another pass, hitting the house and collapsing it. However, SGT Wayne Wood, of B Company, 1st Battalion, 320th Field Artillery wrote "I personally witnessed the destruction of that house within moments of my arrival by one of our howitzers in direct fire." Once source write that the first howitzers from the 82nd Airborne arrived at 0100 on October 26th, but this was incorrect

Wood also wrote "I looked up at a little house situated about halfway up the mountain, which had just been hit by three explosions and was now billowing with thick black smoke. The Grenadian PRA (People's Republican Army) had been using it as a sniper's nest since the landing. They had been having a great time harassing our troops on the airstrip below. I was told later the house had been knocked out by our third gun section, chiefed by SSG Jimmy Hunter. It was the first artillery mission against a hostile force since Vietnam. He and his crew put several rounds of high explosive 105mm rounds into the house."[160] No matter which story is correct; the house was destroyed and the Rangers were able to evacuate to safety.

At some point I had lost CPT Jacoby, so I figured I needed to get back to the Company CP. I told the Ranger grenadier "good luck" and then moved back up the ridgeline to the white house being used as the CP.[161] I found my rucksack in front of the house by a low stone wall where I had dropped it earlier and laid my M-21 on the rucksack to use as a rest while firing.[162] The fighting had slowed

---

[160] Facebook post from Wayne Wood on 27 October 2013.

[161] I have had many civilians over the years ask me if I was afraid in combat. I never was. There were moments of intense surprise, such as when the Cubans ambushed us on the trail, but it really wasn't fear. During combat I was just focused on what needed to be done and nothing else seemed to matter.

[162] In the XVIII Airborne Corps Sniper school the students were taught to use their ALICE packs as a support for the sniper rifle, in lieu of sandbags or other support. When

to just scattered sniper fire coming from the Cuban barracks. I finally had a chance to put my ART-1 scope on the rifle and I also clipped on an M-16 folding bipod that I borrowed off of one of the automatic riflemen.[163] This helped steady my aim.

Though my rifle had been zeroed previously at Fort Bragg, I had been taught in Sniper School that the weapon needed to be rezeroed after the scope was re-attached. The Rangers had all left the perimeter to conduct further missions, so I was the only sniper there. I decided to take advantage of the situation. In front of the wall where I was set up the terrain dropped down about five feet to a dirt trail. One of the machinegun teams had moved there, about ten meters away. I yelled down to them that I was about to start firing and to not panic. I then remembered what the Rangers had told me about the rules of engagement. I needed to see a weapon in the hands of the enemy before I could fire. I sighted in on the Cuban barracks, and waited.

There were about a dozen buildings that looked like double-wide trailers or warehouses about 500 meters in front of my position. Beside these buildings were construction vehicles and military trucks parked in a motor pool. These were all down in the "valley", below where I had my position. I did not see any Cubans in those buildings, but from time to time I could see green tracers come from there, reaching up to the hillsides aiming at anyone the Cubans could see in our Battalion.

There were three long rectangular buildings about 800 meters from my position, higher up on a hill. These buildings looked to be made of cinder block and were painted white or light blue. The Cubans in the compound were openly running back and forth in front of those buildings. There was a taller rectangular building in the back, which appeared to be a warehouse or a garage to fix

---

the M-21 fired part of the flame would travel down the gas tube, pushing back the bolt to load the next round. In the front of the gas tube was a hole, which was used to lubricate the piston. Whenever the rifle fired a bit of the flame would come out of that hole, leaving a blackened burn mark on the rucksack. It became a matter of pride as a way to identify the snipers by having this burn mark on the rucksack.

[163] Unlike the movies made in Hollywood, soldiers are taught to not fire on full automatic unless it is absolutely necessary. This is especially true if the soldier is in a light unit, such as the Rangers or Paratroopers. There is only so much ammunition that can be carried, so each shot must count. In an infantry squad there are two Fire Teams of four to five men each. Within that fire Team there is a Team Leader, a Grenadier with an M-203 grenade launcher, a rifleman firing semi-automatic with the M-16, and there is an automatic rifleman firing full auto with an M-16. In the Rangers there is an additional second rifleman. In 1984 the automatic rifleman's M-16 was replaced by the M249 Squad Automatic Weapon (SAW).

vehicles. The Cuban flag was flying attached to an air conditioner on the larger building. In front of three lower rectangular buildings was a cinder block wall, covered by a corrugated sheet metal roof. This was a covered parking area containing a Mercedes automobile and a Soviet style jeep.

In between the two buildings I could see one of the motorcycles that the Rangers had been riding when they were ambushed and wounded. The Cuban compound sat in a "U" shaped bowl. Behind it was a tall hill that the Cubans still controlled. A thousand yards behind that hill was the Grand Anse beach, where more medical students waited to be evacuated. To the right of the compound was a ridge that continued until it ended at my position at the "white house". A Company occupied this ridgeline, and our CSC Company was to our right on that ridge. The opening of the "U" was to our left, and contained a road that led to the airfield. To the left of the Cuban compound was another tall ridge line occupied by our B Company, under the command of CPT Michael Ritz.

We had the advantage of height. I could see down inside the whole compound, but I could not see behind the buildings and vehicles. However, due to the Cubans being surrounded, both B Company and CSC Company could see behind any buildings that were blocking A Company's fire.

___

After the invasion Cuban LTC Matamoros said "Our buildings were in a hollow, so the defense was organized on the hills around the Military Mission and the new and old camps."[164] As the Rangers drove the Cubans and PRA soldiers back on October 25th their defense collapsed into the camp until there was nowhere to go except the large hill to the rear of the compound.

___

I watched the compound, looking for a target, and saw one. One Cuban, dressed in a khaki outfit, walked casually to the vehicles parked behind the cinder block wall that was 800 meters away. He was carrying an AK-47. There was no wind as far as I could tell, since down in the compound the Cuban flag was limp. I thought that it was great for the Cubans to put out a wind gauge for me. I clicked the scope to 800 meters and put the point of aim on the Cuban's chest as he opened the Mercedes door and leaned down. The shot was loud and it echoed around the ridgeline. I did not know where the shot went. By the time I was able to put the scope back on the target, the Cuban wasn't there anymore.

[164] Adkin, *Urgent Fury*, pg. 227.

Another Cuban looked out of the door of the barracks to see what had happened and I fired at him too. He also disappeared, but I didn't know if I hit him. That was a long shot, with a rifle that was not re-zeroed, so I most likely just scared them both when the round zoomed nearby.

I was a little frustrated that I had no way of knowing where my shots were going. In a perfect world I would have been able to zero my sniper rifle before I went to war, but we were the only company to actually do what the 82nd Airborne says is "the mission". We deployed within 18 hours right into a war. Actually, we deployed in 14 hours, taking whatever equipment and training we had.[165]

Normally a sniper would work as a two-man sniper team with a spotter. However, I was the only school trained sniper in the Company and I had no one to spot for me. I saw our Company First Sergeant, 1SG Bernard Polachowski, watching me. He was probably wondering what I was firing at. I looked up at him and grinned. "I'm trying to zero this weapon, First Sergeant. I'm not having much luck. Want to help me out?" He laughed out loud and came over. I set him up beside me with the M-49 spotting scope.

I told him that I needed to zero the scope, so I chose something closer to aim at. There was a large tank; it was either water or propane, behind one of the closer warehouses 400 meters away. It was white, so I figured that any hits on it might actually show. I fired a few shots and eventually the 1SG was able to see a hit. From there I narrowed the shot group until I was hitting within a few inches of the aim point. I then turned my attention to the Mercedes by the Cuban barracks, 800 meters away. I fired a few shots, with the 1SG determining where they went, until I was able to knock out the front window of that Mercedes.[166]

Once I had established my zero, I began to fire on any Cubans that showed themselves. Green tracers would zip up towards the hill trying to find me. Each of those streams of green tracers gave me a new target to look for. The sun was

---

[165] LT GEN Jacoby wrote to me in an email on September 8, 2007 "As far as I know we are the only unit that actually executed a no-notice N-hour sequence into combat…I have often thought the battalion deserved a unit award, (recall A Co was the Initial Ready Company) and LTC Hamilton deserved far more credit for pulling it off."

[166] I have had a recurring dream since Grenada, and I usually have it at least once a week. In the dream I have a rifle of some type, or a pistol. Someone close to me is threatened, or else I am threatened, but whenever I try to fire the weapon, it doesn't shoot. It just hits on an empty chamber and goes click. This is a nightmare for me, and I wake up with a jolt. I think the origin of that recurring dream is Grenada, when I could not tell if my sniper shots were doing anything or not.

going down and it was getting harder to see, so I did not know if I hit any of the enemy that I fired at or not. The 1SG told me after each shot that the Cubans had fallen down, but this may have been them taking cover as the bullet "snapped" nearby. At the time I thought my harassment fire of the compound must have been effective since they did not fire back again until morning.

When the sun finally set the 1SG left to check the perimeter. I figured that I was only making noise with my rifle, but down in the compound the Cubans most likely thought it was a nightmare.[167] I quit firing once it got dark enough and ate my first meal since the popcorn I ordered at the movie theater on October 24th. I opened up one of the "Bright Star" MREs and leaned back on the stone wall. A small puppy came over and I fed him some of the beans from the MRE. That puppy would be adopted by our platoon and would go with us through the invasion. The squads gave him different nicknames. He was called "Fritz", "Flash", and "SGT Mitchell."[168]

As I sat there eating the MRE I wondered if the 12.7 mm anti-aircraft gun that was a few feet away could be used against the Cubans. There were cases of ammunition for it all around the gun, some spilled out on the ground. One of the magazines of the weapon was lying there also. A couple of guys from the M-60 machinegun came over and we tried to figure out how to make it go "bang". After awhile we figured that it would more likely hurt us than it would the enemy, so we left it alone.[169]

82nd paratrooper Harry Shaw wrote of landing in the evening "No one gave the order to dig in, the fact that tracers were flying overhead was enough to set

---

[167] The Army at that time only had limited night vision devices, and I don't know of anyone that had any type of NVG (night vision goggles) or scopes in Grenada. In training, the only ones who would use such a thing were the key leaders.

[168] SFC Mitchell was one of the oldest paratroopers in our battalion and he was a bit of a legend. He also acted like he was senile or crazy all the time. He had white hair and mumbled a lot. Supposedly he was in Vietnam and had been shot up several times and the legend was that he woke up in a body bag, spit in the face of the chaplain giving him last rites, and crawled out. Supposedly he was also put in for a Medal of Honor, but due to discipline problems he never received it. The legend lived on in Grenada and there was word that each night he would leave the perimeter and hunt down the Cubans with just a knife. It was all rumors, but the Army thrives on these things. SGT Mitchell was ancient, at least 45 or so! One squad called the dog SGT Mitchell because of the white hair.

[169] After I returned home I tried to learn everything I could on how to fire an anti-aircraft gun. It seemed logical that anyplace where there would be an airborne operation, there would be a lot of these guns laying around. Due to my own research, when I went to war the next time, during Desert Storm, I knew how to fire these weapons.

everyone scraping out a firing position in the rock infested soil at the far southeast portion of the airfield where we had taken up positions. It was then that the endless months of training paid off. Dig in, stay down wait for orders! Badda boom! Badda bing! The Grenadian militia and their Cuban allies put on a fantastic show of what not to do on a battlefield. The (Little Birds) would swoop in at treetop level and sure enough a stream of tracers would follow well behind their wake. It was then that the Marine AH-1 Cobra Gunships would pop up from their positions just offshore and a stream of 2.75-inch rockets would come raining down on the fool who had dared to fire."[170]

That night I slept beside my M-21 rifle. I kept it on its bipod, facing the direction of the Cuban compound. The full moon lit up the area, so it would not be easy for an enemy to sneak up on our perimeter. I unholstered my .45 pistol and went to sleep with it in my hand. Sleep that night was sporadic, since from time to time someone would fire into the darkness. I don't think that it was the Cubans and most likely it was our own paratroopers who were seeing things. To add to the noise the CSC gun jeeps fired bursts from their machineguns at suspected targets. To make sleep even harder to find, I continued to wake up throughout the night hearing the A-7s firing their cannon at targets in the distance at St Georges.

After the Marines had secured Pearls Airfield in the north, they planned to land helicopters into St. Georges to relieve the trapped SEALs at Sir Paul Scoon's house. This plan was quickly shelved because of the intense anti-aircraft fire that met earlier helicopter flights. LTC Ray Smith commanded 2nd Battalion, 8th Marines, 22nd MAU and now had to determine where to land his Marines near St. Georges.[171] He also had to keep Pearls airfield secure. About five miles north of St. Georges there was the beach at Grand Mal Bay. Smith decided that one company would land there, while he would fly another company from Pearls airfield to rendezvous with them.

The Marines of G Company began landing there at 7:00 on the night of October 25th with no opposition. These Marines had to wait nine hours on that beach before they could continue. The delay was due to the Marines of F

---

[170] Extracted from *Hardcore Harry's Blog*, http://hardcoreharry.wordpress.com/2010/05/21/d-day-grenada-urgent-fury-part-ii/ on 04 November 2013
[171] MAU stands for Marine Amphibious Unit.

Company flying from Pearls Airport on CH-46 helicopters that eventually landed near the beachhead.

Our 3rd Battalion had yet to arrive and after flying for hours in the crowded airspace over Grenada, they returned to Barbados to refuel. They would not begin landing until almost 0300 of October 26th. When they did begin landing MG Trobaugh gave COL Stephen Silvasy, the 2nd Brigade Commander his mission. LTC John Raines and his 3/325 would move east and clear the True Blue peninsula. LTC Hamilton was to move our 2/325 north to Morne Rouge Bay, overlooking the Grand Anse beach, to support a rescue of the Americans there. To do this we had to go through the Cuban positions at Little Havana. The two Ranger Battalions would be in reserve to support the effort. Though Trobaugh wanted more battalions from the 82nd Airborne to come to Grenada, none would arrive until October 27th.

"Mad Jack" Hamilton designated B Company, 2/325 the assaulting element against Little Havana, while A Company, mine, would attack towards the village of Frequente. After B Company assaulted through the compound, they were to push up the hill to the rear and onto the Morne Rouge high ground. On order, they would then clear the Grand Anse campus. CPT Rocke's C Company would be held in reserve in case the Grenadians counterattacked.

The Commander of B Company, CPT Michael Ritz, did not know enough about his target and decided to do a leader's reconnaissance of the Cuban positions. When the Battalion had conducted the EDRE at Camp MacKall before the invasion, CPT Ritz performed a leader's recon and had been "killed" by the enemy. This was an omen of things to come.

At 0400 Ritz passed through the LP/OP position of his 3rd Platoon.[172] With him were SGT Terry Guinn, LT Steve Seager (the 3rd Platoon leader), 2nd Lieutenant Fidel Perez (1st Platoon leader), his RTO and two others.[173] Ritz

---

[172] LP/OP stands for Listening Post / Observation Post.

[173] RTO stands for Radio Telephone Operator and was the radioman. The radios used by companies were the PRC-77. Terry Guinn gave me some of this information when I contacted him in 2007. Guinn wrote that there were only four people on the leader's recon, beside CPT Ritz. It was two sergeants and two lieutenants. Guinn carried an M-203 grenade launcher, and wrote, "I had the grenade launcher for several reasons. My ETS (end time of service) date for my first three years was just four days prior to the Grenada expedition and I didn't take my oath until that day on which I could have been headed home. I was sworn in to my reenlistment for an MOS in the radar field by Captain Ritz five days before he died. As it turned out in the last month, I went from Squad leader to excess baggage because they had to make sure my position was filled. A grenadier in our platoon had a broken leg and therefore couldn't deploy. No one wanted the grenade

chose where the support position would be for the morning attack and left behind the LT Perez, his RTO and the other two paratroopers, taking LT Seager and SGT Guinn with him. The Cuban perimeter could be seen from that support location.

Ritz continued with Seager and Guinn to a saddle in between the B Company perimeter and the enemy's location.[174] Ritz found an abandoned Cuban position that contained crates of hand grenades and some communication wire leading up the hill.[175] Ritz drew his .45 pistol and made a "follow me" gesture to Seager and Guinn. His choice was not a good one, since Ritz was moving towards the Cubans with only minimal firepower. He was also moving away from his own support.

As they moved forward Ritz held onto the wire, letting any enemy on the other end know that it had been picked up. Guinn later wrote that he saw a flare go off and CPT Ritz was standing there, holding the wire, looking like "Iron Mike".[176] Ritz was hit several times by machinegun fire and killed instantly.

Guinn wrote "I took a bullet through the right arm opening of my flack vest and it exited through my chest just below my left collar bone. It felt like I had been hit by a truck. My very first thought was to hit the dirt (like hitting the dirt at this point was even a choice) and take cover. I immediately returned fire as I felt my life draining away. As I was blacking out, I could only aim in the direction of the tracers that were coming from 10 to 15 yards away. There was no pain, panic or fear of dying at this point, only the thought of returning fire and going out in a blaze of glory. I remember hearing the 2nd Lieutenant calling out for "Captain Ritz" and thinking how bad an idea that was; in essence he was

---

launcher and they were going to leave it behind. I wanted all the fire power I could get, so I talked them into letting me have it." Guinn was not the RTO, but the other SGT in the group was. Guinn could not recall his name. Guinn was with CPT Ritz for security.
[174] The low ground in between two hills is known as a saddle. Basic infantrymen are taught to remember what it looks like by comparing it to the cleavage of a woman's breasts.
[175] The two methods of communication there were transmitting through radio waves and "land line". The land wire was the most secure, since the enemy could not pick it up though radio waves. However it was susceptible to be cut by artillery, or from a clumsy soldier tripping over it.
[176] Iron Mike is the nickname of a giant bronze statue of a WWII paratrooper located on Fort Bragg.

letting the enemy know there was an officer among us. I was in and out after that."[177]

Seager wrote "When we approached the top of the hill, covered with some thicker foliage, the world went nuts around us. I was anticipating some possible engagement so when the first round was fired, I was on the ground returning fire in seconds. Many things happen at this time in a quick fashion. Sgt. Quinn let out a blood curdling scream, bullets flying, etc.... I did fire off a magazine and reload and was a bit surprised I was still living as we were so close that I extended my rifle with my hand and was within feet of the Cubans.

"I figured I was going to take a bullet and die at any moment. I too knowing this remember how I had this wash of peacefulness rush over me and I was alright with this being my last moment on earth. However, which is kind of strange I had this thought, not a feeling, that said "I know I am going to die but I am going to take as many of you son of a bitches with me as I can." Not exactly a Guidepost story moment but a thought non the less I will always recall and never forget."[178]

Upon realizing that I would have an opportunity to make it back I took off my flashlight to get me closer to the ground, it was attached front of LCE,[179] and moved back towards the platoon perimeter."[180]

"As I moved away, having been a lifelong hunter, I gravitated towards thicker, denser vegetation to conceal my escape. Much the same way all those rabbits and deer would do I hunted growing up on our Michigan farm. I also had the luck of a C-130 Spectre firing large rounds of some type and as the rounds hit, I would use that noise to help conceal my next move."

"For the record after I had moved, I am guessing approximately 20-30 yards away from the initial ambush site I was fired on again. This new engagement

---

[177] Guinn wrote "I volunteered to go on the reconnaissance because I was on an adrenalin rush like nothing I had ever felt. I was carrying an M16 with an M203 grenade launcher attached. The Grenade launcher did me no good in the ambush as a weapon, but it made a big difference when it came to balance in a prone fighting position. I didn't have much strength and if it weren't for the grenade launcher, I would have probably been firing into the dirt.".
[178] Facebook post by Steve Seager on 21 December 2020.
[179] LCE stands for load carrying equipment. It consisted of a pistol belt, two 1 quart canteens and two ammo pouches, hung on a pair of nylon suspenders.
[180] Retrieved from www.ranger.org/rangerHistoryGrenadaPersonal.html on 11 September 2007. Seager misspelled Guinn's name and thought it was Quinn.

surprised me but after a few shots from the enemy this quickly subsided, and I went about moving back towards my platoon line. Btw, based on how the initial attack took place I could not return the way I came. I was cut off and had to move what I have come to believe was my left facing the ambush site, westerly and eventually south."

"After being fired upon twice and knowing that the Third Herd would probably be coming to save the day, I was plenty nervous about linking up with my platoon and not getting shot up by enemy fire. As it turned out Third Platoon was moving forward and I saw them before they saw me. I waited in the brush until the outside left platoon wedge was even with me and quickly, and with very strong articulation let a trooper I was very familiar with know my presence. I ended back up with platoon, gave a quick sitrep and proceeded up the hill to take out the enemy position." [181]

Guinn wrote "I came too at one point, when it was still dark, with all my gear gone and I could hear the Cubans nearby. I was told days after that the Cubans had used, among other things, the cellophane from a pack of cigarettes as a makeshift field dressing to stop my bleeding."

The killing of CPT Ritz woke me up with a start. I could see tracers flying into the sky from where B Company was located, but I wasn't able to see any enemy in the early morning light. I was never told that there was to be an attack by B Company. The other platoons may have been informed about the attack, but since I was pretty much on my own, attached to the CO, I was out of the loop. I could see the buildings of the compound through my 10-power scope, so I fired more harassment fire to keep their heads down. To get a better view I moved away from my rucksack and got up on the front porch of the white house that was being used as our CP. Overnight condensation had built up in my scope, so everything I saw was fogged. I wasn't able to detect anything very clearly until it was lighter.

As the sun came up, I was able to make out an intense firefight between B Company and the Cubans. Green tracers going uphill and red tracers going down.[182] The distance between the two groups was less than 50 meters. Some

---

[181] Facebook post by Steve Seager on 21 December 2020.
[182] The Soviets used green tracers and we used red ones. A tracer is a bullet that is filled with phosphorous material. It ignites when it is exposed to the air. As the phosphorous burns, it leaves a red trail behind it, showing where the bullet is going. Normally a

of the B Company paratroopers fired their LAWs at the Cubans, since they had no hand grenades, but the LAWs rocketed through the Cubans, exploding behind them to no effect.

I could see the Cubans in the compound clearly through my scope, but I didn't have orders to shoot. I kept yelling out to CPT Jacoby, asking permission to fire, but he thought I would hit our men by accident. He ordered me to hold my fire. Finally, as the Cubans began withdrawing from the B Company attack, he gave me permission to engage the enemy.

The first Cuban I put my sights on had on a blue work shirt and a pair of blue jeans. The distance was about 600 meters. Due to the long distance, I was able to fire, and then bring the sight back down onto the target to see if it was a hit or not. When I fired the Cuban in the blue shirt stumbled, and then looked right at me with a surprised expression on his face. He then fell down and did not get up. I later found out that I had hit him in the side of his chest, the bullet exiting out the other side.[183]

I fired at other Cubans running, but I don't know if I hit them or not. Though I was the only sniper in my company, a couple of CSC snipers showed up at my position to see if they could hit any targets. They came up on the porch with me and soon we were all firing away at anything that moved. We were not even trying to conceal ourselves and it seemed like every Cuban in the compound began to fire in our direction. When the bullets started slamming into the tin roof and the walls of the house where we were located the CSC snipers dove off the porch and laid low in the bushes, seeking cover from the rain of fire. I stayed on the porch, continuing to fire at any enemy I could see.

SP/4 Ray Meier watched me on the porch and years later wrote, "The next morning (our first morning there), you and Howard were positioned in front of a white house on the front porch. Haring, Zamora and I were at the first position to your left. I was carrying two LAW's and Howard didn't have one, so he asked me to bring him one of mine. This is before any of the firing started. I went over to you guys with just the LAW, leaving my M-16 at my position. (Dumb, even only being 10 yards away). Anyway, I was half way back to my position when all hell broke loose. I dropped right where I was, in tall thick grass. Listening to the rounds fly over my head, I thought, I can't stay here much longer. It was then that I saw a thick blade of grass get cut in half by a bullet

_____

machinegun would have one tracer round for every four regular rounds. So for every tracer that is seen, there are four bullets that aren't detectable to the human eye.
[183] I know he could not see me on my porch, almost a half a mile away, but he looked right in my direction. Through the 10x scope it appeared that he looked right at me.

and fall right in front of my face! I immediately jumped up and dove to our position. We then started receiving fire near our position, and I remember Haring yelling at me; "Thanks a lot Meier!" Marborough our M-60 gunner was on our left. He wasn't firing, so we started yelling at him to fire his weapon. He yelled back that he had no targets. We told him to just spray the area in front of us, so he did. At the same time all this is happening, we looked over at you and Howard. You were locked in on someone taking shots with your sniper rifle, and we could see that you were also getting shot at, because right above your head and behind you, that white house was getting nailed. After that fire-fight, we knew you had a big set of Kahuna's!"

Enemy tracers are pretty scary to watch as they seek you out. They start out slow, just barely moving, and then as they get closer, they scream in faster than the speed of sound. If it is a machine gun firing at you, there are four unseen bullets chasing after that tracer. I figured there were at least nine Cubans with AKs or machineguns firing at me on the porch. The tin roof on the house sounded like someone was throwing rocks on it. I continued to fire at all the enemy firing towards me. At one point I felt a pain in my hand and looked down. A splinter had lodged into the back of my hand and I pulled it out with my teeth, and then continued to fire.[184] Blood trickled down my hand, decorating the front porch rail. I saw another target and pulled the trigger, but nothing happened. Unfortunately, something was wrong with my rifle. It wouldn't fire anymore!

It took me a second to realize that my magazine was empty and I had just run out of ammunition. I reached down to get another magazine, but the other three magazines I had were also empty. I only had four magazines and I had emptied one at the hidden ambusher the day before, one at the BTR-60, and between last night and this morning I fired two more magazines. I had more ammunition, but it was in my rucksack, about twenty yards away, near the anti-aircraft gun. This was a stupid cherry thing to do, and I should have reloaded my magazines after I had fired them.

I left my M-21 where it was and I jumped down off the porch. I ran to my rucksack, opened it up and grabbed the boxes of MATCH ammunition. As I ran back to the porch our Company XO yelled at me to get down.[185] He said

---

[184] After the invasion I heard of the huge amount of medals given out to units that had not even been there on the first two days. I also heard of Purple Heart medals given to paratroopers that never saw action, and only had a minor superficial injury. I joked that I should get one for the big splinter that went into my hand. It had to have come from a near miss when a bullet hit the wooden rail on that porch.

[185] XO = Executive Officer, the second officer in command of a company.

that they were shooting right at me and the rounds were about two feet above my head. I laughed and then yelled back "When they get down to about six inches, I'll duck! Right now, they can't hit shit!" I quickly loaded a magazine and then rammed it in my weapon. I took up a good kneeling position on the porch and picked up where I left off, firing at whatever Cubans I could see. If I couldn't see them, I fired at the location where the green tracers originated.

The M-60 machinegun to the left of the porch opened up on the Cuban compound, laying down suppressing fire. The bullets were striking everywhere, with no real pattern. It had to be frightening to be in the bullet's beaten zone. Whenever I saw a Cuban head pop up, I fired, making it duck down again. The noise was deafening!

Sean Bermingham, with an M-60 crew, later wrote "I always remember you on the porch of that house on the first day, since I was only 20-30 feet to your right, popping off round and getting hits! Colonel Silvasy, the new Brigade Commander, had to hit the dirt when the rounds started flying in our direction!"[186]

Over in LT Nicholson's 1st Platoon perimeter they were firing mortars and machineguns down into the Cuban compound. I did not know of the mortars until after the invasion and thought the explosions in the compound came from naval artillery. I also did not know about our own 105mm artillery that was at the west end of the runway, firing over our heads and into the compound. Before the day was over, they would have fired 600 high explosive rounds.[187]

John Sporgitas was part of the artillery that fired into the compound and he wrote "On the morning of the 26th we received word through FDC that the infantry was receiving heavy machine gun fire from the compound and they called a fire mission. The FOs and FDC were having a difficult time plotting the targets because of the poor-quality maps they were using. Our section sprang into action. At the time, I was one of three SP/4s in the section, so I was the #1 man or loader. SP/4 Lemons was gunner and SP/4 Westover was AG."

---

[186] Letter received from Sean Bermingham in September 2013 right before the 30th anniversary reunion.
[187] The artillery howitzers arrived at 0100 on October 26th. Initially the artillery had set up south of the runway, but the high explosive rounds would then fire over the aircraft landing on the runway, making a potential disaster possible. During the night the artillery moved to the north west end of the runway. For the Rangers waiting on the runway to go home and for incoming 82nd Airborne paratroopers, the pounding of the guns was a constant noise.

"We could hear small arms fire in the distance to our front on the other side of the airstrip and behind the ridge. Heredia and Dyment were busy prepping the ammo. I went to my position in the trails but saw that Westover was already there yelling at the guys to get the ammo going and bring him a round. So, I had to go to the AG position. I was a bit worried because I didn't have much experience at that position and I didn't want to screw it up, especially on that mission. I tried not to rush entering the elevation numbers into the sight and elevating the tube. Westover loaded the first HE round and I closed the breach and grabbed the lanyard handle, standing by. Then SSG Hunter yelled "fire" and I pulled the lanyard to fire the first round down range."

"At that point, we became the first army howitzer to fire in combat since Vietnam. Our training had kicked in and everything we did was like second nature. We fired nine more rounds with the FO sending adjustments to FDC. After the final round we were told that we scored a hit through the roof of one of the barracks. I don't know what we hit or if we hit anything at all because we could not see the target because of the ridge in front of us. Once we stopped firing, we heard the A-7s starting to make strafing runs at the compound flying directly over us. We could see dark smoke come out of the front of the aircraft and then a couple of seconds later we heard the sound of its Gatling gun. They made multiple passes over us coming in from the sea to our rear. I couldn't imagine being on the receiving end of all that fire."[188]

SGT Gerald Bannon, from A Company's 1st Platoon, saw movement to his front during all this chaos, and rose up on one knee to get a better look over the grass. He saw a Cuban at the same time, doing the same thing that he was. The Cuban fired first, hitting Bannon in the arm and knocking him backwards. He yelled out "I'm hit", but a nearby paratrooper didn't believe him. Since there was so much noise no one heard the shot fired at Bannon. The paratrooper yelled back, "You're kidding!" Bannon answered, "NO! I'm really hit!"[189]

SP/4 Ray Meier also saw Bannon go down and later wrote to me, "We were all yelling for him to get down, and he was just to your right. Just to the right of the porch you were on. I guess you were too busy sending rounds down range. Anyway, Haring, Zamora and I thought they got him good because he went flying back. However, by the time the fire-fight ended, the medic and a couple other guys had him on a jeep, and he was extracted fast. We didn't hear

---

[188] Facebook post by John Sporgitas on 27 October 2013.
[189] This story was told to me by Bannon after we returned to Fort Bragg. He was shot in the upper bicep. He wanted to keep the bloody shirt, but his wife washed it and sewed up the hole.

until an hour later, or so, that he was shot in his upper left arm. Next time we saw him was in pictures at Walter Reid Med. Cen."[190]

Brad Gallardo wrote "I HAD Cubans shooting at Me, John Zamora and SFC Howard. Tony Peebles and his squad were a little further up the trail and captured the guys hiding under a crawl space under a house. I got a call over the radio telling us to come get flak jackets. I informed SFC Howard and he tasked me to collect a team of guys and go get them. I made contact with SGT Gerald Bannon, and while I was asking him for one of his guys a bullet tore thru his tricep and exited out his bicep.

I checked it out and called for SP/4 Clancy (medic) and proceeded on with my task. The Cubans followed my movement down the tree line until we hit the opening and they cut loose on us. I returned to cover and tried again. Same thing again, so I chose an alternate route."

82nd trooper Walter Hall wrote "Early in a.m. hell broke out... it was freaking loud! I remember being in the high prone position looking for the enemy, when bullets started whizzing by my head. I got down on my stomach. He (the enemy) was locked on me! I started digging into the ground. Somebody from my unit took him out. MY GOD, I ALMOST PISSED MYSELF!!"

Gary Marsh was with the A Company mortar section on Goat Hill and saw the Cuban who had wounded SGT Bannon, he wrote "If we would have had rounds for our mortars, I could have took out that sniper that got Sgt Bannon. I had him marked on his position but no ammo; it was to arrive any time after unloading it off the plane."

---

[190] Email from Raymond Meier on September 11th, 2010.

Give me my god,
what have you left?
Give me what no one else
would ever ask.
I dont want riches, not success, nor even health...
I want insecurity and unquiet,
I want torment and chaos.

Battle of Calliste

One paratrooper, PFC H___, was so shaken by Bannon being wounded that he feigned being injured and limped to the airfield. We didn't see him again until after we returned to Fort Bragg. Gallardo wrote about the incident "When Bannon got hit and I called Clancy, H___'s eyes about popped out of his head from fear. He was just on the other side of Percy, got scared and bailed. After I got Clancy on Bannon, I grabbed Percy and Cambell and went to the TOC. H___ was still there with the rest of 1st Squad but when we got back, he was gone. I asked Sengbusch what happened to H___, and he just smiled at me, shook his head and said he twisted his fucking ankle or some bullshit...then gave me that look like "What bullshit!"[191]

Meanwhile, Cuban bullets were still coming at me from the farthest compound 800 meters away. It would have been easy for the Cubans spot our position from there due to the M-60 machinegun firing below me, spitting out a tracer every fifth round. I also had myself and the other two CSC snipers on a porch, on what seemed to be the only white house in the area. After the initial flurry of bullets that had come my way, the CSC snipers returned to the porch. I could see the enemy tracers race up to us. They would hit the ground in front of us, spraying dirt and rocks around us, or they would slam into the walls and roof of the house. I couldn't see any Cubans that were shooting my way though, just their tracers. To keep them suppressed I fired into the windows of the barracks and at any heads that would pop up.

Rounds from the company mortars began raining down on the hill to the rear of the compound. Each round would explode a little closer to the compound as they were walked into the target. One round finally hit one of the buildings that looked like a warehouse and set it on fire. Our Company M-203 grenadiers fired at the closer buildings, about 400 meters away. There must have been Cubans in them, but I couldn't see anything due to the trees in my way. Mark Banker, one of the grenadiers, later told me they were aiming for a gasoline truck in the Cuban motor pool.

---

[191] Facebook message from Brad Gallardo on 30 November 2011

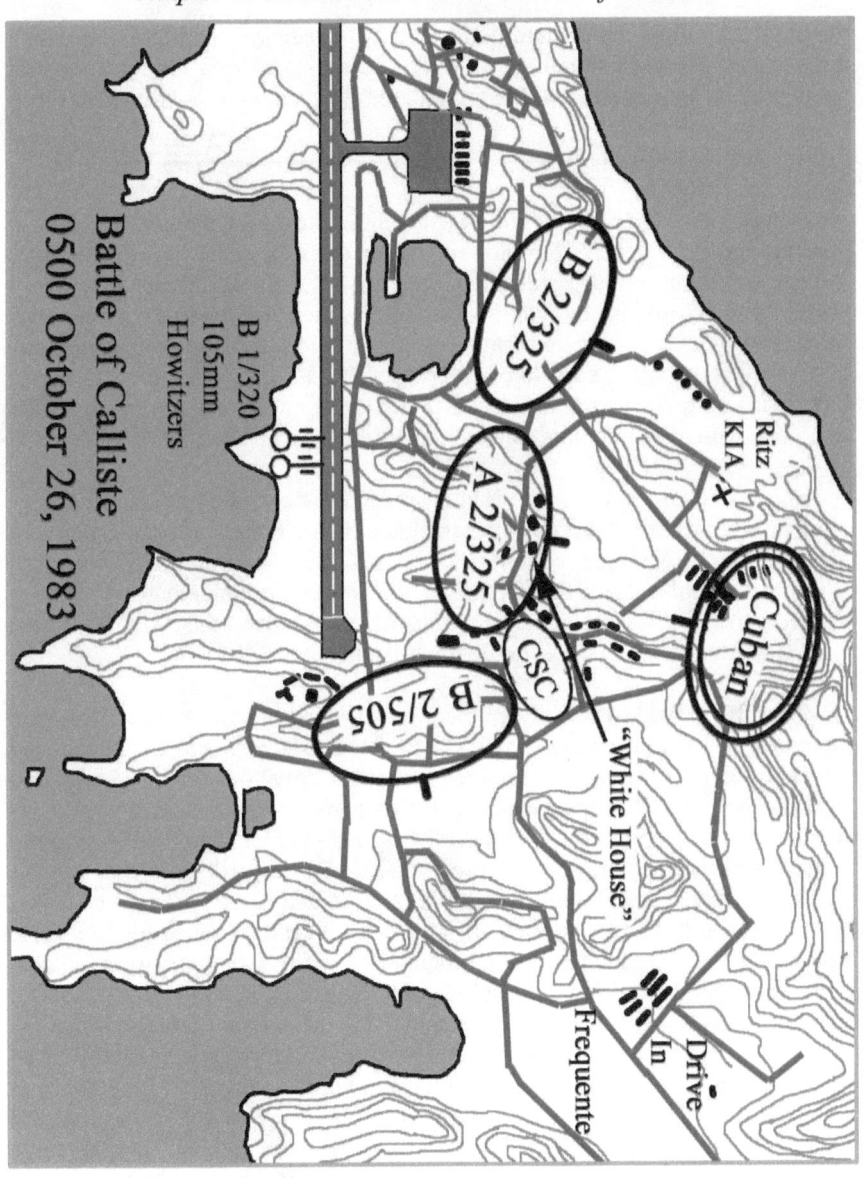

Battle of Calliste
0500 October 26, 1983

B 1/320
105mm
Howitzers

B 2/325

A 2/325

B 2/505

CSC

Cuban

Ritz
KIA

"White House"

Frequente

Drive
In

The truck was at the extreme range of the grenade launchers, so they weren't able to hit it. One of the grenades landed on a warehouse and set it on fire. As the building burned the ammunition stored inside began to explode. The small arms ammo popped continuously, while tracer rounds from the machineguns sputtered and spun around on the ground, setting even more buildings on fire.

Bowen later wrote, "It was like getting hit in the gut. Up to that point I had been talking to LTC Jack Hamilton on the radio because CPT Ritz hadn't made it back, but in my mind, I still thought he, 2LT Steve Seager, and SGT Terry Guinn were coming back. Top Madsen telling me I needed to give that order made me suddenly realize that they weren't coming back and I was in command. Let me tell you, there is a world of difference between saying "Follow me, men," and giving an order to a unit that you're not leading. The latter is much harder to do than you might think, especially under those conditions."[192]

Bill Blaine was "Mad Jack" Hamilton's S-3 officer, and he wrote that they knew Ritz was dead, "because after the burst of fire there was no contact with Cpt Mike Ritz. My recollection was that Jack Hamilton told me he was dead and to get a replacement CO for Mike. No indecision Immediate. Matter of seconds, likewise the movement of 1st Lt Mike Okita and the order from then LTC Hamilton for the 1SG to go up the hill and get him."[193]

Capetillo moved cautiously up the hill and found two Soviet recoilless rifle positions that had been abandoned. There were numerous AK-47s abandoned on the ground and ammunition scattered about. Capetillo left his 1st Squad, led by SSG Gary Epps, to secure the area. B Company never received any grenades or mortar rounds before they left Fort Bragg, but they had been able to acquire some grenades after arriving on the island. SFC Capetillo did not like his situation and suspected another ambush. He decided to do a technique he had used when he had been in Vietnam known as "reconnaissance by fire". He brought up an M203 grenadier and ordered him to fire a grenade twenty meters to the front, and keep firing them as he moved forward. Capetillo moved forward with one squad, while another squad covered them. When Capetillo's platoon was almost to the top of the hill, the third grenade fired by the grenadier flushed out the ambush site and the Cubans opened fire prematurely.[194]

The 3rd Platoon rushed forward and captured the hill, but continued to receive a high volume of fire from the Cubans. Capetillo's men had thrown all their grenades and their M16 ammunition was running low, so he ordered them

---

[192] Facebook message from Jim Bowen on October 8, 2019.
[193] Facebook message from Bill Blaine on 31 December 2020.
[194] Raines, Edgar. *The Rucksack War*, pg. 374.

to not fire unless they actually saw a target. One of his men had been wounded in the firefight. When they ran out of grenades one of the troopers found several cases of Soviet grenades. Capteillo was wary of the foreign grenades and threw one of them towards the Cubans, counting the timing of the fuze. When it exploded after four seconds Capetillo told them "They were good to go" and his men began throwing the Soviet grenades at the Cubans as fast as they could. The Cuban firepower slowed when they realized they were not going to take back the hill from the 82nd paratroopers. As they searched the hill one of Capetillo's men, Kevin Carlisle, found SGT Guinn.[195]

Guinn wrote "I came to again after daybreak to the sound of gunfire and voices. Someone yelled out 'There's a dead body over there!' All I could think of was that they were talking about me. I screamed, 'No'... ok, so at that point it was probably more of a thought than a scream. The medics started working on me and asked, 'Can you walk?' I think they were serious! They tried getting me out of the area that would soon be the target of an air strike, by carrying me in a sitting position between two soldiers on a rifle. That didn't work and with the air strike about to begin, a machine gunner threw me over his shoulder and carried me back to the Company CP. I had lost so much blood that my own Platoon sergeant didn't recognize me, 'Who the &%$# is that?' In the medical tent at the air strip, I heard the doctor tell someone to get the chopper that had just left, and that it would return in 20 minutes. The doctor said it had to come back because 'This man doesn't have 20 minutes' as he worked to try and stabilize me."[196]

Capetillo was ordered to withdraw, but not before they found CPT Ritz. The Cubans saw them leaving and put up a half-hearted attack, but Capteillo was able to get his men, the wounded Guinn and the body of Ritz back to B Company. As he fell back to the position held by 1st Squad, he discovered the body of SSG Gary Epps. Epps had attempted to unload one of the Soviet

---

[195] Raines, Edgar. *The Rucksack War*, pg. 375; Facebook message from Kevin Carlisle on 26 August 2019.

[196] Guinn wrote that after he was extracted "I was kept away from the other soldiers most of the time, which was probably due to the risk of infection. The only soldiers I came in contact with, other than top brass making the rounds (including Secretary of Defense Weinberger), were a PFC who had a bullet bounce around in his Kevlar helmet and another soldier who shot himself in the foot. I flew by transport helicopter with them to Walter Reed Army Hospital. Immediately upon landing, the PFC and I were greeted by a full bird colonial who apologized to us for us having to fly with the soldier who shot himself in the foot. We were treated like royalty."

recoilless rifles when the round went off, killing him and wounding five of his squad members.[197]

Steve Logan, one of the paratroopers with Epps wrote, "Third platoon had no grenades until after Epp's death. I found two cases of Russian grenades and was showing my men how to check the fuses for "zero time" grenades when Epps was killed. The recoilless rifle 'exploded' when Epps opened it; (his body flew too far for it to have been a grenade in the tube). I carried someone to the beach after that attack, but I thought it was an enemy soldier. Doc might remember. My memory is a wreck from that blast (don't remember much after that, for eight days)."[198]

Kevin Carlisle was also there, and he wrote "I caught some shrapnel and some singed parts during the blast that got Epps. I was taken down with Guinn. I remember trying to help when he got to medical tent. They had me holding up an IV bag and squeezing trying to get fluids in as fast as possible. I did so till I started getting light headed I guess after the adrenaline starting wearing off."[199]

John Clift was one of the paratroopers in Epp's squad, and he wrote "I was way too close to SSG Epps when the round went off, I remember the heat and the bright white flash and flying backwards When I hit the ground there was a lot of blood and I heard someone hollered that there was a dead man over here, I thought to myself dam I am not dead. Later to find out that it was SSG Epps.

---

[197] Epps is one of the only two paratroopers killed from the 82nd Airborne in the invasion of Grenada. The Ritz/Epps Gym and Ritz/Epps Field on Fort Bragg are named in honor of them. There are differing versions of what happened to Epps. We were told that he had been killed by a defective Soviet style grenade. One version of his story was that he was trying to place the grenade in the breach of the recoilless rifle, when it went off. Terry Guinn wrote "during the pullout from the attack in which I was found wounded, when the air strike was coming in, SSG Epps tried to destroy some Cuban equipment with some hand grenades he found. One of those hand grenades exploded when he pulled the pin. We were not issued hand grenades." Brad Gallardo wrote "His squad was in pursuit of the PRA that was dug in behind the compound on the hill behind it... As his squad was moving thru a Recoilless rifle position with bad guys in front of him...He didn't want to leave the recoilless rifle behind him without doing something to keep it from being used...They had no time to break it down so he took out a grenade to booby trap if. He pulled the pin and inserted it in the breach then shut the breach door...As he walked away it opened and the frag fell out behind him detonating and killing him and wounding a few in his squad".
[198] Facebook message from Steve Logan on December 21, 2020.
[199] Facebook message from Kevin Carlisle on December 21, 2020.

I also remember the air strike starting pretty close to where medics were working on us."[200]

Lieutenant Bowen remembers, "An hour or so later, after SSG Epps was killed and 3rd Platoon had several wounded on that hill, I sent my platoon under SFC "Bull" Harris to reinforce them in the support position, then began to move out with 1st Platoon to get on line for the assault when all of a sudden we got both artillery and air support, and the Cubans started surrendering. A long day in a young 2LT's life. I will forever be grateful for the leadership of 1SG Madsen and the platoon sergeants - I realized later that they knew a long time before I did that CPT Ritz wasn't coming back." [201]

During the attack Lieutenant Mike Menu was near a house where a Cuban was holed up. Menu was attached to Bravo Company from the 1/320th Artillery on their FO Team. As the troopers moved past the building a grenade was thrown in. When it exploded it killed the Cuban "blowing his knee off" and Menu was wounded by the shrapnel coming through the walls.[202]

SSG Lalone with the gun jeeps in CSC Company, had spotted several Cubans and he pulled out his sniper rifle. He wrote "I was able to shoot 3 of the enemy as they came over the hill, but my scope was screwed on the M21. I carried two weapons in, folded my M16 w/203 in my ruck."[203]

CPT Ritz's body was carried back to the airfield on a stretcher by paratroopers Michael Maris and J.D. Shipp. As it passed through the paratrooper's defensive line, each man was able to look on the face of their fallen commander. Anthony Peebles wrote "I remember seeing Capt. Ritz lying on the ground and Lt Col. Hamilton was talking to some men around him, and seeing him and knowing he was KIA, made tears of anger come to my eyes and all I wanted to do was kill somebody."[204]

Brad Gallardo wrote "As LT (Nicholson) and I were walking up, I saw both Ritz/Epps under ponchos with their feet sticking out from underneath. I looked to the right where 50-60 Cubans sat with their hands tied behind their backs. I rotated my selector to full auto, then saw Mad Jack talking to the 1st Ranger Battalion CO and XO", which changed Gallardo's mind about committing a war crime. In all situations where soldiers decide to cross that line, there is the

---

[200] Facebook message from John Clift on January 13, 2021.
[201] Facebook message from Jim Bowen on October 8, 2019.
[202] Facebook message from Joe Muccia on 13 August 2013.
[203] Facebook message from Charles Lalone on 30 September 2014.
[204] Facebook message from Anthony Peebles on 30 May 2011.

Triple Canopy

leadership that either stops it or allows it to happen. Mad Jack's presence was enough to stop the slaughter of the Cuban POWs. [205]

Ritz would be flown back to Fort Bragg, where his wife was waiting. She was seven months pregnant.

The artillery rounds were used to mark the targets for pilots flying the A-7s. Normally white phosphorous was used to mark targets, but the 82nd howitzers did not have any WP rounds. John Sporgitas, the assistant gunner on the only firing 105mm at that time, wrote that his artillery "had HE and no smoke. Our Goats and most of our gear and ammo arrived later that day." [206] Jim Bowen, the acting Bravo Company commander, wrote "the purpose of the fire mission was because the A-7s couldn't ID the target and they needed a spot. One round did hit the flagpole... the FO, a buck sergeant... walked the rounds down the hill to the compound. The fire mission was HE quick – I was told there was a restriction on using WP because of the fire hazard.[207] But HE did the trick."[208]

The artillery and mortar barrage were stopped so that two A-7s could strafe the compound. When the A-7s dove on the Cubans the cannon rounds ripped up a baseball field 100 meters in front of the buildings. I saw bodies flying through the air. At first, I thought that it was cows or goats, but I later discovered that it was Cubans who were in a trench. The way the bodies just flew up and were torn in half just didn't look like people to me anymore.

The A-7s continued to dive on the Cubans, and with each pass they were closer to the main building where the Cuban flag was still flying. This was the building that had been firing at us the most and the flag was an easy identification mark for the pilots. I stood up and yelled out "This Bud's for

---

[205] Facebook message from Brad Gallardo on 31 May 2011.
[206] Facebook post from John Sporgitas on 23 July 2014. The Goats he wrote about was the M561Gama Goat.
[207] HE quick meant High Explosive rounds that would explode when they made contact with the ground. Other types of HE rounds were "delay", which meant they would explode a fraction after making contact... so that they could penetrate building or bunker walls, and "air burst" which meant they would explode in the air above the target and rain shrapnel down onto the enemy.
[208] Facebook post from Jim Bowen on 23 July 2014.

you!" when one of the aircraft made a direct hit on a warehouse.[209] Chunks of concrete flew in all directions.

The Cubans in the far compound were being suppressed so much that they were just raising their AK-47s up over a concrete wall and spraying the hillside. I looked through my scope and saw a Cuban pop his head up behind a concrete wall. One of the CSC snipers fired and the head dropped. I saw the same Cuban pop up again, so I fired. The head disappeared and did not rise up again. I could see at least four other Cubans behind that wall all "spraying and praying" in our direction.[210]

Hugging the ground behind that wall was Colonel Pedro Comas Tortolo. He had been sent by Fidel Castro to take charge of the mission in Grenada and had arrived on October 24th. Laying there with him behind the wall was LTC Matamoros, the commander of the Cuban defenses, and Carlos Diaz, the senior civilian Cuban official.

Matamoros wrote "When the sun came up, they concentrated their fire, throwing mortars, planes, cannon and machine guns against us. I was wounded by the fourth mortar shell. Tortollo wanted to see what the matter was with me. I shouted I had been hit in the waist but was OK. He asked if I could crawl over to where he was, and I said no, but that we couldn't hold that position, and that we should leave it and go to the hill behind the offices to be protected from a direct hit." [211]

I saw one of the Cubans behind the wall get up and start running to the door of the building. Judging the shot I saw that there was no wind and he was running directly away from me. This is what is known as a zero-lead shot. I did not have to compensate at all for windage. I fired. Due to the long range, I was able to come back onto the target after the recoil and watch as the bullet hit the Cuban in the back of the head, blowing it apart and spraying a red mist

---

[209] I have no idea why I yelled that. Youthful enthusiasm I guess. The Budweiser slogan was fairly new and must have stuck in my mind. After we returned from the invasion one of the local T-shirt shops printed a picture of a mortar, with the round flying out of the barrel, and the words "This Bud's for you!" underneath. I guess I wasn't the only one who thought it was a neat saying.

[210] "Spray and Pray" is the nickname for not aiming when you fire on automatic.

[211] Adkin, *Urgent Fury*, pg. 262.

towards the barracks. Another Cuban began to get up when he saw the A-7 coming in for another attack. As he ran the 20mm cannon rounds tore into his upper body, blowing him apart and the momentum threw him through the door of the barracks.[212]

Another Cuban, wounded in the leg, was dragging himself towards the same open door. His comrades in the door waved at him to hurry. I yelled to the M-60 gunner that there was a target right in front of that door. The M-60s weren't able to see anything that far away and were relying on the snipers using their scopes to guide them onto targets. The machinegun bullets impacted all over the wounded Cuban's body. The CSC snipers also fired into the body, riddling it. Amazingly the Cuban continued to crawl, and made it into the doorway of the barracks.

The final pass of the A-7s hit the side of the building with the Cuban flag. The wall fell outward and the flag tumbled to the ground. B Company had advanced close enough to the buildings to fire M-203 grenades and LAW rockets into the concrete walls. One rocket exploded on the side of the building, sending the wall crashing inward as dust blew out the windows.

Suddenly I heard the unmistakable sound of a DRAGON being fired.[213] There was a lull in the fire as everyone recognized the sound and turned to watch what would happen. The missile hit the largest building making a huge explosion, smoke and dust covered the area. The Cubans finally had enough and a white flag waved frantically out of one of the windows. The order to cease fire was echoed around the ridgeline. Cubans came out of the buildings waving white bed sheets, letting us know they weren't going to fight anymore. They gathered in front of the smoking buildings, carrying out their wounded.

CSC Company was ordered to take one of the gun jeeps down to the compound and accept the surrender of the prisoners. Several of the Cubans ran out the back of the buildings and up into the hills. We weren't allowed to fire due to honoring the white flag, but CPT Jacoby gave the order to fire into the hill in front of the Cubans to discourage them from trying to get away. When the M-60 let out a long burst several of the Cubans in front of the building ran back inside the barracks, but when the CSC gun jeep showed up, they came back

---

[212] In Adkin's book Matamoros states that Diaz and another Cuban were killed by a grenade or a mortar, however no one was close enough to fire a grenade and the mortars had also stopped firing. I think what Matamoros witnessed was Diaz being hit by my shot, and then the other Cuban being hit by the 20mm round from the A-7.

[213] The DRAGON launches with a large explosion and then a series of popping noises from smaller charges are heard as the optical sight guides the missile onto the target.

out again. As the paratroopers searched the surrendering Cubans, more of the enemy came out from the bushes and trees on the hill. It looked to me to be about fifty prisoners.

During the intense firefight Cuban LT Salina Garcia, the only woman in the compound, received a radio message from Castro ordering them to continue fighting and not surrender. Most of the defenders did give up, but Tortolo, Garcia and a small group of Cubans made their way out the back and up the steep hill to Morne Rouge. Their goal was to reach the Soviet Embassy on the other side of the ridge.

Down in the compound the prisoners and the dead were counted. There were 86 Cubans who had surrendered and sixteen dead were found in and around the buildings. After the firefight was over, I was able to relax for about an hour. I continued to overwatch the barracks, but the fighting was over for now.[214]

The 1st Platoon moved down into the valley and checked out the warehouses for any survivors who wanted to surrender. As the paratroopers moved between the burning buildings tracers popped and sputtered in the fires, sounding like popcorn cooking, only much louder. LTC Hamilton moved with the 1st Platoon and as they passed by the bodies of the Cubans on the ground, they fired into them with their rifles. Some of the M-203 Grenadiers fired at the corpses, to make sure they were dead.[215] Brad Gallardo wrote that "There were some blown up Cubans and PRA's down there, mangled beyond recognition."[216] Terry Book said that he saw one of the enemy without a face, and was "obsessed" with finding it. He wrote, "wasn't much blood splatter, but he still had part of his forehead and lower jaw. He was next to an old 1940's car. His chest was rising and falling much more calmly than mine." [217]

Brad Gallardo wrote of the carnage by the wall, under the carport that I had been firing at earlier. He took a seat by the wall to rest and saw "two KIAs at my feet on either side of me...One I could see he'd been hit in the leg, it just exploded. The other was clearly hit with 20mm... It went thru a 6" steel post then thru an 18" brick wall and cracked his head open like you took a black diamond watermelon and cracked it across your knee. Split open completely thru to his chin."[218]

When the 1st Platoon searched the buildings, they found cases of ammunition and weapons stacked to the roof. There were bayonets still packed in Cosmoline, which was quickly snatched up as souvenirs by the paratroopers. All of the soldiers who were armed with just a .45 pistol uncrated the AK-47s

---

[214] During this time I pulled out my camera and took a picture of the compound, with smoke rising above it. I also had time to write in my journal what I had witnessed.

[215] Firing upon the dead was taught as the proper way to clear an objective. As the attacking troops move across they fire into each of the enemy, since it could not be certain if they were faking or not. It is better to fire extra rounds into a corpse, than to have that same "dead" body come back to life and attack from the rear. Once the objective was cleared it was then considered improper to fire at the bodies. When 1st Platoon went across the compound the objective was not considered cleared yet. Except for the few CSC gun jeeps that had accepted the surrender, none of the American units had gone across the objective yet.

[216] Facebook message from Brad Gallardo on 25 October 2010.

[217] Facebook message from Terry Book on 25 October 2021.

[218] Facebook message from Brad Gallardo on 25 October 2010.

and carried them for the rest of the invasion.[219] It quickly became obvious that this was not a stockpile just for the Cuban workers. Evidence would show that it was going to be used to spread communism throughout Latin America.

SGT Sengbusch had carried his personal .357 magnum revolver with him and as Gallardo cleared the bodies, Sengbusch covered him. Gallardo wrote "We spied out a dead Cuban laying in front of a small building. I didn't want to just go past it without checking it out for booby trap (paranoid something might have been under it) and someone behind us would get blown up. So, I decided to do the old EIB body check.[220] Once I got all wrapped up on him and began to slide my hands under, I thought "Oh man, what if his guts and crap are all out?" I changed my mind and said "to hell with it" and Sengbusch popped the body with his pistol."[221]

There were also tons of food, such as tins of sardines, which were packed into the cargo pockets of the paratroopers and eaten later. Many of the soldiers had left their food back at Green Ramp to lighten their load. Among the neatly stacked piles of supplies were cases of Cuban cigars and cigarettes, which was a huge relief for the smokers in the company. Many of them had not packed any cigarettes during the hasty invasion preparations and would literally kill for a smoke. Since we still did not have any grenades the cases of Soviet grenades were broken into and passed around, though after the rumors of Epps being killed due to booby-trapped Soviet grenades, we were wary about using them.

After 1st Platoon had moved through the warehouses the rest of the company was ordered to move around the compound, searching the houses there. The hills on the sides and rear of Little Havana were steep and men began going

---

[219] As mentioned earlier, the men on crew-served weapons, like machine guns and mortars, did not have rifles and only carried a pistol.

[220] The EIB method of searching a body was to have one soldier cover the body with his weapon, just in case the enemy was only playing dead. The other soldier would get on top of the body, wrap himself around the body and check underneath for booby traps. If there were any, the body would absorb much of the blast. Next the soldier on top of the body would hold onto the body and roll it over. The soldier covering the body would then visually check for anything underneath. If there was, he would yell "bomb!" and the body would be thrown back down on the bomb, absorbing the blast.

[221] From a Facebook message from Brad Gallardo on 05 April 2012. There were a lot of personally owned weapons on Grenada, though it is not allowed by regulations. LT Nicholson had his 9mm, I had my 1911A1 and one Ranger company commander carried his M1 Garand rifle.

down due to the heat and lack of water.[222] When 3rd Platoon was ordered to search a small village, I was put on a steep hill beside the village, to provide overwatch security as a sniper. Providing the security for me was SP/4 Bush, an M-203 Grenadier in the HQ section that was attached to me by 1SG Polachowski after the firefight. Bush would continue to be my spotter and security for the rest of the invasion.

Bush moved out to find us some water in one of the nearby houses. There was a 55-gallon oil drum beside a building being used to catch rain water off of the roof of the house. Though I suspected the water was filled with all sorts of vile bacteria, I still had him fill up both my one-quart canteens. I dropped in an iodine pill for good measure.[223]

After the village was cleared, we continued to move up that steep hill toward Morne Rouge. It was slow moving due to the heat and our extremely heavy rucksacks.[224] The paratroopers began shedding weight as we climbed. The hill was littered with all sorts of equipment and clothing, but not the ammunition. After that morning's firefight we all knew that this would not be an easy mission. I struggled to move up the hill and then would stop when my heart felt like it would explode out of my chest. I would then just sit there, resting on my rucksack, looking down into the valley we had fought through, and when I regained my strength, I would struggle on and do it all again.

Halfway up I took my "alert box" out of my rucksack and put it up in the limbs of a tree. I had a Peak One Coleman stove that I had carried for years and I contemplated throwing it out, but it had too much sentimental value. I also carried around a small stuffed puppy that an old girlfriend gave me. I had made the puppy a camouflage uniform, and it had gone on all the jumps with me, in my cargo pocket of my pants. Though it was a silly toy, it was a talisman for

---

[222] In the 2/325 twenty-nine soldiers suffered from a heat injury. When the 3/325 arrived they suffered forty-eight heat casualties and the battalion aid station used up the entire supply of their IV solutions on the injured soldiers.
[223] Small bottles of iodine tablets were given to each soldier and would be carried in a small pouch on the side of the canteen cover. These were good as long as they were gray in color, but when they turned a reddish-brown they would no longer be effective. The troops who were there on the first day had an outbreak of hookworm infections months later, due to sleeping on the ground, or from drinking water that had not been treated.
[224] According to the Centers for Military History, the weight carried by the paratroopers on Grenada was the heaviest average loads ever carried by soldiers, in history. The average ruck weighed around 150 pounds. This is the average, not the heaviest. The second closest weight in history was the weight carried by the British soldiers in the Falkland Island War, with rucksacks averaging 140 pounds, or 65 kilograms.

me, so I didn't throw it out either. He was called "Line Doggy" and had been to Panama, Egypt, Italy, and Sudan so far.

I stopped again when I got close to the top of the ridge to catch my breath. I could see some of our paratroopers dragging out the Cuban bodies from the buildings down below me. As I sat there, I saw eight dead Cubans dragged out by their feet and stacked by the road. The soldiers who were tasked to do that later told me that the 7.62mm bullets went in one side of the cinderblock walls and out the other. So, all of my shots that I had fired at the buildings, to keep the Cubans heads down, had punched through the walls.

While I sat there, trying to slow my heart down, I heard some troopers saying that the SEAL Team holding out in St Georges had been slaughtered. Another rumor said that DELTA Force had also been in the city and had taken heavy casualties. After I quit feeling like I was going to have a heart attack, I got back up on my feet and continued the slow climb up the hill. While the Special Operations were being killed by the enemy around St. Georges, we were losing our soldiers to the heat. Five of the company was down and CPT Jacoby ordered any men wearing a flak jacket to take them off and tie them to their rucksacks. I had never even put mine on.

Though the fight at Sir Paul Scoon's house had been intense, the men of SEAL Team 6 had not suffered any casualties. LTC Ray Smith of the 8th Marines moved out with a company and five M-60 tanks early in the morning of October 26th to relieve the SEALs. The tanks did not have any ammunition for their main guns because it had been stored underneath other equipment on board their ships. The task force commander, Captain Carl Erie decided not to move the cargo piled on top of the shells.[225]

The Grenadian soldiers at Queen's Park and Green Bridge heard the loud rumbling of the approaching tanks. A few brave souls fired upon the Marines, but the Grenadians lost heart and quickly departed. The Marines' G Company would occupy Queen's Park for the next few days. Since they had not encountered any resistance, the Marines pushed on towards Mount Weldale and Sir Scoon's house. At 0730 they linked up with the SEALs, amazed that there were no casualties on the American's side.

---

[225] Raines, Edgar. *The Rucksack War*, pg. 333.

Sir Paul Scoon and his wife were flown by helicopter out to the *USS Guam* around noon. On the ship Sir Paul was shown the letter from the OECS requesting that the United States convene militarily on the island. There was also a letter, backdated to October 24th, from Sir Paul Scoon, requesting military intervention. Scoon signed the letter, authorizing the invasion.

At Fort Frederick, Grenadian General Hudson Austin and Ewart Layne realized that the situation was hopeless. After the intense fighting of the first day, the Grenadians had lost the will to fight. Many of the PRA soldiers were taking off their uniforms, putting on civilian clothes and blending into the population. Austin gave the word for the rest of the PRA soldiers to cease fighting and avoid arrest. Before the sun came up on October 26th Austin and the other RMC leaders left the tunnel at Fort Frederick and slipped into the dark.

CPT Robert Dobson and his G Company, 2nd Battalion, 8th Marines, approached Fort Frederick slowly, expecting resistance. Instead, all they found was an abandoned fort. Inside the tunnel where the RMC had been hiding, they found a stash of classified documents, and piles of ammunition. Among these weapons were three brand new 82mm mortars.

After arriving at Point Salines MG Trobaugh learned that there were American students at the Grand Anse campus, 2 ½ kilometers north of Point Salines airfield. Through a HAM radio operator inside the Grand Anse campus, Mark Barettella, it was discovered that there were no PRA soldiers in the dormitory. They also learned of enemy positions around the student's location. With the assistance of Barettella the military was able to tell the students what to do when the rescue mission arrived. Unfortunately, because the 82nd Airborne had to leave its vehicles behind at Fort Bragg it meant there were no resupply convoys, and more importantly, no long-range radio communications. MG Trobaugh had to have Barettella relay the information through the USS *Guam*. To make matters even more screwed up whenever the *Guam* changed course Trobaugh was no longer able to receive any messages.

Around the time that our company was moving up the steep hill, Barettella radioed information that several PRA soldiers were near the campus, but they were not in the buildings. Trobaugh was ordered to secure the campus by nightfall. Initially the plan was to have Hagler's 2nd Ranger Battalion move by foot to the left of our battalion and seize the campus. The students would then be brought back to the airfield by confiscated trucks under cover of the night.

On board the *Guam* MG Norman Schwarzkopf changed the plan and decided that the Rangers should go in by Marine Corps helicopters that were not being used on the ship.[226] The Marines were not in favor of the plan, since they had lost so many helicopters that day, but Vice Admiral Joseph Metcalf approved the plan.

Nine CH-46s carrying three under strength companies of the 2nd Ranger Battalion would secure the compound and set up blocking positions so that the PRA would not be able to penetrate the campus. Four CH-53 Sea Stallion helicopters would then arrive and take the students back to the airfield, where they would be flown back to the United States. Prior to the rescue the area around the Grand Anse campus was to be bombarded by A-7 Corsairs, Spectre Gunships, Naval artillery, the 82nd Airborne's howitzers at the airfield, and the mortars of the Rangers and the 325th Infantry in the hills around the airfield.

The students were told by telephones and HAM radios to place white sheets out on their buildings so that they would not accidentally be targeted. The students were also told to wear a white armband, so they would not be mistaken as the enemy. Until the Rangers arrived all the students were to lie on their floors and told to not move. Anyone running around outside the buildings would be considered hostile and would be fired upon. Unfortunately, it was learned that there were additional students that stayed 800 meters away, and they had to be given time to move to the campus. Due to this the bombardment was postponed until 4:00 p.m.

I finally got to the top of the hill and had a fantastic view of the island. I could see the airfield to the south with smoke still pouring out of the buildings at Little Havana. Transport aircraft continued to fly in our 3rd Battalion.[227] To the east I could see the island, and I dreaded the steep hills covered in jungle that loomed all the way to the horizon. To the north I could see the ocean and a beach. This was the Grand Anse beach.

The 3rd Platoon was placed into a hasty defense on top of the Morne Rouge, facing out towards the steep ridge. While I had been climbing the hill, they had spread out and searched the area for any signs of water. They found several rain barrels beside the empty houses and they filled their canteens. CPT Jacoby

---

[226] Schwarzkopf was commander of the 24th Mechanized Infantry Division, but had been attached to Metcalf for the operation as the Army advisor to the Admiral.

[227] David Yeager wrote in a Facebook post on 30 October 2013 "Still amazed some wild little revolutionary didn't take out a C141 on the runway with an RPG-7 - there were enough of both on the island. We had no capacity to put out a fire - the crash rigs had been disabled. A flaming hulk on the runway would have hampered things a might."

found me and told me to follow him as his security element. We met LTC Hamilton by the deserted houses and he told CPT Jacoby that our Bravo Company would be attacking a warehouse to the east, and he wanted Jacoby's Alpha Company to support them if they got pinned down. This warehouse turned out to be a supply post in Frequente.

CPT Jacoby had all the platoons rotate through 3rd Platoon's position so that they could fill their canteens. We did not know when we would get another resupply. If this next attack was like the assault on Little Havana, we would need it. Bravo Company moved to their attack position by going down to a road that paralleled our company on the Morne Rouge ridge. We moved adjacent to them, down the ridge towards our support position, the highest point on the ridge. There we waited.

Bravo Company had a hard time of it. So far, they had suffered numerous killed and wounded, to include their company commander. The loss of such a key leader will devastate the morale of any company. Bravo Company's objective was Radio Free Grenada 1,100 meters away, but to get there they had to move across terrain so thick that the soldiers had to cut their way through the jungle. Forty-seven-year-old CSM Catalino Barajas moved with the company and the Vietnam veteran described the movement as "rough". The company was still wearing their flak jackets and the heat took a devastating toll. Thirty soldiers went down due to heat and had to be evacuated. When LTC Hamilton learned of the heat casualties he ordered every man in his battalion to take off their flak jackets.[228]

The CSC gun jeeps drove to where Ranger jeep Juliet-5 had been ambushed the day before and they saw two BTR-60s on the road. The paratroopers fired their LAWs at them but the armor vehicles ended up being abandoned. However, as the LAWs exploded, the CSC troopers began to receive fire from around the knocked-out vehicles. The PRA had waited until the first two American gun jeeps had drove past them and then they fired on the last three jeeps. The M-60 gunners in the back of the CSC jeeps stood up, entirely exposed, and returned fire on the enemy positions. One of the gunners had his Kevlar hit by an AK-47 round, ripping the helmet off his head. The enemy bullets punched holes through his cargo pockets on the sides of his pants.

---

[228] Raines, Edgar. *The Rucksack War*, pg. 379.

While the gun jeeps returned fire on the Cuban ambush CPT Rocke's Charlie Company also fired upon the enemy from a ridge to the east of some warehouses. The PRA did not even know they were there when they had ambushed the gun jeeps. LTC Hamilton was with Rocke and gave him permission to fire his mortars.

Danny Davis was there and he wrote "2d platoon leader Mark Fields adjusted our company mortars onto the vehicles. When the first 81mm rounds impacted near the vehicles, whoever was manning them, decided it was time to leave. They abandoned the vehicles and fled on foot." [229]

Twenty-nine rounds slammed around the PRA causing death and chaos and setting the BTR-60s on fire. Since the 82nd Airborne did not back down from the ambush the enemy withdrew into the surrounding jungle leaving behind four dead and an unknown number of wounded.[230]

Grand Anse Rescue

I sat on the ridge hearing the CSC gun jeeps firing in the distance, while an AC-130 Spectre gunship began the bombardment of Grand Anse. The 20mm and 7.62mm Gatling guns sounded like a thunderous chainsaw. When the automatic 105mm on board fired, the round would go over my head with a whoosh, then I would hear the cannon firing off in the distance from the aircraft, then I would hear the louder explosion as the artillery round struck the target. Whoosh-boom-BANG! Whoosh-boom-BANG! Off shore artillery screamed in right beside our ridge and detonated below on the houses lining the beach. We could see the rounds flying through the air. The A-7 Corsairs streaked in, firing their 20mm cannons at the bungalows, while Cobra gunships weaved in and out of the area to fire upon any suspected targets. To add to all this noise our Company's 81mm mortars fired down onto the beach, directed by LTC Hagler in a helicopter above the carnage. They were striking a police training college and the transmitter building for "Radio Free Grenada". It was a hell of a show!

---

[229] Facebook message from Joe Muccia on 06 March 2013
[230] This information was told to me by SSG Charles LaLone of CSC Company after the invasion. LaLone had the knack to be in the right place at the right time when any photographer showed up and he is in many of the more famous pictures of the invasion.

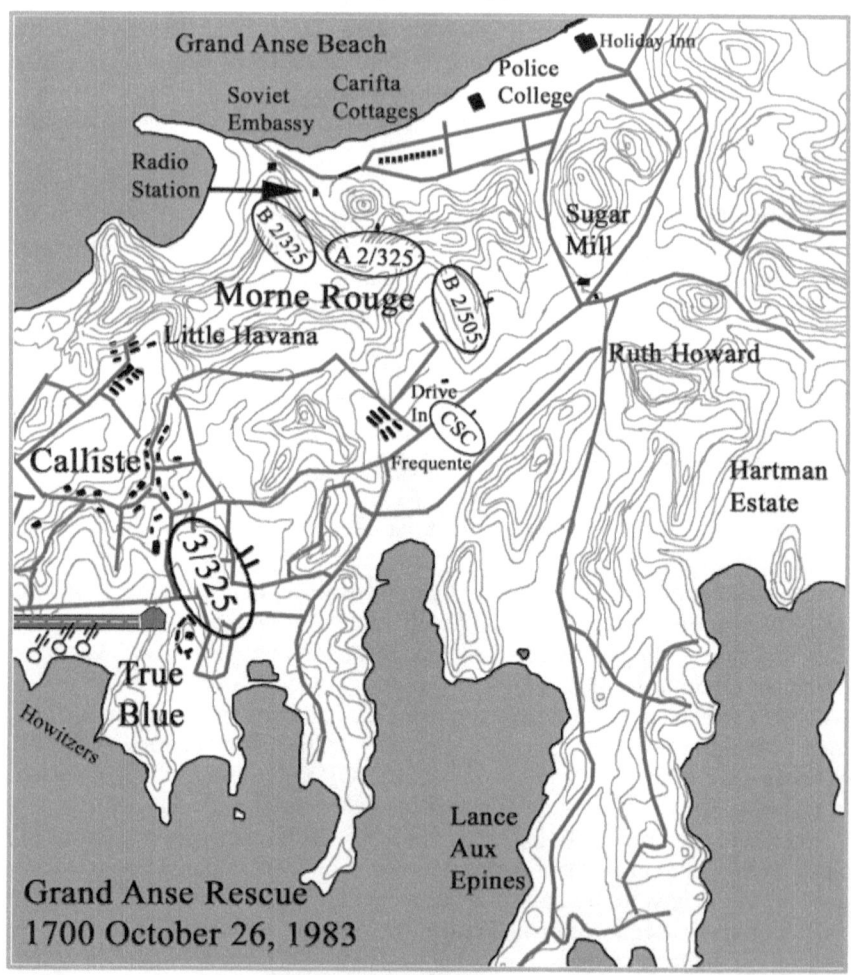

Grand Anse Rescue
1700 October 26, 1983

Clausewitz once said that no plan survives the first contact with the enemy. Every war, since the beginning of recorded history, is full of mistakes and bad planning. The reason for that is simple. The planning is good, but there is an enemy trying to do everything to defeat that plan. Those who criticize the military for not being perfect, or for not being prepared for everything that will happen in a war, are obviously not soldiers and do not know history.

The attack on Grand Anse began to go wrong from the start. Alpha Company of the 2nd Ranger Battalion were in the first three helicopters and landed 500 meters southwest, near the radio station that was being bombarded.

Luckily the bombardment ended 20 seconds before the first helicopters touched down.

Though the Rangers came under small arms fire, bullets did not take out any helicopters... a palm tree did. "Kurt Sturr's bird, numbered "03", entered the zone, it was hit by a palm tree that had been hanging over the sliver of beach they were to landing on. The fronds, whipped around by the rotor wash, smacked into the fore pylon, disabling the system. The pilot, immediately feeling the vibrations in the stick from the damaged rotor, triggered his intercom switch and yelled to the crew to abandon the aircraft. The Marines piled out of the aircraft and onto the beach, and subsequently moved up to the sea wall. The Marine crewmen only had the briefest of moments to yell a warning out to their passengers, to the consternation of the Rangers of 3rd Platoon. Since the landing occurred during high tide, water started pouring in the aircraft and the Rangers scrambled to exit, many losing critical equipment."[231]

"LCpl Marty Dellert, the crew chief of Chalk 7, decided to check out the rotor and pylon to see if the aircraft was flyable, so he bolted from cover and clambered up the aircraft and began inspecting. The rotors were damaged but not destroyed. Believing that the aircraft was airworthy, he signaled to the other crewmembers to reboard. They did so and after he climbed down, the pilot fired up the Phrog. "03" vibrated terribly but the pilot was able to nurse the stricken craft back to point Salines."[232]

The Grenadians had been facing our company, ready for the impending attack, and were surprised by the Rangers landing in their rear. The enemy quickly left the area. Ranger Scott Underlook later wrote "Our CH-46 had a .50 cal on the right side of the AC going into Grande Anse. I was near the window closest to the gun when the gunner opened up on the threat on the beach. Still remember the red tracers going down toward the buildings. We had to tailgate jump into the surf and run the 100 meters to our blocking pos to get the students out."[233]

After the Ranger's Alpha and Bravo Companies had secured the objective, Charlie Company on board the CH-53s came in to get the students. The Rangers grounded their rucksacks near a wall with the intention of returning to them. One Ranger, Kurt Sturr, wrote "when we extracted, it was chaos, of course, and

---

[231] Facebook post from Joe Muccia on 09 November 2015.
[232] Ibid.
[233] Facebook message from Scott Underlook on 04 March 2013.

no one got on the right birds. We were extracted far from our insertion point, so our rucks stayed there." [234]

Sturr wrote that a Cuban mortar was firing upon them from the Soviet Embassy "we blew the roof off the Soviet Embassy. The Sov's bitched in the UN about it, and Jean Kirkpatrick, then US Ambassador to the UN, and a chick that I would never want to rumble with, completely denied it. 'Never happened,' she said."

Many of the students were terrified during the intense bombardment. They lay on floors and placed mattresses over the windows to protect against flying glass. Some of the students were so scared that they did not come out until it became quiet again, but by then the Rangers had left. These students would later be rescued by the 82nd Airborne soldiers as they moved northwards to St Georges. Though the CH-53 can only hold 55 people, the Rangers jammed over 60 students in each of the helicopters, rescuing 233 students with no injuries. The rescuers only had one Ranger wounded due to friendly mortar fire.

During the rescue the crew of the CH-46 that had been abandoned in the surf returned to try to fix the helicopter. Finding nothing seriously wrong, except part of the rotor blade missing, the pilot was able to fly it back to Salines. After the students had been extracted the Rangers threw out a yellow smoke grenade, the signal for the orbiting CH-46s to return to retrieve them. "Chalk 8, piloted by Dick Gallagher, approached the LZ and as the veteran pilot tried to settle into the zone, the unexpected occurred. The rotor wash had caused a palm tree to arch back and when the aircraft settled, the rotor wash subsided, releasing its hold on the tree allowing it to fall forward, smashing the rotor system. Like his fellow aviator in Chalk 7, Dick immediately called for his crew to abandon the aircraft. They were able to board a CH46 as passengers and fly back to Salines."[235]

Pilot Al Mirabella wrote "I am not sure what a/c (aircraft) it was either 07 or 11 but when we picked up the Rangers, I made sure the Msgt (Master Sergeant) was on the right gun. When we started into zone the Msgt open up. My window was open and he scared the shit out of me. The next time we went in the zone I think everybody was shooting something out the right side. We

---

[234] Extracted from http://www.socnetcentral.com/vb/archive/index.php/t-43622.html on 27 September 2007.
[235] Facebook post from Joe Muccia on 09 November 2015.

picked up the crew of 00 and some Rangers. I had to land in the water and when we came out, we were so heavy with water we bounced off the water 3 times."[236]

The CH-46 that was left behind was nicknamed "Double Nuts" due to it having the number "00". The CH-46, or known as a "Phrog" to the Marines, was a Marine helicopter from HMM-261. The helicopter was abandoned and the Rangers were flown back to the airfield.[237] The entire operation took 26 minutes.

Years later Jim Bowen wrote of the Grand Anse rescue mission "Our platoon (2nd PLT, B 2/325) broke out of the jungle coming down the hill just as the raid was going down. We watched from up the hill and saw the Marine helicopter rotor hit the tree (and subsequent gun runs to destroy the helicopter) and the AC-130 taking out our objective, the radio station (Radio Free Grenada). We had a Cuban POW with us that we picked up in the movement to contact over the hill from the Cuban compound, after clearing a house on the cliff overlooking Grand Anse. Later that afternoon, moving toward the Soviet embassy, I had the ambassador at gunpoint for a brief time until he identified himself."[238]

In the chaos of the extraction eleven Rangers from 2nd Platoon, C-2/75, led by 2LT Robert Dorsey and SSG Barry Shughart, had pushed north up the beach into a blocking position on the flank, missed the call to extract, and had been left behind. Charlie Company's Commander, CPT Mark Hanna, tried desperately to have a helicopter extract the abandoned Rangers, but he was unable to contact them with his radios. The Army and Marines were using different radios and frequencies. Hanna told the Squad leader to try and make his way to our position on the ridge southwest of Grand Anse. Unfortunately, the Rangers did not have the frequencies that the 82nd Airborne was using and they did not want to try and conduct a linkup in the dark with soldiers who were tired, hungry and on edge.[239]

---

[236] Facebook message from Joe Muccia on 06 March 2013.

[237] One of the more famous photographs of Grenada is of that CH-46 sitting in the surf, covered in bullet holes. The damage was not caused by the enemy, but was caused by the AC-130 gunship attempting to destroy the helicopter.

[238] Facebook message from Jim Bowen on 01 August 2014.

[239] This would have also been a linkup with forces that the Rangers were thought inferior to them and so they did not trust them. My Special Forces team had the same problem during Operation Desert Storm, when we were operating behind Iraqi lines. At one point we thought we might have to withdraw, and we did not want to go through the elements of the 101st Air Assault Division or the tank units to our rear, because we thought they would be scared and shoot us by accident. So, we actually planned to head west, and try to make our way to Israel, instead of going through our own lines.

Before the main force was extracted, they had attempted to destroy the helicopter with gunfire. Luckily the Squad leader searched the downed CH-46 and found three inflatable rubber rafts. The Rangers were able to work their way out to sea using helmets and rifles as paddles. Things got worse when it was discovered that two rafts had also been hit by the friendly fire that had peppered the helicopter. All the Rangers piled their gear into the remaining raft and swam alongside until they were finally rescued at 11:00 p.m. and taken to the destroyer USS *Caron*.

As for Double Nuts, the order was given for a AC130 to shell the aircraft and render it and it's cryptological communications equipment unusable.[240] "According to Spectre crewman Bill Walter, "there was no real reason to shoot this bird...though I was one of the guys who did it (by direction of Fires Network). The reason we were given was to destroy the Crypto, but as everyone who was there knows secure com was whacked anyhow and units were loaded with a combination of day 23, 24, 25 and 26 codes. That just won't work, so if the radios were useless to us, they were useless to the PRA also. We had fire control problems engaging Double Nuts, so we missed (a lot) with 40mm and 105mm. The 105mm gun was fired caveman style with Jeff Drummond on the lanyard. We fired on it for about 30 minutes and never hit directly with the 105mm gun (but did frag it as you can see). We also hit a small building on shore adjacent to the Spice Island Inn where Dr Robert Jordan had taken cover in a walk-in cooler.[241] Nobody had any idea he was there. We were called in after the Rangers were paddling out to sea...would like to hear what they saw from their end."[242]

After the bombardment of Grand Anse, and the rescue of the students by the Rangers, we were told that we would be spending the night on the Morne Rouge ridge. As the sun sunk down into the ocean our platoons moved into a cigar-shaped perimeter along the steep ridge. The ridge had a few trees, but the terrain was mainly tall grass, with steep sides dropping off to cliffs. We were told that all the firing we had heard was due to a Ranger assault on the beach and that they had five men killed.[243] I was ordered to go make an OP on the south side

---

[240] The pilot of "Double Nuts", Dick Gallagher, retired after 20 years in the Marine Corps as a lieutenant colonel and then taught at the high school and elementary school in Jacksonville, NC. Gallagher died in August of 2021.

[241] Dr. Robert Jordan was still living on Grenada in 2021, and hosting the annual reunions of the Grenada veterans.

[242] Facebook post from Joe Muccia on 09 November 2015.

[243] Like most of the rumors, this one was false.

of the ridge, with my security being SP/4 Bush and PFC Stillson, from the HQ section.

In the twilight the move to the OP was treacherous. The hill dropped off to an extremely steep angle that was almost a cliff. When we got about 50 yards away from the perimeter, I told Bush and Stillson that we would be there for the rest of the night, and we had to make do the best we could. One of us would stay up for an hour, while the other two would sleep, then we would rotate. Sleeping was almost an impossibility due to the steepness of the hill. The only way I could prevent myself from rolling down the hill was to stab my Randall knife into the ground, and lean against it. Throughout the night my ALICE pack would occasionally roll down the hill, and I would have to wake up and get it.[244]

Our entertainment for the night was Bravo Company attacking the warehouse. There was a short intense fight that ended with the warehouses catching on fire.[245] All night long ammunition popped and exploded in the buildings. On the Grand Anse side of the ridge helicopters and artillery continued to fire upon hidden targets.[246] The A-7s would roar in, fire at some enemy, and then roar away. We had no food, since we had only been given three C-Rations back at Fort Bragg. My only food since I arrived was the one MRE that I ate on the night of October 25th. We expected to get resupplied, but what we didn't know was that there was no food coming. The aircraft that landed at Salines throughout the night were filled to capacity with 82nd paratroopers from the 3rd Brigade.[247] None of these had any type of vehicles to resupply the thousands of soldiers on the island.

---

[244] Brad Gallardo wrote about where he slept during the first two days, "Day one slept on Goat Hill in front of that little blue house at the end of the trail which is about 40m or so down west of that house you were at. Day two after the fire fight we moved up and around the mountain behind that other compound and were supposed to assault it, but we had too many heat casualties and they reassigned the assault to the 505th. So once we recovered we moved on down to Grande Anse and searched the student billets, and the helicopter. Then moved a little ways searching homes in around Grande Anse and then stood down for the night, right there when the rain came" Facebook message from Brad Gallardo 21 November 2011.

[245] This "warehouse" may have been the Radio Free Grenada building, that was stockpiled with ammunition.

[246] I later learned that a lone PRA anti-aircraft gun had dueled with the Spectre gunship. For most of the night the gun would fire, then move, while Spectre searched it out. The AA gun was never hit and the extremely lucky crew got away unharmed.

[247] The 3rd Brigade was commanded by COL Steven Scott and consisted of the 1st and 2nd Battalions of the 505th Infantry, and the 1st Battalion, 508th Infantry. On the night of October 27th only the Brigade headquarters and the 1/505 had arrived. Back in Fort

Triple Canopy

By the end of the second day of the invasion the Point Salines airfield was secured and the Cuban barracks at Little Havana pacified by the 2/325th Infantry, with the loss of only two paratroopers.[248] Sir Paul Scoon had been rescued in St. Georges by the SEALs, while the PRA supply base at Fort Frederick had been captured by the 8th Marines. The Rangers had rescued 230 students and faculty at Grand Anse. The 3/325th Infantry had arrived and occupied the True Blue peninsula with only a single sniper to resist them. After numerous attempts to get the sniper to surrender the paratrooper finally killed him. The battle for Grenada was over except for a few holdouts, though none of the U.S. forces on the island would realize this for a few more days. Due to the intensity of the fighting during the first two days by the Rangers and the 2/325, no one wanted to underestimate the enemy and most expected renewed resistance.

Before the sun came up my "sniper section" was told to return to the perimeter. My scope had so much condensation in it that there was a puddle of water inside it. I couldn't see anything out of the fogged lens. I unscrewed the ends of the scope, dumping out the water, but it didn't do any good. If any firing started, I would have to use the iron sights.

Our company moved in a single file down the steep ridge to the Grand Anse beach. The paratroopers were slipping and falling down due to the heavy rucksacks. As we neared the bottom of the ridge, we saw a group of soldiers crossing our front but we didn't know who they were. We immediately prepared to ambush the unknown soldiers, but soon discovered it was our own Bravo Company crossing our path.

At 0630 we moved from the jungle covered hill and out into open onto flat ground that was a residential area. Our company led the way, moving onto a road that had burning buildings on either side of the road. Ammunition cooked

---

Bragg all units that remained were put on alert and told they would be going, but the 504th Regiment, in the 1st Brigade, were never deployed. This led to the nickname of "No War 04" regiment, and the title continued to haunt the 504th Regiment until the invasion of Panama. One company, A Company, of 2/504, led by CPT Humble, did deploy as a detachment.

[248] I have never been able to accurately determine how many wounded the 82nd Airborne had during the invasion. There may have been as many as 140 wounded. Almost all of these wounds were inflicted by friendly fire or by accidents. The United States has the most deadly military in the world, even to ourselves. Due to better and more realistic training the number of accidental deaths has dropped dramatically by the 21st century. More of the military died by accidents in the peacetime years of the early 1980s, than in the entire Iraq War of the early 21st century.

116

off in the burning Police Training College. When we first heard it all of us hit the dirt as the ammunition exploded, thinking it was a sniper firing at us. This soon got old and after awhile all I did was fall back on my heavy rucksack and sit there. Soon the sudden noises didn't faze me and I walked on through the devastated and burning Grand Anse beach.

Not all the buildings were on fire. There was a row of beach cabins with orange tiled roofs that had been shattered by machinegun and artillery fire.[249] The Spice Island Inn on the other side of the road had been hit also, with ripped mattresses lying out on the road. Gaping shell holes still smoked in the street, while pieces of the blasted roof tile covered the street and surrounding roofs. We continued our slow forward movement, stopping occasionally to check out the buildings for any PRA or Cuban holdouts.

As I passed by one side street, I saw "Double Nuts", the destroyed CH-46 helicopter sitting in the surf. American equipment was scattered around the road. We moved past the downed Chinook helicopter and stopped the column, each platoon searching the area adjacent to them. The 1st Platoon searched the area where the helicopter had crashed. I was told that they found a body that they thought was an American pilot. They also told me that paratroopers from the HQ section put the corpse in a body bag and moved it out to the road to be picked up by following units.[250] The 1st Platoon also found ALICE packs with M18A1 Claymore mines.[251] The platoon took the mines and put them in their own rucksacks. One of the paratroopers found an UZI in the sand, and he stuck it in his rucksack. I found a black beret with the Ranger crest on it in an ALICE pack. I took off the crest and put it in my pocket for luck.

After we passed by the helicopter our Bravo Company followed. They also searched the helicopter, throwing the large equipment, like the machineguns,

---

[249] This was the Carifta Cottages.

[250] This may be yet another one of the many rumors being told during the invasion. No American military personnel were killed during the Grand Anse rescue and to this day I have not been able to identify who the body was or if there was one. It may have been a Cuban soldier or one of the other Eastern European military advisors.

[251] The M18A1 Claymore mine is not one that is detonated by stepping on it. The Claymore is set up and aimed and when the detonator is pushed 1 ½ pounds of C-4 explodes, sending 700 steel ball bearings through the target. It has a kill radius fan of 50 meters out and 50 meters wide.

into the surf. We were later told that one of their paratroopers tried to move a loaded .50 caliber on the Chinook and it fired, wounding him.[252]

We continued to move forward until 2nd Platoon stopped at a crossroads that had a restaurant, a travel bureau and some other shops on one side of the road, on the other side was a Holiday Inn. Outside the restaurant was an abandoned anti-aircraft gun. Behind the buildings were PRA defensive positions dug into the hillside. Abandoned cases of RPG rockets and AK-47 ammunition were strewn about the area, along with uniforms and equipment.[253] It seemed that the PRA soldiers had changed into civilian clothes and blended into the civilian population.

I moved to the front of the column and dropped my rucksack. I tried to open the restaurant to find food, but it was locked. I didn't think it would be right to break the window and go in. The place was abandoned and we had yet to see any of the population of Grenada yet.

I climbed some stairs to look into the second floor of the restaurant. When I reached the top of the stairs, I laid my sniper rifle down and pulled out my .45 pistol. The door on the second floor was not locked and I went inside by myself. It was quiet. For the last two days I had been around other soldiers, heard the crack of nearby bullets and heard the explosive force of our artillery land nearby. Now it was very quiet and I couldn't hear anything at all. The room looked like any modern apartment, with a living room and a kitchen. The lights were all off, so I looked around and found the refrigerator. As I opened it, I heard a noise come from the living room. This scared the heck out of me and I dropped down on one knee, aiming at the back of a couch. I heard movement on the other side of the couch, and it sounded like someone getting up. I aimed at the couch, and had my finger on the trigger, when an old man rose up. I had his head in my sights, but he just stared at me, shocked to see a dirty paratrooper in his home, aiming a pistol at him. He just stared and I continued to point the pistol at his head. I told him "82nd Airborne. I'm from the United States". He whooped and jumped up, and began to tell me how glad he was to see us. I lowered the pistol and told him that he needed to go outside. It turned out that he was the owner of the restaurant and hadn't known the Americans were here.

I came back down the stairs and leaned against a cinderblock wall near the building. SSG Moore came up and told me that I was stupid to go in a building

---

[252] Like most of the rumors, I never did find out if this was true or not. I have never been able to find an exact list of who was wounded in the 82nd Airborne. I have a partial list that was published in a Fayetteville newspaper on November 1st.

[253] The rockets and ammunition we found were from North Korea and the Soviet Union.

by myself. I agreed, but I was also pretty hungry and not thinking clearly. Food was the foremost in my mind. As he quietly scolded me a shot rang out nearby. I found myself behind the wall, searching the jungle through my scope. I must have done a back flip over the wall, but I don't remember doing it. My acrobatics ripped out the crotch of my camouflage fatigues.[254] The shot had come from PFC Vela, one of the M-60 assistant gunners, who drew out his pistol to enter one of the buildings and then fired it by accident. He had a bad habit of walking around with the pistol on full cock.

Mark MacNamara wrote about this time "We were so dehydrated and that was our first chance to get water, since the day before out of an oil drum that was filled with water and little white worms swimming about. I only had my two 1-quart canteens, because my 2-quart canteen came off my rucksack on the plane going to Grenada. Wearing the heavy BDU's I was a thirsty Airborne troop. I half filled my canteen with water from a drainage ditch along the road. I put two iodine tablets in my canteen shook it around and took a drink in desperation and spit it right back out because it tasted like sewage. About 45 minutes later, with the taste still in my mouth, the Grenadians came out of the resort and gave us grapefruit. Then after that Vela went behind the restaurant looking for a water spigot. After filling up on water we started humping down the road and I saw a sergeant with two 2-quart canteens hanging off his rucksack and remembered he was sitting right across from me on the plane down. I told him it was mine and he refused to give it back. Jesus saved his live that day!"[255]

I continued to look for food or water around the restaurant and found a group of Grenadians hiding in a kitchen, scared to death that we would shoot them. They had spent the night in there while the apocalyptic bombardment tore up the area around them. They showed me where to get water behind the hotel. I found some coconuts that had fallen off of a tree and tried to split one open. I couldn't figure it out so Bush gave it a try and threw it on the road several times, but neither one of us were able to crack it open.

While we were attempting to break open the coconut an American drove up on a Moped, and nearly wrecked when he saw the two paratroopers, armed to the teeth, repeatedly throwing a coconut in the road. He jumped off the bike and ran over to us, waving wildly and telling us how glad he was to see us. I

---

[254] Many soldiers do not wear underwear, since after a long period of not bathing in the field the underwear will stiffen, and cut like sandpaper on long road marches. The nickname for going without underwear is "going commando". In Grenada I uncharacteristically wore briefs, which came in handy at this time, or else I would have been more exposed to the elements than I would have liked.

[255] Facebook post from Mark MacNamara on 27 October 2013.

took him to CPT Jacoby and he told us that a bunch of Americans were hiding in a nearby basement, but they were too terrified to come out. They had sent him out on the Moped to find help.

While we were listening to his story a Grenadian ran up and told us that several West Germans were also trapped in a basement nearby. The Grenadian told us that the PRA was killing civilians and gunning them down in St. Georges. That was the first time we had ever heard the name of the enemy we were fighting and asked him what PRA stood for. We now officially knew who the enemy was. Until then we thought that we were just fighting Cubans. A Canadian drove up with his son in the car and asked us if there was some way we could get him off the island. We sent all of these people back to LTC Hamilton and the Battalion headquarters.

CPT Rocke's company discovered twenty students that the Rangers had missed. They were sent back to our 3rd Battalion under escort. Since we had not harmed any civilians so far, the Grenadians started to come out from hiding, giving themselves up to us. All were happy to see us. I had one of the Grenadians crack open the coconut for me and I shared it with Bush. The restaurant owner opened up his shop and began to pass out all the food he had. He gave me some bread, while a Grenadian civilian gave me a fried chicken leg. That meal of stale bread, chunk of coconut and cold chicken leg was the best thing I had ever tasted in my whole life.

CPT Jacoby decided to have us drop all our extra gear at the Holiday Inn and leave some of the company behind to guard it. There were many of our men who were ill due to the effects of the heat and they were chosen to guard the rucks. Whenever we stopped the medics would set up impromptu aid stations and plug in IVs in the worst casualties, but since there were no resupply vehicles, they had to limit who would get treatment.

I had put my M17A1 protective mask in my rucksack when I rigged it to jump and so far, I had not taken it out. Early in the invasion there was a rumor that Cuba was sending MIGs armed with chemical weapons to attack us, but that rumor was laughed at by most of us. The one thing that the United States did have was naval and air superiority. So, my M17A1 gas mask never came out of the ALICE pack.

I also did not wear my Kevlar flak jacket. I didn't trust it and thought that if an AK-47 bullet would hit it, it would mushroom and make the wound worse. I preferred to have one go in quickly, and maybe come out just as quick.[256] That Kevlar jacket had also been strapped to the top of my rucksack and never taken out.

At the Holiday Inn I dropped off my rucksack, the mask and Kevlar vest. I kept my LAW rocket and I put my journal in my pants cargo pocket. We would not see these rucksacks again for three days. After dropping off our rucksacks we continued down the road towards St. Georges. We would move for a short distance, stop, and then search the jungle and houses on either side of the road.

The Battalion headquarters and Bravo Company followed behind us as we moved forward. When they moved through the Holiday Inn where I had eaten the fried chicken, they were hit by sniper fire. One of their men on security at the restaurant had fallen asleep and his platoon left him there, forgetting he was on security, while they searched the area. The paratrooper woke up when he heard a noise and found two enemy soldiers standing near him. They fired, hitting him twice, and then ran off. He was found by his buddies, lying on the ground bleeding to death, when they came back to get their rucksacks.[257] We were told that the soldier had died, but this turned out to be another rumor.[258]

---

[256] It was not until after Desert Storm in 1991 that the military began considering body armor that would actually stop a high-powered rifle bullet. When the new vest came out and we were finally given a vest with ceramic plate, we still didn't trust it thoroughly. To prove to ourselves that it would stop a bullet, my Special Forces Team fired different rounds into one vest, ruining it, but showing us that it would actually work. Today soldiers wear all the armor necessary, but the news media complains constantly if they aren't given enough to protect them thoroughly. I always preferred being able to move fast, to being weighed down. The tradeoff was between speed or protection. I preferred speed.

[257] The soldier was Brent Taylor. Lieutenant Steve Seager said that Taylor had two wounds that day. In addition to the grazing neck wound on the right lateral neck, he had a shrapnel wound right lateral thigh that was caused by the M-203 round he had in his outside pants pocket being hit by a round and shattering. Taylor's Kevlar helmet, with the AK-47 bullet stuck in it, was on display in the 82nd Airborne Museum in the 1980s.

[258] After the invasion I heard another version of what had happened. The trooper was on security, had detected the enemy, and fired an M-203 grenade at them. The grenade went out about 10 meters and struck a tree limb, detonating early and wounding him. A third version of this story was mentioned in *The Rucksack War* "General Trobaugh and his staff ... were convinced that on 25 October an antiaircraft gun mounted on the Soviet Embassy roof had shot down the Marine Cobra gunship. They also suspected that some Cuban fighters had taken refuge in the two compounds and believed that some of those soldiers were sneaking out to waylay individual Americans and then darting back to their sanctuaries. An incident on 27 October about one mile south of St. George's, near the

The restaurant owner's name was Michael Sacremouche and continued to serve food to the passing American soldiers of the 2/325. Bravo Company was treated to hamburgers and cokes from his shop. Sacremouche would yell "I love it! I love it! I love it, Uncle Sam!" whenever he heard gunfire.[259]

Though the news media was not allowed on the island, many attempted to slip through the US Navy blockade of the island. Two Newsweek reporters, Linda Prout and Wally McNamee, were able to race through the blockade in a speedboat and wade ashore at the Holiday Inn. They watched as Bravo Company paratroopers searched young Grenadian men they had found and then would let them go. Sacremouche told the reporters that many of the men were Marxist militiamen who had changed into civilian clothing.

While the reporters were at the crossroad shopping center paratroopers burst into a house near the restaurant and found six Cuban soldiers, along with several AK-47 rifles. The CSC gun jeeps also drove to the medical school to find much needed medical supplies for the 82nd Airborne. The reporters were there when a wounded paratrooper was carried in on the back of one of the jeeps. He had been wounded in the arm and the leg by an unseen sniper. There were no medics nearby, so the soldiers placed the wounded man on a poncho and one of the American medical students that missed being rescued began to work on him.

Though their own supply vehicles had not arrived yet the paratroopers simply acquired vehicles to use. Any abandoned vehicle on the road was hotwired, or "jump" started by popping the clutch. While the reporters watched another wounded soldier was brought in, on a commandeered blue Daihatsu

---

Soviet Embassy, only deepened their mistrust. On that day an infantryman from Company B, 2d Battalion, 325th Infantry, Spec. Brent Taylor, was guarding a road when he was attacked by three enemy soldiers of undetermined nationality. Through good luck, good equipment, and hard fighting Taylor survived to tell his tale, and General Trobaugh and his advisers became ever more vigilant about the potential threat inside the Soviet and Cuban compounds" Taylor's helmet had been hit by an AK47 round and the bullet remained in the helmet. The Kevlar worked though and the bullet did not penetrate. Taylor's Kevlar helmet that had been hit by the AK-47 round was on display in the 82nd Airborne Museum in 1984.

[259] Prout, Linda. "Let's Clean this Up", *Newsweek*. 102, no.19 (7 November 1983).

pickup truck. He had been shot in the neck.[260] Medics worked on him and then all of the wounded were placed on a MEDEVAC helicopter and flown to the airfield. They were flown out by the 82nd Airborne's UH-60s that just recently arrived on the island from Fort Bragg.[261]

More sniper fire made all the Bravo Company soldiers take cover and a patrol was sent into the jungle to flush out the enemy, returning later empty handed. LTC Hamilton sat in the restaurant with his Battalion staff, smoking captured Cuban cigars, and becoming angry at the slow pace and lack of supplies.

After the Battalion had moved on from the Grand Anse beach the 2nd Battalion Aid Station set up there to handle any casualties. According to Edgar Raines in his book *The Rucksack War*, the medics "manning the station were startled when the marines conducted an air assault on Grand Anse. Anticipating possible resistance, the marines charged off their helicopters with their weapons at the ready. It was, reflected an ambulance driver at the aid station, Spec. Denis Deszo, a very dangerous moment. Fortunately, the Marine fire discipline held until Specialist Deszo and his buddies could identify themselves as Americans. The marines, Deszo later recalled, "were not happy with the news."[262]

Due to the continued resistance Admiral Metcalf ordered MG Trobaugh to attack the enemy camp at Calvigny. This had been one of the objectives that the 2nd Ranger Battalion was supposed to have seized on the first day. The fact that it had not been taken did not seem that important on the first day, since all of the fighting was at the airfield. Camp Calvigny was built by the Cubans and used by the PRA as a training base. Once the invasion had started all available PRA soldiers were moved to their combat positions, and Calvigny was deserted. By D+3 though, the camp started to become an embarrassment to the Joint Chiefs of Staff since it had not been cleared.

The Rangers had been at Salines airfield preparing to return home when they were told that they had one more mission. They had already begun to unload their ammunition and leave it for any follow-on forces. When it was learned that the mission was to take Camp Calvigny some of the Rangers had a sense of foreboding, but they methodically reloaded their magazines. The Rangers were told that there may be 600 Cuban soldiers and six anti-aircraft guns waiting for

---

[260] The media published pictures of the medics at the restaurant treating this paratrooper covered in blood. The paratrooper was PFC Kirk Shartzer.
[261] Grenada was the first time that the UH-60 Blackhawk helicopter had been used in war and it became trusted due to its rugged durability to withstand damage from enemy fire.
[262] Raines, Edgar. *The Rucksack War*, pp 464-465.

them. They had less than an hour to get ready to assault the objective, where supposedly they would be outnumbered by the enemy. Greek historian Thucydides wrote, "The Spartans never ask how many there are, but where they are". The American counterpart to the Spartans, the Rangers, also did not worry about how many there might be. Their mission was to take Camp Calvigny.

Colonel Hagler main concern was going into the compound without having artillery prepare the objective. Joe Muccia wrote that Hagler "flat out told the 82nd CO, the JSOC CO and the Admiral who was the task force CO to get fucked because they wanted to put 2nd Batt into Calvigny without prep fire on the objective. Calvigny was estimated to have somewhere between 600-1000 Cuban/Grenadians. The LZ was so small they could only get four Blackhawks into it at one time. Since they were packing 15 Rangers per bird, the odds for 60 Rangers going up against 600 is not good. Hagler knew this and wanted to give his guys the best chance of survival. He DEMANDED prep fires and eventually got them."[263]

Jim Hicks wrote "After a couple days in Grenada we were all pretty much surviving on adrenalin but that was still running extremely high. Our OpOrd for the Calvigny Barracks objective pushed the adrenalin into extra-super-duper-overdrive.[264] Specter and the Navy would prep the objective and lift fire just in time for us to go in. We would be broken into chalks of eight Blackhawks to insert into an open area within the built-up compound. This open area looked awfully small but we knew the pilots were very good and they'd make sure we all fit. We were all rigged up and ready to go in, double-checking everything we already double-checked and then double-checking the guys next to us even though we already double-checked them a few times. When the prep started... everything else stopped."[265]

The plan to take Calvigny involved a half hour bombardment from the seventeen howitzers of the 82nd Airborne, naval gunfire from the USS Guam, and aerial bombardment from A-7 Corsairs and a Spectre gunship. Before it was over the howitzers would fire 500 high explosive shells into the camp. The barrage would begin at 4:30 p.m. The Rangers would then fly from Salines in four waves, with four Blackhawk helicopters in each flight. As they came in the door gunners would pepper the camp with their M-60 machineguns.

---

[263] Email from Joe Muccia on December 23, 2008.
[264] OPORD = Operations Order, or the plan for the attack.
[265] Extracted from http://www.socnetcentral.com/vb/showthread.php?t=32820 on 27 September 2007.

The Ranger air assault was delayed a further 15 minutes so the aircraft could bombard the camp after the artillery strike was over. In his book *Urgent Fury, the Battle for Grenada,* Mark Adkin wrote that the 82nd Airborne's artillery had missed the target altogether, with the rounds falling into the sea. Adkins blames this on the guns not having proper coordinates, and that the 82nd had left their aiming circles back at Fort Bragg.[266]

Wayne Wood, of the 1/320th Field Artillery wrote a review of the book, stating "His assertion we left our aiming circles behind when we deployed is not only untrue, but ludicrous...As a gunner I was responsible for "laying" my piece for azimuth of fire. It was Division Standard Operating Procedure (SOP) for both battery commanders - and Executive Officers of firing batteries to jump in with an Aiming Circle strapped to their legs. I personally saw our aiming circles on the island and used them to lay my piece."

Wayne continued that Adkin "blames the "inaccurate" (fire) on Calvigny Point to the lack of aiming circles on the island. How then, did we do such a great job on Grand Anse? If we did miss the target as he says, it was due to bad maps, not negligence on the part of our commanders...We had fire support teams with the infantry calling our fire. The bombardment lasted over 30 minutes! If we had been missing, our FIST teams would have corrected our fire."

Wood's explanation for Adkin's belief that the artillery fell into the sea was due to later artillery fire, when "the 319th FA who were left on the island as part of the occupation force, had to fire several thousand rounds of ammunition into the sea after the battle because the condition they were in rendered them unsafe to transport."[267]

Whether the artillery was effective or not did not matter, since after the air strike the buildings in Camp Calvigny were reduced to smoking rubble. The Ranger Blackhawks approached from the south over the ocean. The helicopters climbed over the cliffs, only to be surprised by how close the camp was to the cliffs.

In the first wave the first helicopter landed at its designated position. The next two Blackhawks overshot the target and prepared to land farther into the camp. Something happened to the third Blackhawk and it spun forward,

---

[266] An aiming circle is a covered compass, similar to the ones used by surveyors.
[267] Extracted from:
http://www.amazon.com/Urgent-Fury-Grenada-Intensity-Conflict/dp/0669207179   on September 26, 2007

smashing into the second Blackhawk. The fourth Blackhawk saw the carnage and avoided the wreckage, only to land in a ditch, damaging the tail rotor. The pilot attempted to take off, but the helicopter spun forward and smashed into the ground.

Historians disagree on whether or not there was any enemy gunfire during the insertion. Rangers who were on the first helicopters into Calvigny "stated that green tracers from the north side of the compound and the adjacent peninsula swept the LZ as they came in. One of the pilots and several of the Rangers were hit by the enemy fire."[268] The Rangers believed that the enemy troops who were at Calvigny moved to a ridgeline nearby and fired on the helicopters from a distance.

In a matter of seconds three Blackhawk helicopters were down and three Rangers were dead due to the flying rotor blades. Two more that had sought shelter from the blades of the downed helicopters were wounded. 1LT Bill Eskridge had his leg chopped off by the whirling buzz saw.

Warrant Officer Thomas Speaks, piloting the Blackhawk, told the Washington Post, "I saw a soldier with his face, arms and legs chopped off. I remember trying not to step on his brain, which was lying right there. A guy will always wonder, what am I going to do when this happens. There's a guy all mangled. But it goes right past you. Altogether, there were about six wounded on the ground."[269] Speaks heard gunfire, thinking it was sniper fire, and he had lost his gun when the helicopter smashed into the ground. He took a rifle from one of the wounded, and then, with the help of another wounded Ranger, he got up and ran for cover.

SGT Stephen Trujillo ran to the downed Blackhawk, not knowing if there was any enemy fire or not, and pulled the wounded out of harm's way. He would later write "My uniform was stiff with blood as it dried, I smelled of carnage, and plucked particles of brain tissue from my hair. It was futile as I had been immersed. I was glad for the Ranger haircut. I gargled with a quart of priceless water to rinse the flavor of blood from my swollen tongue."[270]

The other helicopters came in, but no enemy was found. There were no bodies, no wounded, no prisoners. The Rangers would spend the night at Calvigny in the smoking ruins of the camp.

---

[268] From a Facebook post by Joe Muccia on 22 November 2013.
[269] *"Hospitalized Locally"*, The Washington Post, October 31, 1983.
[270] Extracted from http://magickingdomdispatch.blogspot.com/ on September 25, 2007.

Tragedy would strike the paratroopers of the 325th Infantry also. Colonel Raines' 3rd Battalion continued to move north on the True Blue-Grand Anse Road, following on the right flank of our Battalion. When the 3rd Battalion was 300 meters from the village of Ruth Howard civilians poured out of buildings and cheered the American soldiers. It was a scene out of the liberation of France in WWII. The paratroopers got caught up in the moment, laughing and talking, and let their guard down. Snipers opened fire from houses in the village. The paratrooper's main concern was the safety of the civilians and they herded them back into their homes. The snipers continued to fire while the paratroopers moved into an attack position. Finally, one trooper fired a LAW rocket in the direction of the sniper fire, scaring off the enemy.[271]

Continuing through Ruth Howard the 3rd Battalion climbed a rugged ridge to a cliff located in what was known as the Woodland Estates. This piece of key terrain would dominate the battlefield. Unfortunately, the movement was through steep jungle terrain and the radiomen were not able to communicate since they mainly relied on the short whip antennas. Due to this the 3rd Battalion soon intermingled with the 2/325th and a short firefight ensued until the two elements made sure they were not the enemy. Some of the soldiers in this fight thought that the Cubans or PRA had set up another ambush to delay the capture of the cliff.[272]

Back at Ruth Howard a Marine Corps ANGLICO team heard the firing from the jungle.[273] They had no radio contact with the 3/325th Infantry so they had no idea what was going on. Like most of the military on the island, the ANGLICO did not have call signs or radio frequencies to contact the USS *Guam*. Suddenly the village came under fire from automatic weapons fire from the east and west, areas already cleared by the 3rd Battalion. More weapons fired from the south, which they thought must have been the paratrooper's response to the firefight.[274]

A Chief Warrant Officer attempted to contact the brigade to conduct an airstrike on the snipers, but he was unable to communicate with them. He was able to talk to CPT William Stephen, the air liaison officer of the 1/505th Infantry, and tell him of the situation. CPT Stephen cleared the chief to have aircraft attack the enemy.

---

[271] Raines, Edgar. *The Rucksack War*, pg. 431.
[272] Ibid., 432.
[273] ANGLICO stands for Air Naval Gunnery Liaison Company.
[274] Raines, Edgar. *The Rucksack War*, pg. 434.

The chief initially wanted to bring in SPECTRE gunships, but there were none available. However, he was able to contact four A-7 Corsairs that had been bombing Calvigny. The chief knew what the target was, a white house with a red roof, but he did not know that Silvasy's 2nd Brigade headquarters had moved near the target building. Fifteen minutes after the artillery barrage slammed into Calvigny, the A-7s came in low, at 200 feet, and dived on the supposed enemy three times, without firing. Each time the aircraft flew over the house the chief would call "mark on top" to let the pilots know where the target was. However, when the aircraft came in for the actual gun run, it was no longer on the right bearing. The chief saw what was about to happen and yelled into the radio microphone to abort the mission. It was too late; the A-7 had already let loose a burst of 20mm cannon fire into Silvasy's command post.

Robert Bence was near the hill and wrote "The bird flew over my position as I reporting into the CP, I was in the open and in the line of fire when all hell broke loose. I saw the impacts and men being hit. The bird passed over me and was turning out to sea. I swore the pilot looked my way; my vibe was that I knew he knew he fucked up... I was pissed, pointed my weapon at the bird and turned to my senior NCO's and coined the phrase "I guess we are not in Kansas anymore." Shit just got real!"[275]

Seventeen paratroopers were wounded in the strafing attack. Radio operator Harry Shaw had both legs torn apart by cannon fire. Over the years many historians have stated that SGT Sean Luketina had his legs blown off, but Harry Shaw wrote "Sean and I were wounded together. Sean's story, and my story, has somehow become a composite memory over the years. Sean had taken shrapnel up higher in his torso and suffered some severe internal injuries that would result in uremic poisoning and the docs having to remove his legs a month later. Captain Jean-Luc Nash has told me that other than the internal injuries, Sean did not appear to be that serious... Captain Nash and Specialist Timothy Andruss were the first ones to enter the barracks after the air strike. They were the ones who took care of Sean and myself."[276]

Harry told of his wounds "I knew three things right then. Number one, I was in a great deal of pain. I have heard it told and often repeated that when someone if injured really severely that that person does not feel pain. I can only assume whoever made this lie up had never really been injured because the pain was immense! Number two: I was thirsty–very, very thirsty. I could not believe that it would be possible to be that thirsty. I was losing a lot of blood fast. Lastly, my legs were gone. If I survived this my life would be forever different. Right

---

[275] *Facebook* post from Robert Bence on 4 April 2014.
[276] Facebook post from Harry Shaw on 27 October 2013.

now, I wanted two things, something to drink and morphine. I would deal with
the missing legs later.... if I survived."[277]

Michael Dodig wrote "The building that the TOC was occupying was on top
of a hill, and the grade was too severe. Also, there was just a dirt road up to the
house, and on the sides of the road the vegetation was about knee-to-waist high
and it was very thick. We had to get the wounded down the hill to a place where
the Black Hawks could land. Harry says he was carried out on a door. I
remember that there were some bed frames -- just bare springs and no mattresses
-- in the building. We pulled them apart and used the frames as litters for some
of the wounded. I remember seeing Shaw and Luketina, but I didn't carry them.
I was working with some others."[278]

Harry Shaw told *Corpus Christi Videos,* "The United States Navy just shot
us and now SGT Steele is beating the crew chief up. The crew chief saw me
and SGT Stewart and saw all the blood and completely freaked out and wouldn't
get off the helicopter to help load us. SGT Steele had gone on board and pulled
him of and started beating the tar out of him, saying you are going to help my
man!"[279]

Mark Adkins wrote in his book that MEDEVAC helicopters were unable to
fly to the USS *Guam* due to a sudden downpour of rain, but Michael Dodig
wrote "The rain didn't prevent the medevac. It wasn't raining then, but rained
later. In fact, I remember looking up at a bright sunny sky to see if the A-7's
were coming for a second pass."[280]

The helicopter first flew to the 82nd MASH unit near the airfield, but realized
that there was nothing they could do for Shaw except "last rites". Once Shaw

---

[277] Extracted from *Hard Core Harry's Blog* at:
http://hardcoreharry.wordpress.com/2010/05/22/all-come-tumblin-down-urgent-fury-part-iii/ on 04 November 2013.
[278] Facebook post from Michael Dodig on 27 October 2013.
[279] Extracted from http://www.caller.com/videos/detail/hardcore-harry/ on 04 November 2013. Harry Shaw later wrote on a Facebook post "I have ever since had a love-hate thing going on with that aircraft. In a sense, the experience of combat, forever creates a duality in us and our existences. I am always here where I am, and, I am, at every moment, there, in Grenada. I always hear that sweet-sickening growl of the gun run. I am, always, at every moment, there in that choking smoke and dust and gore. Sean is still there, and, we are always talking about death, and dying, and God. I am always there amidst a most tremendous tragedy, and yet, always there among an unlikely saving grace as the medical department of the USS Independence, CV-62, heroically refuse to give up on a young paratrooper who refuses to give up.
[280] Facebook post from Michael Dodig on 27 October 2013.

and the others were flown to the Guam they were finally operated on. Shaw said that he never received morphine or anesthetic the entire time and was only given "an ice cube". The surgeons took out part of Shaw's small intestine and he said "somewhere in there I passed out." Though Luketina made it to the *Guam*, he would later die of gangrene in both legs at Walter Reed Hospital less than a year later.[281]

The pilot finally heard the abort order and the screaming of "cease fire" in the microphone by everyone on the ground that had a radio. The flight of A-7s peeled off and flew back to the USS *Independence*. The chief warrant officer did not know of the friendly fire casualties and called in a flight of A-6 Intruders to bomb the original white house with the red roof. The first Intruder dropped a MK-20 Rockeye, but it also missed the target.[282] The second bomb hit near the target, but did not explode. Finally, the chief was ordered to stop the attack by LTC George Crocker, the 1/505 Commander. Crocker decided that they would deal with the snipers without the inaccurate air support.[283]

Amazingly through it all, the Brigade TOC continued to function. Shaw wrote "Seventeen out of 25 troopers hit, yet the TOC continued to function. THAT, more than anything, made me damn proud to be a paratrooper. No single unit suffered a greater loss in Grenada, yet it did not cease in its mission."[284]

---

[281] When President Ronald Reagan gave a speech to the West Point cadets in 1987, he said "As Commander in Chief these 7 years, I have been struck again and again by the professionalism of our military officers and by the dedication of the soldiers I have met in the field. But one who impressed me most deeply is a member of the U.S. Army I never met. His name was Sean Luketina, and he was 23 years old. He did not have the privilege of attending this academy. He was a sergeant, a soldier like those you will command. In this month of October, four years ago, Sean Luketina fought in the invasion of Grenada. He was wounded--badly wounded. He was evacuated to a hospital in Puerto Rico, where his father, a retired Army officer, joined him. He slipped in and out of a coma. During a moment when he was conscious, his father asked him: "Sean, was it worth it?" "Yes, Dad," he answered. Then his father asked: "Son, would you do it again?" Sergeant Luketina looked into his father's eyes and said simply this: "Hell yes, Dad." Duty. Honor. Country."

[282] The MK-20 Rockeye is a 500 pound cluster bomb, that releases 247 anti-tank bombs.

[283] The commander of B Company 1/505th Infantry was relieved the next day due to this friendly fire incident.

[284] Facebook post from Harry Shaw on 23 July 2014. Harry Shaw died in his sleep on August 13, 2021. In his obituary it states "he was hit, first in the arm by a Cuban sniper and then 10 minutes later by "friendly fire" from an A7 airstrike".

Alpha Company continued to move down the road, stopping to search for any enemy or weapons. While we moved it began to rain. Since we were all extremely hot, and had not cleaned up in almost three days, we welcomed the rain. I was with CPT Jacoby and did not have to clear any houses, but I stayed near the road watching for anything suspicious. Though Bravo Company was coming under fire towards the rear of the column, our company at the head of the column was not.

An old Grenadian woman came out of her house and thanked us for being there. She told us that we would not find Bishop, since he had been protected by a magic spell. This is the first time I had heard the name Bishop and did not know who he was. The woman said that Coard tried to kill Bishop, but he could not kill him because he was now bulletproof due to a voodoo ceremony.[285] I had no idea who any of these people were, and just smiled at the woman.

The rain did not last long and the paved road began to steam due to the humidity. It reminded me of rain on hot summer days in North Carolina. While I over-watched I saw a sight that I had not seen since we arrived on the island. One of our own GAMA Goats arrived with C-Rations, Cuban cigarettes and mail.[286]

I sat down in a ditch on the side of the road, wolfing down a C-Ration meal and gulping quarts of water that I no longer had to conserve. As I sat in the ditch, covered with water dripping from the trees, I read my mail. The letters were all dated before the invasion and consisted of a Playboy magazine, with Joan Collins on the cover, and a letter from my mom. The Battalion Chaplain walked down the road, talking to the soldiers, and told us that he would mail any letters we wanted to send. All we had to do was write on the back of the cardboard C-Ration box. I quickly wrote out a letter to my dad that simply said:

---

[285] Years later when I was in Africa I came across soldiers who thought they were bullet proof and invisible due to these voodoo ceremonies. Just to be on the safe side I had one of their priests make me bulletproof and invisible to. You never know...

[286] This was the only GAMA Goat to arrive on the island. The M561 "Gama Goat" was a six wheeled vehicle that was unique due to rear wheels that turned opposite of the front wheels. This "amphibious" vehicle was created for the jungles of Vietnam and continued in military service until the introduction of the High Mobility Multipurpose Wheeled Vehicle (HMMWV) or Humvee. The Gama Goat saw its final combat during Desert Storm in 1991.

Triple Canopy

I'm alive. This is the first time we got resupplied so they're letting us write a quick letter on the back of a C-Rat box. I'm a sniper. Seen some action. Mainly moving inland to get the American hostages. Tell Sean having a wonderful time, wish he was here. The people are glad we're here. Got to go.

Me.

The Grenadians continued to come out and celebrate their liberation once they realized we would not shoot at them. Some of them gave us Coca-Colas and others tried to give us rum. I turned down the alcohol since I did not know what might happen in the near future. While we searched the houses, we heard Radio Free Grenada blasting rock and roll music from radios located inside. The announcer still vainly told the Grenadian people that they were needed to repel the invaders.

As I sat in the ditch leaflets floated down from the sky and landed amongst us. I put a couple of them in my pocket. On one it said "Safe Conduct Pass. To those who are resisting the Caribbean Peace Force – this pass will save your life." Another leaflet told the citizens of Grenada "Do not leave your home. Avoid confrontations and do not interfere with the U.S./Caribbean Forces. If fighting starts in your area stay in your homes and on the floor. Stay off roads and highways. Further emergency information will follow. Please remain calm and no harm will come to you."[287]

We continued to move down the road and searched the houses until the sun began to set. We had moved about two miles since we climbed down from Morne Rouge ridge earlier that morning. As the sky began to darken, we stopped beside a hotel that was on a beautiful beach covered in white sand and palm trees. My 1st Platoon moved inside the hotel, creating an urban defensive position with paratroopers at each window. The rest of Alpha Company spread out around the hotel in a defensive line that linked up with Bravo Company behind us. The hotel had become the company CP. We were told that we had to stop our forward movement since we were the farthest unit in the push to St. Georges. Since we were so far ahead of the rest of the 82nd Airborne, they were

---

[287] During and after each war there are always uninformed critics that talk of how brutal the American military is towards civilians, but what they don't know is the extraordinary care taken to make sure that civilians are not harmed unless it is absolutely necessary. It has gone to the extreme of allowing US soldiers to come under fire by the enemy, without ever returning fire, just to make sure the civilians are not harmed.

worried that we would soon outdistance any artillery support if we came into contact with the enemy again.

I was not with the 1st Platoon in that hotel, but instead was with CPT Jacoby and the HQ element. Major Baine, the Battalion XO, drove up to our command post in a jeep that had a captured RPK machinegun on the hood. The jeep was weighed down with all sorts of captured items, so that the frame of the jeep nearly touched the ground.

MAJ Baine drove down to the 1st Platoon's defensive position at the hotel and told CPT Jacoby that the hotel would become the Battalion CP. He also told us that the 3rd Brigade of the 82nd Airborne had arrived at the airfield. This was the 1st Battalion, 508th Infantry under LTC Hugh Shaw. Since it arrived so late the 1/508 would not see any action and was delegated to guarding Salines airfield.

As the 2nd Platoon moved into their defensive position a shot rang out. Everyone jumped for cover, then began scanning the nearby houses for the sniper. We soon learned that SSG Hudson, of the 2nd Platoon, had shot a dog that looked like it was going to attack him. Mark Banker wrote that he "was so pissed off at Hudson for shooting that damn dog in the leg. He told me he shot it because it was barking...now it was yelping and making all kinds of noise. Nothing like announcing our arrival!"[288]

Back at the airfield our HHC had spent the morning burying the dead from the fight the day before. They dug ten graves and afterwards moved vehicles to the Frequente village. A platoon of engineers set up security over the vehicles on a nearby hill, but soon came under fire from snipers located in the nearby houses. The engineers returned fire until sunset. As soon as it became dark a Grenadian approached the engineers and told them that the Cubans had used his house and his neighbor's house as cover. He also said that they had been caught in the crossfire and several civilians were wounded. The engineers sent a patrol back with the man and escorted 28 civilians, eight of them wounded, to safety.[289]

As it began to get dark CPT Jacoby had to look for a new company CP, since Battalion was taking the hotel. He decided to use the porch of a house that was near the 1st Platoon. A few of us wanted to clear the house, mainly to see if there was any food inside. Though food had been delivered to us earlier in the

---

[288] Facebook post from Mark Banker 27 October 2013.
[289] Raines, Edgar. *The Rucksack War,* pg. 467.

afternoon, it had only been a single C-Ration and we were hungry again. It seemed like I was never going to get enough to eat.

The door was locked so I took out my .45 pistol and tried to smash in the window on the door. Though I hit is several times with the pistol, the window refused to break. It did gouge a piece of metal out of the slide of my pistol. SP/4 Bush got tired of waiting and ran up, kicking the door and slamming it inwards. We rushed in and immediately fanned out into the room, made sure it was clear, and then headed towards the kitchen. We found a case of Coca-Colas and filled our cargo pockets with them. There were also tins of sardines, cans of corned beef and boxes of crackers. As I passed by the bedroom, I saw one of the guys looking into a jewelry box. I told him to not even think about it. He looked embarrassed and quickly headed outside. Taking food was one thing, but no one was going to take anything of real value while I was there. We were foragers, not looters.

As it became dark the streetlights came on around the houses. Almost immediately we heard gunfire coming from near the hotel and I dove to the floor. I crawled towards a window and searched the darkness, looking for any signs of the enemy. All I saw was SSG Moore from 3rd Platoon, shooting out the streetlights. He was worried that they would give away the positions of his squad members so he decided to "turn them off".

After our hearts stopped racing, we took all the food we had found and quickly moved outside, away from the officers. We were worried that they would make us put the foraged food back. In the shadow of the house, we ate everything, washing it down with the warm Cokes. Ironically after we finished the meal, our GAMA goat returned from the airfield and issued out more C-Rations. We ate that heartily also, and I finally began to feel full.

As soon as we finished eating the sky opened up and poured rain on us. It continued to rain the rest of the night. Since we had no rucksacks, and no ponchos, we took cover where we could inside a small house being used as a Post Office. My "bed" was some old cardboard I had found inside the house.

Ray Meier wrote about that night "I remember Gallardo going outside the perimeter to take a leak, and not knowing the challenge/password coming back. He just kept yelling back to us; DON'T SHOOT, it's me Gallardo. I think it was Kelly Harris that said, "Someone shoot him anyway" and we all laughed."

Gary Marsh wrote "we had a mortar position set up not to far from there, McPeak shot at a retarded guy that ran through our gun position at night. I heard halt! Halt! And automatic gun fire and then the retard ran past SP/4 Tolar. He couldn't shoot because he had his rifle apart cleaning it. What a crazy night!"[290]

Mark MacNamara also wrote of the rain "On the night of the pouring rain 2nd Platoon, 2nd Squad was in the Jungle on the right-hand side of the road. We set up for an ambush, but I didn't know what was in front of us. It poured rain at first... I just laid in jungle floor and searched the darkness. Of course, when you're tired its night time and raining and your eyes are straining to see something that you know is creeping up to you. You will see it and I did!!! At first, I thought it was PVT Patallono in 3rd Squad. I asked him where he was. 'Right here' he answered. With the rain and the dark, I couldn't tell where he was and I didn't want to shoot a guy in my platoon. I thought he might be out front taking a piss. So, I asked him again if he was in position. He said he was. So, when I saw that thing in the dark rainy jungle again, I opened up with my M-16. The M-60 also fired, then all was quiet till morning and we never even went forward to see if anything was hit. I think it was Casper the ghost."[291]

I was still worried about a Cuban counterattack or some sort of fight, so the night was long, tense and tiring. Visibility was almost non-existent as the rain poured down. Occasionally a burst of gunfire or the deafening explosion of a Claymore mine exploding would startle us. I could also hear Spectre firing up a golf course that was right beside our position. The shock waves from the explosions would rock us out of any sleep we might have been able to get. We were told that the Cubans had dug positions at that golf course, but no enemy appeared. I wondered where they could have gone to.

After the invasion there were those who criticized the objectives of the invasion, the tactics used, the decisions made, the lack of preparation by the military and the mistakes made by the leaders. In other words, the criticism was the same as every other military engagement the United States has been in since Vietnam. One of the complaints was that the whole invasion moved too slowly. To the north the Marines had seized their objectives fairly quickly and were waiting for the Army in the south to rendezvous with them. However, the Marines in the north had little to no resistance and they also had their tanks and other vehicles. In the south of the island the Army had met stiff opposition on the first two days, which set the tone for the next week. The Army also did not have much of their vehicle support and had to move northwards by foot.

---

[290] From a Facebook message posted on 30 May 2011.
[291] Facebook post from Mark MacNamara on 27 October 2013.

The 82nd Airborne moved slowly northwards in Grenada, clearing any pockets of resistance and ensuring that no enemy would be left behind them to organize a counterattack. Ironically in the Iraq War in 2003 the military leaders were criticized for bypassing pockets of resistance and allowing them to organize a guerilla warfare against the coalition forces. Grenada was also ridiculed for being over too fast with very little casualties, but the Iraq War was criticized for taking too long and having too many casualties.

Grenadian General Hudson Austin, Ewart Layne and Liam James abandoned Fort Frederick and decided they needed to leave the island by boat. To remain would most likely be a death sentence. The trio determined that their best chance for survival would be to escape to Guyana, a few hundred miles to the south. The fugitives armed themselves with some pistols, grenades and an AK-47, then headed south to the Westerhall Point peninsula. There they met Gerhard Jonas, an East German advisor to the PRG. Jonas agreed to help them get off the island, and he hid them in the home of Douglas Stone, a resident who was not in Grenada at the time.

On the morning of October 27th two British officers with the Barbados defense forces, Lieutenant Commander Peter Tomlin and Major Michael Hartland, approached Brigadier Rudyard Lewis and told them that they thought Austin would try to get away by using a boat. Lewis listened to their plan to find Austin, and then gave the two a captured GAZ jeep with a U.S. soldier as an escort. Lewis told them to conduct a reconnaissance of Lance aux Epines. When the small group arrived there the marina owner told them that there had been enquiries about Jonas wanting to hire a yacht to go to Westerhall. The marina owner pointed them toward Dod Gorman's house on a nearby hill.

When the trio arrived there, they found more American students who had not been rescued, along with some British expatriates. One of these was John Kelly, the British representative in Grenada. Kelly did not appear to be happy that the invasion had occurred but Kelly's wife directed Tomlin to a European sitting in the corner. Initially the European man pretended to be drunk and then he pretended to not know English, but after a few hours of interrogation it was learned that he was Gerhard Jonas. Jonas finally told the officers that he knew where Austin was and that he was supposed to get him a yacht to sail to Guyana.

Tomlin devised an elaborate plan to capture Austin by sailing a yacht from Lance aux Epines to Westerhall Bay. There they would lure Austin onboard the yacht and capture him. However, after one failed attempt by the British officer,

MG Trobaugh told Tomlin that the mission was aborted since it was too dangerous and he did not have any available Special Forces that could do the mission successfully. Austin, Layne and James would remain at Stone's home for another two days.

While the lead elements of the 82nd Airborne continued to move towards St. Georges, the Marines of E Company 2/8 were ordered to do a reconnaissance into the Mount Horne area. There the Marines discovered an abandoned PRA command post at the Mount Horne agricultural station. Inside the station radios, maps and other military equipment was discovered.

Towering above the Marines was Mount Saint Catherine, almost 2,800 feet above sea level. The Grenadians at Mount Horne told the Marines that some of the PRA might be found up there. There was no real road leading to the top of the mountain and it was only a muddy track, twisting through the dark jungle. The Marines went anyway with some of the smaller jeeps and a flatbed truck. As they climbed along the road it began to rain hard. Near the top the Marines spotted several PRA soldiers carrying a mortar. As the Marines fired at them from a distance the PRA quickly ran into the security of the jungle. The Marines didn't hit any of them, but they did discover mortar and RPG ammunition.

On the return down the mountain disaster struck when one of the TOW jeeps overturned, injuring two of the Marines. The wounded needed immediate medical assistance, but it was impossible to bring in the MEDEVAC helicopters in that driving rain on the steep terrain. The two Marines had to be carried down the mountain on improvised stretchers. When they arrived at the small town of Paraclete, they were finally able to call in a CH-46 to extract their casualties.

On the morning of the third day, October 28th, I was wet, tired and hungry again. Since we had no rucksacks, it did not take long to start the march down the road to St. Georges. Our pattern of march, stop, and search the houses continued. We were trying to find any sign of the enemy, but it seemed that they had disappeared. Rumors were that the enemy had drifted into the jungle to organize some sort of defense, or to start a Vietnam-style guerilla warfare. At one house 1st Platoon discovered some burned papers on the ground with some military symbols on it. CPT Jacoby told me to move behind the house and overwatch the platoon while they searched. I moved around the house, but there was a thick impenetrable jungle covering a steep hill. I could not watch anything from there, so I returned to the front of the house and helped in the search.

The house turned out to be another hotel and the owner was an old lady who kept asking us when the war would be over. She was excited about the prospect of tourists returning to the island and more customers coming to her hotel. One of my team members found a Soviet F-1 grenade and he gave me it to me to use since I did not have any grenades. I thought it looked "cool" but I was also wary about using it after hearing what happened to SSG Epps. The burned papers with military symbols turned out to be just notes taken by one of the Grenadian militiamen during a class. He had burned them so that he wouldn't be arrested. We continued the mission and moved on down the road.

At 0730 we came across some Marines stopping cars at a roadblock and checking them. This surprised us because we didn't know there were any Marines in our area.[292] The battalion column stopped while the necessary coordination was made between the two units. While we waited Bravo Company at the end of the column was told to search the golf course that the Spectre gunships had been hammering the night before.

I talked to one Marine and asked him if he had seen any action. He looked disgusted and said that he "hadn't seen a damn thing yet". He said that they had landed north of St. Georges and then moved to their present location at Ross Point without any opposition. We were able to get some gun oil off of them, which was great because after the stormy night we were running low on oil for our weapons. Some of our guys didn't have any oil at all and were losing the battle against the rust.

Our appearance contrasted sharply with the Marines. They were all clean shaven and looked sharp. Meanwhile we had four days growth of beard, were muddy, and had not bathed since we left Fort Bragg. We stank. My uniform had the crotch ripped out of the pants and now that initial rip had split down to my knee. Some of our men carried Soviet weapons and equipment. The M-203 Grenadiers carried their grenade rounds in AK-47 chest pouches or tied loose onto their LCEs. Some of our guys had been called off guard duty when they deployed and had been wearing their nicest jump boots. These boots look good, but do not stand up to severe conditions and had broken down. Some of the men resorted to wearing Soviet half boots that were found in a crate at the airfield. The same happened to some of the older jungle boots that dry rotted in the heat and moisture. A local cobbler volunteered to repair some of the boots and was able to do it overnight. Assistant gunners, with no rifles, carried AK-47s or RPKs. A few men had Makarov or Tokarev 9mm pistols shoved into their pistol belts. Soviet grenades dangled from our LCEs. Our Battalion X.O. drove around with an RPK on the hood of his jeep. We looked like a bunch of pirates.

---

[292] These were the Marines of F Company, 2nd Battalion, 8th Marines.

Those Marines at Ross Point must have wondered where we had been and what had we seen during the last three days.

The Marines said that they were going to withdraw now that we were there, and they would only leave a guard at the temporary embassy at Ross Point. Once they got back to their ships, they would continue on to Lebanon to replace the Marines who had been killed in the suicide attack in Beirut.

We cleaned up the best we could at Ross Point, then laid in the grass and let the sun dry us out. We were issued C-Rations for lunch and I traded the Marines my C-Rations for their MREs. The Marines said that they were always issued MREs and were sick of them. I told them that I always was issued C-Rations and was also sick of them, so we were both happy with the trade.

While I laid there relaxing, I saw my first reporter. He wandered over to us and told me that he was from Newsweek magazine. We ended up asking him more questions than he asked us. I asked him what did America think about the invasion, and he replied that the American public was dead set against the whole thing. He also said that there were protests in the streets back home against the invasion because it had no justification. He told us that the Americans back home thought we were just killing civilians for no reason, "stormtrooping" across the island.

He asked me my opinion and I said "Fuck them, what the hell do they know!" I also told him that the Grenadians celebrated every time they saw us, and gave us food and drinks. I did not find out until after I returned home that the reporter had been lying about everything and the American people were almost unanimously behind us.[293] However until we returned home and learned the real truth, we were worried that we would be looked down upon by the American people, the same as the Vietnam vets were when they came home ten years earlier.[294]

Brad Gallardo wrote to me later that "The LT told me to set up a road block and sit tight, so I took Boulderoff's crew and found a nice driveway that overlooked the road on a bend. There was a Marine gun crew there on one corner of it. I plopped Boulderoff down on the opposite corner of the yard and put Walters and Malbrough on the road. Well, the Marine gun crew leader came

---

[293] A CBS television poll had 91% of the American public favoring the invasion.
[294] In my journal I wrote "Fuck them. These people loved us, and I never did see a dead Grenadian, only dead Cubans."

strolling over looking a little perturbed. He comes up to me and says "Well, yalls the Gallant Eagles, eh?" I said yea that would be us."[295]

Gallardo continued "After a couple of hours the Marines moved off the spot.... I had set up on the porch with my radio while Malbrough and Walters were on the road. Mitch Boulderoff was pulling overwatch with the M-60. After a while a Diplomatic Staff car showed up, Boulderoff moved down on the road with them. It appeared they were excited...I asked them who the hell is that? They yelled up "Commies." I said get them out of the car and search them. It turned out to be Soviet, East German and Cuban advisors. I gave the CO a shout on the radio and waited for word on what to do with them. He called Battalion and they said to cut them loose. Well, we did, only after we rousted the hell out of them."[296]

LTC Hamilton drove up in a jeep and issued CPT Jacoby new orders. Alpha Company was to move to a tall hill overlooking Ross Point and secure it. We trudged up the hill and began searching the houses there. We discovered an American, who asked me how he could get off the island. I pointed to the Marines leaving Ross Point and told him he might be able to hitch a ride with them, but they were going to Beirut.

He told me that about 250 Grenadians had come down out of the hills, threw their weapons into the jungle, changed into civilian clothes and then blended back into the population. He also pointed to a white house on a hill above us and warned us about it. He said he thought it was a possible Cuban strongpoint. Other Grenadians listening to us also said that we needed to watch out for that house. None of them gave us details though. The house was pretty far away, and out of our perimeter, so we did not search it, but I constantly turned my eyes that way, wondering what was in there.

While the rest of the company was searching the houses around the hilltop I sat down and watched the surrounding neighborhood for anything suspicious. I had with me PFC Lewis, the Commander's RTO. We were outside a house that had not been cleared yet and there were papers scattered around the door. I picked up one of the pieces of paper and saw something written on it about the PRA. Lewis pointed over at the grass and we saw an AK47 and a cattle prod

---

[295] From a Facebook message from Brad Gallardo on 19 September 2010. I guess this was a reference to our Company jumping into Operation Gallant Eagle a year before, which was pretty famous at the time.
[296] Facebook message from Brad Gallardo on 19 September 2010

laying just ten feet from us. I gathered a group of soldiers attached to the HQ element, created an improvised team and went inside the house to clear it.

Once our eyes adjusted to the dark, we saw papers scattered everywhere on the floor. On the wall were propaganda posters of Che Guevara and posters of a young Fidel Castro. There were also various posters spouting the greatness of Communist life. I picked up some passports from the floor and shoved them in my cargo pocket. Among the items thrown on the floor were manila folders, logs, film negatives, paper money and coins. Weapon racks lined the wall. I picked up some large copper coins that had Queen Elizabeth on them, to keep for a souvenir.

In one of the rooms there was a hole in the floor. PFC Lewis and I jumped down into the hole to see what was there. Inside there were even more gun racks. We continued through the room and down a dark hallway until we came back outside in the garage. There were parts for motorcycles and cars strewn about. The two of us moved around the stripped carcasses of some cars and checked out another room attached to the garage. Inside we found Soviet aid bags and medical supplies all over the floor.

Inside the garage was a refrigerator and my first instinct was to open it, and see what might be in there to eat. I learned a lesson that day that is forever seared into my memory. When you are in a place that hasn't had power for a week, DO NOT for ANY reason open a refrigerator. When I cracked it open the smell of rotting meat engulfed the garage. I slammed the refrigerator door shut, but it was too late. We both began gagging from the smell and had to run outside to throw up. I imagined that this is what dead bodies would smell like if they were left out in the sun for a week. Many years later, in another war, I would learn firsthand about the smell of dead bodies. That was not the same, the dead bodies smelled worse.

I went back to the CP and told CPT Jacoby what I had found. I told him that we needed a team from the S-2 to come and analyze what we found. I also handed over the passports that I had pocketed back in the building.

While we had been searching the building the GAMA Goat arrived carrying our ALICE packs. We had been living out of our pockets for two days and it was a welcome sight to see our heavy rucksacks. Though we didn't want to carry them, they did have some luxury items such as socks, poncho liners and gun oil. The supply sergeant handed out more C-Rations, which we devoured. It seemed we could never be full after the first three days of starvation. He also handed out warm Cokes, which tasted fantastic. We were finally issued U.S. hand grenades that had been taken from the Rangers when they left.

While most of the company celebrated the return of their rucksacks, I didn't get mine. All of the rucks from the HQ section were missing. I figured they were still laying near the restaurant where we dropped them. The mortar section had their jeeps now and I was able to convince one of the mortarmen to take me back to the restaurant to find the missing ALICE packs.

When we arrived at the "rear" we found the rucksacks lying neatly in a row, just as we had left them. The REMFs were back there too, stockpiling the captured weapons, and carrying them around while their buddies took pictures of them.[297] They put on as many grenades as they could to look like real soldiers, then sat around in the shade relaxing and drinking cold cokes. After we loaded up the rucksacks in the jeep we drove back to the "front", where I ate my cold C-Ration and drank my warm coke. The mortar jeep returned to the mortar section that was digging in at the beach, back at the Ross Point Inn. The rest of the company was digging in on top of the hill at a doctor's house.

That night I took the poncho out of my ruck and threw it over my head. I then turned on my flashlight and sewed up the hole in the crotch of my pants. While I was doing it, I accidentally sewed my underpants to it and had to cut a hole out of the underwear. I could have undone the thread, but it was hot under that poncho liner and I didn't feel like sewing it all over again. I spent the night on the front yard of the doctor's house, not in an OP, but taking turns on radio watch with the headquarters section.[298]

When I returned from retrieving the rucksack, I heard a dog barking. This was a constant noise, never ending. It just continued going woof... woof...woof...woof, never ending. I finally decided to kill it so I could get some sleep. I walked over to the dog and found it chained to a tree in the yard of the doctor's house. I pulled out my Randall knife, ready to cut its throat so I could go back to bed. The dog began to bark at me rapidly, realizing that I

---

[297] REMF is an extremely derogatory term to an infantryman. It stands for Rear Echelon Mother Fucker. These are the folks who don't fight the war, but take the credit for it. They usually are the bureaucrats of the military who make the life of the combat soldier a little more hellish.

[298] Radio watch is a lot easier than pulling security or being on an OP. While doing the latter the soldier must remain vigilant and awake. One technique is to rub hot sauce underneath the eyes. The pain will keep anyone awake for a while. However radio watch is one where the soldier might be able to sleep after taking his turn listening to the radio. Neither will give the soldier a full night's sleep though.

wasn't there to be its friend. I knelt down, grabbed the dog by the neck and then put the knife to its throat. I stopped. Obviously, this dog was someone's pet, maybe some child who lived in the house. I grabbed the dog's mouth and held it shut. I then whispered to the dog "I don't want to kill you, but if you don't let me sleep in peace I will!" I walked back to my rucksack to the miraculous sound of silence. The dog had listened to me and there was no more barking for the rest of the night.[299]

On the afternoon of October 28th E Company 2nd/8th Marines were ordered to search a cave near the Mirabeau Hospital. Upon arriving the Marines left their vehicles to search the hospital on top of a ridge. One squad was left behind to guard the vehicles. The Marines moved quietly through a banana plantation on top of the ridge and discovered one civilian carrying coconuts. He was quickly detained so that he would not make any noise. The lead Marine then saw three men standing by a Land Rover. The people of Grenada were predominately black, and these men were not.[300] One of the Marines yelled "Freeze!" and the three suspicious men took off running. The Marines opened fire, wounding two of the men, but the third ran through a house and escaped. The wounded men were soon discovered to be Cubans, one of who had been mortally wounded.

While the Marines were taking care of the two wounded Cubans, the squad that was left back at their vehicles came under fire. The Marines returned fire for a few minutes, then the enemy withdrew. The Marines at the top of the ridge quickly moved to the vehicles to provide assistance, but they came under fire

---

[299] Facebook message from Brad Gallardo on 22 November 2011. Brad Gallardo wrote of another incident at the doctor's house "That night we set in around that mansion where that lemon grove was...there was a bungalow in stilts on the other side of the road with a deck overlooking the road. We took that house over and used it as our corner of our perimeter...That night you and I sat on the deck pulling security over that area...Well SFC Howard was in the house as well...I heard a knock at the door and there was a pretty Grenadian girl who began telling me her house had gotten bombed and she didn't have anywhere to go...... I was stymied with what to do with her so I went and got Howard and turned her over to him...Next thing I know Howard took her in the bedroom.... Later Sengbusch showed up looking for Howard, I pointed the way...He walked in on him...I have no clue what happened after that."

[300] This was a quick way to tell the difference between the PRA and the Cubans. Any person found who was not Black was automatically suspect.

from a squad of men above them. The Marines returned fire with machineguns and a LAW rocket. Though no enemy was found, it was presumed, due to blood on the ground, that some of them had been wounded.

A Grenadian who had been with the PRA told LTC John Raines that President Bishop was buried at Calvigny. A team from the 3/325 was guided to the location and they found five bodies in a pit. The bodies were badly mangled, and did not have any heads or hands. None of the bodies could be positively identified as Maurice Bishop.

More of the 82nd Airborne arrived at the airfield on the afternoon and night of Friday, October 28th. This was LTC Keith Nightingale's 2nd Battalion, 505th Infantry. Still later that night LTC Ralph Newman and the 2nd Battalion, 508th Infantry arrived. The 2/508th Infantry had received the mission to clear the Hartman Estates east of Lance aux Epines. As they did two companies came under sniper fire from the PRA around 4:00 in the afternoon. The two forces fired at each other for about an hour, with little effect, until the paratroopers wounded a Grenadian soldier. The rest of the PRA broke contact and withdrew into the jungle.[301]

As all this was going on, American citizens continued being evacuated with every aircraft that left the island. By the end of the day there had been 662 United States citizens evacuated, and 82 foreign nationals. While this was going on the Libyan ambassador ran from one group of students to the other to make sure that none of his staff defected to the United States. Only 21 students had been left behind, mainly due to any lack of information on their whereabouts. Because of the constant sniping and with the threat of a protracted guerilla war, XVIII ABN Corps commander, Lieutenant General Jack Mackmull, decided to alert the 101st Airborne Division for possible deployment.[302]

The old man who lived in the house where we stayed the night came out and introduced himself. He told me that he had been living in Grenada for the last

---

[301] Raines, Edgar. *The Rucksack War*, pg. 469
[302] Though the 101st Division is known as "Airborne" it is not a division of paratroopers. The 101st had been one of the original paratrooper divisions in WWII, but in 1974 the division lost its airborne status and was redesignated as an airmobile division. The soldiers did not jump into the battle using parachutes, but instead flew into battle on helicopters. The official term for the division at the time of Grenada was the 101st Airborne Division (Air Assault). The only paratrooper division in the Army at the time of Grenada and at the time of the writing of this book is the 82nd Airborne Division.

20 years. He also told me that he had been a Lieutenant in World War I with the British army and showed me his scar where he had been wounded in battle by the Germans. He said that he was glad that we had kicked the communists out and anything on his property was available to us. He offered all of us a drink of gin, but we turned it down. Instead, I asked if we could get water from his garden hose and he gladly showed us where it was.

I filled all my canteens from the hose and then walked down to a grove in his backyard. I picked some grapefruits, lemons and limes and then sat with some of the machine gunners, eating them while we enjoyed an incredible view of St. Georges harbor. I ate the fruits with some sugar from my C-Ration pack. There were ships in the harbor and more ships from the US Navy out on the horizon. It was breathtaking and I could see why the doctor chose this spot for his house.

While I was relaxing, enjoying the view, 3rd Platoon sent word that they had found a weapons cache. I picked up my sniper rifle, grabbed SP/4 Bush, and headed on over to 3rd Platoon's area to check it out. When I arrived, I saw everyone taking cover because they thought they saw someone on the hill with a weapon. I set up a quick sniper position and scoped the hill. I wasn't able to see anything suspicious and told them that I thought that it was clear. We were within view of that suspicious white house on the hill, but the 1st Platoon was supposed to be up there.

The weapons cache ended up being a variety of weapons scattered in the jungle. The 3rd Platoon brought them out and stacked them in the road. There were three RPGs with missile packs, twelve AKS rifles, twenty AK-47 rifles, one RPK machinegun with a full ammo can of ammunition, 6,000 rounds of 7.62mm Soviet ammo in rucksacks, and various other Soviet equipment.[303]

I started back to the CP to let the Commander know that he needed to bring up a vehicle, when a shot rang out behind me. Everyone hit the dirt. I quickly scanned the surrounding area, but found nothing. I asked where the shot came from and someone told me it was a sniper. I doubted it, and figured it was someone accidentally firing their M-16. This happened a lot during the invasion. After a few more minutes I got up and moved back to the company CP.

After I told CPT Jacoby about the cache, he radioed LTC Hamilton. Both came out to see the weapons. The colonel told CPT Jacoby that he also wanted

---

[303] The AKS is the folding stock version of the AK-47.

to see that house that I cleared the day before. Before he arrived at the house, I wanted to go inside one more time to pull out some of the more important looking items. I took two guys from the HQ section and went down into the garage. I told them to bring out the aid bags and anything else that looked important. We started on the top floor and I told them to gather up the cattle prods and put them out on the lawn. Bush did what I ordered, but the other two guys from HQ were too scared to touch anything. They had just recently heard about SSG Epps being killed. The rumors were that he had been killed by a booby trap and it had blown off his face, killing him and wounding six others. I tried to convince them that nothing was booby trapped, but they didn't want to touch anything. The two HQ guys were useless to me, so I told them to get back to the CP.

I went back upstairs and started looking through the papers on the floor. I found some more passports and a book that looked like it said "terrorists of the Caribbean" in Spanish. There were stacks of files stamped "Secret" with a history of all the "terrorists" in the area. There were numerous books on communism, along with films and videotapes. There were also ledgers of arms, ammunition, vehicles and lists of who was issued all of this equipment.

I found a plastic Soviet canteen and put it in my cargo pocket. Once I was outside, I wondered if maybe it had been booby trapped. I really needed another canteen for water, so I "hugged" a tree, by putting my arms on one side, while using the tree as cover in case there was an explosive charge. I figure my hands may be blown off, but I would live. I unscrewed the lid while standing in this odd position, and found the canteen half full of rum. I dumped out the rum and shoved it back in my cargo pocket.

I went back and forth to the house, carrying documents outside for the colonel to look at. One of my squad members from 1st Platoon saw what I was doing and came over to help me. When LTC Hamilton saw the piles of documents he wanted to go in the house too. I told him that the intelligence had been sitting there for the last two days, but no one from the S-2 had showed up yet.[304] By the time we returned to the company CP half of the battalion officers were there wanting to see the weapons. Bush and I escorted them back to the cache. They ogled them for awhile, being more tourists than soldiers, and then we loaded up all the weapons on a Soviet jeep.

---

[304] In a battalion or brigade headquarters the different staff sections are numbered. S-1 is personnel, S-2 is intelligence, S-3 is operations, S-4 is supply and S-5 is public relations.

I drove back to the company CP in the Soviet jeep and then grabbed the two guys who had been too scared and refused to do anything earlier. I told them they were coming with me and no whining this time. Since everyone had been in the house, they were no longer afraid of booby traps and helped me to carry out more documents. The Battalion S-2 finally arrived and took all the documents away. Many of these documents were used to justify the invasion during a Congressional hearing after the war.

For most of the day I was either clearing out the Intell site or acting as a tour guide for the various members of the chain of command who wanted to see everything. On one of my shuttles to the site, a Grenadian girl came to me with a nasty looking dog bite. I took her to one of our medics, "Doc" Blackenby, so that he could fix her. He took a look at her, but figured it might need some stitches, so he took the girl back to the doctor's house to let the old man sew her up.

While I was at 3rd Platoon's cache site, SSG Moore told me that he had heard Reagan on a radio saying that the war was over. There were radios playing in most of the houses, usually playing a song by Bob Marley or Eddy Grant and SSG Moore had heard the news about Reagan on those radios. Moore told us that Reagan said the Marines would pull out and go to Beirut, while the 82nd Airborne would remain there for at least six more months until stability returned to the island. Like most rumors, this one turned out to be not true.

Towards the end of the day, I was told that my 1st Platoon nearly got into a firefight with C Company of 3rd/325th Infantry. We had heard how the 3/325 had called in an airstrike on our own brigade headquarters a few days earlier, and now we tried to stay away from them as much as possible. They seemed trigger happy. Our battalion wasn't doing much better, and there were numerous accidental discharges.

As the sun set, I was surprised by being issued an MRE for the evening meal. 1SG Polachowski had been trading with the Marines by giving them our C-Rations for their MREs. MREs were still a novelty, so we liked any we could get. Within a year the novelty would wear off. I washed down the MREs with more hot Cokes. We had cases of them from the airfield warehouses.

While I had a few minutes, I wrote another letter home on the back of a C-ration box.

I'm still here. I don't know what to do about my bills. I guess I'll skip them till I get back. If I'm here for Christmas don't send anything, I have to hump everything I get, or throw it away. The Grenadians have all blended into the populace again, so we're just fighting Cubans. The people love us, like France 1944, offering us booze and food. I could use some socks and some 35mm color film. A lot of guys have got war trophy's, but they're in the areas after I leave. They're still hot while I'm around so I haven't got nothing yet. I'm still a sniper in H.Q. section, so I'm part of everything that goes down. Got to go.

Me

That night it was nice and quiet. No barking dogs. No Spectre gunships. No quiet intensity, wondering if you would die by a sudden thrust of a knife at close range. All I had to worry about was radio watch, while I looked up at the starlit sky in a tropical paradise.

During October 29th the most recently arrived battalions of the 82nd Airborne pushed out from the southwestern tip of the island to search the interior. The 2/508 under Colonel Newman borrowed some Grenadian vehicles and drove to Good Hope Estate, four kilometers east of Ross Point. With them were elements of the 3/325th Infantry. The motorized group became known as Task Force Newman. The paratroopers discovered a PRA supply dump full of ammunition at Good Hope Estate. There were five trucks loaded with ammo, with seven more truckloads scattered on the ground. As the paratrooper's vehicles broke down the Grenadians gladly lent them replacements and then cheerfully waved them on to the next village. Task Force Newman continued their motorized reconnaissance in force to the eastern coast, and the village of Crochu, meeting no resistance.[305]

The Marines of H Battery, 2/8th learned of Coard's hiding place on the outskirts of St. Georges from his fellow Grenadians. The Marines surrounded the house and using a bullhorn demanded that Coard come out and surrender. There was no response. Finally, the Marines threatened "Come out or we'll blow the place up". Coard's wife and children came out, with their hands up, followed by Coard. They were ordered face down in the dirt and then

---

[305] Raines, Edgar. *The Rucksack War*, pg. 472.

handcuffed. Grenadians gathered, yelling their hatred. One man screamed "We want to take him apart, piece by piece".[306] The prisoners were quickly flown out to the *Guam*.

On Sunday morning the local Grenadians came up to our perimeter and gave us some hot coffee in china cups. I gave them some of the items out of a C-Ration meal. I don't think they were hungry, but the gift was a souvenir to them of the American's invasion and they were happy. Down in the town we could hear the people singing, and the church bells ringing. There was a rumor that the 101st Air Assault Division had arrived to reinforce us.[307]

The Grenadians told us that the Cubans had come up to the perimeter, in civilian clothes, to check us out, and then they went back into the jungle. An AK-47 had been laid on the ground outside the perimeter, but the owner had drifted back into the night. During the night 3rd Platoon had found more weapons. Also, during the night a local police commissioner identified one of the PRA, and 1st Platoon captured him.

While I was drinking the Grenadian coffee there was a burst of gunfire that came from the direction of the cave. I assumed it was 1st Platoon doing another recon by fire. No one seemed alarmed in the perimeter, only curious. A few days earlier we would have all jumped for cover, but we were now able to tell the difference between hostile fire and that of our own forces.

I didn't have much to do while the platoons patrolled in the surrounding city. I wrote C-Ration box letters to several people back home. I wrote to a girl I had met at a Revolutionary War reenactment, hoping that it would lead to more romantic interests, but it never did. A lady who worked in the doctor's house came out and gave us all the "goodies" she could find in the house. This consisted of Vienna sausages and chewing gum. Some of the HQ guys had found an abandoned building and discovered that the electric power was now on. Inside the building was a shower that worked and they took turns taking hot showers. I wandered around the grove, eating lemons and enjoying the view. It was a pleasant existence.

Later in the afternoon CPT Jacoby came over and told me "Let's go" so I followed him down to the Spice Island Inn at Ross Point. We were supposed to

---

[306] Adkin, *Urgent Fury*, pg. 301.
[307] Only the helicopters of the 101st Aviation Group participated in the operation.

replace the Marines guarding the embassy there, but they knew nothing about it, so we went back up the hill again. When we got to the top, we learned that 3rd Platoon had found more documents in a different house. The 1st Platoon had found more weapons and two footlockers full of uniforms, equipment and two radios. Everyone grabbed something for a souvenir. SP/4 Bush grabbed a Makarov 9mm pistol. Some of the 1st Platoon were wearing the British style half-boots that they had gotten from a shoemaker in town. The shoemaker told the paratroopers that they could have whatever they wanted. During their patrol the Grenadians had also given them food and cokes.

Since hostilities were almost ended, President Reagan lifted the ban on reporters. We started seeing more of them roaming through the town. A reporter from Newsweek magazine followed 1st Platoon into the CP. He was wearing a camouflage jacket, a pair of black pants and a Kevlar helmet without a camouflage cover. If I had seen him during the first few days I most likely would have shot him. We learned from him that Reagan had initiated the War Powers Act, and that we would be there for 60 days.

Two French filmmakers with a video camera showed up and I was able to borrow some black and white film from them for my pocket 35mm camera. These cameramen were able to get the only video of the invasion that I ever saw on television.[308]

That night we got the word that we would be doing an airmobile in the morning to the interior of the island. Brad Gallardo wrote to me years later that "I got word over the radio that 1st Platoon needed to check out the golf course. So, I checked in with the 2nd Platoon RTO SGT Brown, as he was standing near me. I verbally made sure he was to tell his platoon leader to hold tight there, pulling rear guard. While this was going on a little euro flat bed truck was passing by turning up the hill. Boulderoff and his machine gun crew with a few other guys from 1st Platoon jumped on this tiny little truck to get a ride up the steep tightly turned hill. This thing had a short wheel base and high off the ground, lightweight and now it was loaded with an M-60 team and 3rd squad trying to pile on. It putters up the hill. Well about the 4th tight turn it turned over spilling troopers everywhere, it was hilarious. A keystone cop moment if I ever saw one. Intel had it there were BTR's somewhere around in the woods

---

[308] The reporters gave a copy of the video to 1LT Nicholson, and he allowed me to borrow it and make copies. I put the copy of this video on YouTube with the title of "Grenada – Operation Urgent Fury – Part 1" and "Part 2".

on the edge of the course... That night Spector rolled on the scene and hosed the Golf course down." [309]

1SG Polachowski did some slick trading with the Marines and was able to get each of us two more MREs. We moved the platoon's defensive positions into the company CP, and the HQ section combined with the 2nd Platoon and the Mortar Platoon to do security on the perimeter. That night I had radio watch again. After I pulled my shift, I woke up the next man on watch. He took the radio from me and I tried to get some sleep.

On October 30th Gerhard Jonas, the East German advisor, had been questioned by the CIA on the location of Hudson Austin. The mission to capture Austin fell to the 2/505 of the 82nd Airborne. Jonas led the paratroopers to Austin's hideout, and by 3:30 the soldiers had cordoned off the area. The three RMC leaders gave up without a fight and were flown out to the *Guam*.

General Vessey and Admiral McDonald drove through St. Georges, inspecting the progress so far. Vessey was disappointed with the slow advance of the 82nd Airborne and pushed MG Trobaugh to move inland. This also led to one of the most uninformed quotes of the operation, that many in the Marine Corps like to use to show superiority to the Army. Vessey stated, "We have two companies of Marines running rampant all over the northern half of this island, and three Army regiments pinned down in the southwestern corner, doing nothing. What the hell is going on?"[310]

However, the Marines in the northern half of the island had little to no opposition, while the airfield to the south was heavily defended with anti-aircraft guns, Cuban forces, and PRA armor vehicles. All of the casualties happened in the southern half of the island, to include the Marines pilots who were killed, and the Army had not been "doing nothing", but had just conducted the first hot drop zone airborne operation since the Korean War. So, to say Vessey made an ignorant comment is an understatement.

In the center of the island was Grand Lake Etang, which was reported to be a PRA supply base with arms and equipment. If the Cubans and PRA were

---

[309] Facebook message from Brad Gallardo on 19 September 2010.
[310] Taken from Heritage Press, USMC Quotations,
http://www.usmcpress.com/heritage/usmc_quotations.htm

reorganizing in the interior of the jungle, this would be a prime destination. MG Trobaugh gave this mission to LTC Hamilton and the 2/325.

At 0300 CPT Jacoby woke up and found no one on radio watch. He was mad as hell since we had to get ready for the day's mission. The guy I woke up for radio watch had fallen asleep, and didn't even remember me waking him up.[311] I quickly packed up my rucksack and threw it on the GAMA Goat. While we were getting ready, the company received orders from Battalion that we had to relieve the Marines at Ross Point. The Marines were finally getting ready to leave because they had received orders to do another assault landing on the northern island of Carriacou.

CPT Jacoby grabbed the 2nd Platoon leader and told him to get his platoon down to Ross Point. He then grabbed me and we both went down to the Spice Island Inn to make sure that we really were relieving the Marines. He wanted to make sure it wasn't another mistake like the day before. It turned out that the Marines were actually leaving, and the CO called up the Mortar Platoon to come over to Ross Point and fill in for the Marines. It took our company a while, but we finally took over responsibility for the security of the Embassy.

Hugh Simmons wrote "My squad got to guard the US ambassador to Grenada just before we were relieved by the Marines. I will never forget walking the perimeter behind his room at night and I could see him in his room scrambling to load his suitcase with his clothes and who knows what else. He was definitely freaking out though and I just chuckled at him to myself. "Damn civilian" I thought." [312]

Gary Marsh, the squad leader of Simmons' squad, wrote "There was a drunk CIA agent that fell down and his weapon, a Mac 10, went sliding across the concrete. I went over and picked up his weapon and we carried him to his room and put his drunk ass to bed. What a piece of artwork he was." [313]

Though we had the mission to provide Embassy security, we also had to prepare for the air assault into the interior of the island. The rumor was that MG Trobaugh had told General Vessey that he would send us into the interior for

---

[311] This is actually quite common. To make sure the next man is definitely awake it is best to make them stand up, walk around, and then pull the guard. If the soldier just lays there, he may fall asleep and never remember being awakened at all.

[312] Facebook post from Hugh Simmons on 29 October 2013.

[313] Facebook post from Gary Marsh on 29 October 2013.

three days to determine if there was any resistance left. If we found nothing we would be going home.

Ray Meier years later wrote to me that "Capt Jacoby took me and a couple of guys for his security when he went to the American embassy. He went and met with Schwarzkopf and an Admiral, along with Mad Jack (Hamilton) and the Marine's battalion commander. Well, it went to shit for us, needless to say. Hamilton was making a bid to let us stand down and pull security in St George and let the Marines go and conduct mop up since they didn't do anything. Well, the Marine commander threw a fit. He said we were scaring the hell out of the locals with our damned German helmets and that we were looting. Well Hamilton came UNGLUED and reached out and grabbed LTC Ray Smith, USMC, and threatened to beat his jarhead ass... all the while Schwarzkopf and the Admiral were all saying JACK settle down!!" [314]

We moved out to a golf course, at 0400, under cover of the night. We picked up 2nd Platoon and the Mortar section along the way. Another unit filled in as embassy security when we left, but I never did find out who they were.[315]

It seemed that everyone had their own vehicles that they had commandeered or captured since we had arrived. Since none of their own vehicles arrived, they made do with what they had. Our Mortar Section had the "Gandhi Mobile". This was a military VW-looking micro bus that carried the mortar ammunition. The side door had been pulled off, and the front doors were bent backwards towards the front, and held there by a rubber inner tube. Rucksacks, C-Ration boxes and equipment were tied to the roof and the outsides of the van. "USA" and "All The Way" had been written on the side of the vehicle with a camouflage stick, so it wouldn't be mistaken for a PRA vehicle. It was called the Gandhi Mobile, because the driver, PFC McPeak, had shaved his head before the invasion. He looked like Mahatma Gandhi. [316]

---

[314] Our Battalion commander was known as "Mad Jack" because of incidents like this one, but we thought he was a great commander. Facebook message from Ray Meier on 18 September 2010.

[315] In my journal I wrote "the peacekeeping forces of Zaire and Britain are here". I did not know what I was seeing and this was our Caribbean allies, along with their officers, who appeared to me to be British.

[316] Facebook message from Gary Marsh on 14 November 2010. Gary Marsh later wrote "Mcpeak took that van from the Cuban motor pool in the compound. we tried to get the front doors off of it. we beat on the hinges with a hammer ,pulled up to a palm tree with the door open and backed up trying to rip the door off that didn't work, so we kicked out the windshield and used the window rubber seal to tie the front doors open just in case we got hit. That van had no brakes but the emergency brake did work. We also wrote

The PZ for the helicopters was the golf course that Spectre had fired up a few nights earlier.[317] I quickly ate a cold C-Ration for breakfast and then camouflaged my face with a cammie stick. The Blackhawk helicopters came in at sunrise and landed on the golf course. The golf course looked normal, but when the Blackhawks landed, they sank up to the bellies of the aircraft due to the soft, spongy muddy ground. We ran for the doors, ducking lower than ever, since the rotor blades seemed so much closer to the ground due to being stuck in the mud. The doors were fixed open and the seats had all been taken out, so there was room for everyone.

We flew in a formation of a dozen Blackhawk helicopters, escorted by Cobra attack helicopters, with Spectre gunships overhead. The door gunners watched intently for any signs of enemy fire coming from the jungle. The terrain looked like pure torture to all of us infantrymen. The hills and mountains went straight up and then straight down, surrounded by triple canopy jungle.

The helicopter formation arrived at a small village and began circling, as the Cobras went down to draw any possible fire from the enemy. There was no sign of any hostility though. We hovered over a soccer field on the edge of the village and everyone jumped out before the helicopter could touch down. We quickly ran to cover before the next bird could come in. In less than two minutes the entire company was on the ground, in a 360-degree perimeter around the LZ. After they left, we laid in the thick red mud, waiting for any signs of the enemy, and listening to the helicopter support hovering overhead.

A few of the Grenadians came out to see what all the commotion was and they soon discovered that the Americans had arrived. We got up and marched down the small road through the village, but the village soon became a large town. We stopped and searched the houses along the way in the same manner when we had approached St. George days earlier. The Grenadians followed us and gave us fruit, cigarettes, and Cokes. The similarities between this and the images I had seen of the liberation of France in WWII kept flashing through my mind.

Once we moved through the town the Grenadians quit following us and we continued up an incredibly steep mountain. The company would stop about every 500 meters so that we could rest. We would catch our breath and then

---

all over the van in camo stick USA and USA PARATROOPERS with jump wings on the sides of the van. Dave Thuma was having a blast with the camo stick. also Tim became a master driver with using an e-brake on the hills, only a couple times did Thuma and I think we might have to bail out in case Tim didn't get it stopped."

[317] PZ stands for Pickup Zone. LZ stands for Landing Zone.

continue on up the mountain, sweating heavily and sucking in air like we were drowning.

After a climb of two kilometers, we finally made it to the top of the mountain. LTC Hamilton was there waiting for us. He had landed his helicopter there instead of climbing that steep mountain. With the Colonel was a CNN camera crew, who followed us everywhere we went. The company moved down a red dirt road to a large lake that was called Grand Lake Etang. From the look of it, the lake was part of an old volcano and the steep hill we had just climbed was also part of that volcano.

There was supposed to be a PRA training camp at the lake, so we were told to search the area. There were trails all around the lake, but they were all old. The red mud was everywhere and there were no fresh prints in it. The HQ section set up a CP near the lake, while the platoons were sent out to patrol around the lake. The HQ section rotated pulling OP security, which really annoyed the RTOs. They wanted to stay at the CP and not have to worry about pulling security. Everyone started becoming relaxed and letting their guard down. SP/4 Bush was smoking everywhere we went, which is a cardinal sin in the world of infantrymen. When I told him to put out his cigarettes, he would just look at me like I was an idiot, and tell me "There's nothing out there to worry about". However, I grew up on a diet of John Wayne movies, and whenever you let your guard down was when the Indians would attack. So, I continued pulling security, though it seemed as if I was the only one.

The 1st Platoon had found a tunnel system down by the lake and they sent some men down into the tunnels with a flashlight and a .45 pistol. They didn't find anything though. As the sun began to set the company moved back up the road to where two houses sat on either side of the road. HQ and 3rd Platoon formed a perimeter around one house, while the 1st Platoon formed around the other house. The 2nd Platoon had a defensive position off in the jungle.

While we prepared our defenses a Grenadian in a uniform walked up to us and surrendered. He acted very strange, like he was mentally disturbed. We later found out about the bombing of the mental hospital, and we assumed he was one of the patients. He was given food and placed under guard. One of the Platoon SGTs had broken open his claymore mine and we boiled some water for coffee by lighting the C-4 plastic explosive inside.[318] That night was peaceful, like most of the nights since it had rained.

---

[318] C-4 would not explode if it caught on fire, and was an excellent way to heat coffee. The veterans from Vietnam knew this, while us new soldiers had not learned it yet.

Triple Canopy

In the morning Lieutenant Tiernay, the S-2, arrived to question the prisoner. Instead of taking him back to Battalion headquarters he just let him go. Some of the soldiers said that they wanted to take him into the woods and shoot him, but it was only talk. Most didn't understand that he was just a mental patient from the bombed hospital. As soon as the sun came up, we were on the march. We marched through B Company's perimeter, which was around a landing zone, and then continued on down the mountain. The going was extremely steep, and though we were on a paved road it took us almost two hours to get down the mountain. Due to the steep grade, we got blisters on the tops of our toes during the march.

When we got to the bottom of the mountain we marched into a small town and then fell out on each side of the road to take a break. As soon as we sat down, we were assaulted by dozens of reporters. After a few minutes listening to their uninformed questions and statements, we continued our march into the center of the town, and set up a company CP beside a schoolyard soccer field. The three platoons went into the towns and surrounding hills to search for any enemy.

A Grenadian citizen approached our group and told me that he would take us to where a PRA officer was hiding. CPT Jacoby sent a squad from the 1st Platoon out to find him, along with an M-60 machinegun for support. A short time later the squad returned with the prisoner. The man told us that he had been in the PRA, but he had quit. He told us that he would be glad to take us to where weapons were hidden. He guided the 1st Platoon to a weapons cache, but it did not have many weapons.

In the late afternoon 3rd Platoon radioed the HQ and reported that they had found a BTR-60. CPT Jacoby grabbed me and the RTOs and drove down there in a captured jeep. It took awhile to get to where the 3rd Platoon was located, since they had covered a lot of ground on their patrol. The road was also one of the worst I had ever seen, but at the end of it was a beautiful, almost pristine, BTR-60PB. LTC Hamilton had come with us and asked if there was any "brave troop that wanted to clear this thing out?" SP/4 Bush and I looked at each other, and shrugged "what the hell", then jumped up on the armored vehicle.

As we crawled into the vehicle, we carefully checked it for any booby traps, or trip wires. As we cleared the vehicle, we would throw things out the side doors. We soon had cleared everything on the inside, except the guns. The colonel wanted us to clear them too, but I had no idea how the big gun operated. I was able to clear the 7.62mm machinegun fairly quickly, throwing the ammunition belt out of the turret hatch. The big 14.5mm cannon had a belt of ammunition hanging from it, but I had no clue as to how to clear it. No one else

156

did either. I looked around for some sort of button, or latch, and found a lever on the side of the weapon that looked like it would open the feed tray cover. I told the colonel that I would give it a shot. Both of us in the vehicle put on our Kevlar helmets and hoped for the best. I aimed the gun upwards, and then pulled the latch.

BOOM-BOOM-BOOM-BOOM-BOOM!!!

The heavy machine-cannon fired a five round burst into the hills. I looked out of the vehicle and saw everyone on the ground. My ears were ringing so I had no idea what they were saying. I told them that what I found was obviously not a latch to open the cover, but must have been the trigger. Everyone in the chain of command yelled at me to leave it alone, then they all got up off the ground.

The colonel decided that he wanted to have this vehicle taken back to the airfield. He asked us if we could drive it back there. All of the controls were written in English and Russian, so it was easy to figure out what the dials meant. Unfortunately, we didn't know how to start it up. A handful of Grenadians had been watching us analyze the BTR-60 and two of them approached me. They said that they could show me and Bush how to drive it. They pointed out how to start it up and where to put the fuel. I asked them had they ever driven one of these before and they told me that they had never seen one before. They were obviously PRA soldiers, but didn't want to admit it.

Bush and I backed off and let them try to start it up. They tried to crank it up and it did start, but then it died a short time later. The colonel told Bush to try to drive it back to the airfield and ordered the 3rd Platoon to guard it. The colonel, CPT Jacoby and I then returned to the company CP.

When we arrived back at the soccer field the CO requested that trucks be sent out to our location so that we could return to Bravo Company's position on the mountaintop. We needed seven trucks to carry the whole company, but only two confiscated vehicles showed up. One of the trucks broke down as soon as it arrived, leaving a single civilian dump truck to shuttle the entire company up to the mountain.

CPT Jacoby sent me with the jeep to return to 3rd Platoon, and give them a radio in case they had to stay with the BTR-60 for the rest of the night. About halfway there I stopped the jeep, amazed at what I saw in the road. Half of the townspeople were pushing the BTR-60 down the road, with Bush in the driver's seat steering the vehicle. A GAMA Goat was pulling it from the front. The 3rd

Platoon was relaxing on the side of the road, watching all this free entertainment, as they drank cokes and ate food the Grenadians had given them. One Grenadian offered me some hot sauce in a wine carafe. I sampled some of it and it nearly set my entire face on fire. It was the hottest thing I had ever tasted in my life!

I had to get back to the CP before dark, so I dropped off the radio, grabbed a warm coke, and drove back to the HQ. When I returned to the company CPT Jacoby told me that the single dump truck used to shuttle the company had also broken down. Due to this our section spent the night beside the soccer field. CPT Jacoby left me there, and went up to the Bravo Company defensive position to spend the night. As the sun set 3rd Platoon arrived back at the soccer field, but without the BTR-60. They had dropped it off with our 3rd Battalion. Later we heard that the 3rd Battalion set it on fire, but this was typical of the rumors during that time. None of us had linked up with our rucksacks that night, so I took off my shirt, draped it over my face, and slept on a folded-out C-Ration box.

I learned after we returned to Fort Bragg that the Mortar Platoon leader was relieved that night for dereliction of duty. Earlier in the day the mortar platoon had set up near the Company CP by the soccer field. PFC Gary Marsh and SGT David Thuma had taken two small teams around the field, making sure there were no threats. The only thing they had found was "a bolt action rifle in the bottom of an outhouse buried in crap." After the recon the mortar section set up their 81mm mortars, while the Mortar platoon leader, 2LT F_____, went into the village to question the people. The lieutenant returned with a young Grenadian woman and let her stay at the mortar position. Gary Marsh wrote that this was a "big time No-No. Sgt Thuma and I went to SFC Bush (mortar PSG) and told him that we can/t allow people in our gun position, but he didn't say shit to F_____."[319]

That night the mortars had one guard roving between the three positions, so that they could get some sleep. At two o'clock in the morning Marsh woke up to check the gun positions, but found no one on guard. He checked his roster and PFC Bettis should have been on duty. Marsh woke up Thuma and they both searched the area for Bettis, but he was missing. They both went to the FDC position and told SFC Bush that Bettis was missing. When they went to find LT F_____ to let him know, they discovered that F_____ was also missing.

---

[319] Facebook message from Gary Marsh on 17 October 2011.

Thuma was now worried that the men had been captured and said "What the fuck is going on?"[320]

The night was dark and the men couldn't see anything at all, but Marsh said they could hear noise "below our gun position, banana leaves and shit snapping." They tried to get an illumination mission authorized, but LTC Hamilton denied the request. The Battalion Commander told them to wait until first light and send out recon patrols into the village and the surrounding jungle. Thuma and Marsh were able to convince Bush to let them run several two-man teams around the soccer field, looking for any sign of the missing men. They took radios with them in case they needed support.[321]

When Thuma neared one of the huts by the village, he saw someone sleeping on the porch that looked like PFC Bettis. Using the radio, he told Marsh to link up with him quickly. Marsh sprinted across the soccer field, staying low so he wouldn't be seen. When he linked up with Thuma they both moved closer, leaving the other two men in the patrol behind, to cover them. They quickly discovered that the man on the porch was Bettis and woke him up. After telling the sleeping private to move back to the other men on security Thuma looked through the window and saw LT F_____ in bed with the young woman, with his M16 leaning up against the wall.[322]

Marsh wrote "So I kicked the door open and Dave covered me while I grabbed the LT. I wanted to kill the fucker; Dave kept saying "don't hurt him Gary". So, we placed him under arrest and took his weapons." Both of them took the lieutenant back to the FDC position and called the Battalion HQ to let them know they had found the men. LTC Hamilton told Bush to put a guard on the lieutenant and he would be there at first light. Marsh wrote "I just remember hearing the chopper coming up the valley. We were told to stay at our gun position so I didn't hear what Mad Jack had to say. Never saw LT F_____ again. LT F_____ had ordered Bettis to go with him and not say a word to anyone so Bettis was not charged with anything because he was ordered to go" [323]

Early the next morning I was startled awake by CPT Jacoby telling me that we needed to get ready to do an airmobile in 15 minutes. I only had what was on my back so it didn't take long to get ready. We lined up everyone into "lifts"

---

[320] Ibid.
[321] Ibid.
[322] Ibid.
[323] Ibid.

and waited for the helicopter to arrive. We waited, and waited, and waited some more. As we waited there were rumors being spread around. The word was that we would be going back to the airfield at Point Salines for some "R+R" and then we would return to the jungle to hunt down the remaining PRA.[324]

Another more promising rumor began to circulate that we were going to go home. This rumor started because the Mortar Platoon had turned in their mortar rounds before they left the hill. They wouldn't have done that unless they no longer needed the ammunition. While we sat there, waiting for our helicopter, the Mortar Platoon drove by in their GAMA goats, and the "Gandhi Mobile". They were heading to the airfield. A few minutes later we received the word also… we were going home!

After sitting around all morning, the Blackhawks finally arrived sometime after lunch, but they couldn't find us. We were a small group of camouflaged soldiers beside one of many soccer fields and they didn't know which one it was. We popped smoke grenades; then talked to the helicopters, trying to get them to see the smoke. Finally, they recognized us and landed on the tiny field.

On the flight back we flew over St. George's city and it was a beautiful sight. When the helicopter touched down onto the airfield, we all jumped off and ran to a group of guys under an American flag. We began cheering, and yelling that "we're going home". We finally noticed that the soldiers near the flag were not from our company. I spotted the "Gandhi Mobile" in a valley by the airfield and had our troops move in that direction.

Once we finally found our company, we moved to an area to turn in all of our ammunition. I had hollow points for my .45 caliber pistol and the scuttlebutt was that it was against the Geneva Convention. I never knew if this was true or not, but I shoved the bullets under the pile so no one would see them.[325]

After turning in the ammo, I took off my shirt for the first time in nine days. We stank, but we enjoyed baking in the sun! Our company moved to a grass

---

[324] R+R stands for "rest and relaxation".

[325] The United States is one of few major powers that did not agree to IV-3 of the Hague Convention of 1899, and thus is able to use this kind of ammunition in warfare. However, for years the United States military respected this Convention and refrained from the use of expanding ammunition, but after announcing consideration of using hollow point ammunition for side arms, with a possible start date of 2018, the United States Army began production of M1153 special purpose ammunition for the 9mm Parabellum for use in situations where limited over-penetration of targets is necessary to reduce collateral damage.

covered peninsula and we just laid there all day, relaxing, sleeping, and wearing only our pants. I took my boots off letting my feet "breathe" for the first time since I had arrived on the island. All of our mail had been accumulating back at the airfield, so each of us had a small pile to read. We savored every word, even the junk mail. I shaved for the second time since the invasion began. When it was dark, we threw chem lights into the water, and watched as they slowly moved with the tide.[326] Everyone slept soundly on the beach, snoring loudly, with no worries.

On the morning of November 3rd, the Battalion had a memorial service for CPT Ritz and SSG Epps. We all put our uniforms back on and gathered around two M-16s jammed into the ground on that grassy peninsula. There were two helmets on the rifles representing the two fallen soldiers. The chaplain said his remarks and some men led the group in prayers. A final roll was called. The men of the company would be called, and they would answer "Here, 1st Sergeant". Finally, the names of the fallen would be called, answered by silence. The names would be called again and still it would be silent. The names would be called a third time, and then the 1SG would say that the names were to be dropped from the rolls. When it was all done a firing squad fired over the water with three volleys. Normally a firing squad would have blank rounds, but all we had were live rounds.

Afterwards we returned to our company areas and were ordered to lay out all of our equipment, to make sure nothing had been lost. We also were told to turn in all of our war trophies, so that the items could be inspected. Anything we weren't allowed to bring back would be confiscated. The inspectors told us that everything could go back, except anything that could be fired, or might blow up. We were disappointed in this. Throughout history soldiers were allowed to bring back the weapons taken from the enemy. We all understood that we couldn't bring back any fully automatic weapons, but we thought that pistols and semi-automatic rifles were legitimate war trophies. With Grenada a new rule was created. Soldiers were no longer trusted to bring back any weapons. I didn't quite trust the inspectors, who were not from our battalion, so I put my name and unit on every item that I had picked up.

I had a Cuban LCE with a Makarov 9mm pistol in the holster. With much disgust I turned in the pistol. Some of the men in my company were so insulted

---

[326] Chem lights are Cyalume night sticks. These are plastic "sticks" that contain peroxide and a glass vial of phenyl oxalate ester. By bending the stick, the glass vial breaks, combining the two chemicals and making them glow. During the time of the Grenada invasion, I only saw their use with the military, however today they are used by children as toys, especially around Halloween.

by the lack of trust that they threw the pistols into the water instead of turning them in. I also had a sling from an AK-47 and a cloth Soviet vest that I had taken from the BTR-60 that Bush and I had searched. All of the items, except the pistol, were returned to me later on. [327]

We repacked everything in our rucksacks just in the nick of time. The clouds gathered overhead, the wind blew, and the rain came down in sheets. None of us ran for cover. We just sat there in the driving rain, letting it rinse out our clothes, our hair, and our bodies. The rain didn't last long, but when it stopped everyone stripped off their clothes and hung it over bushes and rucksacks. We grabbed bars of soap that were in some "care" packages and jumped into the ocean, totally naked and not caring at all.[328]

The "care" packages had arrived that day and had been packed by the wives of the paratroopers. The packages were filled with shaving cream, razors, toothpaste, toothbrushes, envelopes, candy bars, pens, chewing tobacco and cigarettes. There was enough for everyone. After we had washed up in the ocean and swam around for awhile, we laid out in the sun, still naked, letting our clothes and our bodies dry.

In the afternoon all of the snipers of the Battalion were told to gather around an officer from the division. He asked each of us if we had actually shot anyone. Those who said they hadn't were sent back to their companies. The officer then questioned the three of us who were left. He wanted to know where the enemies were located when we had shot them. I think the reason he was doing this was to try to locate any enemy bodies that were not discovered yet. The other two snipers told him that they had shot at the Cubans inside the compound. The officer dismissed them, since any bodies found in the compound had multiple wounds, and were already discovered. He then asked me about what I had done.

I was the only remaining sniper from the group. I told him about firing on the enemy when the BTR-60 attacked. He wasn't interested in that incident. I

---

[327] The Marines did not have this policy in 1983, and they let their men bring back weapons that were semi-automatic or bolt-action, but no automatic weapons. In 2019 I was finally able to acquire a Czech VZ-52 rifle that had been used on Grenada.

[328] Though there may have been female soldiers on the island, we never saw any. Since women were not allowed in the combat specialties, they were not seen much by the frontline soldiers. In recent wars the news media focused on women soldiers in the war zone, so that much of the American public thought there were women in the front lines with the infantry, Rangers and Special Forces. Though by the time of this re-release of the book, 2021, there are women who have graduated from Ranger and Special Forces, so far none have been used in the front-line combat.

also told him of firing on the Cubans behind the wall in the compound. I estimated that I had shot at five to seven of them there. He had found the bodies of those enemy soldiers and wasn't interested. I then told him about shooting the one Cuban who ran for the door of the barracks and hitting him in the head. He wrote this one down, and then told CPT Jacoby "That's one we found". I also told him of shooting the one Cuban on the side of the hill, who was firing on B Company. The officer wrote that one down, and told CPT Jacoby that they had found that lone body, shot through the chest. He told CPT Jacoby that I would get credit for those two, and then walked away.

That night LTC Hamilton ordered CPT Jacoby to have us get dressed and go through customs. We grabbed our rucksacks and moved to a building on the edge of the airfield. Inside we were all subjected to a customs screening by an officer not from our Battalion. For a few minutes I thought that my personal .45 pistol was going to be confiscated. The customs officer saw my shoulder holster, and then noticed that I was carrying a pistol that was not Army issued. It had Pachmeyer rubberized grips. The customs officer held up the pistol to the Company Commander and asked him "Is he allowed to have this?" CPT Jacoby just saw a .45, and not the custom grips, and he told the customs officer "Yes, he's a sniper." The officer handed it back to me and told me that I was cleared. I walked away, relieved. I had been worried, since I did not want to be caught carrying my own pistol, but I also did not want to have it confiscada.

After clearing customs, we lined up in chalks on the runway. We slept on the runway where nine days earlier we had fought, and some of us had died. At 4:00 in the morning on November 4th we were told to wake up and move to the hangers to await the aircraft. As the sun was beginning to rise a C-141 landed on the airfield. We crammed the whole company into a single C-141. When the plane lifted off the ground everyone on board cheered. The Air Force guys gave us MREs and sip-ups.[329] They told us that when we got back to Pope Air Force Base there would be a ceremony and a presentation, and that maybe even President Reagan would be there. Like most rumors during the invasion, this proved to be false. Our "C Company" would not be going with us, but returned to their Battalion. They would redeploy back to Bragg in December with the rest of the 2/505.

The flight was long, the hours dragged by, seeming to take an eternity. Most of us tried to sleep anyway we could, but there were too many paratroopers scattered about to be able to get comfortable. As we began the final approach

---

[329] Juice Boxes. "Sip Ups" was the name brand for the only ones at that time.

towards Fort Bragg everyone kept asking the men by the windows if they could see the ground. When the plane finally touched down everyone cheered again.

We taxied down the runway for what seemed another eternity. We just wanted to get off, but the plane continued to roll very slowly. The back doors opened up and we could see the runway rolling by underneath us. The C-141 stopped, and we were told we had to wait.

I remembered the Newsweek reporter telling us in Grenada that the Americans were against us, and the war was not supported by the United States. As I looked out upon the crowds and crowds of waiting people, I knew that the reporter was full of crap and had been speaking out his ass. There were banners hanging everywhere, civilians, soldiers and news reporters all on hand to greet us. We continued to wait inside the aircraft, in the shadows, not visible to the people outside. Finally, we marched down the ramp in a single file. The crowd cheered, the bands played and it was pure pandemonium! We marched up to two men behind a podium and formed a twelve-man front. I asked who they were, and someone told me it was John O. Marsh, the Secretary of the Army, and a 2-star General of the Air Force.

The Secretary of the Army gave a speech as we stood in the drizzling rain. No one wanted to hear the speech. We wanted to get this over with and the crowd wanted to see their loved ones. The Secretary moved to three men, an officer, a sergeant and a private who had been selected from our Battalion. He then pinned on the Combat Infantryman's Badge on each of their uniforms. It was then that we learned we would all be getting this award. The three chosen paratroopers would go to Washington and have breakfast with President Reagan in the next few days.[330]

---

[330] The Combat Infantryman's Badge, or CIB, is a blue rectangle piece of metal, with a musket inside of it, surrounded by a wreath. This award was created in WWII and given to those infantrymen who participated in actual ground combat. Initially soldiers awarded the badge received an additional $10 a month. According to the regulations (AR 600-8-22) "A recipient must be personally present and under hostile fire while serving in an assigned infantry or special forces primary duty, in a unit actively engaged in ground combat with the enemy." It also states "Battle or campaign participation credit alone is not sufficient; the unit must have been in active ground combat with the enemy during the period." When we were awarded the CIB, I felt guilty, since my father had been awarded the same award for being under fire for 30 days. I had only been under fire for 3 days. I do not feel so guilty anymore, since by the time my 20 years in the Army was over, I accumulated the other 27 days and more. I was awarded the CIB three more times in my 20-year career.

Someone in the formation complained "I hope they get some 80 paks here" as we began to march off. When we rounded the corner, and as the band played on, we saw what appeared to be every 2 ½ ton truck and military dump truck in Fort Bragg waiting to take us back to our barracks. Soldiers climbed on board and the long convoy of returning warriors slowly made its way to their barracks. Crowds of people lined the streets, cheering us on. When we hit the 82nd Airborne Division area all the trees had yellow ribbons tied onto them. Banners were strung from the light poles on either side of the streets, welcoming us home! The units that did not go to Grenada were lining the streets, welcoming us home. The Riggers had placed a giant American flag on the parachute shake-out tower, and were on the roof holding it down.[331] There were camera men everywhere. I held my M-21 rifle high and gave a "peace sign" to those looking up at us.[332] Many of the people reached up to give us a "high five" as we drove by.

We drove into our own Battalion area and got off the trucks. Wives and girlfriends ran to their men and hugged them. I walked alone to the dayroom, where the wives had put up posters, streamers and yellow ribbons. We lined up to turn in our weapons, all while being interviewed by television reporters. It was good to be home.

Aftermath

The United States officially lists 19 killed during the invasion, however if the Special Forces soldiers are counted, and those mortally wounded and died elsewhere, the number could be as high as 29 killed. Initially the military stated that 87 men had been wounded, but later the number was adjusted to 152 wounded.[333] During the invasion 29 Cubans were killed and 59 wounded, with 602 unwounded prisoners being taken. There were also 67 Grenadians killed and 368 wounded. A list of all of the American wounded that I could verify is at the end of this chapter, along with the names of those who died.

---

[331] Riggers are the soldiers who pack the parachutes.

[332] I wasn't doing this to show my feelings of peace, but it was my way to show that I had gotten two of the enemy.

[333] During the invasion the Navy's Seal Team Six had four killed. The Marine Corps had three killed and 15 wounded. The Air Force had two wounded. The Army had 12 killed and 108 wounded, this includes the Rangers, who had eight killed and 69 wounded, the 82nd Airborne which had three killed and 36 wounded, while the rest of the Army casualties were from the Delta Force. In addition to the combat wounded the 82nd Airborne also suffered one killed and 25 wounded in non-combat conditions.

Colonel Pedro Comas Tortolo returned to Cuba, but did not receive a hero's welcome. He and several other officers were court martialed for dereliction of duty, and reduced in rank to private. Tortolo was sent to Angola, when he returned to Cuba he vanished from public view and was last seen as a 70-year-old taxi driver in Havana.[334]

Fourteen of the Grenadian leaders who fired into the crowds at Fort Rupert, and who executed Bishop were sentenced to death. Among these were Austin and Coard. The death sentences of all were later commuted to life in prison. In 2007 they were still in the Richmond Hill prison.

After the invasion there was criticism that the Army took too long to take the island, and that the invasion was not planned or executed properly. Historians know that no war in the history of mankind has been executed properly. No matter how well you plan, the unknown will create situations to which troops can only react. The army that reacts the quickest, and without too many mistakes will win. General Patton summed it up when he said "…an imperfect plan implemented immediately and violently will always succeed better than a perfect plan."

Another criticism of the invasion was that there were too many awards given out to the few soldiers who were on the island. On this point I would have to agree. Initially the dates for the award of the CIB were October 24th to November 4th, 1983. This was the time that our Battalion was on the island, and a soldier had to be there on those dates to receive the award. After we returned home, I discovered that in the other battalions everyone over the rank of SFC (or E-7) received a Bronze Star medal, or some other significant combat award. This shocked and angered most of us because it cheapened the awards.

The only significant action against the enemy happened within the first two days of the invasion. During that time the only units in the Army to see any action were the two Ranger Battalions operating at less than half strength, the operators of Delta Force, the helicopter pilots and crewmen of Task Force 160th, and the three companies of my own 2/325th Infantry of the 82nd Airborne.[335] The total number of all these soldiers who saw combat was less than 1,000.

---

[334] Extracted from *US Invasion of Grenada, 30 years later*
http://www.miamiherald.com/2013/10/24/3709792/us-invasion-of-grenada-30-years.html on 25 October 2013
[335] Even Adkins, who writes critically of the invasion in *Urgent Fury*, wrote "the only airborne unit to see … infantry combat was the 2d/325th at Little Havana on the morning of the second day. Their ponderous performance during Urgent Fury was well rewarded."

However, the Army awarded 812 Bronze Stars (59 with "V" devices for valor), and 5,079 Army Achievement Medals (99 with the "V" device for Valor). After a few months the dates for receiving the CIB also changed from 22 October to 21 November, 1983. A total of 3,530 Combat Infantryman's Badges were awarded to those soldiers who never heard a shot fired in anger.

Besides the CIB, every soldier involved in the invasion received the Armed Forces Expeditionary Medal. This award was a campaign medal given to any soldiers, sailors, airmen or Marines who were on or near the island during the invasion.[336]

When it came time to determine who would receive awards in our company, I was proud of the direction we took. Our company was just as guilty in giving everyone who was an E-7 or higher the Bronze Star medal.[337] However, only one Bronze Star with a "V" device was awarded in our company, and it was received by my platoon leader, 2nd Lieutenant John Nicholson. A "V" device is a small copper "V" that is worn on the medal or ribbon to designate specific acts of valor and heroism. LT Nicholson's citation reads:

For exceptionally heroic actions against a hostile force in the country of Grenada on 26 October 1983. As a Rifle Platoon Leader assigned to Company A, 2d Battalion (Airborne), 325th Infantry, 2LT Nicholson's platoon came under heavy enemy fire which resulted in the wounding of one of his Squad leaders. With total disregard for his own personal safety, 2LT Nicholson moved up and down his line giving encouragement and personally directing the timely and accurate firing of his mortar gunners and machine gunners. After the surrender of the Cuban compound, 2LT Nicholson led his platoon into the compound to secure prisoners and aid in the evacuation of the wounded. 2LT Nicholson's actions in combat reflects distinct credit upon him, the 82d Airborne Division and the United States Army.

There were only seven Army Commendation Medals with a "V" device awarded in our A Company for actions taken during the Battle of the Calliste

---

[336] The AFEM was created for the Cuban Missile Crisis in 1962, and is awarded for "any military campaign of the United States for which no other service medal is authorized."
[337] The Bronze Star medal can be awarded for bravery, acts of merit, or meritorious service. It is only given out during war time service and is not a peace time medal. However, it can be given for non-heroic actions. If the award has a "V" device for valor, then it is given for heroic actions during combat. The Bronze Star is the fifth highest combat award. The highest is the Congressional Medal of Honor, followed by the Distinguished Service Cross, the Silver Star, the Distinguished Flying Cross and then the Bronze Star. The Bronze Star Medal was created during WWII.

area. Six of these medals were awarded to soldiers who were below the rank of SP/4. Had they been a higher rank they would have most assuredly been awarded the Bronze Star medal. These paratroopers were SP/4 Edmonds and Schofield, PFC Messenger, Lowe, Freimuth and Callaway. All of the awards were given to the M-60 machine gunners and their crews who were in a position near me during the fight for the Cuban barracks on day two of the invasion. All six citations state the same thing:

For exceptionally heroic action against a hostile force in the country of Grenada on 26 October 1983. As a machinegun crewmember assigned to Company A, 2d Battalion (Airborne), 325th Infantry, PFC Edmonds, while under direct enemy fire from Cuban positions, manned his gun and delivered timely and accurate fire against the enemy. PFC Edmond's actions in delivering devastating machinegun fire was instrumental in the victory of the Calliste area. PFC Edmond's personal courage and professional skills provided a valuable service in achieving victory in this decisive battle and reflects great credit upon him, the 82d Airborne Division, and the United States Army.

The final "V" device was awarded to me, but I was a bit of an enigma to CPT Jacoby. I was put in for a Bronze Star, but I had also missed the initial callout. So, I would be awarded and punished, on the same day. My Bronze Star Medal was downgraded to an Army Commendation Medal, with a "V" device for valor. I also received an Article 15 under the Uniform Code of Military Justice. My punishment was to supervise other soldiers while they did extra duty punishment, for a period of 14 days. The citation on my award reads:

For exceptionally heroic action against a hostile force in the country of Grenada on 26 October 1983. As a Sniper assigned to Company A, 2d Battalion (Airborne), 325th Infantry, SGT O'Kelly positioned himself under direct fire and delivered timely and accurate fire against the enemy. SGT O'Kelly's actions resulted in at least two confirmed kills and was instrumental in the victory of the Calliste area. SGT O'Kelly's personal courage and professional skills provided a valuable service in achieving victory in this decisive battle and reflects great credit upon him, the 82d Airborne Division, and the United States Army.[338]

Though there were many problems with Operation Urgent Fury, what is amazing is that a military invasion with over 20,000 servicemen from all four services was planned and executed in less than four days. I have heard critics

---

[338] My name was misspelled in the citation, but this was common. Half of my orders in the Army had my name misspelled.

say that Grenada wasn't a "real war" and therefore deserves no recognition. For those who died on Grenada, it was real to them.

Though the enemy defending Grenada were facing a huge invasion force during the entire operation, on the first day of October 25th the Americans were outnumbered. In the south of the island, and in St. George's, the enemy defending the airfield had between 1,500 and 2,000 soldiers, equipped with the best anti-aircraft artillery in the world, and moving in armored vehicles. That anti-aircraft was deadly, with nine helicopters destroyed during the Operation. One Task Force 160th Blackhawk and two Marine Cobras were shot down during the Richmond Hill Prison mission. One Marine CH-46 Sea Knight was disabled during the Grand Anse mission, and three Blackhawks of the 82nd Airborne crashed during the Camp Calvigny mission. The defenders also had the strength of knowing the terrain, and having unlimited resources of ammunition and supplies.

On the first day the United States Army faced them with less than 500 Rangers, 300 Paratroopers of the 2/325, and less than 100 operators from the Special Forces and Navy SEALs. Less than 1,000 men, or about one battalion, were used to take an airfield, rescue hostages, and destroy any resistance. Though we had air and naval superiority, it was a daunting task to those on the ground. Add to this that we had to limit any civilian casualties. Amazingly all the objectives were completed 72 hours after notification on October 24th.

Now, compare this to the fighting in Somalia during the "Battle of Black Sea", that became famous due to the movie "Blackhawk Down". In the two days of fighting there they had 19 killed, 73 wounded, and one captured. They had two helicopters shot down and three damaged. So, there were more killed, and more helicopters shot down in Grenada than Somalia. Now, to be honest, I think the fighting in Somalia was a lot more intense, mainly because it was only about 160 Rangers and Delta against about 4,000 enemy. But Grenada is looked upon as a joke, as if it was nothing at all, while the other is looked upon as an epic fight. The difference is how the media treated both. Since Reagan did not let the media on the island, they treated the operation as an embarrassment, however the newspaper series that created the book "Blackhawk Down" was written by one of their own, so the media treated it like it was an epic fight (and, for my brothers who were there in Somalia, it was an epic fight, do not take this the wrong way).

Grenada broke our passive international policy after the Vietnam War. At the beginning of the 1980s we were losing the Cold War to the Soviet Union. Many in the United States believed that it was futile to continue to resist against

Soviet communism. Grenada was Ronald Reagan's first link in a chain that led to the wall coming down in Berlin, less than a decade later.

Operation Urgent Fury provided a testing ground for the United States Army under Reagan. During the administration of Jimmy Carter, the military suffered in training. During the Carter administration I conducted a parachute operation on a drop zone in Fort Bragg, but there was no fuel for the aircraft. Instead, we drove down the drop zone in a 2½ ton truck, dropping off paratroopers at different intervals. We would then pull our parachutes out of the pack, and when the truck blew the horn three times every paratrooper would simulate landing. We would then roll up the chute as if it was a real airborne operation. This lack of training led to many peacetime deaths. In 1980 the military suffered 2,392 peace time deaths. In the following year of 1981 the lack of training and experience led to 2,380 deaths. Ironically the number of deaths at the height of the Iraq and Afghanistan wars in 2004 was 1,887.

The final 82nd Airborne paratroopers returned home from Grenada on December 13, 1983. Afterwards those of us who actually fought on the island were debriefed so that any lessons learned could be incorporated into future training.

I was interviewed by the Inspector General of the Army as to how sniper operations could be more efficient. I told his staff that there needed to be a better scope, since the ART 1 scope let so much condensation in.

I suggested that sniper rifles be fitted with night vision scopes. I also told them that there should better training for snipers, such as what the Marine Corps sniper school teaches. The snipers should work in sniper teams and not just have a group of HQ soldiers thrown together with a sniper. I recommended that the Army go to a bolt action rifle, such as the Marine Corps Remington 700, and that snipers be authorized to carry an additional weapon, such as a pistol or a small automatic weapon like an MP-5K. All of the mistakes in Grenada were corrected and it paved the way to the quick victories in Panama in 1989 and Desert Storm in 1991.

When I returned home from Christmas vacation, I tried to put all my experiences down onto paper. I typed out thirteen pages of what I remembered and submitted it to the 82nd Public Affairs Office. After we had returned from Grenada some soldiers wrote stories for the local paper, the *Fayetteville Observer*. The soldiers who wrote those stories were not in any of the actual fighting, and they grossly exaggerated what happened there. Due to this, those soldiers were punished. I wanted to make sure that did not happen to me, so I had my story cleared through the PAO. I never found out what happened to the story until many years later, when I saw it published in a Japanese magazine that I found being sold in the 82nd Airborne Museum. I never received any payment for the story, but I thought it was great anyway. I didn't write the story for money. I just wanted to get the real story out there. The PAO cleaned up the story quite a bit, deleting much of what I saw.

After the invasion training in the 82nd Airborne seemed to have lost its urgency. Everyone acted as if another war would not happen for awhile, and many of the veterans treated each exercise as being futile.[339] The combat veterans ETS'd and new "cherries" took their place. When it came time for me

---

[339]   The 82nd Airborne would not go to war again for another six years.

to reenlist, I decided to leave the 82nd Airborne and move on to a group that seemed more professional.

The war against communism was picking up and the Special Forces were actively involved fighting against Nicaraguan rebels in Honduras and El Salvador. I reenlisted for Special Forces training, but I was not a good swimmer. The Special Forces swim test required a candidate to swim 100 yards. Unfortunately, it was not 100 yards straight, but the candidate had to swim 50 yards, then turn around and swim back. I practiced at the Division pool, but when it came time to actually test, I was not able to turn around and swim the rest of the distance. The Special Forces candidates left me behind and I had to await new orders. I did not want to be assigned to some regular infantry Division, so when I heard of a new Ranger battalion being formed in Fort Benning, Georgia I quickly volunteered. This new battalion was the 3rd Ranger Battalion, and becoming a Ranger would affect me for the rest of my life.

Killed during OPERATION URGENT FURY

PO1C Kenneth Butcher	SEAL Team 6	24 OCT – HALO accident
SGT Randy Cline	1/75 Rangers	25 OCT – Killed in gun jeep ambush
SSG Gary Epps	B 2/325	26 OCT – morning fight in Cuban Compound
CPT John Giguere	USMC Cobra	25 OCT – shot down trying to rescue Howard
SP/4 Philip Grenier	2/75 Rangers	27 OCT – Calvigny helicopter crash
SP/4 Kevin Lannon	2/75 Rangers	27 OCT – Calvigny helicopter crash
CPT Keith Lucas	TF 160th SOAR	25 OCT – killed flying DELTA to Richmond Hill
SP/4 Sean Luketina	82nd Signal	30 JUN 84 – wounded TOC strafing – later died
PO1C Kevin Lundberg	SEAL Team 6	24 OCT – HALO accident
PFC Marlin Maynard	1/75 Rangers	25 OCT – Killed in gun jeep ambush
PO1C Stephen Morris	SEAL Team 6	24 OCT – HALO accident
SGT Mark Rademacher	1/75 Rangers	25 OCT – Killed in gun jeep ambush

PF Dinesh Rajbhandary	B 2/505	23 NOV – killed by another soldier cleaning rifle
CPT Michael Ritz	B 2/325	26 OCT – morning fight in Cuban compound
PFC Russell Robinson	1/75 Rangers	25 OCT – Killed in gun jeep ambush
SC Robert Schamberger	SEAL Team 6	24 OCT – HALO accident
CPT Jeb Seagle	USMC Cobra	25 OCT – shot down – killed by PRA as surrender
1LT Jeffrey Sharver	USMC Cobra	25 OCT – shot down trying to rescue Howard
SGT Stephen Slater	2/75 Rangers	27 OCT – Calvigny helicopter crash
SP/4 Mark Yamane	1/75 Rangers	25 OCT – killed while clearing airfield

Wounded during OPERATION URGENT FURY [340]

CPT Jimmy Alexander	USAF	27 OCT – Brigade TOC strafing by A7 Corsair
SGT Randall Arnold	USMC	Smashed ankle between two jeeps during HELO move
CPT Robert Awtrey [341]	USAF	27 OCT – Brigade TOC strafing by A7 Corsair
SGT Gerald Bannon	2/325, 82nd ABN	25 OCT – wounded during fight at Calliste
SGT Jeff Beatty	SFOD-D	25 OCT – wounded Richmond Hill Prison attempt
MAJ William Boykin	SFOD-D	25 OCT – wounded Richmond Hill Prison attempt

---

[340] Though there were 152 Americans listed as wounded, these 91 wounded are the only names I was able to track down.

[341] A third airman, Airman Daniels was wounded slightly with rock and debris shrapnel, but apparently no fragments from the 20mm, but did not seek medical attention and was never awarded a Purple Heart.

LCPL Brian Birth	USMC	
CWO Dave Bramel	TF 160th SOAR	25 OCT – wounded Richmond Hill Prison attempt
SSG Richard Campbell	2/75 Ranger	
SGT S.M. Campbell	Unknown, Army	
SP/4 Kevin Carlisle	2/325, 82nd ABN	26 OCT – wounded with Epps during fight at Calliste
PFC Michael Cate	82nd MP CO	Accidental Discharge - wounded by .45 pistol
PFC John Chiasson	313th MI	
SGT John Clift	2/325, 82nd ABN	26 OCT – wounded with Epps during fight at Calliste
1SG George Conrad	HHC 2/75 RGR	Accidental M203 round fired by another Ranger
SGT Tony Davis	1/75 Ranger	25 OCT – wounded, paralyzed south edge of runway
SP/4 Thomas Denney	82nd ABN	
SP/4 Edgar Dick	2/75 Ranger	
SSG Michael Donovan		
SSG Robert Edmonds		
CPL Troy Ennis	3/325, 82nd ABN	
LT Donald Erskine	SEAL Team 6	25 OCT – Wounded in Radio Free Grenada mission
1LT William Eskridge	2/75 Rangers	27 OCT – Calvigny barracks helicopter crash – lost leg
SGT Kevin Evans	B 1/75 Ranger	25 OCT – shot in hand on top of Goat Hill
1LT Sydney Farrar	A 1/75 Ranger	25 OCT – wounded during BTR-60 attack

Chapter 1: AIRBORNE - The Invasion of Grenada

CW3 Bill Flannery	TF 160th SOAR	25 OCT – wounded Richmond Hill Prison attempt
SGT Dana Foley	2/75 Ranger	
CPT August Fucci	FIST 1/320, 82nd	27 OCT – Brigade TOC strafing by A7 Corsair
SSG Eric Gardner	1/508, 82nd ABN	Accidental Discharge
SGT Jeffery Garrett	Unknown	
SGT Craig Garrison	2/325, 82nd ABN	Powder burns
SP/4 Gary Genovese	B 1/75 Rangers	25 OCT – wounded riding motorcycle at Salines
SSG Jose Gomez	2/75 Ranger	
SGT Terry Guinn	2/325, 82nd ABN	26 OCT – wounded with Ritz during Calliste fight
PFC John Haas	2/325, 82nd ABN	26 OCT – wounded with Epps during Calliste fight
SP/4 Harold Hagen	B 2/75 Ranger	25 OCT – 90mm gunner broke leg on jump
BM2CL Van Hall	SEAL Team 6	25 OCT – Wounded Radio Free Grenada mission
SSG Roy Hardin	2/75 Ranger	
SGT James Harlow	2/75 Ranger	
LCPL Keith Harter	USMC	Finger shot off by a sniper at Pearls Airport
SGT Harry Hennemann	USMC	
1LT Patrick Higgins	82nd ABN	
SP/4 John Hiles	2/325, 82nd ABN	26 OCT – wounded with Epps during Calliste fight
CPT Timothy Howard	USMC Cobra	25 OCT - Shot down - lost arm – Richmond Hill

175

SGT Michael Hummel	2/75 Ranger	
SGT Ronald Johnson	82nd ABN	Shot in foot
MAJ Robert Johnson	TF 160th SOAR	25 OCT - wounded flying to Scoon's residence
SGT Ron Johnson	B 1/75 Rangers	25 OCT – wounded riding motorcycle at Salines
SGT James Keen	B 1/75 Rangers	25 OCT – wounded riding motorcycle at Salines
LCPL Don King	H 3/10 USMC	28 OCT - sniper gunshot wound to arm
Danny Knoft	SEAL Team 6	25 OCT – Wounded in Radio Free Grenada mission
WO2 Daniel Kuchenberg	82nd Combat AVN	27 OCT – Calvigny barracks helicopter crash
SGT Steven Kurlowicz	1/75 Ranger	Wounded in right arm
SP/4 Jay Lee	Unknown, Army	
SGT Michael Luciano	2/75 Ranger	
SFC Francisco Magana	2/75 Ranger	
SSG Alfred Manso	2/75 Ranger	
SP/4 Ronald McCall	Unknown, Army	
SP/4 Scott McCormick	82nd Signal	27 OCT – Brigade TOC strafing by A7 Corsair
CPL Kenneth McGuigan	USMC	
1LT Michael Menu	B 1/320	26 OCT – wounded by grenade on morning Calliste fight
PVT Rand Miller	1/75 Rangers	25 OCT – wounded in Ranger gun jeep ambush
SSG Gary Minerve	TF 160th SOAR	25 OCT – wounded Richmond Hill Prison attempt

Chapter 1: AIRBORNE - The Invasion of Grenada

PVT William Mozingo	2/325, 82nd ABN	
SP/4 Bradley O'Bryan	2/75 Ranger	
SGT Frederick Olmstad	2/75 Ranger	
SP/4 John Pelott	82nd ABN	27 OCT –Brigade TOC strafing by A7 Corsair [342]
SFC Scott Perry	SFOD-D	25 OCT – wounded Richmond Hill Prison attempt
SGT Sean Powers	B 1/75 Ranger	Broken bones
LT Paul Price	TF 160th SOAR	25 OCT – wounded Richmond Hill Prison attempt
PFC Michael Reed	82nd ABN	
SP/4 Loren Richards	TF 160th SOAR	25 OCT – wounded Richmond Hill Prison attempt
SGT Jeffrey Rogers	C 2/75 Ranger	27 OCT – thrown out of helicopter at Calvigny
PVT Timothy Romick	1/75 Rangers	25 OCT – wounded in Ranger gun jeep ambush
SSG William Sears	A 2/75 Ranger	25 OCT - paralyzed at Calvigny helicopter wreck
1LT Scott Schafer	82nd Combat AVN	27 OCT – Brigade TOC strafing by A7 Corsair
PFC Kirk Shartzer	2/325, 82nd ABN	27 OCT - wounded fight at Grand Anse intersection
SP/4 Harry Shaw	1/320 ARTY	27 OCT – Brigade TOC strafing by A7 Corsair
SGT Al Siciunas	82nd ABN	Wounded by shrapnel
MAJ Larry Sloan	TF 160th SOAR	25 OCT – wounded Richmond Hill

---

[342] Says he was in "Brigade Intel" and that his Kevlar saved him.

		Prison attempt
1LT Scott Snook	307th Engineers	27 OCT – Brigade TOC strafing by A7 Corsair
WO1 Thomas Speaks	82nd Combat AVN	27 OCT – Calvigny barracks helicopter crash
SGT Joe Stewart	FIST 1/320	27 OCT – Brigade TOC strafing by A7 Corsair
2LT Michael Talbot	LNO, 313th MI	27 OCT – Brigade TOC strafing by A7 Corsair
SP/4 Brent Taylor	2/325, 82nd ABN	27 OCT – wounded fight at Grand Anse intersection
SGT Steven Todd	C 2/508	31 OCT – Accidental Discharge – wound in leg
SSG Gary Toth	82nd ABN	
SGT Thomas Walker	Unknown	
PFC Jason Woodsford	2/325, 82nd ABN	Accidental Discharge
SP/4 Brice Vivan	2/75 Ranger	
SP/4 John Williams	2/75 Ranger	
SFC Stan Wood	SFOD-D	25 OCT – wounded Richmond Hill Prison attempt
SP/4 Curtis Young	2/75 Ranger	

Chapter 2: RANGER – A Death in the Family

A Company, 3rd Ranger Battalion
1984-1986

Throughout American history there have been Rangers. One of the first documented units with the title of "Ranger" in their name was Church's Rangers, that first saw combat during King Philip's War in the late 17th century. Church's Rangers were part of the Massachusetts Militia fighting the Wampanoag under "King Philip". The term "Ranger" came from official reports that told of the scouts ranging so many miles during patrols. This was strictly an American term, since there wasn't a lot of land to go "ranging" on in Britain.

Since that time there have been Rangers in every single war fought by the United States, except World War I. During the War Between the States there were Rangers on both sides, and the most famous, and successful Ranger unit of that war was Mosby's Rangers. The Rangers of WWII, Korea and Vietnam became the masters of light infantry warfare, but the modern Ranger Battalions began right after the 1973 war directed against Israel. [1] The Chief of Staff of the United States Army, General Creighton Abrams, realized that a highly mobile unit was needed that could be rushed to any hot spot in the world quickly.

---

[1] Known in Israel as the Yom Kippur War, and in the Arab world as the Ramadan War.

Though there had been Ranger Battalions in WWII, they were disbanded at the end of that war and had never been reformed.

In 1974 General Abrams ordered the formation of two Ranger Battalions. Abrams felt that having an elite unit within the military would raise the standards of the entire United States Army. Since the end of the Vietnam War both the morale and the standards had dropped and Abrams believed that once a Ranger had finished his time in the Ranger Battalion, he could then be disseminated throughout the ranks of the regular Army, raising the standards of all. Abrams wrote "The Ranger Battalion is to be an elite, light, and the most proficient infantry battalion in the world; a battalion that can do things with its hands and weapons better than anyone. The Battalion will contain no hoodlums or brigands and that if the battalion were formed of such, it should be disbanded."

"The organization of the Battalion must be done right, there (is) no timetable for this effort, (that) it must be determined first what has to be done and with what equipment and facilities. Wherever the Ranger Battalion goes, it is apparent that it is the best."

After Israel successfully conducted an airfield seizure to rescue hostages in Entebbe, Uganda, the Rangers focused their training on similar techniques. During Operation Urgent Fury the 1st and 2nd Ranger Battalions conducted an airfield seizure in Grenada under enemy fire, with very few casualties. Due to the actions of the Rangers on Grenada they were awarded the Valorous Unit Award, one of the few units that received any recognition for Grenada.

President Ronald Reagan realized that the United States Army needed to be increased in order to win the Cold War and "leave Marxism-Leninism on the ash-heap of history". [2] Light mobile forces would be needed to deploy quickly in the world, so President Reagan redesignated the 7th and 25th Infantry Divisions as light infantry, and reactivated the 10th (Mountain) Division. The 1st and 3rd Special Forces Groups were reorganized along with a new Special Operations Command (SOCOM). Due to the actions of Rangers in Grenada a 3rd Ranger Battalion was reauthorized in 1984.

Prior to Grenada I had never known much about the Rangers. Since it was a peacetime Army, there wasn't a lot of action after Vietnam. Grenada changed all that and brought the Rangers into the spotlight. When I failed the swim test for Special Forces training, I was up for "worldwide assignment". This meant

---

[2] President Ronald Reagan's 1982 speech given to Parliament in Westminster, London.

Chapter 2: RANGER – A Death in the Family

I could be assigned anywhere and might no longer be a paratrooper. I didn't want to sit still and wait for orders, so I called the Department of the Army and asked for a Ranger assignment. They told me there was a new battalion starting up in Fort Benning, the 3rd Ranger Battalion so I told DA to send me there. [3]

"The Soldiers Of The Wilderness,
A Rough And Hardy Band,
In Woodland Garb, With Woodland Arms,
We Guard This Forest Land.
Tis Ours To Breathe The Battle Smoke,
To Range The Trackless Wood,
To Struggle With The Howling Storm,
And Swim The Flashing Flood.
Deep In The Gloomy Forest,
Unseen By Human Eye,
We Track The Foe, We Strike The Blow,
And, Nameless All, We Die.
I Love The Savage War Whoop,
And Whistling Of The Ball;
The Woods, The Rocks, The Boiling Streams,
I Love Them One And All.
And Yet Their Memory Is Entwined,
With Thoughts Of Sore Distress,
Of Famine, Grief, And Danger,
And Bitter Weariness.
For The Ranger's Gun Has Echoed,
From A Thousand Pathless Mountains;
And The Ranger's Blood Has Stained With Red,
A Thousand Limpid Fountains."

Cpt. Jonathan Carver, Jr. Ranger, French & Indian War (1755-63)

I arrived at the 3rd Ranger Battalion in June of 1984 though the companies in the Battalion had been forming since April. I was assigned to A Company and I was the only NCO there who was not already Ranger qualified. Right away I noticed a different attitude. In the 82nd Airborne I would talk to Platoon Sergeants and the 1st Sergeant in a casual way, standing in no particular manner. However, in the Rangers I had to change to meet their standards. When a Ranger talked to anyone that was a higher rank, he stood at parade rest or

---

[3] I later received a welcome letter from the unit that I would have been sent to. It was the 6th Battalion, 31st Infantry in Fort Irwin, California. This was an OPFOR unit, and pretended to be Soviets to allow other units to attack them. My chances of doing something "real" were zero in a unit such as this and I was lucky to dodge that bullet.

attention, depending on the rank. I was assigned as the A Team Leader in the 2nd Squad of 1st Platoon.

Rangers all wore "high and tight" haircuts, like the Marines, and the haircut had to be fresh each Monday.[4] Even the uniform was different. The rest of the Army wore the camouflage BDUs but the Rangers wore the old green cotton "Jungle fatigues", so immediately everyone knew who you were.

The commander of the 3rd Ranger Battalion was LTC William Ohl, and he was a "lead from the front" officer. He had fought in Vietnam with Ranger units, like many of the other platoon sergeants in the Battalion. When the Battalion first started it resembled a foreign legion. We had soldiers from Australia, Columbia, Northern Ireland and Thailand. One of our squad leaders, SSG Kelso, had served as a mercenary in Rhodesia.[5]

Prior to General Shinseki's usurpation of the Black Beret, the only soldiers authorized to wear that color beret were Rangers. Each Ranger had to earn the beret by making it through RIP (Ranger Indoctrination Program) before they could join the Battalion. However, we were truly a Ranger Cohort Battalion. Everyone arrived at the same time and the privates were all fresh out of Infantry school. Due to this we ran a Battalion level RIP program. This consisted of a lot of physical training, several airborne operations, and some small unit tactics. I was in charge of the platoon, since I was a SGT. Everyone higher rank than me, SSG and above, went to different type of training for the leaders of the Ranger Battalion. Our training lasted until September, when the Battalion was awarded the Black Beret.[6]

The Battalion next went through a mini-Ranger school. This was an immersion into patrolling, such as the type that is taught in the Ranger School. The whole Battalion went to the Mountain Ranger School in Dahlonega, Georgia, then the following week we were immersed into weapons training and airborne operations. In between missions we would all go out and party hard in Columbus or Atlanta. Since we were something new and intimidating to the

---

[4] The high and tight continued to be worn by the Rangers until the war in Afghanistan began. Due to the Rangers working so close with "Delta Force" they were allowed to grow their hair out to blend in.

[5] Kelso would later become the Command Sergeant Major for the Ranger School, and today there is an award called the "CSM Michael Kelso Enlisted Leadership Award" that is awarded to the highest ranked enlisted Ranger as selected by their peers for demonstrating outstanding leadership, initiative, and motivation.

[6] The beret ceremony for the battalion was conducted in the rain. The Battalion Commander and CSM put on their berets, and then they went and put them on each Company Commander and First Sergeant. Both the officer and NCO chain of commands then worked their way down to the Team Leaders, and then we put it on the heads of each of the Privates.

"legs" on post, there were constant fights.[7] LTC Ohl threatened to kick out the offenders, but he never did.

I would do road trips to outlying towns in Georgia, such as LaGrange, with my whole squad, that created an even closer bond. One of our wildest road trips was during the last year that the Army was authorized to wear the Khaki uniform. We took four carloads of Rangers to New Orleans, put on the Khaki uniform, and partied over Labor Day weekend. The purpose for the trip was a bachelor party for PFC Walters, on my Team. In the end we were broke, hung over, but had the time of our lives. The bachelor party was pretty effective, since Walters called off the marriage and returned to New Orleans the next weekend to be with a girl he met.

The day after the New Orleans Road Trip the Battalion jumped into the Florida Ranger School. We learned how to fight in the swamps as Rangers and how to do rubber boat operations in the surf. We also lost our first Ranger. Private Wesley McDavid of C Company had returned to Dahlonega to do some mountain training on his own. He was killed while climbing.

The cycle of hard, realistic training and harder partying went on throughout the summer until the activation ceremony in October. We were becoming tougher, stronger and it was a tighter unit than any I had served in. When the 3rd Ranger Battalion was officially activated the last Battalion commander from WWII arrived to present the Battalion colors to LTC Ohl. Rangers from Korea and Vietnam were there to do PT with us and party with us afterwards. The 1st and 2nd Ranger Battalions were there, and received their Valorous Unit Awards from Secretary of the Army John O. Marsh for their actions during Grenada.

After activation each of the levels of the company were tested in an intense series of field problems known as an ARTEP.[8] Each squad, platoon and then company would be tested to see if they were combat ready. Each evaluation had a surprise, to make us think out of the box on ways to solve the problem. An example of this was when our helicopter set down in the middle of nowhere, and the pilot said "get out, we've been shot down". We then had to E+E back to safe lines, though we only had a vague idea of where we were.[9]

Our Platoon SGT was an extremely experienced Ranger, SFC Akuna. He also had served in Vietnam with one of the Ranger companies. During the platoon ARTEP we did not do so well, but this was mainly due to the inexperienced Platoon Leader we had. He was a Lieutenant who had no clue, and would not listen to the advice of any of the NCOs under his command. Since our Platoon didn't achieve their goals, an example had to be made.

---

[7] The most derogatory thing you can call a paratrooper is "leg". This is someone who is not airborne qualified and is only "straight leg" infantry.

[8] ARTEP stand for Army Training and Evaluation Program.

[9] E+E stands for Escape and Evade.

Unfortunately, the platoon leader, Lieutenant Kan____, was not relieved, and instead the Company Commander, CPT Minton, relieved SFC Akuna. The platoon was amazed and angered by this, but I realized then that officers will protect their own, instead of doing what is right. SFC Akuna was thrown to the wolves, and SFC Maddison became our new Platoon Sergeant.

The Rangers train in an extremely realistic manner. Due to this there were numerous injuries and sometimes fatal accidents. An example of this was when the 2nd Ranger Battalion was training in Honduras and their mortars fell short, landing in the midst of a company CP, killing the commander. We were told about that when we learned of the 2nd Battalion commander being relieved of his command.

Unfortunately, my lack of a Ranger Tab in a Ranger Battalion slowed my professional progress. The 1SG of the company, a slightly psychotic Ranger, First Sergeant Jones, did not want to promote me until I went to Ranger school. Unfortunately, the first school slots weren't going to happen until the next year. Due to the 1SG's suggestion, the Company Commander, CPT Minton, continued to deny me a Ranger school slot. I never did know why, but he felt I wasn't ready to go to Ranger School. To add to all this, my girlfriend in Shelby, North Carolina had left me for a guy just out of prison. I was so demoralized at one point that I was thinking of getting out of the Army and becoming either a Sherriff's Deputy in Hillsborough County, Florida, or going to South Africa and joining their army.[10] The wars against Communist rebels in Africa were numerous and it seemed like a chance for adventure.[11]

The main thing that changed my mind about getting out was finally going to Ranger School. Though the Ranger Regiment had been created after the activation of 3rd Battalion, they still were not fully functional. Each Battalion still ran its own RIP and Pre-Ranger course. The 3rd Battalion didn't have a Pre-Ranger course, so it created one in January 1985.

---

[10] I had lived in Tampa, Florida at the time that I joined the Army. My brother, Sean, had joined the Marines, done a four year tour, and then returned to Florida to work as a ship welder.

[11] I got the idea for this from Mike Kelso, who had been in Rhodesia in 1978, and from another Ranger, PFC Dengate, who was from Australia and had been in the Selous Scouts in Rhodesia. At this time Angola and Rhodesia had been lost to the Communists, so South Africa seemed to be next. When I was thinking of going to South Africa my brother Sean told me that if I went, he would go too. SSG Kelso talked me out of the idea. He left Rhodesia when it all fell apart and he had a bad experience there. In 2005 Kelso would become the CSM for the US Army Infantry Center in Fort Benning.

Chapter 2: RANGER – A Death in the Family

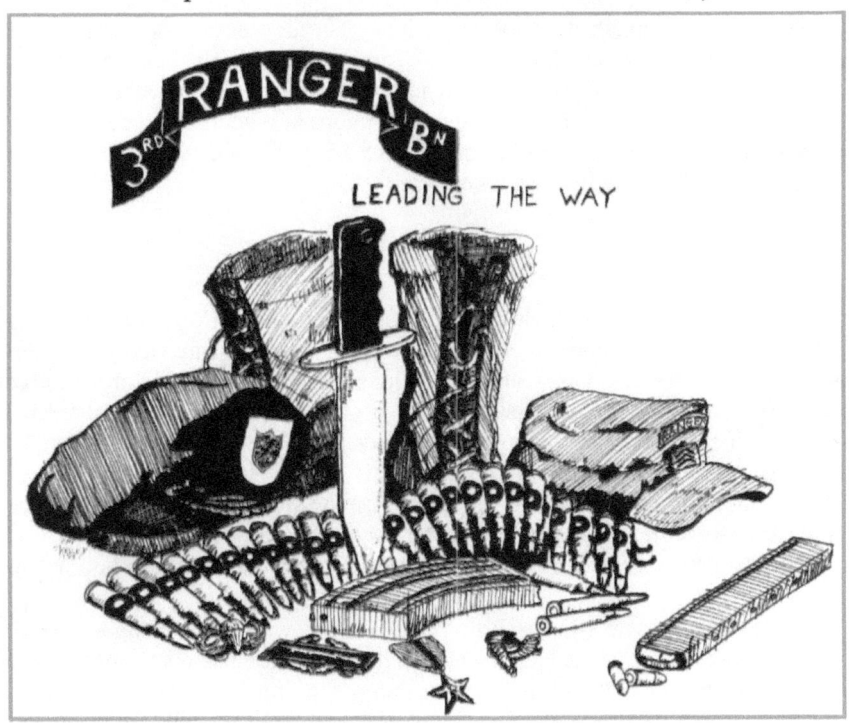

RANGER
3RD BN
LEADING THE WAY

I was in the first 3rd Ranger Battalion Pre-Ranger Course. Since it was the beginning of winter, everything we did was in the cold, and due to this many Rangers quit. Some with frostbite, and another situation I have never seen, blisters appearing on hands and cheeks, almost instantly, due to extreme cold. If a Ranger quit the course, he would no longer be in Battalion.[12] I refused to do this and excelled in the Pre-Ranger course, and I soon became the man each patrol leader chose to be the point man. I was able to see things others missed and I did not get lost.

After the Pre-Ranger course the chain of command, mainly 1SG Jones, still didn't have faith that I would be able to make Ranger School. I imagine this was the bias against non-tabbed NCOs.[13] Since I was raised in the 82nd Airborne and not in a Ranger Battalion, I was perceived to be flawed. I was knocked off the primary list for Ranger School and put on alternate. I was not thrilled about this since without completion of Ranger School I would not be promoted and

---

[12] BDUs were known as Battalion Departure Uniforms in the Rangers. Once you saw a Ranger out of his Jungle Fatigues and into BDUs, you knew he was leaving for good.
[13] Those who graduate Ranger School get to wear the black and gold Ranger tab. If you did not have this, you were "non-tabbed".

would never advance within the Battalion. The day before the first course started something happened to one of the primary Rangers, and I ended up being chosen to go. I was concerned that I wouldn't be able to pass the swim test, but the Ranger swim test is only 15 meters in fatigues and boots. I was able swim at least that far. I would be in Ranger Class 5-85, the first class that 3rd Ranger Battalion attended, and it was a winter class. [14]

Ranger School

In 1985 Ranger School was 58 days long. It consisted of basic patrolling procedures in Fort Benning, mountain patrolling in Dahlonega, Georgia, desert patrolling in Fort Bliss, Texas, and finally jungle and surf ops in Eglin Air Force Base, Florida. Though the course teaches patrolling, it is merely the tool needed to determine leadership. An infantry officer must be Ranger qualified if he wants to succeed in the military. Due to this the course is mainly filled with officers, but no rank is worn in the school by the students, so no rank can be pulled on a lower ranking enlisted Ranger Instructor (RI). You are only as good as your last patrol. You must earn respect, and it is not just given to you due to your rank.

Ranger School was run by COL "Tex" Turner, who was a Ranger legend. The first week is known as Hell Week and it weeds out those who cannot make it. [15] On the third day the class ran three miles, at a little over 7 minutes a mile. We started with 245 men and lost 75 students on that first run. When I first showed up, I was worried that I might not make it, but I soon discovered that Ranger School was not that hard, especially after being in a Ranger Battalion. [16] All of us Battalion Rangers would clique together, and we would joke aloud, wondering when it was going to get hard. It was almost like a vacation from the Battalion.

---

[14] A winter Ranger course is perceived to be much harder. More gear is carried, which means the rucksacks are all heavier. The chance of getting frostbite or hyperthermia in the mountains is much greater. Due to this some Rangers sew the Ranger Tabs on their uniforms with white thread, to signify that they went through the course in the winter. I never did though.

[15] The Hell Week cadre is known as the Morgan Team, named after Daniel Morgan, the victor of the Battle of Cowpens in the Revolutionary War.

[16] Not everyone who wears a Ranger Tab has served in Battalion. In the Ranger School not everyone is airborne qualified. The Ranger School is only a couple of months long, and many of the graduates will never do anything "Ranger" again for their entire military career. However if you are in a Ranger Battalion you must pass the Ranger School to remain. Every day in the Ranger Battalion you are put to the test. The unit patch for the Rangers is a black, white and red scroll. There is a saying amongst the Rangers, "The Tab is a badge, but the Scroll is a way of life".

186

Chapter 2: RANGER – A Death in the Family

I was the only enlisted soldier in my platoon during the Fort Benning phase. Almost all of the other students were 2nd Lieutenants right out of West Point. They knew virtually nothing, and I helped them out as much as I could. For example, when we were given an M60 machinegun for one of our patrols, one of the officer students asked me how I put the M60 on semi-automatic. The M60 machinegun doesn't have a semi-automatic setting, but since he was really new to the Army, he did not know this.

After Hell Week the survivors were sent out to Camp Darby, in Fort Benning to learn the basics of patrolling and reaction drills.[17] At Camp Darby the students lived in "pup tents" out in the elements. It either rained or snowed on us for about half of the time we were there. Small painful cracks appeared in our hands from the cold and our faces always felt sun burnt due to being in the cold all the time. Food was limited to only two meals a day and we received about four hours of sleep each night. In the mess hall the Ranger students only had three minutes to eat and then they had to get out. Eating fast became a habit.[18]

Though the food was limited, it was the best food I had ever had in my military career. This wasn't just due to being hungry all the time, which we were, but it was also due to the caliber of the cooks and the mess halls. Mountain Ranger phase is known more for its blueberry pancakes and muffins than anything else. Ranger students also excelled at the art of hoarding food and would collect cream and sugar packets from MREs to eat dry later on.

Since I came from a Ranger Battalion the officer students would use me for the critical positions on every mission. This meant that I would be the compass man, or point man, surveillance leader, POW and search, or be one of the Team Leaders. I didn't mind, since I didn't trust the green officers to know what to do.

Throughout the Ranger course the airborne qualified students would parachute into the different missions, while the non-airborne students would be trucked in. Due to this the non-airborne students were looked upon with disdain, since the chance of being hurt and not making Ranger School was not as likely for them.

At the end of Camp Darby, the students had a five-hour break. We treated that break like a four-day pass. The "real" Rangers went and got their traditional

---

[17] These drills were for things that needed to happen instantly, and with little thought. For example if there was a near ambush, everyone would automatically assault into the enemy's line. To stay in the ambush area would mean death. If there was an artillery attack, everyone would have to run out of the impact area, which may be 500 meters away.

[18] Once I retired from the Army I gained a lot of weight. This is mainly because I ate like I did when I was on active duty, and I ate fast. This doesn't work in a more sedentary civilian lifestyle.

haircut and then went to Pizza Hut. The key was to find a place where it was buffet, all you can eat. After eating, the 3rd Battalion Rangers went back to the Battalion and washed clothes. The final thing, before going to mountain phase was to eat at Burger King.

Though no food is allowed at all in Ranger School, it was a challenge to slip something in. Some Ranger students would put honey in shampoo bottles, or put powder sugar in baby powder containers. I filled a two-quart canteen with M&Ms, and I filled up a flashlight with Slim Jims. In the mountain phase I would eat a few M&Ms each day as my treat. If we were caught with these treats, we would be thrown out of Ranger school, so it was a stupid thing to do.[19]

The Mountain Phase of Ranger School is in Dahlonega, Georgia where Mount Yonah is located. The main difference is that we were able to eat three meals a day and we weren't rushed. The increased food was due to the cold weather, which was lucky for us because in summer classes that would not happen. Students that had failed the Mountain Phase, but were recycled, joined us, bringing our numbers back up to what we had before we suffered our losses due to attrition. Our platoon gained a Thailand exchange officer called Kulkhanchit. We called him Top and he ended up being my "Ranger Buddy".[20] We stayed in small wooden shacks that were decorated with graffiti from all the other Ranger classes. I added my own graffiti by drawing a picture on the wall for the next class.[21]

After learning how to climb and rappel down Mount Yonah we began the graded mountaineering patrols. The terrain was brutal, and when combined with the cold weather it was the toughest time I had in Ranger School. A few of the officer students began to show their true colors and didn't like the idea of an enlisted man besting them. They became a thorn in my side, but due to the peer evaluation system in Ranger School, many of them were recycled. During the Mountain Phase the students were not allowed any snacks or candy, but they

---

[19] Rangers became food fanatics and would write fan letters to the makers of the MRE candies and cakes. One Ranger in Battalion received a case of chocolate nut rolls (one of the better cakes) due to his fan mail. Rangers in Battalion would also receive boxes from their buddies, meant to annoy them. The box would be labeled "To my big bad Ranger" or some such on the outside. A box sent to me was addressed from Seka Lovelace, 6969 Straddlemee Drive, Climax, PA. Inside was empty beer cans, pizza crusts, Snicker wrappers, and a Polaroid photo of all my buddies eating these items. Due to this, I do the exact same thing to anyone I know who goes through Ranger School.

[20] Supposedly no exchange officers would fail the Ranger School. We were told that two African officers had failed the Florida phase, and when they were returned home they were executed due to the embarrassment. Whether or not this actually happened is not known.

[21] When these old wooden shacks were torn down in the early 2000s, they let one remain, with all its graffiti, for historical purposes.

were allowed to chew gum. Due to this I consumed mass quantities of Hubba-Bubba Raspberry bubblegum. The sugar in it was what I craved.

After the last mission we all were able to go to the NCO Club and pig out on all sorts of junk food, before we prepared to conduct an airborne operation into the desert. All of the non-airborne students would take a bus out to Fort Bliss, so once again they were looked down upon. The good thing about being airborne is that a certain amount of sleep is required before a training jump. This backfired on us at Fort Bliss though because the RIs ran us harder. Whatever rest we had acquired was burned up quickly. After we landed on the drop zone and rendezvoused with our RIs, we ran all over the desert to our first target. The RIs did this by throwing artillery simulators the entire way there, and because of the immediate action drill you have to run from the explosion. The RIs said that we didn't need any sleep, since we got so much before the airborne operation. Our rucksacks were extremely heavy and we missed our objective that night due to all the chaos caused by the RIs.

The next day I was made the PL (Patrol Leader), which is a critical position. In each phase the student has to hold a major role on the patrol. This could be the PL, or the assistant, the RTO, or the M60 machine gunner. The most important was the PL. Throughout Ranger School the students dreaded when the new RIs would switch out in the morning. The first thing they would do is yell out "Roster Number XX, you are the PL!" All the students waited to see if their number was called. If you were chosen as a PL, you had a good chance of messing up and being recycled. Anyone that failed a patrol twice would be recycled. Two recycles and you failed Ranger School. My new "Ranger Buddy" an officer named Oldre, was recycled after Desert Phase for failing two patrols.[22] Two other 3rd Ranger Battalion Rangers were recycled in Desert Phase, which left me the only survivor from the original 3rd Battalion Rangers.

When I assumed the role of PL, I refused to panic and start "pinging".[23] I had the students continue improving their positions in the patrol base, and then had them eat breakfast. Normally a PL would have all the squad leaders come in and begin barking out orders, so that it would look good to the RIs, but didn't really serve a purpose. It was all for show. Instead, I wrote up the Operations Order, and when I gave it to the men the RI did not make any corrections. That was a first. I received an enthusiastic "go" on my patrol.[24]

---

[22] Each Ranger had one other person that they did everything with. This was mainly for safety so no one would be fatally hurt or lost during Ranger School. This person was the student's Ranger Buddy.

[23] Pinging was the nickname for chaotic panic. The term came from the old cartoon *Ricochet Rabbit*, who would go "ping, ping, ping" and then bounce off the walls.

[24] During a patrol, the PL would be switched out three times. One for the planning, one for the movement, and one for the actions on the objective. This way more students could be tested. So, all I had to do was the planning and I passed easily.

The Desert Phase was located just north of Fort Bliss. The desert was a combination of sand dunes and bushes. The lights of El Paso could be seen in the distance, and this included the lights of all the restaurants and steak houses. It was just one more way to torment us. The weather was another bit of torture, and it bordered from hot desert sun to snow flurries. Yep, snow flurries in the desert on the border of Mexico.

Sleep during that phase was about three hours a night and food was back to two MREs a day. In my journal I wrote "I found the best thing in MREs. Take cocoa powder, add just enough water to make a paste, and then squirt in a peanut butter package. It's better than sex!" After each phase the students would write a peer report that would rate other students on whether or not they pulled their own weight. The officers came up with a system where one would write a glowing report on one officer, and then write a not so good one on another officer. Each officer would switch off, so all officers would get a great report and a mediocre one. This way it would all be fair and no one would fail. I refused to do that and I told the officers that I would write exactly what I see. If they are squared away, the peer evaluation will reflect that. If they are a "soup sandwich", the peer report will also state that.[25] I did not want to let any "spotlight Rangers" get through the course.[26]

The Desert Phase of Ranger School was only twelve days long, but it was constant patrolling. Once we finished the Desert Phase, we prepared to jump into Florida for the last phase of Ranger School. Like most deserts I have been in, it rained on us pretty much the entire time. When we jumped into Florida it also rained. At the beginning of each phase of Ranger School it rained on us. We were pretty used to being wet and cold all the time.

Florida Phase of Ranger School is where all the lack of sleep and food accumulates to its highest level. The reason Ranger School does this is to create stress and fatigue, so they can evaluate how the students react under real pressure. Florida Phase is where all the great Ranger hallucination stories come from, such as the story of a Ranger trying to put quarters in a tree because he thought it was a Coke machine. I did not see any hallucinations like this, but a huge problem was when the patrol would stop, and everyone would get down on one knee. As soon as you saw the shadow of the man in front of you get up, you would then continue to move on the night patrol. However, every bush and tree looked like a kneeling man to me, so I would not know if we had moved unless I listened really hard. I hallucinated that they were people, kneeling, waiting to move. My Ranger buddy for Florida Phase was *Top*, the Thai, and he was just short enough to look like a palmetto bush. Each night movement I

---

[25] A soup sandwich is an Army term, which means to be totally screwed up, such as making a sandwich out of soup.

[26] Spotlight Rangers are those who perform well when the RIs observe them, but turn into whiners and liabilities when no one is watching them.

had a bad habit of walking into trees, and then move again, and smack into the next one, like a pinball. Each movement was long, usually all night and freezing cold as we waded through the swamps. Exhaustion was the norm.

The aggressors for Florida Phase were the 3/325th Infantry from the 82nd Airborne. They knew what we were going through, so the "bad guys" would have MREs in their pockets. When we "killed" them in an ambush, we would then search the bodies, and anything we found was OK to eat. When we flew in the helicopters the pilots would do the same thing, and covertly point at a bag on the floor, which would be full of Snicker bars and other candies. We would take them, and then pass them around to the rest of the squad. The RIs would all go blind during those moments.[27]

On one mission we ambushed a truck that had vegetables, live chickens and goats. Inside the truck was ammo cans that we could use for cooking pots. We moved away, about 700 meters, and killed the animals. I was the cook for that detail, and put everything into the ammo cans. Each student got around three cups of my stew.[28]

The final mission in Ranger School was run on nothing but adrenaline, MRE coffee and cake. We had to row the Zodiac rubber boats to an LCM (Landing Craft, Medium) and then ride the LCMs a few hours to Santa Rosa Island. At the island we unloaded the Zodiacs and rowed them to the island. Our objective was to find a guerilla leader, Jose Jiminez, and raid his camp. We pulled off a raid that got praise from the RIs, but we just missed the target of Jose Jiminez. We then had to carry the Zodiacs across the mile-wide island on our shoulders, put the boats in the sound on the other side, and row another five miles to the mainland.

The mission to find Jose Jiminez continued with a swamp movement through chest deep mud. I was a squad leader and my squad had the blocking position on a road leading to the objective. When the main company attacked the guerilla camp, our job was to make sure that no one entered or escaped. The attack went off at the right time, but "Jose Jiminez" escaped, and came running down the road into our blocking position. The only thing between us and graduation was him, so everyone dived on him and duct taped him until he couldn't move anymore. We brought Jose to the guerilla base camp, where a

---

[27] There were some RIs who had fun with this, and made dog food sandwiches, and left them on a truck at an ambush site. He would later ask the Rangers what they did with his sandwiches he made for his dog.

[28] For some reason, on almost every mission if a cook was needed, I ended up with the job. I just knew how to cook better than most and I credit my mother, Joyce O'Kelley for this. Also, I had some help from my step-grandfather, Grover Finney, that played in a Big Band in the 1930s, owned a restaurant and served in WWII as a cook. He went ashore at Normandy, but not until D+3.

huge iron pot of refried beans had been cooking. Everyone got out their canteen cups and ate all the beans they could stomach.

After the missions were all over, we cleaned up and prepared for the return to Fort Benning for graduation. There were 24 recycles from the Florida Phase that had to do it again. Some recycles had to go all the way back to Hell Week. The head RI told me that I had rated highest on patrols, spot reports and on peer reports. He told me that I would be the Distinguished Honor Graduate for Class 5-85.

My mom, dad and brother showed up for the graduation. No one in the 3rd Ranger Battalion attended this very first Ranger School graduation of their Battalion, because they were in Panama, training at the Jungle Expert School. While I had been in Ranger School, they moved the Battalion from "Starship" barracks at the main post, to older barracks out near Harmony Church.[29] The Battalion returned the next day and I learned that one of the youngest Rangers on my Team, Russell Hobgood, had gotten married to his High School sweetheart. I was not too thrilled with any women, since they kept leaving me for "Jodies" who lived closer, but I thought Hobgood and his fiancé were a perfect match. They were both innocent and naïve.

I became an acting Squad Leader, since SSG Carey, my squad leader, was on profile for an injury he sustained in 1981.[30] Since I was finally a Ranger School graduate the Battalion sent me to the E-6 board, which I passed easily. My squad was down to five men, but we were tighter knit than ever.[31] This was because we had all shared the same experiences and pain with the creation of the Battalion. All the privates had become SP/4s and in the past year they had become Rangers. On our off time we would hang out at Hobgood's house, or Walters and Wilroy's trailer, to drink, shoot guns or chase women. Hobgood's wife, Laura, loved firing my M1A, but wouldn't touch anything else.[32] Brett Beaudette was my corporal, so as a fellow NCO we hung out more than anyone else. On a few occasions I would reenact the Revolutionary War, or WWII, but we all mainly stayed anchored to the Ranger Battalion.

---

[29] The new barracks in the 1980s were self-contained buildings that had everything in them. They were nicknamed Starships, since they were huge modern looking units.
[30] Being on profile means that a soldier is limited in what he can do until he recovers from his injury.
[31] A Ranger Squad is larger than a normal squad. It is two five man fire teams and a Squad Leader.
[32] The Springfield Armory M1A is the civilian version of the M21 Sniper Rifle. I had bought one with my reenlistment bonus after Grenada.

The Battalion went through their ARTEP in May of 1985. Each of these field problems were one more step to becoming fully combat ready. The Battalion ARTEP was one of the most intense training missions I have ever been on. This was mainly because our aggressors were the 1st Ranger Battalion out of Savannah, Georgia. The 3rd Battalion conducted a night jump onto Hunter Army Airfield, in Fort Stewart, Georgia, right beside Savannah, so the terrain is all low country swamp. A good portion of the Battalion landed in the trees on the edge of the airfield. After eliminating the enemy on the airfield, the Battalion then moved all night through the swamps to get out of the area. We were exhausted, and at one point when we stopped and got down on one knee, everyone fell asleep while they were still kneeling. I woke up and was startled to see that the sun had come up, and all around me were kneeling, sleeping Rangers.

When we moved into a company sized patrol base, we were attacked by three truckloads of Rangers dressed as Nicaraguan rebels. The fight was intense, with my four-man squad fighting a withdrawal all the way back to the main company CP, and then doing hand to hand with the 1st Battalion Rangers who broke through the perimeter. We finally beat them back, but there were 1st Battalion bodies everywhere.

We continued to move out of the area, collecting water on ponchos whenever we could since our canteens were all low. Helicopters searched for us in the swamps, so we had to hide whenever they came over. From time to time we would ambush a patrol looking for us, or let it go and slip by them. The whole company had to cross a river, but we had to do it without being detected. We put up rope bridges and over three hours moved everyone to the other side.

While we moved, we could hear the PSYOPs trucks driving around, loudspeakers blaring about how we should give up, and how our wives are sleeping with the neighbor next door.[33] Since most of us were under 25, and single, it was pretty funny.

SFC Maddison, the platoon sergeant, took some men out on the road to ambush the PSYOPs truck, but instead they captured a 2½ ton truck. The Rangers jumped on the truck as it drove by, swarmed over it, and pushing the driver out of the seat, yanked the shift into neutral. There was a lieutenant on the truck, who objected, but he was duct taped to a tree with the driver. We used the "Deuce and a half" to move our platoon quickly to the next objective. It was a Battalion level attack on a resupply base, and each company had a different objective.

---

[33] PSYOP stands for Psychological Operations. They are the unit that drops leaflets on the enemy, and broadcasts propaganda to make him not want to fight.

That attack was another extremely realistic fight, with hand to hand going on. Soldiers are not supposed to actually harm another in training, but since this was Rangers going against other Rangers, it was survival of the fittest. At the end of the attack the company moved quickly to a PZ and boarded CH-47 helicopters. Our helicopter landed on a dirt ramp built on a swamp. After we boarded it began to lift off, and the crew chief yelled "HANG ON!" The Chinook then just stopped, and dropped about 20 feet back into the swamp. We all landed hard, but no one was hurt, and the helicopter didn't appear to be broken. We left the helicopter where it had crashed, and took the next one that arrived (on the following page is a drawing that I made after the ARTEP and depicted the "bloody" fight between the two Ranger Battalions).

Our return to Fort Benning was a celebration of a conquering army. The Battalion jumped onto Lawson Army Airfield during daylight, and then marched to the hangers under the admiring eyes of the wives and girlfriends.

Now that the Battalion was qualified under conventional methods of warfare, we had to learn the unconventional side of the Rangers. The senior NCOs were going to be the instructors on these unconventional methods, so each day we studied up on the modules we would teach to the Rangers, while they recovered from the Battalion ARTEP. Each night the materials we studied had to be locked up in the Battalion safe, since it was considered SECRET at that time.

Since we all had trained hard during the ARTEP, we were given a three-day pass. I drove down to Florida to stay with my brother Sean, in Tampa. Sean had been in the Marine Corps as a Sniper in H+S Company, 1st Battalion, 2nd Marines from 1982 to 1985, but he had gotten out and became a below waterline ship welder.

During the day we would go out shooting, and during the night we would chase after women in the bars. It was very similar to being in the Ranger Battalion. The day before I was to return to Battalion, he took me out three-wheeling on his ATV. I had never been on an ATV before, and I ended up driving it off a small cliff. When I landed the ATV rolled over me, knocking my wind out. I tried standing up straight, but I couldn't. After 15 minutes I still couldn't breathe right, so my brother drove me to the hospital. It turned out that I had broken three ribs and punctured a lung. The civilian doctors were confused about what to do with me, so I told them to send me to MacDill Air Force Base. I know that they knew how to handle a chest wound.[34]

---

[34] MacDill AFB was the home of SOCOM (Special Operations Command).

WE KILL MORE
BEFORE NINE O'CLOCK,
THAN MOST PEOPLE
KILL ALL DAY!

At MacDill they cut into my side and put in a tube to inflate my lung and drain out any liquids. I remained in the hospital for a week, while they monitored the progress of my lung. I was worried that I would not be able to stay in the Ranger Battalion due to this injury. In my absence there was a change of leadership positions in the platoon. SSG Carey was promoted to SFC and became the 2nd Platoon Sergeant. I then became the 2nd Squad Leader.

During my time in the 3rd Rangers, I still did some reenacting, but most of the events were over a seven-hour drive. When I first started participating in Revolutionary War reenacting, I met Bert Puckett. He was only a kid, but after his parents were divorced, Bert, his younger brother Bryan and his sister Betsy would always hang around me. Bert would listen to my "war stories" and due to this he decided to go into the Army when he turned 18. Since I got into the Rangers, I had been writing to Bert Puckett's younger sister Betsy quite a bit. She had "developed" early in life, and I took notice. Unfortunately, she was 14 and I was 23, so I felt odd about it, but we were just writing letters back and forth. I mainly did it because I was tired of having girlfriends that left me for the first "Jodie" at the first opportunity, and it was nice to write to someone who wouldn't be running off anytime soon.

During this time the Battalion had been training hard learning how to do Jump Clearing techniques on an airfield. I still was not fully recovered, so I was not able to train with them. I felt that I had let them down. I was always embarrassed about what happened and for years afterwards I would tell girls that the scar on my side was due to a gunshot wound. However, when Bert Puckett became a Ranger, I told him the truth. This is because Robert Rogers wrote in his rules that you should never lie to another Ranger. Since I had graduated Ranger School, I had immersed myself into the Ranger culture. Each morning the Ranger Battalion would say the Ranger Creed before PT. This was not merely lip service to know some lines on a piece of paper, but it was a Creed that we all truly lived by. If any Ranger violated the Ranger Creed he would be removed from the Battalion.

Around this time, in the "real world" Islamic terrorists hijacked TWA flight 847 and kept it on the ground at Beirut airport. All the Rangers assumed we would deploy there eventually. When Rangers go on leave, they do it all at the same time in what is known as "block leave". In June of 1985 the block leave was cancelled due to Flight 847. President Reagan was considering using us as an option and we could not take leave until the situation had settled itself. This was the life of a Ranger then, and now. This is why girlfriends all tended to be temporary. It would take a special woman to put up with this lifestyle.

While we waited to go on block leave, we continued to train in the unconventional techniques of the Ranger Battalion. We learned how to assemble a "Little Bird" after taking it out of a C-141 aircraft. The Little Bird is an MH-6 helicopter that could carry six Rangers on the outside on a small seat that looked like bench. The AH-6 Little Bird was the attack helicopter version. Both were carried inside a cargo plane, and then were quickly assembled upon arriving at the airfield. The Little Bird and its capabilities were considered SECRET in the 1980s, and during our training we would have to conceal the helicopters whenever the Soviet spy satellites would come over head. Today the image of the Little Bird is no longer secret, and you can see the helicopter in movies such as "Blackhawk Down" or on FPS (first person shooter) computer games.

Bike Teams

During this time the Rangers also trained with their Bike Teams, Jeep Teams, and the Jump Clearing Teams. On the way to the training at the airfield on Fort Benning, the Jeeps and Bikes would drive on the back roads, so no one would see them. When we were seen by non-Ranger soldiers on Fort Benning, they would stare at the odd sight of heavily armed, black painted jeeps driving by, wondering who we were.

I was on the Jump Clearing Team (JCT). Our job was to be the first onto the runway and clear off any obstacles. We jumped in "light", which meant no rucksacks. All the tools we used, such as crowbars, saws, bolt cutters were all taped to our weapons. Each section of the Jump Clearing Team would jump at intervals along the airfield. The Section Leaders would spot their area from the ramp of the C-130 and then go. Each section would follow their leader down, landing as close as possible to the objective on the field. If the parachute landed off the strip, you left it where it lay. If it was on the airstrip, you packed it up quickly, threw it off the strip, and then started running down the runway clearing it.

The first person on the strip throws down a chemlite, signifying that a JCT team is there, and then they run and link up with the JCT team in front of them. He bypasses all obstacles, and mainly looks for a flour line on the airfield where the other JCT team was. If there is no flour, he will spread out a line of flour on the airfield, and then run back to his team. The chemlite is picked up and the team will then move down the runway, clearing all obstacles. They run until they get to the assembly area, then pull security on the perimeter. The bike teams would finish the clearing by riding down the runway verifying that it had all been cleared (the drawing on the next page is one I made of the Bike Teams, and it had to be cleared through the Battalion S2, for me to sell it in the local shops. I had drawn "GLINT Tape" on their arms, and the Battalion CSM said that it had to be removed due to OPSEC).

After the JCT teams did their job, the aircraft would land and then the jeep teams would move out to blocking positions. The rest of the aircraft would land, and the Little Birds would be assembled to go after their targets. Once the airfield was seized the Rangers would move on to their secondary targets, or would hold the airfield until larger units, such as the 82nd Airborne, would arrive. The Rangers did this both in Grenada in 1983 and Panama in 1999.

We tried some new techniques and even jumped with a small bulldozer that could be used to remove any larger obstacles, such as the steel rods that the Cubans had drove into the airfield in Grenada. Some officers wanted to outfit an old WWII Stuart tank located on post with a bulldozer blade so that we could have firepower too, but it was determined it wouldn't survive the jump.[35] We would run back and forth on Lawson airfield during training, sweating out buckets in the Georgia heat.

Each night my original squad, Wilroy, Beaudette, Walters, Olson and myself, would go out on the town. We usually went to strip clubs, such as Little Richards.[36] I didn't want any more real girlfriends, since they caused me nothing but heartache. The temporary ones that would take one dollar at a time was good enough for me. Hobgood didn't go out with us much anymore, since he was now married. He was such an inexperienced guy, so we figured he needed to catch up on all the loving he could from his new bride. [37]

The "final exam" for our Airfield Seizure training was when we flew to Wright Army Airfield in Savannah and took down the whole airfield in 12 minutes during the daylight practice. We ran out of water during the mission and were all dehydrated, but we continued to train.[38] We then did the same mission, but at night. Afterwards we flew back and were allowed to sleep late, all the way until 0830.

The day after that night mission we learned how to do "fast roping". In 1985 the fast rope technique was classified SECRET, but today that technique is taught to all soldiers and can be seen in many movies. How Fast Roping worked was that each Ranger would grab a thick rope and slide down it, like a fire pole. Each Ranger is not hooked into this rope, like rappelling, but just slid down the rope. Once a Ranger hit the ground he had to quickly move away, since the next Ranger is coming down fast. An entire platoon could fast rope down four ropes on a CH-47, and be on the ground in 30 seconds.

---

[35] I always wondered what that Stuart tank, outfitted with a modern 23mm chain gun, would have looked like.

[36] When I returned to Fort Benning in 2013 all of these strip joints were long gone. Where Little Richards was located now sits a hotel. The Traffic Light Lounge became a small park with flags from all the different branches of the military.

[37] The following print showing Ranger motorcycle jeeps had to be cleared through the 3rd Battalion S-2 and the CSM, for me to release it for sale. Anything the Rangers in the drawing were wearing that was considered SECRET at the time had to be erased from the print

[38] This would be considered a dangerous career ender in today's military, but it was just the way it was back in the Reagan time frame.

198

It was July in Fort Benning, and it was hot! We continued to train, wearing rucksacks, flak jackets, and helmets.[39] We had started dehydrating since the day before, and it was taking its toll. My squad was slowly being whittled away. SP/4 Walters and another squad leader sprained their ankles badly on one of the fast rope runs. On the fourth fast rope run one of the Rangers didn't let go of the rope at the bottom, and instead ran with it, carrying it away from the helicopter. I was the last one out of the helicopter on that run, and one of my squad reached to grab the rope, but it wasn't there anymore. He started to fall out of the helicopter, so Hobgood tried to grab him, but he went out the back, so I grabbed him and we all fell ninety feet to the ground. If one goes, we all go. All I remember was thinking "this is going to suck" and when I hit, it did. I had the wind knocked out of me, and it felt like my lung had collapsed again. Another Ranger ran over to my body and covered me with his so that the dirt flying off the ground from the rotor wash wouldn't sandblast me. I was still trying to get my breath while I looked for the rest of my squad. They were all around me, also not moving. Hobgood was not able to stand, and a new Private

---

[39] This was the days before body armor, so the closest we had to armor was the Kevlar flak jackets.

to our squad, Terry, was not moving at all. It turns out that he was paralyzed. He had come down head first and snapped something in his spine.[40] The seriously injured were taken away in a medevac helicopter. I stayed, since I was able to get up and walk around and not considered seriously injured.

The company commander, Captain Minton, told me to fly back to the Battalion and take his van to the hospital so I could take another Ranger who had wrenched his knee out of place during another Fast Rope run. There were also three heat casualties that needed to go. At the hospital I was X-rayed for any damage, but nothing was broken. They gave me some Motrin and told me to report back to duty.[41] From that day on I hated fast roping, and it was the only thing that really scared me.

The long summer of Jump Clearing training continued. We would jump into Northfield, or McIntyre Air Force Base in South Carolina.[42] We trained while wearing whisper microphones and used suppressed MP5 submachine guns. I didn't like the whisper mikes, which were attached to the headphones we wore. When I used them, I couldn't hear anything except what was on the radio. As we ran down the runway, clearing it, we were ambushed and I never heard the shots due to the noise in my ears. Regimental commander Colonel Wayne Downing watched us clear the field, and then had us do it in a different direction. We ran our ass off all over that field, but we did it successfully each time.

The day after that mission we received a "Bravo" alert. This was a real-world alert, and the Battalion kicked into overdrive. SSG Croft, my roommate and I, were issued M-21 sniper rifles, since we were the only snipers in the platoon. We soon learned that we would be going to Uganda.

In 1971 Apollo Milton Opeto Obote had been overthrown by Idi Amin. When Idi Amin was exiled by Tanzanian forces in 1979, Obote was re-elected as President. Many Ugandans thought that the vote was rigged, and civil war erupted in the African nation. On July 27, 1985 Obote was overthrown by his

---

[40] I thought that Terry had been medically discharged from the Army, but I bumped into him in Germany when I was on a LRSU team. He had recovered, no fully, but enough to be back on active duty. He told me his left side still felt numb most of the time.
[41] The 800 mg Motrin is known as Ranger Candy, since it was considered the cure for all injuries.
[42] Northfield was one of the airstrips used by Colonel Doolittle to train his men on how to take off from aircraft carriers in preparation for the Doolittle Raid in WWII. The marks, showing how long an aircraft carrier was, is still painted on the airfield.

own military, led by the Okello brothers. American, British and French nationals fled the fighting in the capital and headed towards Kenya in a convoy. The Ranger Battalion was alerted when contact was lost with that convoy. The mission was called Operation Castle Hellion. Our Battalion was to seize Entebbe and retrieve the "precious cargo".[43]

We stayed at the Battalion area, waiting for word to go; waiting for President Reagan to unleash his Rangers. However, by the afternoon we were told that contact had been reestablished with the convoy and we would not be going after all. We were all disappointed. With all of our training we knew we could take Entebbe quicker than the Israelis did in 1976, and we wanted to try.

The Battalion suffered another blow in August. The 3rd Ranger Battalion commander, LTC William Ohl, lived up to the Ranger Creed, when he addressed the whole battalion one morning before physical training. The Colonel said that there had been an investigation against him, but the charges proved to be false. He said that he had personal problems, and he would not be able to live up to the standard of the Ranger Creed. He told us that he had filled out his reassignment orders and he would not be able to continue on as the Battalion commander.[44] The Executive Officer, Major John Vines, assumed command until another Battalion commander would be appointed.[45] A lot of the Rangers had tears in their eyes because LTC Ohl had been there with us since the beginning.

Honduras

Though this was a major blow, life continued on in the Battalion. In September the Battalion deployed to Honduras for a "show of force" mission. Communist Nicaraguan rebels had been fighting our Special Forces in El Salvador and Honduras, and a show of force would let Nicaragua know that the border was not an easy target. I had just graduated from jumpmaster school, and it was my first jumpmaster duty with the Battalion. The drop zone was an extremely small one, covered in trees, barbed wire fences, and cattle. The jump

[43] If something had to be removed, whether a hostage, or a nuclear device, it was known as the PC or precious cargo.
[44] We never did know officially why COL Ohl resigned, but rumors surfaced that he had been having an affair with his secretary, and had taken her on official trips using government funds.
[45] In 2007 Lieutenant General John Vines was the commander of the XVIII Airborne Corps and the Multi-National Corps – Iraq.

in was not that bad, but the walk to the assembly area wore me out. The terrain in eastern Honduras is extremely mountainous, and it seemed like I had to walk up a hill to get anywhere. PFC Hobgood and another one of my Rangers went down due to heat casualty in the assembly area. The 3rd Platoon never came into the assembly area on the drop zone, and no one had seen them. However, we continued on with the mission without them.

We were in Honduras to conduct training, but everyone over the rank of SFC had a magazine of live ammunition. We carried C-4 plastic explosives, and the mortars all had live rounds. We conducted some blank firing assaults on "enemy" positions, but the heat was taking its toll. Six NCOs from my platoon went down. The Rangers were giving themselves IVs of lactated ringers solution, sticking needles in their own arms like heroin junkies, or were just drinking their IVs when they ran out of water.

On the second day one of our Honduran guides pointed to a mountain in the distance, and told us that was the way to the coast. He also said that no army had ever gone over that mountain range. That was almost a dare for us, and the Company Commander told use to strip our rucks to the bare minimum. All we needed was food, water and the explosives we all carried.

As we traveled, we refilled our canteens from the mountain streams along the way, purifying them with iodine tablets, then continued to move up, higher and higher. The men were exhausted, but they continued to climb. On one mountain we had the point men hacking with machetes, while the men behind climbed up, hand over hand, on the stumps. We ate the sugar cane on the mountain tops, and would blow holes in the jungle to get resupplied by helicopters. The choppers that brought in food and water would take out our heat casualties. All the while it rained on us, making the jungle very cold. We never really knew where we were, we just knew that we had to keep moving to the ocean.

We lost our lead squad, and there was no radio contact in the thick triple canopy jungle. We continued on the mission without them. SSG Flowers, the 1st Squad Leader, and I were put in the lead and given machetes. We hacked our way through the jungle until we came to a mud trail in the jungle. It was dark, really dark. I felt the trail in the inky blackness, and could tell that jungle boots had walked over the ground.[46]

---

[46] This sounds like a bullshit story, but jungle boots have a specific "waffle pattern" to the soles of the boots and vary from the smooth sandals or sneakers used by the Hondurans. Some of the Rangers wondered if I was part Indian due to my ability to

Chapter 2: RANGER – A Death in the Family

We continued down that trail, sliding down the mud in the dark to a small village. I was able to find a road in the dark that led to the village. We asked a villager how far away was the ocean, and he said it was about an hour away. It was the road to Trujillo, the town where we could be picked up by aircraft. SFC Maddison asked him, "is that an hour away just walking, or carrying pineapples." For some reason I thought this was the funniest thing I had ever heard at the time and busted out laughing.

Trujillo was called a resort town, but it consisted of huts with thatched roofs on the outskirts, and the town made up of cinder block buildings with tin roofs. We marched to the beach, collapsed on the sand and slept until the sun came up. Once it was light enough, we jumped into the ocean and bathed. We had not been able to clean up since we had arrived, a week earlier and we most likely left a "ring" around the ocean.

The kids in the town came out and gave us coconuts, limes and pineapples. We did not have much money, but we paid them with MREs. They loved them. I sent one kid out to bring some cokes, and he brought back a case, and also brought a red headed American Peace Corps girl, Carolyn. She didn't believe the boy when he told her that the beach was covered with Americans with guns. She asked what we were doing there, and I told her we were engineers, sent there to build a road.

While I was talking to her, our lost squad and the lost 3rd Platoon both arrived within a few minutes of each other, so we finally had the whole company together again. The Company Commander told us not to leave the airfield, but on the airfield were two bars and a whorehouse.[47] So we were content with our life. The beers in Honduras are extremely potent, someone told us that was due to them being made with formaldehyde.

While we were at the airport bar Honduran troops stood guard at the doors. That night we staggered back to the beach and slept on the sand again. The ocean looks very different in a country with hardly any power and lights and I imagined this is what it looked like at the beginning of time.

Our Bravo Company was jumping into Honduras that day, and it was determined we would use their aircraft to return home. Unfortunately, one of their C-130s broke down, so we would have only two aircraft to fit the company

---

track. I used to tell people I was from the Galway tribe (Galway is in Ireland) and some believed it.

[47] The most expensive girl was $7.00.

in.[48] As we waited on the side of the runway a civilian rusty DC-3 flew into the airfield. It barely missed the palm trees on the end of the runway, and then it almost ran out of room to land. Hondurans came out the side door of the DC-3, some carrying goats, and one had a crate of chickens. We all wondered how a C-130 would be able to get in there, but more important, how would it get out again.

---

[48] A C-130 could hold 64 paratroopers with all their equipment. A company normally consists of around 150 men, so it would be a tight fit, but not impossible.

Chapter 2: RANGER – A Death in the Family

Our aircraft came diving down to the runway, and just missed the trees. The palm trees swayed in the backwash of the propellers. The mountains around the area were high enough that the aircraft had to do a sharp descent to land. After it had come to a stop, we were told we had to quickly load up, so the next C-130 could come in. We strapped into the seats, and were told to hold on. The pilot stood on the brakes and revved the engine to the maximum. We all said a prayer and crossed our fingers. When the pilot let the brakes go, we jerked forward with a start, did a "wheelie" and zoomed down the runway. As soon as we had airspeed the C-130 went straight up to clear the mountains. Once we finally leveled out, we started breathing again.

We landed at Palmerola Air Base, where the covert war against the Nicaraguan rebels was conducted. We were told not to take any pictures and to not wander away from the plane while it refueled. AC-130 Spectre gunships sat waiting around the airfield. AC-130s were considered TOP SECRET in the mid-1980s and you were not allowed to take pictures of them.

Once we returned to Fort Benning training was conducted at a slower pace for the next few months. My squad shrunk down to three people for one mission due to injuries, and due to those going to Ranger school. However, by November I had one team that was all SP/4s and all were Ranger qualified.[49] We were highly trained and experienced. Wilroy broke up with his girlfriend, so he, Kirby, Walters, another squad leader, SSG Pat___ and I went out a lot to the strip joints.[50] I didn't want to go to any "normal" places, because I just didn't fit in anymore. Though I liked Betsy, I also realized she was too young and wrote her telling her that I couldn't be in love with her anymore. The Rangers had taken the place of everything else for me.

Hobgood was our only "normal" Ranger, and had a good home life. Though their trailer hardly had any furniture, the him and his young wife were really happy together. Whenever I wanted something besides a strip joint experience,

---

[49] Ranger qualified specialists were known as the SP/4 mafia. They didn't have enough power of an NCO, but they were the demi-gods to the privates.

[50] The reason SSG Pat___ is not identified here is not because he was cowardly or incompetent, but the opposite. He was a close friend of mine, but he got into some trouble when he joined a Ranger motorcycle gang known as The Wingmen. After I left the Ranger Battalion he was arrested when he was caught stockpiling arms and ammunition in an off post storage unit. He spent some time in Leavenworth, but he is free at the time of the writing of this book. I concealed his name in case he wants this part of his life to remain hidden.

I would go over to their house and cook up steaks, have some beer, and enjoy their home life.

In November 1985 we had to play OPFOR to the 2nd Ranger Battalion's ARTEP.[51] We knew this would be good, since there is nothing quite like real combat as a Ranger-on-Ranger field mission. We were issued AK-47s, which was a novelty, and we wore BDUs. Every Ranger had two simulators and a smoke grenade.[52] Backhoes came in and dug trenches on a hill for the 2nd Ranger Battalion to attack. At that time Hurricane Juan poured rain all over us for two days, so the area was covered in thick red mud. For three days we manned the trenches, doing patrols and getting a lot of sleep. Finally, Regimental Recon got ambushed by SSG Patton's squad, which let us know they would be coming soon.

The first OP to get hit was manned by SSG Flowers' squad. Flowers didn't fire, but had his men throw their simulators to disorient the attacking Rangers.[53] As the attackers came onto the objective, we threw our smoke grenades. It was dark to begin with, but with the smoke no one could see anything at all, except for flashes in the night from the AK-47s firing. It looked like strobe lights in a disco. The fight in the trenches was furious, and a few Rangers ended up with broken bones. One Ranger was slashed in the arm by an entrenching tool.

My squad had bunkers in the middle of the perimeters and I watched the battle's approach. Doc Cabbarone, the medic, protected the hospital tent, and he put up a hell of a fight, taking out six of the 2nd Battalion Rangers before he went down too. My squad was told to fight and die when overrun, except Hobgood, who was supposed to surrender. The 2nd Battalion Rangers fired him up when he tried to surrender and "killed" him. I took the remainder of my squad and went out into the road, finding 1SG Jones. He took control of us and put us in a hasty ambush. Two 2nd Battalion Rangers came out of the trenches with a red lens flashlight, following some commo wire. When they got close,

---

[51] OPFOR stands for Opposing Forces, and are the "bad guys" in a field problem.
[52] There are two types of training simulators, artillery and hand grenade. The artillery simulators whistle before they explode; the hand grenades do not. They make a huge explosive noise and a gigantic flash, but there is no shrapnel to harm anyone.
[53] In 2010 Richard Flowers was the Brigade Command Sergeant Major of 4th Brigade of the 82nd Airborne Division. He suffered a heart attack after one of his Afghanistan deployments and retired out of the Army shortly afterwards.

we fired up all the rest of our ammunition. The AK-47 blanks were as loud as the live ammunition. The 1SG saved his ammo and ambushed a company that was down the road in reserve.

With all the real casualties the "real world" ambulances were called in. I got into one of the medic trucks and disguised myself as a medic, folding my boonie hat up so it looked like a patrol cap. I was able to get back on the objective, let the CO know what happened, and then cut loose some of the POWs there. There were about thirty real world casualties due to the intense fighting. Hobgood was taken prisoner after all and was able to escape later that night and link up with us at the rally point.

Ranger Hobgood

Towards the end of November, we did platoon live fire exercises. November 22nd started like any typical day in the Ranger Battalion. First call in the NCO barracks was 0430. In the enlisted barracks the CQ would go through the building shouting "first call" at 0515, but the NCOs got up on their own. I stumbled down the hallway in my underwear, rubbing sleep out of my eyes. I shaved in the 2nd floor bathroom of the old WWII wooden barracks that I called home. My roommate was SSG Croft, and we lived in an extremely small room. No air conditioning in the summer, except an open window. If I extended my hands either way, I could touch the bunk beds or the wall.

The morning had a chill, and I could tell that winter was coming. I walked over to the platoon barracks, a more modern cinderblock building from the time of the Vietnam War. SGT Sheehan, my A Team Leader, had his Team up, but SP/4 Sharp, the other Team Leader was not there, and his men were still sleeping. I banged on the doors and told the B Team to get up. As I walked back down the hall, I saw Sharp finally arriving. I figured I'd let it ride for now, and told him to make sure there were no sick calls today. When I said that, Hobgood walked in the door and said "Ooh, I'm going to be sick!" I told him to cut the crap and go find out if there was going to be a company formation or not.

The company ate breakfast in the mess hall, drew weapons, and then was trucked out to Cactus-Woodson Range. We formed a "horseshoe" around LT Varney, our Platoon Leader, and he briefed us about the range. The range was pretty rough looking, with chest high grass, scattered pine trees and scrub oaks. Each squad grabbed their cans of ammunition from the ammo point, and then moved to the assembly area. The ammo cans were dumped in a pile, and then each of the Rangers moved to their positions in the perimeter, facing out. SFC

Maddison called the Squad Leaders to the center of the perimeter and told the leaders to have their men camouflage. From this point they would remain tactical. This meant that talking was non-existent, or extremely low, and no Ranger would be smoking or making any noise.

The live fire attack onto the platoon objective was supposed to happen at 11:30, but when that time arrived and nothing happened, the Rangers began to relax. Something was wrong, and everyone might as well conserve their energy. SP/4 Kittle, the SAW Gunner in Sheehan's Team commented that the platoon in front of us hadn't even gone yet. Sheehan sat down beside me and told me that the range was a "bad luck" range. I asked him what he meant.

He told me "Remember the last time we were here? We had them damn fast movers buzz us. Hell, we nearly shot them down. It took us two hours to reopen the range." Kittle spoke up, "First platoon is finally moving."

I told the squad to wake up; we would be moving as soon as the firing stops. I told Sharp that he would be moving on a 134-degree azimuth until he hit a numbered panel on the range. Sharp made a smart-ass comment that we could do it in our sleep by now. He had joined the squad recently and I was still trying to figure out if he would fit in.

First Platoon began firing on the objective, but then the noise abruptly stopped. We all looked around. Something was wrong. The word was passed down that there was a "check fire" on the range. Sheehan looked up and said "told you, bad luck range."

What had happened was that when 1st Platoon had begun to fire, some soldiers from the 197th Infantry, the non-Ranger unit on post, took down the barriers and started doing a police call on the range. The 1st Platoon couldn't see anything down range, since the grass was so high, and shot up the truck of the 197th "legs". Some of the bullets punched holes through their pants too. Since there was nothing to do until it was cleared up, the 2nd Platoon was told to eat their MREs.

The check fire was lifted, and we heard 1st Platoon begin to fire again. For a second time it was cut short. SFC Maddison called in the Squad Leaders and told us that two civilian hunters had wandered onto the range and we would begin when they cleared them out.

After Grenada I had bought a Colt 3x scope for my M-16 rifle. Whenever we did a live fire, I would take the scope out of a padded case and attach it to the carrying handle of the rifle. SFC Maddison saw the scope and asked if it was effective or not. I told him that I was able to acquire targets quicker with it, and hit them faster. He told me that he wouldn't want to have one, since it would probably break. I showed him that you could still use the iron sights with the scope mounted by looking underneath the scope. He thought it might be OK then.

First Platoon moved out again, and this time they were able to finish their attack on the objective. Our platoon started moving, each Fire Team getting into a wedge. My squad was the lead squad in the platoon, with Sharp as the point man. We moved through the woods and out into the open field, when an ear-splitting roar made everyone jump to the ground. Two F-16s had buzzed the range and then were circling back. Another check fire was called, while the platoon laid back and watched the air show.

I looked at Sheehan and I told him, "Don't even say it... I know. Bad luck range."

Once we received the all clear, we began moving forward into the open field. As we approached some dirt mounds a pneumatic machine gun opened fire on the platoon.[54] I moved to Sharp and told him to have his team move up to the dirt mound on the right, while Sheehan's Team was ordered to move to the dirt mound to the left. I crawled back to LT Varney and told him "We got contact, 200 meters to the front." The squad had already begun returning fire and the LT had to shout over the noise. "Hold your men in place, and set up a base of fire!"

I crawled back to the squad and told them to continue to fire at the pop-up silhouettes, but they needed to save some ammunition for the next objective. The rest of the platoon deployed to the left to flank the "enemy". I could see several staff officers and CPT Minton on the mound behind us, watching the attack. The M249 SAW was deafening, and I knew I should wear earplugs, but I didn't want to miss something that might be said. As new silhouettes popped up, I fired tracers into them, signaling the squad where to fire.

I saw a yellow smoke grenade billowing over to the left, signaling that the rest of the platoon was in position, and was beginning their attack. I yelled "shift fire! Shift fire!" and the squad began firing over to the right, away from the platoon. In a real situation this fire would be directed towards any enemy trying

---

[54] The pneumatic gun uses gas and a spark plug to fire. It has the same noise as a car backfiring.

Triple Canopy

to get away or trying to reinforce the objective. Sharp yelled back to me "I see them on the objective".

I ordered "Cease Fire! Cease Fire! Cease Fire!" and ordered the Team Leaders to make sure all the weapons were on safe. I also called for an ACE report.[55] SFC Maddison had been with the M-60 machineguns, and ordered my squad to link up with the platoon. We ran to the platoon, but they were in the wood line, and not on the range. The LT told me that the Company Commander had lost contact with the control tower on the range, and we were to hold there until the CO told us to move out.

I moved among the squad, critiquing what they had done. Hobgood was in Sharp's Team and I told him that he needed to stay down, and quit popping up on one knee. If there were real bullets being fired back, he would be dead. Hobgood told me that if it was real, he would be a lot lower. The grass was too high, so he said he had to rise up to see the targets. If he lay down, he couldn't see anything at all. I joked to Hobgood that if he kept getting up, I would put him in the Weapons Squad. They never got down either. He fired back that I would never get rid of him, and he would be in the squad until he died.

The platoon was ordered to move onto the range again. The grass was even higher on this field. I could barely make out some rusted out hulks a few hundred meters away. The pneumatic machine gun opened up, and all of us hit the dirt. I ordered Sharp to put his men on line with Sheehan's Team, but no one was firing yet. They couldn't see any targets at all. I crawled back to the LT and told him that I thought the targets were 100 meters to our front. He told me to keep my squad on line, and then lay down a base of fire like before. The platoon would move up to the left again.

As I crawled forward the silhouettes popped up just in front of the squad. The two teams began firing point blank into the targets. I yelled back to the LT that the objective was right to our front. Sheehan's men were too far back, and I was worried that his Team might not see the other Rangers to their front. I ordered Sheehan to have his team move on line with Sharp's. Both Teams were on their knees, or crouched down, so that they could see above the grass. I kept yelling that they needed to stay down, but if they were down, they couldn't see the targets. I then compromised and yelled out to kneel when they fired, but then get down in between shots.

---

[55] ACE stood for ammo, casualties and equipment. The Report would sound like "3, 0, up!", which would mean the soldier had three full magazines of ammunition, no casualties and no equipment was lost.

210

Chapter 2: RANGER – A Death in the Family

I look to Sharp's Team, and I saw Sharp kneeling over Walters and his M249 SAW. They were trying to get it to work. I yelled over to Sharp to quit messing with the SAW and watch his Team. The M-60 machinegun opened up, so I don't think he heard me. Now that the M-60s were firing I figured the platoon must be linked up. There was no way to tell though due to the tall grass.

From off to the left I heard someone shout "cease fire!" over and over. Every Ranger echoed this command when they heard it. This was a command that was only used if there was an emergency, or if the mission was over. The attack had just started, so something must have gone wrong. I stood up to see what had happened. The spectator officers behind me yelled at me to stay down.

I asked my Squad what happened. Did anyone see anything? Kittle told me that someone was moving out there to our front.

"Where?" I asked, looking over the objective area. The officers still yelled to get down, but most of them were looking too. "There!" Kittle pointed, "Right in front of me!"

To the front was a large cement block, and beside it I could see someone kneeling. The body fell forward, slumping to the ground. I ran forward and one of the officers yelled to stay where I was. I ignored him and kept running. As I got closer, I could see what looked like a person. The officers were still yelling, but I sprinted forward. I saw the body and figured it had to be dummy, there was no reason for someone to be this far forward of the line. I had been in live fires before where dummies were used, dressed up in old uniforms.

It wasn't a dummy, it was a Ranger, and there was blood splashed all over the cement block, the Ranger, and the ground. I figured it had to have been one of the damn staff officers, getting in the way and getting shot. I stopped about ten feet away and dropped my rifle. I knelt down beside the body, and then crossed myself. This was odd, because I was not even Catholic, and I had never done this before in my life.

The blood covered Ranger was still holding an M-16. It couldn't be one of the officers then. Blood was pouring out of his mouth, nose and eyes in a steady stream. He had on glasses, and there was a bullet hole in one of the lenses. I pulled the M16 out of his hands and rolled him onto the ground. The back of his head was gone, and there were pieces of his skull on his shoulder. I took a few of the pieces of his skull, and fit them back into his head, and held them in place. Blood poured over my hands and into my lap.

211

Triple Canopy

I could hear someone yelling behind me "It's only a dummy! It's not real!" A medic ran up and tried to push me out of the way, but if I let go the pieces of skull would fall out. I yelled at the medic and told him that I was holding his head together. The medic, Cabral, looked at me in an odd way. CPT Minton ran up and stood there with his mouth open. I held out one bloody hand to him, and said "It's no fucking dummy. I don't know who it is. He doesn't have a face."

CPT Minton told me to go back to my squad and have the men clear their weapons. I wanted to tell him that I had to stay. I was the only thing keeping the skull together. The medic grabbed my face and turned it to him with both hands. He said, in a quiet voice, "you don't need to hold his head together. I have it. Go to your men."

I walked over to my M16 laying on the ground, and picked it up. I held it over my head, and then yelled "Lock and clear". As I pulled out my magazine Kittle asked me who it was. I told him "I don't know." I then pointed to my face and said choked out, "It's gone."

In the back of the Ranger Handbook were the Catholic last rites. Most of my squad was Catholic, so I whispered the prayer. "Hail Mary, full of Grace".

SFC Maddison yelled out for everyone to form up. I told the Team Leaders to account for all their people. Sharp looked at me, and then said "Hobgood, he's gone!"

I felt like someone had just punched me in my stomach. Of all the Rangers, it just had to be him. He was one of my original Team. His wife, Laura, was only 19 years old, and she was 7 months pregnant with their first child. I walked up to SFC Maddison and said "It's Hobgood! It's Hobgood!" He stared at me, and then said "I know. Don't lose it!" I turned around, remembering the last rites and whispered, "Pray for us at the hour of our death."

I felt someone put a hand on my shoulder, and I turned. It was SSG Pat___. "It's Hobgood", I told him. "I can't remember the last rites".

SFC Maddison yelled "Fall In!" He turned over the platoon to another sergeant, and then had us march back to the control tower. It was a fast march, as if we all wanted to get away from the area. "Slow it down!" the sergeant yelled. Someone else yelled back "They are all in shock, let them wear themselves out."

Chapter 2: RANGER – A Death in the Family

As we walked along, I heard two of the older sergeants talking. One asked "Did you ever have anyone killed?" The other sergeant answered that he had, back when he was in 1st Battalion. "You remember Lau?" he said. "Yeah," the other sergeant said, gazing off at the trees, "Lau."

We were winded when we got to the tower. We were told to unload all our magazines and collect up the ammunition for turn in. A lone helicopter landed out where Hobgood was, blowing red smoke from a signal grenade all around. The helicopter was only there for a few minutes, and then it took off again. All of us were silent, following it with our eyes. SFC Maddison walked up from the range and told all the squad leaders to come to him.

We gathered around, and he said "Listen up! I want you to take your men off somewhere, and train them! Train them hard! I don't care what you do, just work them. Get their minds off of this!"

I walked away, wondering how we could take our minds off of this. I moved the squad to some trees and then we sat down. I told them "You guys were slow taking apart your weapons after the last road march. I want to see what you were doing wrong. You got two minutes to disassemble your weapon." I glanced at my watch, "Go!"

The men stripped down their weapons, mechanically, all within a minute. Kittle looked up, "Is he alive?" I told him that I didn't know. It was pretty hard to tell. I said, "The helicopter didn't stay out there very long, so maybe he is still alive."

I looked at the squad, "You have two minutes to put your weapon back together again. Go!"

When they had all finished, we were told to reform the platoon. A military police car had arrived, with a civilian car behind it. The medic, Cabbarone, was with the MP. The front of his shirt and pants were covered in blood. SFC Maddison said "We're going to ground our gear, right here", pointing to a spot on the ground. "Squad Leaders, I want you to take your men over there, and do remedial training. The CID wants to question some of you.[56] They will call you up, one at a time." He pointed to me, "First they want to see O'Kelley, Kittle and Doc".

---

[56] CID stands for Criminal Investigation Division, and are the Military Police Detectives of the Army.

I walked over to the MP who was questioning Doc. I could hear Cabral as he told them what happened, "…and we pulled two bullets right off of him, both were 5.56 rounds. One went in the back of his head, and came out his cheek. There was a big hole in his neck too."

The MP glared up at me and asked "Are you the man's platoon sergeant?" I told him no, I was his Squad Leader." He asked me his name and serial number, and I told him. I had the serial numbers for all my squad memorized. "Russell D. Hobgood." The MP then asked me what I saw, and I told him. When he was finished, I walked over the CO.

"Sir?" I asked, "What about Hobgood's wife? Who's going to tell her?"

He just looked at me and said, "It's being covered, Sergeant." I walked back to my squad. They all wanted to know if Hobgood was OK. I told them that I didn't know. I told them what I heard Doc say to the MP. I told them that the Doc said he still had a pulse when he was put in the chopper. I figured he would come out with some brain damage, he had to. His head had a huge hole in it. I tried to fix it. I hoped it worked.

I then got angry, "Why was he out there?! Kittle, you saw him first, what the heck was he doing?"

Kittle looked back, "I don't know, SGT. I was shooting at the targets I saw someone move in front of me. I saw him try to stand up, and then he fell down. He got up one more time, on his knees, and then he fell over."

I wondered what he felt when the bullets hit him. Did he know it? What was he thinking? I asked the squad, "Did anyone see him crawl out there?", but they were all quiet.

Kittle said "He used his sniper training really good. He got out there with no one seeing him." I heard someone in the squad mutter, "I bet Toy's gun killed him." I yelled back, "At ease that noise! We ain't blaming anyone!"

Toy was in the weapons squad and had been on the M-60 machinegun. I told the squad that Doc had said that they pulled 5.56mm bullets from Hobgood, so Toy was clear.

Kittle realized what I just told the squad and said "I bet it was my SAW. He was right in front of me." I fired back, "I said at ease that shit!" Sharp spoke

214

up, "Yeah, no one can be blamed. There were too many of us. Think of all the bullets flying down range."

The squad all started talking. "I wonder what he was thinking." Someone else said, "Laura gets 30 thousand dollars now." I told him that the money wouldn't last a year. I told the squad to stay where they were, and then walked over to the 3rd Squad. I signaled SSG Pat___ away from his squad and asked him if he heard anything new. He told me he hadn't heard anything. I told him that he may still be alive, "He still had a pulse when he went into the medevac chopper. I've seen guys take a lot of stuff, and still walk away."

SSG Van Heuten walked over and asked if we had a smoke. He was the Weapons Squad leader. I told him "Don't tell Toy, but my guys think his gun did it." Van Heuten told me, "The Doc says it was a 5.56. I don't want to blame anybody." Doc Cabral walked over, looking tired. "Got a smoke?" he asked.

"What did you see?" I asked him. He looked around and said "I'm not supposed to talk about it." He then ignored the order to not talk and said, "When we picked him up, the bullets were falling off of him." We all heard the word to reform the platoon because the CO wanted to talk to us.

"How is he, sir?" someone asked. "I don't know yet," he said, "he was still alive on the medevac. I don't know why he was out there, or what went wrong, but this isn't going to stop us! If I had my way, I'd run it again right now. This is a dangerous job. That's why you get paid 83 bucks more each month. I doubt if we will be able to run it again today, but we'll definitely run it again. Right now, we are waiting for an Army investigation team out of Fort Rucker. As soon as they get here, we'll proceed on. Until then, just go back to what you were doing."

The rest of the company, those who were not involved in the shooting, was allowed to go back to the barracks. As they filed onto the busses, we looked at each other with hollow eyes.

The sun was beginning to set at the range. Chow had been mermited out to the range for the platoon.[57] While we waited, we all questioned each other about what we saw. The Battalion Commander, LTC Hale, drove up in his jeep and

---

[57] Mermites were green metal containers that kept food hot or cold.

called all of us around him. He stood up in his jeep and told us that Hobgood was pronounced dead at Martin Army Hospital.

The blow, like a shock wave, went through our ranks. Up until that moment we all had hopes that Hobgood would come out alive, like we did when we were all hurt during the fast rope training. Some of the Rangers stared at the colonel; others just looked at the ground.

"I know what a unit goes through when this happens," he said. "I've seen some units fall apart and I've seen other units get tougher. This event should not break you down. You're Rangers!"

Two vehicles drove onto the range with their police lights flashing. It was the investigators from Fort Rucker. We were told who would stay to be questioned, and who would be allowed to go back to the barracks. All of our weapons would remain on the range. The investigation team also wanted to go down range and see where it happened. For some reason I was not one of those chosen to stay and talk to the investigators, so I got on the bus to go back to the barracks.

Someone asked, "What did they want the weapons for?" Another answered, "I bet the bastards are going to match the rifling to the bullets, just so they can blame someone." Someone asked me what did I think happened, but I just told them that I didn't want to talk about it. I leaned back in the seat and closed my eyes.

When I got off the bus someone told me that the chaplain wanted to talk to me. I waved them off, and then went to the orderly room. I went to the supply room; however, the Supply sergeant wasn't there, but a SP/4 was. I told him to get some inventory sheets and some boxes and meet me in the barracks. Since I was his Squad Leader, I had to inventory Hobgood's wall locker.

The chaplain walked into the orderly room, wearing his dress uniform. "SGT O'Kelley", he said, "Mrs. Hobgood wanted you to call her." I asked him how she was. He said that Toy's wife was with her, and that she was "OK". I told him that I would call her after I did the inventory.

The supply SP/4 was in the barracks. "Cut it" I told him. The bolt cutters snapped Hobgood's lock off. I saw Hobgood's wallet and car keys on the top

216

shelf. I pocketed them, and told the SP/4 that I would give them to Hobgood's wife later that night.

The squad had brought in some beer. A lot of it. Cases were stacked in the hallway. I had already drunk three beers, and decided I had enough courage in me. I dialed Laura Hobgood's number, but no one answered the phone. I dialed Toy's number, and a girl answered the phone. I asked if Laura was there, and she sounded relieved that it was me.

"I'm glad it's you", she said, "Laura's not here. She took off to the hospital. We couldn't stop her. The MPs are following her, but they can't stop her either. She's outrunning them!"

I laughed, "How can a pregnant girl outrun the MPs?"

"I don't know," she said, "but she's doing it!"

I told her that I would be bringing Hobgood's truck over there later that night. LT Varney told me I had to go to main post first, and see the investigation team. The LT said he'd drive me over, since I already had some beers in me. After we got in the car, I pulled a beer out of my cargo pocket, and handed it to him.

"Here, sir", I told him, "We've been baptized."

I walked into the building behind the LT. The room looked like a doctor's office. Some Rangers were sitting there, waiting to be called in. I sat down beside Walters and questioned him on what the investigation team was asking. Walters didn't know, since he hadn't been in yet. He asked me how Laura was taking it. I told him that she was running away from the MPs, on foot, and beating them to the hospital. I told him that I would be going over there after all this was over.

Walters said, "I'm going with you." Walters was Hobgood's best friend.

A man in civilian clothes motioned me into the room and closed the door behind me. He sat down behind a typewriter and began banging out my name, serial number and other bits of information. As he typed I looked around the room. There was a table beside the typewriter desk, and on the table were several bags of brass shells from the range. In small specimen bottles were chunks of hair and blood.

The CID agent read me my rights, and then started questioning me about what happened on the range. I was getting tired of telling the same story again and again, but I knew that this one counted. I had to stop occasionally to explain to the CID soldier how to maneuver, or how an infantry platoon formed for an attack. When I was finished the CID agent read over the report, then looked up.

"You say that Hobgood was laying beside a cement block?"

I told him "Yeah, a big concrete thing."

He looked like he had caught me telling a lie, "It was a metal structure. Do you want to change the report?"

I didn't understand what it mattered, but I told him "No, I saw a concrete block."

He then asked me if Hobgood had any troubles at home. "Was there any reason he may want to kill himself?"

I looked shocked. "Hell no! His marriage was a perfect one!"

The CID agent then asked, "Was there anybody in the unit who would've wanted to kill him?"

I snapped back, "NO! Rangers don't kill Rangers!"

"Who do you think killed him?" the CID agent asked.

I must have stared at him like I wanted to kill him. "I said, Rangers don't kill Rangers!"

Hobgood's truck was an old pickup that began to vibrate when it hit 30 mph. Walters drove, while I tried to get the defroster to work. It was a piece of junk. We drove into the hospital parking lot, parked the truck, then asked the night nurse where Laura Hobgood might be. I had to describe her to the nurse.

Chapter 2: RANGER – A Death in the Family

"She's petite, dark hair, thin face, but she's also eight months pregnant."

The nurse told me that a CPT had driven her home. We figured it must have been the Company Commander. We both ran back to the parking lot, and drove to Hobgood's house. The house was dark, but there was a note taped to the door:

Tiger –

    Don't get excited. I'm over at Nelly's.

        I love you,

            Laura

Tiger Lilly was Laura's nickname for Hobgood.

Unfortunately, neither one of us knew where Toy and his wife lived, so we drove around the neighborhood, looking for signs of a Ranger. Walters said "There she is!"

Laura was running across the yard of one of the houses, running towards the truck. I didn't want to be there. I lived in a world of men. It was a world where women were not invited, and were only there for occasional diversions. I didn't know how to handle women. Luckily, I had changed my shirt back at the barracks. The one I had been wearing had been covered in Hobgood's blood.

Laura stood in front of the truck, her hair was messed up and her eyes were red from crying. She turned to Nelly Toy, "I told you he'd come back! He's always late!" She looked at me, then the truck, then took off running back to the house.

I introduced myself to Nelly, and then asked how Laura had been doing. "Not too good," she answered, "She keeps saying that the Army made a mistake. She's been spit shining his jungle boots for the inspection Monday."

"Great", I replied. I went into the house and saw Laura sitting on the couch, shining a boot. Toy sat in the corner, looking helpless.

He asked me if I wanted a beer.

"Hell, yes!" I opened the beer and downed half of it in one swallow. Walters came in, looking as helpless as Toy. I figured we all looked that way. No one trained you how to deal with this.

Laura stood up and said "I want to go for a walk." Nelly started to get up to go with her, but Walters told her "I'll go with her."

I asked Nelly what had happened so far. She told me "She wants to see Russell's body, so she can prove he's dead. That's why she went running to the hospital. The MPs finally tackled her, and got her in the car. They almost weren't able to do it though. The hospital gave her some sleeping pills. CPT Minton brought us home."

I asked her if Laura had said anything to her yet. Nelly went on, "She's been off and on about accepting it. Sometimes she thinks that he's alive, and then she'll look at me and say, he's never coming home again. She almost had a fit when she found out. She's an epileptic."

I had known that, but forgot about it with everything else that had happened. Nelly continued, "She's also been talking about killing herself, after the baby's born."

"She's a big time Catholic, right?" I asked.

"She's Catholic", Nelly answered, "but not as religious as Russell. When she called me on the phone, I had thought Will had died. Forgive me, but I was relieved when she said it was Russell."

Laura walked in, carrying a cat. "We found Garfield", she said, "If he stays out all night, he becomes a bad cat." She laid the cat down and then said "I have to get cat food" and headed for the door again.

"I got this one" I said, and followed her out the door. She walked down the road, with just a few streetlights lighting the way. I caught up to her. "I've been getting his uniform ready," she said weakly. I didn't know if she had been getting it ready to bury him in, or if she still thought he was coming back home. "It's hanging up," she said. "I also put a note on the door so he'll know where I am."

We walked into her house. I could smell the cat box odor. "I made some Kool-Aid for him" she said. "I need to get his PT uniform for him. He'll need it tomorrow."

220

Chapter 2: RANGER – A Death in the Family

I told her, "Don't worry about it. Tomorrow's Saturday. You know he won't be doing PT tomorrow."

She stopped and looked at me, "Oh yeah." She began crying, "He's gone, isn't he?"

"Yeah," I told her, "He's gone."

She quickly ran by me and went out the door. I caught up to her and grabbed her arm. "You've got to have that kid!" I told her, "You've got to now!"

"Oh, I will," she said calmly, "I'm going to call him Russell. He always hated his first name, but I think he would like that." She looked up again, "I think I'll join the Army. Maybe as soon as the baby is old enough. I think you can have a baby and be in the Army too."

"Sure" I told her, "Sure you can." We walked back to Toy's house. I told him that I needed a lift back to the Battalion. I saw Nelly holding onto Laura. "You OK here?" I asked her.

"Don't worry" Nelly said, and held onto Laura like a protective mother. I think Nelly was 19 years old.

When I returned to the Battalion, I walked over to the payphones outside of the mess hall. I called my dad. He had been in the Army for almost 30 years. He went to Vietnam twice with the Special Forces. He had friends who had been killed. I too had known men who were killed, but none had been friends.

Sergeants are not supposed to become too close to those they command, but we formed this Ranger Battalion from scratch. We all had been through the same trials. We were all young, under 25, and most of us were single. We had been close. There was no way around that. The Battalion had become our family. The Rangers were our brothers. The Ranger Creed states, "Never leave a fallen comrade". How could we leave Hobgood?

I talked to my dad, not really knowing what to say. I told him of the training, and the range, and other mundane things. I talked of everything, except what I had called him for. I finally blurted it out, and I began crying. I tried to describe what had happened, but all that came out was crying. I don't even remember what he said. I just needed to get it out.

Triple Canopy

That night I lay in the bunk. SSG Croft was not there. He was at BNCOC.[58] I lay alone in the room, staring at the dark ceiling. I thought of the men that I had known who had been killed in the military. I thought of the enemy that I had killed on Grenada. I then remembered trying to put Hobgood's skull back together. Maybe, if I had held on longer, he may have lived.

I cried again in the dark.

The squad had to be at main post at 0700 for further questioning. Each one of us was brought into the interrogation room, one at a time. While we waited, we drank up the coffee and cocoa in the kitchen area. CPT Minton was also there, and let Toy go first. He knew that Toy had been up all night with Laura.

Walters said to the CO, "We want to go wherever Hobgood's body goes, so we can be an honor guard, or whatever." CPT Minton told him, "Don't worry about that."

I told the CPT, "He's been with us since the beginning. Remember how you guys put ashes in his low quarters during 1SG Jones's inspection?" We all laughed.

"Remember Walters' stag party?" someone else said, "We all went to New Orleans. We had a great time. It was so good that Walters didn't get married!"

Walters spoke up, "Yeah, Hobgood pulled my CQ that weekend so that I could go."

Kittle said "Remember that time you got him a whore?"

CPT Minton looked surprised, "A whore?" he asked.

"Yes, sir" Kittle replied, "We all knew Hobgood was a virgin, so we bought him this hooker for $30. Hobgood didn't know that she was a hooker, so she takes him home, and you know that he backed down? He said that he had a girl back home waiting for him."

"Yeah," Walters muttered, "I didn't even get my $30 back."

---

[58] BNCOC stands for Basic NCO Course.

Chapter 2: RANGER – A Death in the Family

CPT Minton grinned, "With friends like you, he didn't need many enemies."

Walters grinned and looked around, "Remember when 2nd Batt captured him? He escaped by taking one of their own weapons, and shooting up their formation. He said SGT O'Kelley told him to never get captured, so he wasn't going to stay captured!"

We all laughed.

Sharp spoke up, "He made the news last night."

"What did it say?" Kittle asked.

"Not much," Sharp replied, "just that he got killed on the range. He'll probably make the paper today.

Walters looked up, "It shouldn't have been him. One of us should've gone instead. Stupid dumb ass."

When I entered the interrogation room, I saw that there was a long table with seven officers seated behind it. I stood at attention until a major told me to have a seat. An officer read me my rights again, and then had me swear an oath that I was telling the truth. Another officer turned on a tape recorder and asked me what had happened on that range. For what seemed the hundredth time, I told the story again.

On a wall there was a drawing of how the squad had been lined up on the range, and the route that Hobgood had taken to where he had been killed. There were also maps of the range and aerial photographs spread out on the table.

The officers asked me if there had been any rehearsals prior to going to the range. I told them that we had rehearsed two days prior. I told them that Hobgood wasn't there during the rehearsals. He had been training on the Dragon missile. The officers then asked me if there had been Safetys on the range, or if we had been given a safety briefing before going onto the range.[59] I

---

[59]   A Safety is the Safety Officer, and ensures that the range is being safe. A Safety is mainly used on a static range, to observe what is going on.

told them that you couldn't have Safetys on a live fire. I did tell them that there were plenty of officer "observers" behind the squad.

One of the officers looked at me, and asked me "Why do you think he did it? Why did he crawl out in front of the squad?"

I looked back, "If I knew that, sir, we wouldn't be here. I figure he was doing what he was taught. He had been in the open, so he crawled to a better covered position. The grass was so high that he didn't know the guns were behind him. He was a good troop, and he was always really gung-ho. He didn't have a death wish. He had more to live for than all of us."

During the previous night Laura had escaped from Toy and his wife while they were sleeping. She had gone back to her house and had set fire to a wooden table that Hobgood had been making. Nelly threw water on it to put it out, but now the whole house smelled of burnt wood. After that Laura was taken to the hospital, for observation.

CPT Minton put everyone who had been on the original Team on SD, or Special Duty, until the funeral. We were to stay with Laura, if she wanted us to. We were also to take care of her, and not let her try to harm herself anymore.

Walters and I drove my truck to the hospital. I was expecting to find her still crying, and looking crazed. Instead, I walked in and found her and Nelly sitting in the hospital bed, giggling like they were at a slumber party. The 3rd Platoon Leader was there, LT Preysler, and the girls were having fun making him blush. They kept calling him "Elvis" due to his name.

The LT went out into the hallway with me, and I asked him what Laura's condition was. He told me "She hasn't been talking like he is alive anymore, so I think this is sinking in. I figure that we ought to go back to her own house. All of the relatives will start arriving soon, and they won't know where to go."

He also told me that the doctors had put her on some tranquilizers, and she seemed better now. We walked back to the maternity ward, hearing the girls laughing. When we looked in, they had the hospital bed folded all the way up, and Walters was squished in the middle.

Chapter 2: RANGER – A Death in the Family

After Laura was released from the hospital, we all drove back to her home. One of the Rangers had bought some hamburger meat at the commissary, while Toy brought a case of beer. We drank while I grilled the hamburgers. Neighbors dropped by, delivering flowers, and telling Laura how sorry they were for her. Hobgood's mom called while we were cooking, and Laura took the phone into another room to talk.

The family relationship was strained, and Laura didn't get along very well with Hobgood's stepmother, father or brother. She didn't mind his mother though.

Later that night we planned our defense. We needed to make sure that Laura did not go running out into the night again, or set anything on fire. I told Nelly that I would sleep on the floor. There was a fold out couch, but I figured that if I slept in the doorway, I would then wake up if Laura tried to run for it again. Wilroy slept in a chair.

The next day was an important one. The Battalion chaplain would be picking her up, and they would go see Hobgood's body. The funeral home had to have time to fix it up first. I didn't know how they would be able to fix up the body. The Kevlar helmet, which would stop a bullet, had worked, but it worked the wrong way. The bullet had gone under the lip of the helmet and bounced around on the inside. I couldn't picture what Hobgood would look like now, only what he looked like on the range with blood pouring out of him.

Nelly came back to the kitchen and told me that she had drugged her kool-aid with sleeping pills. She then told me, "CPT Minton wants Laura to go back to her own house. I don't think it's a good idea. There's too many memories there, and she will be all alone with those memories." She looked at me, "She can't sleep in that bed. They made love in that bed the day before he was killed. He had waited up for her to return from the Ranger Wives meeting. It's like he knew."

I shook my head, "No one ever knows. Ever."

"Get up!" Wilroy yelled, "She's gone again!"

I got up startled, from the floor. "Jesus Christ! It's 2 a.m.! Toy you check the house, me and Wilroy will cruise the neighborhood!"

225

Walters ran to the front door, and then turned and looked at me. Laura was standing in the middle of the yard. She turned her head to us slowly, and then collapsed to the ground.

"This is just bloody great!" I said, and told the other Rangers to follow me. We half dragged; half carried Laura back to the couch. We dropped her a few times, but Toy carried a pillow in front of her as we moved. Every time we dropped her, Laura's head would hit that pillow and not the ground.

"Damn! She's heavy!" Wilroy grunted. "I can think of better things to do on a Saturday night than dragging around a fat lady!"

Once we got her back to sleep, we all stayed up, waiting for it to happen again. All of us knew it was only a matter of time before Laura would head out again, trying to find her husband.

We slept through the morning, exhausted from the night's vigil. Laura and Nelly had gone with LT Preysler to see the Chaplain. Laura didn't remember taking off the night before, and apologized to everyone there. The Chaplain talked to Laura about the funeral. I think Laura finally realized that Hobgood was not returning, and had accepted it. She didn't want to go to a funeral though. She wanted to bury him somewhere in the woods, where no one would find him.

Walters didn't return that afternoon, and no one knew where he had gone. He probably didn't want to return to his best friend's house anymore. Though we were SD, we were told that we had to report to SFC Maddison on Monday morning. We spent the rest of the afternoon watching videotapes of movies that we had seen hundreds of times before.

I told the other Rangers that I was tired of just sitting on my butt. Wilroy and I would go and get some more beer, and something to make for dinner. Toy told us to get some cokes too, since the relatives were coming.

"How many?" Wilroy asked.

Toy counted them off on his fingers, "Hobgood's real mom, and his sister, his grandmother, his dad, and his dad's wife, and both of his brothers. The one in the Army is flying in from Germany. The one in the Air Force is coming from England."

Chapter 2: RANGER – A Death in the Family

Wilroy wondered out loud, "Where the heck are we going to put them all?"

I told them we would have to put them in Hobgood's house. This meant that it needed to be cleaned up. We now had a new mission that needed to be accomplished before we got the beer.

No one had been in the house since Laura tried to set the furniture on fire. The smell from the fire was strong, but the smell from the cat box was stronger. The door had also been smashed in when Laura snuck in to burn the table. I figured that their landlord would freak out when he saw it.

In the middle of the living room was charred wood, surrounded by a puddle of water. Laura had used heat-tabs to set the wood on fire.[60] When they first got married Hobgood and Laura couldn't afford any furniture. All they had was a television and a bed. Hobgood began getting pieces of wood he had found at a construction site and making his own furniture with it.

I looked at the pictures on the bedroom wall. There were the standard soldier photographs of Hobgood in basic training, and Airborne School. There was also a picture of a long haired Hobgood and Laura, taken at their High School prom. Beside that was a picture of their wedding. Stuck in the corner of one picture was a recent photo, taken when we were playing OPFOR against the 2nd Rangers. In the picture Walters and Hobgood had AK-47s, and tried to look mean for the camera. However, Hobgood had busted out in a boyish grin.

I couldn't stay in the house. I told Wilroy and Toy to finish up, and I went to the commissary to buy dinner for Hobgood's family.

When I returned to Hobgood's house, Laura was in the living room with an older man and woman. I held up three boxes of spaghetti for Nelly to see and asked her if it was enough to feed all of us.

When she rolled her eyes I said, "Too much, huh?" I then asked who all the people in the living room were. She told me that it was Hobgood's real father,

---

[60] Before the creation of MREs soldiers would cook their C-Ration cans with heat tabs. A heat tab comes in a green foil wrapper, and it looks like a blue piece of sidewalk chalk. It is a hexamine fuel tab that burns with a blue flame. It is associated with all heat and the Rangers nicknamed the sun "the great heat tab in the sky".

and his wife. Hobgood's grandparents had already arrived and were staying at the Holiday Inn. I took a beer from the refrigerator and opened it up.

"Is she going to see the body today?" I asked.

Nelly shook her head, "They say it isn't ready yet."

I didn't think it would be. I remembered the piece of skull on the back of Hobgood shirt. The blood pouring over my hands. His head feeling like a deflated soccer ball.

I took a long drink of beer, "Did she say where they were going to bury him?"

Nelly looked over at me, "Finally! She got over the idea of burying him in the woods. She's going to bury him at the post cemetery. Full military honors. She wants you, Will, Wilroy, Walters, Kittle and Olson to be pallbearers. The Battalion is going to have a memorial service early Wednesday morning. None of us will be there though."

"Why's that?" I asked, "What's up?"

She told me, "They are going to have the church service at the same time, at the main post chapel."

The Battalion had five days off, starting at noon on Wednesday. I asked Nelly, "Where is she going for Thanksgiving?"

"Me and Will are going up to Tennessee," she replied, "we decided she shouldn't be alone. She'll go with us. We'll stay here with her over Christmas leave. She should be expecting then."

"Hell," I told her, "We'll all be here for Christmas. We're going to be on alert status."

Nelly looked little annoyed, "You know what I'm talking about. Block Leave, after you guys get off alert."

"Laura should know," I told her, "She'll never have another thing to worry about. I know that she doesn't have much for family, but the Rangers are her family now."

228

Chapter 2: RANGER – A Death in the Family

Nelly looked at me, and asked, "SGT O'Kelley, why don't you tell her this?"

I looked away, "I don't know. I keep thinking it was my fault. I should've been watching him. I can't talk to her, knowing I killed off the only thing in her life. I think that's why Walters isn't here either. I don't think he wants to remember. Him and Hobgood were real close."

I faced her, "You don't understand what we are. We know each other better than wives or mothers will ever know us. It may sound gay, but it is a type of love."

I blurted out, "My God, when he was lying out there, I wondered, why! Why him? He had more to live for than all of us. I have nothing. If I died no one would give a damn. I keep wondering, why not me. We all live through something, and that something binds us together. That's what makes a soldier dive on a grenade to save his buddies! You just have this love. I guess that's what you would call it."

I threw the empty beer into the trash can, and grabbed another from the refrigerator. "You know, I always wondered why do the good always go first, and not the slugs? I'm not real religious, but I figured the good always go first, because they are ready to go."

I could feel the tears welling up, so I told her I had to get something from my truck, and left the house.

Three Rangers were the cooks, and the kitchen began to look like a war zone. I had tried to put all three boxes of spaghetti into the pot, but it only needed about half of that. The girls looked in on us, covered their eyes, and stayed away. The empty beer cans were stacked up on the table in a pyramid as we finished them.

Wilroy pointed to the stack, "One of us has to stay up all night."

"Yeah," I agreed, "I'm getting tired of staying up, waiting for her to bolt out of here."

Wilroy asked, "Did you bring anymore videotapes?"

229

"Yeah," I answered, "It was hard as hell trying to find a movie without any killing in it. We figured no one needed to watch that."

"What did you bring?" Toy asked, "*Debbie does Dallas?*"

"Nah," Wilroy pointed to the bag of tapes, "We got *Breakfast Club*, *Star Wars*, the *Green Berets*, and the *Sands of Iwo Jima*. The Duke is in those. There's killing in them, but its old Hollywood, no blood, lots of heroes."

"Good choice." We finally defeated the spaghetti dinner, but the loser was the kitchen. When the girls saw the spaghetti piled high on plates, they considered ordering pizza. As we finished dinner LT Preysler arrived.

"Hey, sir!" Toy said, "We've got about twenty pounds of spaghetti left, if you want any." The LT declined, and reminded us that we needed to be back in Battalion the next day. I asked the LT did he know why we were needed.

"It's a rehearsal for the funeral," he said. "Also, tomorrow we'll finally get to see Hobgood." When the LT said that both girls looked up.

Laura Hobgood pointed at me and said, "I want you there." She walked over to me and looked me in the eyes. "I only want you, Kittle, Wilroy, Walters and Olson to see him. I don't want strangers gawking at him!"

She looked at me with such intensity. I told her, "I'll be there." I felt like a liar though. I wanted to be anywhere but there.

LT Preysler continued, "Tuesday the relatives will be allowed to see him." The LT then looked at me, "I brought Walters with me. He's outside."

Wilroy and I both went for the door. Walters was outside, pacing back and forth with a beer in his hand, beside the LT's staff car.

"Where the hell have you been?" I asked.

Walters lied, "I didn't get the word." He looked down at his feet, "I really didn't want to be here."

"Why's that?" I asked. "It's SD! No duty for three days, and you get away from SGT Maddison."

Chapter 2: RANGER – A Death in the Family

"I know, I know," he said, "it's just… did you see her the first night? She's thinking he is still alive. She put notes all over the house telling him where she would be! I know he's dead. Ain't nothing bringing him back either!"

"She accepts it now," I told him. "She knows he's dead. She doesn't talk too much about it though. Our only problem is at night. We have to track her down when she runs away. I don't think we'll have a hassle tonight though. She sees his body tomorrow, so I figure she'll quit looking for him."

Walters still looked wary. "Come on in fat boy!" I slapped him on the back, "We've got a ton of spaghetti!"

Laura was happy to see Walters, and he fit right into the group by eating about ten pounds of spaghetti, and downing a six pack of Budweiser. Toy had gone to Hobgood's house, picked up his dress greens and was putting the medals and ribbons in their appropriate places. Wilroy and Walters were spit shining his boots so that they looked like glass. Only paratroopers were allowed to wear jump boots in a dress uniform. The set of jungle boots that Laura had shined the night Hobgood had died would be used at the Battalion memorial ceremony.

For the memorial ceremony an M-16 would be stuck into the ground using a bayonet, between the boots. A black beret would be placed upon the boots, while a Kevlar helmet would be placed on the butt of the upright weapon.

After the uniform was prepared Toy hung it on the back of a door and everyone gathered around the TV to watch videos. Late in the night Toy put on *Walking Tall*, even though we didn't want him too. There was a scene in the movie where the hero and his wife were ambushed by a machinegun while riding in a car. I didn't want Laura to watch any movies with bloody violence. The girls were off on their own, talking, and not paying attention to the movie. We all glared at Toy, hoping that he would get the message and turn it off, but he was oblivious.

I stood up and said "This is boring, let's watch something else."

"Wait till after this part," Toy said, and leaned in to watch the ambush, "I haven't seen this in awhile."

I didn't want to make a scene and draw the attention of the girls, so I sat back down. I kept thinking "Don't look Laura. Don't look. It looks just like Russell's head splattered up there."

The machinegun erupted on the screen and both girls looked to see what the noise was. The hero's wife was shot and killed; blood flew out of her head.

Laura stood up, and weakly said, "I'm awfully tired. I want to go to bed." She left the room. All of us glared at Toy, who still didn't get it.

We watched more movies until late in the night. Nelly got up to go to bed, but she first checked on Laura. She cracked open the door, but didn't see Laura anywhere. She looked under the bed, then looked up at the window. The screen had been ripped out. Nelly ran into the living room and told all of us, "She gone again!"

Wilroy and I searched the neighborhood in one truck, while Walter and Toy took their truck to the funeral home, twenty miles away. We figured that was where Laura was heading. I told Wilroy to turn back to Nelly's house. Nelly was standing in the driveway when we arrived. She told me she had called CPT Minton and asked him if she should call the police. He had told her to hold off, but she decided to call them anyway.

"I called the MPs" she told me, "It took awhile, they're not too bright there, but they finally got the message."

I asked her if Walters had called from the funeral home. She said "They called about thirty minutes ago. They searched the building, but it's locked. They got a hold of the funeral director and he's coming to unlock the building."

"OK," I told her, "Call them back and tell them to stay there, till we tell them otherwise. Me and Walters are going tracking."

I walked over to the window that Laura had ripped open. "Strong girl!" I thought. I looked around the ground until I could make footprints out in the dew. I pointed in the direction of the footprints. It headed in the direction of the post cemetery. "Let's go!" I told Wilroy.

We drove a half mile to the post cemetery, looking out the windows for any clues. Once we got to the gate, I stopped the truck. I whispered to Wilroy, "I'll take the left side of the cemetery, you take the right. Look behind all the gravestones." We both climbed over the gate and took off at a jog.

The night was dark and it had a bit of a chill. The moon cast a faint light over the head stones. I ran down each row of headstones, feeling uneasy about walking across all the graves. I reached the end of the headstones and hopped over a wall. There was a building on the other side of the brick wall that was used to hold bodies until the funeral. I looked in the windows, but there was nothing inside. There were small footprints in the sand around the windows though. I followed the footprints until they came to a cement sidewalk and was lost.

Wilroy walked over and told me he hadn't seen anything. I told him there were a set of footprints, but I don't know where she went after she left the cemetery. Wilroy pointed to the woods at the back of the cemetery. "If she went in there, we'll never find her."

The woods loomed ahead of us, dark and ominous. We realized it was futile and went back to the truck.

When we returned to Toy's house LT Preysler was there with Nelly. I asked her if there was any news. She shook her head "no". I told her about the foot prints at the cemetery. "If she keeps this up," I told them, "She's going to lose that kid! Hell, if she keeps this up, she'll die herself! It's getting cold out there. I don't want to see another body, not this soon!"

LT Preysler told me to not be so morbid. "We'll know soon enough what's going on." I looked at my watch. "4:30. It's been five hours."

Nelly spoke up, "Laura said that a week before he died, she saw him asleep on the couch, and she thought she saw blood covering his head!"

"That's an afterthought", the LT said, "You can think of a lot of things that happened after the event and try to attach meaning to them." He turned to us, "Let's go check out the cemetery again."

We all walked to the truck, when Wilroy stopped and pointed. We all looked to what he was pointing at, and saw Laura, staggering up the sidewalk. "Where in the hell were you!" I shouted. Laura looked at me, her eyes glassy. She looked like a zombie.

"I was in the woods" she said tiredly, "I talked to Russell. He told me it was alright; everything would be alright." She then fainted. I grabbed her, but her weight knocked me over. I fell, but turned so she would fall on me and not the ground.

We were told we had to be in the Company area by 0900. I was tired and needed sleep. As I drove through the Battalion gate, I saw the flag at half-staff. After I parked the truck, I walked to the platoon area. Everyone I passed asked me about Laura.

The platoon had been broken down into different sections for the memorial ceremony. There was the Honor Guard and the Firing Squad. The Firing Squad was issued M-21 rifles and moved out to practice firing volleys. Hobgood had been a sniper, so the firing squad had been issued the sniper rifles. All morning the "crack" of the seven rifles could be heard.

I was part of the Honor Guard, and we would be carrying Hobgood's casket, folding the flag and presenting it to Laura. Olson, another friend of Hobgood's, was also part of the pallbearers. Olson was a 90mm Recoilless Rifle gunner. We rehearsed over and over until we got it right, then both our groups came together and rehearsed again. We also had to rehearse what we would do at the chapel ceremony.

SFC Maddison was beginning to piss us all off. He acted cold, like he didn't care that a fellow Ranger had died. When we were practicing the chapel ceremony, he told us, "Walk straight! Remember that's just a hunk of meat in there! It can't be that heavy! You're going to push this box on a roller thing, then the Catholic priest will throw water, and blow smoke, then you'll push him out again."

When we practiced the graveside service he said, "When they play taps, I don't want to see a sniffle, not a single tear! If I hear one sniffle, I'll have you all here over the holidays!"

We practiced until lunch, then we were allowed to go back to Toy's house. Before we left, we were told that there would be a formation at 10:00 the next day. The investigators wanted us to reenact the live fire on the range. All our weapons were still impounded and we all wondered what we would do if we had to go on alert.

At Toy's house we warmed up last night's spaghetti for lunch. LT Preysler arrived to take the girls to the funeral home where Hobgood's body lay waiting. Toy and Walters didn't want to go see the body. They saw it the other night when the funeral director had unlocked the building. They didn't want to go back. Wilroy and I followed the girls in my truck, while they rode with the LT in an Army staff car.

I looked at Wilroy, "You ready for this?" I asked.

"Ready as I'll ever be," he replied.

First, we had to go to main post, to the Finance building. Laura had to sign the paperwork for the government insurance claim. She would get $30,000, the maximum insurance paid for a soldier's life at that time. We then drove on to the funeral home. It was in a rundown section of the town.

Wilroy saw it and uttered, "What a pit!"

All of us went into the funeral home, where the Chaplain waited for us. He wanted the girls to go in first. I had butterflies in my stomach. I really didn't want to be there. I really didn't want to see the body. I knew Hobgood when he was alive, I didn't want to see him laid out in a box. I wanted to remember him the way he was.

One of the girls cried out, jogging me back to what was going on. It was Nelly. She walked out and held onto the wall. She turned her back to me, her shoulder's shaking. I imagine she was wondering if that was how Toy would look if he was dead. The Chaplain walked over to her and said something. He slipped his arm around Nelly and guided her back into the room with Laura. He probably told her that Laura didn't need to be alone right then.

Wilroy looked at me, and we both thought the same thing, "What the hell are we getting ourselves into?" Wilroy shrugged, turned away from me and looked out the window. The Chaplain came out and motioned for us to come inside. We both walked into the small room like we were walking to our own execution.

The coffin was on a small table, raised up off the floor. Both the girls were at the head of the coffin. I stood off to one side, so that both the girls would block my view of the body. I didn't want to see it. I thought, "If they move, I'm going to freak out!"

Laura was talking to Hobgood. She was whispering to him. "Your mom is coming tomorrow. So is your brother and sister." She looked up at Nelly and asked, "Does his tie look straight? I think it is crooked." Nelly didn't speak and just nodded to her.

"Wilroy is here," she said, "So is SGT O'Kelley." She leaned back and motioned to me, "Come on over SGT."

I move forward, like an old man. Hobgood was in his dress green uniform, with his black beret on. He didn't look like he had a neck because the funeral home had pulled the collar high to cover the bullet hole there. His cheeks were puffed out, from the cotton in his mouth. His eyes were closed, but his lashes looked painted on. I fought back a shudder.

"Well," Laura told the body, "We've got to go. I'll bring all the family over tomorrow." Wilroy and I were the first ones to head outside. The girls went into the director's office, with the Chaplain following them.

I stopped the Chaplain, "Sir, is that it? Can we go? We've got stuff to do."

"Sure," he said, "go ahead. They are going to be in there for a little while anyway."

We both rushed to my truck and got in. "Man!" I said.

"That ain't Hobgood!" Wilroy said.

After a few minutes I told Wilroy that I don't ever want anyone seeing me when I die. "Weld that lid shut!"

Wilroy agreed. "I hope I get blown up into a million pieces" he said.

We stopped off at a gas station and picked up a 6-pack of beer. It was gone by the time we reached Toy's house.

I was cooking dinner that night so I ordered everyone out of the kitchen. I was making fried chicken, the way my mom showed me how.

Chapter 2: RANGER – A Death in the Family

Everyone else was in the living room along with Hobgood's mom and sister. Everyone in that family was tall, at least 6 feet. Almost all the siblings were in the military. Hobgood's sister was in the Army Reserve as a nurse.

I finished up another beer and put the chicken on the plate. I carried it into the living room and said "chow's on!" No one went for the food at first, until they saw me eat a piece. Since I didn't fall down gagging, they figured it was OK, and they all began to eat.

The phone in the kitchen rang and Nelly answered it. "It's Hobgood's brother" she said, "He's at the airport!"

"Is this the good brother, or the other one?" Wilroy asked.

"The other one," Nelly said, "the one in the Air Force."

Wilroy and Toy decided they would pick him up. Walters was passed out in the bedroom from the lack of sleep and the many beers we had finished off. I remained, sitting in the living room with Nelly and Laura. A slow love song came on the radio.

"This was our favorite song," Laura said sadly, "They played it at our prom. He asked me to marry him then." Laura looked at Nelly and cried, "He's gone! It's not fair!" She began to sob.

"He lied!" she screamed. "Stupid idiot! He told me he'd never leave me. NEVER!" She picked up a pillow and screamed into it. She threw the pillow aside and buried her head into Nelly's lap. "He's gone," she cried, "he's never coming back."

Nelly looked at me with tears in her eyes. She stroked the back of Laura's head. I got up and went out the back door. I walked into the dark until the tree branches brushed my face. I had gone into the woods. The place I know. Home. I continued walking until I came into a clearing and looked up at the stars. I was frustrated and angry. I raised my fist into the air and shouted. I dropped down to the ground and pounded the grass over and over. The shouted echo drifted through the woods until there was only silence.

Laura slept on a foldout couch in the living room. I slept on the living room floor with my head against the front door. I had stacked chairs on the back door so that she wouldn't be able to get out that way. Wilroy slept in the bedroom with Walters. Both of them had taken so many No-Doz pills, that it made them sick, and then they passed out.

Before Nelly went to bed, she asked if I was going to stay up all night. I told her no, but I would be remaining in the living room with Laura.

"You don't have to be the iron man", she said.

I told her I wasn't; I just didn't want to go running around after Laura that night.

Later I woke up feeling that something was wrong. I quickly looked over to where Laura had been sleeping, but she was gone. I didn't want to wake up everybody, so I calmly got up and opened the front door. Laura was on the lawn, playing with someone's dog.

"Laura", I asked, "What are you doing out here? Come inside."

She looked at me, smiling with an empty stare. "I wasn't tired," she said, "I'm only playing."

"Come on" I said. I reached down, picked her up and took her to the couch.

The next day we went to the Battalion for the 10:00 formation. I saw that the flag was still at half-staff. I figured it would remain that way until after Hobgood was buried. We loaded onto trucks that took us back to the range where it all happened. On the way all the Rangers asked us about Laura, and how she was taking it. We were tired, but we answered them the best we could.

When we arrived at the range the CID (Criminal Investigation Team) was there, along with several officers. We were told we would start the reenactment from the second attack line. We all lined up where we had been on that day. I could see my tracks of where I had walked. We moved to the area where we started and then kneeled down.

I told an investigator that I had put the A Team on line, and then told the B Team to go to the left. I told them that I went with the Lieutenant, then the whole platoon rose up and went to the left. When we were all on line, we stood up to see what had happened.

Chapter 2: RANGER – A Death in the Family

The line of Rangers was relatively straight. The investigators walked up to the different Rangers and asked them questions as they stood there. Finally, the head investigator, a colonel, said "Thank you men, that'll be all."

I looked at the spot where Hobgood had been killed. It was a metal box, not cement like I thought it had been. I could see the bullet holes in it. I also saw a dark stain splashed all over the box and ground.

I turned around and walked back.

We were told that we would have funeral practice again that afternoon. I went into the NCO barracks and grabbed a sandwich out of my refrigerator. On my wall were different awards and plaques. There were pictures of me in Panama, Sudan, Egypt, Honduras and Grenada.

I looked at a small black flag pinned to the wall. On the flag were the words "Ranger, E Company, Class 5-85". Beside the flag was a knife on a plaque. The brass plate on the award said "Distinguished Honor Graduate, Ranger Class 5-85, SGT Patrick O'Kelley". On the corner of the flag was a black and gold Ranger Tab, pinned there with a safety pin. It was the one that I was awarded in Ranger School. I glanced at the clock and thought, "I have time." I unpinned the Ranger Tab and then walked to the parking lot to find my truck.

The funeral home was dark inside. No one was there. I figured that was due to lunch. I stood in the waiting room in my jungle fatigues and dusty jungle boots. I walked into the room where Hobgood's body was. I could make out the shape of the casket in the dark. For a second, I imagined that Hobgood would open his eyes, crawl out of the coffin and point an accusing finger at me. I shook it off and turned on the light switch.

Hobgood hadn't moved at all. He was still dead in his coffin, looking like a mannequin. I walked up to the coffin and took the Ranger Tab out of my pocket. Hobgood had not gone to Ranger School yet. He never would now. You are never really a Ranger until you get that Tab. I pinned the tab beside Hobgood's head, on his pillow.

"You never went to the suck" I told him, "But you earned this."

I backed away from the coffin and turned off the lights. At the door I stopped and looked at the coffin. I then came to attention and saluted.

I dropped the salute and walked out.

The funeral practice was the same as before. We carried the heavy box, loaded with sandbags, and listened to the crack of the rifles. At the end of the practice SFC Maddison called me into his office. I stood at parade rest, wondering what he needed now.

"What do you need, SGT?" I asked him. He looked up from his desk and said "I don't think you SD men should go over there tonight."

"At night is when they need us most," I told him. "We have to track Laura down every night."

"Well," Maddison said, "I don't think they'll need you men tonight. She has family."

"No," I corrected him, "she doesn't have family. Only Hobgood's family is there. We are her family."

Maddison was persistent, "Someone else can take care of her. You can be doing more important things around here."

I didn't want to be insubordinate, but he was beginning to piss me off. "Like what," I shot back, "more funeral practice?"

"Well," Maddison continued, "I figure Mrs. Toy doesn't want a bunch of G.I.s over at her house messing it up."

"She doesn't mind," I argued, "besides, it's just for one more night."

SFC Maddison threw his hands in the air, "Go ahead! I think you are getting too involved with your troops though!"

"Is that all, sergeant?" I asked.

"That's all" he replied.

Chapter 2: RANGER – A Death in the Family

I did an about face, and then walked out. He was right; I was too involved with my troops. I always had been. They were my men, now one of them was dead. The Rangers took care of our own. The Ranger Creed stated "I will never leave a fallen comrade." We weren't going to leave him or his wife.

LT Preysler had taken the girls out to an expensive restaurant, while we stayed at Toy's house. We ate the leftover spaghetti and had restocked the beer. Toy had bought a bottle of Jack Daniels. This was our last night together. Tomorrow we would bury Hobgood, and then we would go our separate ways for the Thanksgiving Day pass.

We were preparing our uniforms, spit shining our jump boots and polishing the brass. Hobgood's mother and sister were at the airport, picking up his brother that was in the Army. He had flown in from Germany.

The girls burst through the door, giggling, with LT Preysler following them in. He was blushing. "And what have you been up to?" I asked.

"You don't want to know!" Nelly said. She tried to suppress a giggle, but then she busted out laughing.

"Never will know," Walters asked, "will we?"

"Sir," I asked the LT, "What have you been doing to these girls?"

The LT was red-faced and he stammered, "You mean, what have these girls been doing to me!" The girls busted out laughing again.

The LT said he could only stay a little while, since he had to get ready for tomorrow. As he left the house Hobgood's mother and sister arrived. They told us that they had left Hobgood's brother over at Laura's house.

"Laura's house," I thought, "It's no longer Hobgood's house."

That night Hobgood's mother and sister sat in the living room listening to the Rangers tell the different stories about him. Walters asked me, "Remember that time when he thought that your girlfriend was dead?"

He was talking about the girlfriend I had in North Carolina who broke up with me. I told Hobgood's mom, "I had this guy in the platoon, Phytheon, who

Y

kept bugging me about my girlfriend. I had just broken up with her and I didn't want to talk about her. So, I told him that my girl had been killed, shot in a 7-11 in a robbery shootout. I acted totally serious. Well," I continued, "the next day Hobgood comes in, saying 'Gee SGT, how's your girl?' Pytheon tells him that she got killed, and Hobgood goes running around the room, banging himself on the head, saying 'I'm so stupid! I'm so dumb!"

We all laughed, then Laura turned to me and said, "Some of the guys had pinned a Ranger Tab to his pillow. I put it on his shoulder. He would've liked that."

"Who did it?" she asked.

"I don't know," I lied, "probably all the guys in the platoon."

"How about the time me and Beaudette took him to *W.D. Crowleys*" Wilroy said, "We gave him a Long Island iced tea and he says 'these aren't so bad'. So, he ends up drinking five of them. He starts shouting that they aren't nothing, and to bring him more! So, we drag him into the car and take him downtown. We go into a Korean bar and put him inside. We then ran out the back door."

Wilroy continued, "We went to a McDonalds and got something to eat. About thirty minutes later we go back to get him and there's Hobgood, passed out on the stage, with some Korean girl dancing over him. We get him back into the car and he threw up at every stop sign! You could've followed our trail back to the barracks."

All the Rangers were laughing and trying to top each other's stories. Hobgood's brother, who was in the Army, came in and listened. He then asked "How did this happen?" We didn't go into detail, but we told him about what happened on the range.

He shook his head, "I still don't understand how it happened! I've been on ranges, and no one was ever killed! Someone wasn't doing their job. Someone was fucked up!"

I told him that Ranger training was more realistic than the rest of the Army, and due to this people sometimes got hurt. He didn't want to hear it, and stormed out of the room.

The beer was gone and the bottle of Jack Daniels was finished. Hobgood's mom said goodbye to everyone, and then went to Laura's house. Hobgood's sister stayed. She talked to Laura about Hobgood's life before the army. Wilroy and Walters had gone on to the bedroom and passed out again.

Hobgood's sister told Laura that he had always wanted to be like his brother. "That's why he joined the army."

Laura said, "He wanted his father to be proud of him too, but he always got the short end from him."

"When he was growing up," his sister said, "our dad used to beat him. I never knew why. He treated him like dirt." She started to cry. Laura began crying, too.

"I hope that bastard's proud," she said, "I hope he feels great!"

Nelly took Laura in her arms, while his sister looked at me, tears on her cheeks. "I screwed up," she said, "I shouldn't have said anything." She lowered her   head   and   cried.
     "Don't worry," I lied, "It'll be all right." I didn't know if I should hold her in my arms while she cried, and try to comfort her, or let her be. I stood there awkwardly, watching all the girls cry.

The memorial service for the Battalion was at 8:00. The Catholic service was at 9:00. The Honor Guard milled around outside of the post chapel, checking out the different people going inside. There were Rangers we hadn't seen in awhile, the ones who had moved on to other units. There were two flags outside the chapel at half-staff, the American flag and a blue flag with a white cross on it.

We watched as people drove by the chapel. They stared out the windows as they drove slowly past, wondering what was going on. They normally never saw us. The men in the black berets didn't come on main post much. The hearse arrived and we arranged ourselves. We lined up behind the hearse and brought the casket out. It wasn't as heavy as the one SFC Maddison had made with plywood and sandbags. We marched slowly into the chapel, carrying the coffin. The coffin was then laid on a roller and covered in an American flag.

A Catholic priest stopped behind the casket and took out a glass of Holy water. He splashed it on the coffin. He said the words, "Hail Mary, full of grace. The Lord is with thee, blessed art thou among women, and..."

I wondered if it was right to throw water on the United States flag.

"...of God, pray for us sinners now and at the hour of our death. Amen."

The plan had changed. The funeral caretakers would push the coffin into the chapel and the Honor Guard would march in behind it. We slowly marched in and then sat in the front row of pews.

The priest continued to pray over the coffin while the Battalion chaplain went up to the altar and spoke to everyone about Hobgood, about his military career, his marriage, his life. He then introduced the Company Commander.

The CO stepped up to the altar and said, "Hobgood was only 19 years old, but it was a full 19 years. He had been to more places and done more than many men would do in a lifetime. I remember Hobgood well; he always had a bright outlook."

The CO looked down, "I want to forget, but I'll never be able to. We'll all remember him." He left the stage, his head bowed down. Some of the Honor Guard wiped their eyes.

We carried the coffin to the hearse and placed it inside. Afterwards we marched slowly to the van that would follow the hearse. We sat inside the van until everyone was ready to move. As we drove to the cemetery Wilroy said, "Look at that geek".

He was pointing to an MP who stared dumbly at the traffic with his mouth open. We all suppressed a giggle.

"Man!" Walters said, "I want to be a priest when I get out! Did you see the wine that guy was downing!" We all started laughing.

"I didn't understand the part about the cookies." Wilroy looked confused.

"It's not a cookie" I told him, "It's a Holy cracker, dumb ass!"

Everyone busted out laughing again.

"I still don't get it," Wilroy muttered.

The van drove down the road, everyone inside was cracking up.

"Can you picture this," Wilroy said, "Here we are, in a funeral, and the lead vehicle is full of laughing Rangers!"

We all laughed even harder.

When we entered the post cemetery our sides were hurting. "Here we go," I muttered.

The van stopped on the opposite side of the cemetery. The Honor Guard stepped out and formed on the road. We slowly marched to the hearse, stopped, and then pulled the coffin out. We carried the coffin to a tent that had been set up in the cemetery. Once we set the coffin down, we all reached down, in unison like we practiced, and picked up the American flag. We stretched the flag tight over the coffin and then stood that way, at attention.

The family and friends were seated under the tent, while around the tent the Company stood at attention. The sun was shining bright upon them. The priest spoke over the casket then stepped away.

"Have the firing squad fire three volleys!" someone ordered.

In the distance I heard the orders, "Present, FIRE!"

CRACK!

The Honor Guard jumped slightly.

"Present, FIRE!"

CRACK!

A woman cried out.

"Present, FIRE!"

245

CRACK!

Other women started crying.

"Present, ARMS!" The Firing Squad saluted with their weapons. The command to present arms was then given to the Company.

Near the flagpole a lone bugler began to play *Taps*. The bugler stood by a single gravestone at the base of the flagpole. It was the headstone of a Medal of Honor recipient, another soldier of 19.

I felt a chill in my spine when the notes rang out from the bugle. A lump formed in my throat. "I will NOT cry!" I ordered myself.

I stared defiantly across at the other Rangers holding the flag. Toy was standing across from me, holding the flag. A tear rolled down his cheek.

As the last notes echoed away, we folded the flag in unison. We folded it into a triangle shape, leaving the end open. SSG Van Heuten, the SGT in charge of the firing squad, marched up and placed the 21 pieces of brass from the expended ammunition into the flag, and then folded the last fold.

CPT Minton stepped forward and took the flag from SSG Van Heuten. Toy let go of the flag and saluted the CO. When he dropped his arm, we all marched off. The CO walked to Laura and held out the flag. He told her "This is presented on behalf of a grateful nation and the United States Army as a token of appreciation for your loved one's honorable and faithful service." Laura took the flag and cradled it in her arms. She looked up at him and smiled, tears rolled down her cheeks.

We sat in the Honor Guard van.

No one laughed.

After the funeral we all went our separate ways for Thanksgiving pass. When I returned to the Ranger Battalion things were not the same. We did missions and training exercises, but there was a sadness to it all. There was also a seriousness that we didn't have before. At any time one of us may die due to the actions of the others.

Chapter 2: RANGER – A Death in the Family

CPT Minton held a meeting of the company and told them what the investigators had concluded. No one Ranger was held responsible. It was deemed an accident. However, during his talk to the company he became angry that one of his men had been killed. He said that the chain of command was at fault, and he also told them that I had a scope on my rifle and I was too busy "playing sniper shit" instead of watching the squad.

This shocked and angered me. I could not believe that because I had an optical scope on my weapon it would cause Hobgood to be killed, but since it was not SOP, he blamed me. Ironically in today's Army every soldier in the United States Army has a scope on their weapon for the same reason that I had one on that range. It allows you to identify targets and engage them quicker. Unfortunately to this day there are Rangers who blame me for Hobgood's death, due to CPT Minton's remarks.

I was now marked. A few days after that the 1SG pulled me into his office and told me that he had never had a man killed in his command, and he wanted me gone. When I went back to my platoon, I told my platoon leader LT Varney what the 1SG had told me and he was amazed. He got angry, but nothing came of it. SFC Maddison didn't bother to stand up for me, and just told me to pack my things and get out. I pulled my squad into a room and told them what had happened. They were pissed and said they were going to leave the Battalion. I told them to stick around and don't leave just because of what happened to me. As I left, they all dropped down and did pushups for me. Though it may seem odd, it was a sign of respect in the Rangers.

Since they could not kick me out of Battalion I was assigned to the S-5. In 1985 the S-5 didn't do much at all. They sold T-shirts and Ranger coins to the Rangers. They took pictures of events. The S-5 consisted of me, a SFC and a PFC. I hated it. The SFC kept telling me that being in S-5 was great because it was the easiest job in the Battalion, I didn't have to do anything. He didn't get it. I didn't want it easy. I wanted to be back with the men. I accepted my punishment though because I felt it was my fate. Hobgood died because of me, so I must deserve what was happening to me.

This purgatory lasted for three months. Finally, Battalion CSM Fox brought me in his office and asked me if I wanted to go back to a platoon again. I jumped at the chance, however I still felt marked.

My brother Sean must have been fascinated by my stories of the Ranger Battalion. He had joined the 12th Special Forces Group in Tampa, Florida, and had become Airborne qualified. However, he did not become fully Special

Forces qualified because he went active duty and joined the 3rd Ranger Battalion in 1986. He was assigned to Bravo Company. I would see him from time to time, but we never would hang out together.

I tried to find something that would replace the camaraderie I had felt before with my squad, and joined the *Wingmen* motorcycle club in Columbus. This was a predominately Ranger group. I bought a Harley and rode with them for awhile, but after there was a shootout with another bike gang in town, the *Iron Cross*, Regiment gave us an ultimatum. You can be in the *Wingmen*, or you can be in the Ranger Battalion, but you can't be in both. I never really fit into the *Wingmen*, so I walked away from them.

When it came time to reenlist I did so for a different Ranger unit. The Army was starting up a series of LRSU (Long Range Surveillance) Units. I was told by Department of the Army that this unit would be manned by Rangers, commanded by Special Forces, and it would be Airborne, possibly HALO. This unit became F Company, 51st Infantry (LRSU) that was stationed in Ludwigsberg, West Germany.

Things changed for me after Hobgood's death. I was in Germany for over two years and in that time, I never wanted a girlfriend, or even a romantic interest. Instead, I would devote myself to the Team I was assigned to. I would make sure no one died again. I would also make sure the Team was never screwed over. Due to this I would make lifelong friends of those in my Team, but those who were not in the Team usually hated me.

When the NCOs of F Company went to Fort Benning for the Long Range Surveillance Leader's Course (LRSLC) I visited Hobgood's grave at the post cemetery. I visit that grave every time I go to Fort Benning. I would leave a 3rd Ranger Battalion coin, or a Ranger Tab. Around his grave were tokens left from other Rangers, such as Ranger scrolls, bottles of beer and flowers. I knew that his death had affected more than me.[61]

None of us took the LRSLC course seriously.[62] All of the NCOs were from the Ranger Battalions and so were the instructors, so we knew each other. There

---

[61] The last time I went to Hobgood's grave, in 2013, I took my 18 year old daughter, Cailin. I told her about what happened. On the grave was a metal 75th Regiment scroll that someone had glued onto it. There were flowers still there. I put a 3rd Ranger Battalion decal on his grave because the grave never mentioned he was a Ranger.
[62] Pronounced Lurs-lick.

was no attempt to "smoke" us because many of the NCOs of F/51st were legends in the Rangers.

The LRSLC course was meant to teach basic reconnaissance skills and how to do long range communications, but we were familiar with all of this since we had trained on it back in Germany. Many of us had gone to the International Long Range Reconnaissance Patrol School (ILRRP) in Weingarten, and had been trained by the SAS.[63] However, we looked at the trip from Germany as a free vacation back to the United States.

My brother was still in the 3rd Ranger Battalion and had married a young girl that he had "stolen" from her Marine husband. When the LRSLC course conducted their survival course and E+E, me and another Ranger NCO, SSG Peterson, evaded to my brother Sean's trailer, drank beer and watched TV. He wasn't there that night, since he was deployed with the Rangers.

---

[63] SAS stands for Special Air Service and is the Special Forces of Great Britain. The SAS wear the tan colored beret, and when the US Army commandeered the Ranger's black beret for the entire US Army, the Rangers then switched to the tan beret.

Triple Canopy

Sean did redeploy before I left the LRSLC course, and this was the last time I saw him alive. The young girl that he married would eventually lead to his death in 1987. She left Sean a few months later for another man. My brother tried to do everything to get her back. He drove to Tampa and tried to talk to her, but she wouldn't even come to the door. He even bought a wig and dressed up like a pizza deliveryman to get close to her. He was on leave from the Rangers when he did this, and he didn't want to return until it was settled. He remembered the time I had been in Tampa recovering from my accident, so he faked that he had been shot by a bike gang so that he could stay and recover in Tampa. He shot himself in the calf with a .22 pistol, and then filed a police report.

This still did not buy him enough time, and since he was no longer devoting time to the Rangers, he was asked to leave the Battalion. This may have been the final straw for him. He had never had luck with women. His first wife, a much older woman, left him, and now his second wife had done the same thing. When the Rangers left him, I think it was too much. He shot himself in the heart in 1987.

I did not want to admit that my older, better looking, stronger brother, would do such a thing. When I came back from Germany on emergency leave, I conducted my own investigation. I talked to the military investigator at Fort Benning, and the detectives in Columbus, Georgia. There were many things that did not match up, and I looked at the case as a murder. When I wanted to get some items from his trailer the authorities would not let me. They told me that all of the items in his trailer belonged to his wife. This angered me, and I knew there were certain things of his that my mother and father wanted.

I broke into the trailer, using all the stealth that the military had taught me, and removed everything of sentimental value. I did not want to be seen, so as I moved through the trailer, I high crawled back and forth. My brother's blood was ground into the knees and elbows of my clothes, but I didn't care. Sean was a Ranger, and my brother. I refused to let his memory be thrown out by an uncaring girl who only knew him for a year. The authorities in Fort Benning later called me and told me they suspected what I had done, but they had no proof.

Sean's funeral was held in Tampa, Florida, where all of his friends were. When he was buried his honor guard was made up of Rangers that knew him. Though he had been thrown out of the Rangers by the higher chain of command, his fellow Rangers did not leave him. They remembered the Creed, "Never leave a fallen comrade."

Chapter 2: RANGER – A Death in the Family

Before Hobgood's death I drank occasionally, but not to excess. However, when I returned to Germany I drank heavily. Because of what happened to my brother I no longer cared about women and it was over three years before I had another girlfriend. I no longer trusted my superiors, especially officers, to back me up. I began to tell lies. I would exaggerate my accomplishments, or make them seem larger than life. I couldn't stand people knowing the truth of what happened. Lying became second nature, but I would never lie to a Ranger. With all I had been through, I was still a Ranger and wanted to live up to the Ranger Creed.

In 2004 I was sitting in my Junior ROTC classroom waiting for the bell to ring and for the students of my first class of the day to arrive. As I took a sip of coffee, I heard a large crash, the sound of breaking glass and I then heard a long piercing scream. I quickly walked around the corner and saw a young girl sitting on a bench; her classmates and her teacher were backing away from her with a look of horror on their faces. They were backing away because she had severed her femoral artery by a piece of glass and blood was shooting out, spreading across the floor towards them. She had been leaning up against a plate glass window and another student smacked it from the outside, just playing around. Between her leaning back on the glass, and the pushing on the glass from the other side by the student, the glass shattered. She fell through the window, unharmed, but then a long piece of glass, shaped like a knife blade, had fallen down from the top frame, slicing through her jeans and her femoral artery.

I ran over to her and saw the blood spreading everywhere. She was still screaming, but I couldn't tell why. I looked around her head, and arms, looking for the wound, and then I noticed she was gripping the glass in her leg. I pulled her hands away as one of the other Junior ROTC instructors, SFC Hall, ran up. At that point the girl fainted. I yelled at SFC Hall to grab her head so it wouldn't fall into the glass and cut her throat. When she leaned back the glass "blade" popped out of her leg and the blood began to flow out quicker. I yelled to another ROTC instructor to call 911, and then yelled to one of my students to get the first aid bag.

I squeezed her thigh, putting pressure on the artery so that it would stop bleeding. I then kneeled down in the blood and waited. The student arrived with the first aid bag, but it was a practice bag for training. I grabbed a used Army bandage and tried to make a tourniquet. When it snapped, I cursed. I tried another one, but it snapped too.

I couldn't let this girl die. I pulled my dress belt out of my pants and then wrapped it around her thigh. When I cinched it down, the buckle snapped off. I began cussing even louder. Nothing was working, but I couldn't give up. If I could keep her alive, then maybe I could pay off my debt so many years ago. I thought of Hobgood as I saw all the blood pouring out of her. It was happening again. I grabbed both ends of the broken belt and then twisted it hard. My hand was still caught in between her bloody thigh and the belt, but I didn't care. I couldn't feel my fingers anymore, but I didn't care. The school nurse arrived, and looked shocked. I told her to get an ambulance here, fast!

The girl, Hailey Jones, opened her eyes. She looked up at me and asked if she was going to die. I told her, "Not if I have a damn thing to do about it! No one dies today, especially you!" I held her hand and told her it would all be alright. I continued talking to her, while my fingers turned blue and I couldn't feel the pain anymore. I continued to hold on to her so that a deeper pain would go away.

Hailey Jones graduated High School in 2007. I felt that I had paid off some of the debt that was put on me so many years ago. Due to my actions, my 19-year-old friend was killed. However, also due to my actions a 16-year-old girl lives.

Is the debt paid in full? I don't know, but I hope so.

Russell Hobgood, Jr. was born in January 1986. Laura went back to California and married again. She had a total of seven children.[64]

---

[64] While I was writing this story in 2008, I received an email from both Russell, Jr. and Laura. I had not written anything about what had happened since 1986, and now both of them contact me as I started again. Russell, Jr. wanted to know about his father. He said that his mother didn't talk much about him. Laura wanted to make sure I was who I said I was. She asked me one question to verify who I was, "Who gave Russell last rites?" When I answered correctly she told me what had happened in her life since then.

Chapter 3: SPECIAL FORCES – A Storm in the Desert

A Company, 3rd Battalion, 5th Special Forces Group
1990-1991

During WWII small groups of men conducted missions behind enemy lines, not to attack the enemy by themselves, but to organize the local indigenous people who knew the lay of the land. These soldiers would train these farmers, tribesmen, and shopkeepers on how to fight. In modern terms these men were known as a combat multiplier, where a handful of men could "become" hundreds of hunters, creating a nightmare for the enemy soldiers. These were not the regulars, but the irregular soldiers, or what the United States call the Special Forces.

One of the more successful special forces used against the Germans were the OSS Jedburgh Teams. These were three-man units that parachuted into Holland, Belgium and France, trained the partisans and then delayed or destroyed the enemy supply system during the Normandy invasion. In the

Pacific Theater, in Burma, OSS Detachment 101 organized 11,000 Kachin tribesmen and led them against the Japanese. The Japanese suffered over 10,000 killed, while the Kachins only lost 206.

After WWII the OSS was disbanded, but many of their agents formed the cadre of the CIA in 1947. Colonel Aaron Bank and Colonel Russell Volckman were two of the OSS operators who remained in the Army and tried to convince the military that the Army needed to create its own unconventional guerilla warriors. Once the Soviets gained nuclear weapons the Army realized that the conventional war of Korea or WWII would not be practical. However small bands of men creating confusion and destruction behind Soviet controlled Eastern Europe would be extremely effective. In 1952 this gave birth to the Special Forces at Fort Bragg, North Carolina.

The new unit was designated the 10th Special Forces Group and was made up of men who were former paratroopers, Rangers and OSS agents that had gained combat experience in WWII and Korea. Many of the men were from Eastern Europe and had fled the communist expansion. Unlike the Rangers and paratroopers of WWII, their missions would not be quick violent raids, but instead these men would be required to live deep within hostile territory for months or years, speaking the language of the people, with little or no resupply from friendly forces.

After the 1953 uprising in East Germany half of the 10th Special Forces Group moved to Bad Tölz, Germany, while the other half stayed in Fort Bragg and became the 77th Special Forces Group. [1] Soon the 1st Special Forces Group was created and sent to Okinawa, where some of the teams took part training Korean guerillas during the final year of the Korean War. Though small, the Special Forces soon proved that they were the elite of the military, much to the annoyance of conventional Generals. When President Kennedy visited Fort Bragg in 1961, he inspected the 7th Special Forces Group. He was impressed with the unconventional warriors, who wore an unauthorized and unconventional green beret on their heads. Kennedy realized that this unit is what he needed for any counterinsurgency operations and he authorized the 5th, 8th and 3rd Special Forces Groups to be created by 1963. He also authorized the funny looking hat, stating that he considered the Green Beret to be "symbolic of one of the highest levels of courage and achievement of the United States military." Soon the men became known by their headgear, the Green Berets.

---

[1] The 77th Special Forces had become the 7th Special Forces.

Chapter 3: SPECIAL FORCES – A Storm in the Desert

My father, James O'Kelley, was born and raised in the mountains of North Carolina. He was raised by his grandmother, but I did not know this until I was an adult. I always thought she was his mother. When he turned 15, he wanted to join the Army to go and fight in Korea, but he was underage. His grandmother, Laura O'Kelley, signed the papers by stretching the truth and saying he was old enough and with that my father became a soldier in the Army. He never went through Basic Training, but was taught his job by the older soldiers who had served in WWII. However, they soon discovered the truth about his grandmother's lie and he was discharged because he was too young. After he married my mother, she convinced him to go back into the military to pay for hospital bills when my brother Sean was born. He enlisted in the 518th Airborne Infantry Regiment of the US Army Reserve. In 1959 he became a paratrooper in the 82nd Airborne Division but he didn't go to the Airborne School in Fort Benning, Georgia. He instead went through Airborne School on Fort Bragg because the 82nd Airborne ran their own. He was assigned to the 325th Regiment that was still using the same gliders from WWII. Unfortunately, he was only one flight away from earning his Glider Wings when the aircraft was considered obsolete and removed from the inventory.

He wanted to join Special Forces on Fort Bragg, but they were a very small group and he told me that "you had to know someone to get in." His job in the Army was classified as a 051, Intermediate Speed Radio Operator (Morse Code), and he saw an opening for this job in Europe, so he volunteered to be reassigned to a Military Police company in Mannheim, West Germany. He did this so that he would have a better chance of getting into Special Forces that was located in Bad Tölz.

Finally, in 1962 he volunteered for Special Forces and was assigned to the 10th Special Forces Group in Bad Tölz. He wrote "In those days the 10th Group was called the "First Company" because that was only a company size unit - the whole group. After I got in the BIG SF expansion began. One day I was in ODA SFOD-6 and the next day the same detachment was named SFOD A-13." [2] My father never went through Special Forces Training though he wanted to, instead he was made an instructor because of his communications knowledge in Special Forces Training Company located at Lengries, Germany. Later he would write the training manual about SF Communications. [3]

---

[2] Email from James O'Kelley received on September 15, 2010.
[3] So my father may be one of the few people in the military who did not go to basic training, did not go to regular jump school and did not go through the Special Forces course, but he was in all these units. Email for James O'Kelley received on January 22, 2012.

The primary unit in Special Forces is the ODA (Operational Detachment – A Team) or simply, the "A Team." It is the same size as a squad in the conventional army. The B Team would be at the company level, while the C Team would be at the battalion level. The A Team is commanded by a Captain and at that time the second in command was a 1st Lieutenant. Today the A Teams have Warrant Officers in the secondary slot. The Team Sergeant on an A Team is a Master Sergeant. There is one Intelligence Sergeant while the rest of the Team consists of two Weapons Sergeants, two Communications Sergeants, two Medics and two Engineers. Since there is two of every specialty the Teams can run "split team" operations of two 6-man elements.

My father became a Communications Sergeant and had to learn all the aspects of long-range radio communications in the 1960s. I remember many nights of him staying up, studying Morse code or writing coded messages on paper, then burning them in the ashtray. To me this was what normal fathers did.

Though the Vietnam War officially began for the United States in 1964, we had been involved in a war there since the Japanese invaded in 1941. After the French withdrew from Vietnam in 1956 the US Army Special Forces sent one A Team there to train the South Vietnam Special Forces. By 1964 the Special Forces advisors increased until Vietnam became the operational domain of the 5th Special Forces Group.

My father deployed to war for the first time with ODA-323 in February of 1965 to Trang Sup. His base camp was in the shadow of Nui Ba Den, or the Black Virgin Mountain, towering 4,500 feet above the jungle. He was assigned to the CIDG (Civilian Irregular Defense Group). The Viet Cong harassed the area around the mountain but there was a radio relay site for all of III Corps on top of the mountain. My dad wrote "The VC never took the top of the mountain but they tried often enough. We could never take the rest of the mountain even though we tried many times." [4]

He never spoke much about what he did there, but sometimes a story would slip out. He told me that on one of his missions he had to lead a squad of Chinese Nung mercenaries to determine if there was a 500-man Chinese unit trying to get into Saigon. They found the enemy and then had to fight their way back to the camp, over a period of three days. During some missions in classified areas,

---

[4] Email from James O'Kelley received on January 22, 2012.

he was given a temporary rank of Lieutenant Colonel by the CIA so that he could advise the "out of country" soldiers.[5] It was due to this temporary battlefield commission that he decided he wanted to become an officer.

My father served one year in Vietnam and then returned home to Fort Bragg. A one-year tour was typical at the time and is similar to how the United States deploys the military during the war in Afghanistan today.[6] When he returned from the war my father went to Officer Candidate School in Fort Gordon, Georgia, was the Honor Graduate, and was commissioned as a 2nd Lieutenant in the Reserves in 1968. Normally the Honor Graduate would get a Regular Army commission, but he was considered too old at 33.

He returned to 10th Special Forces Group in Bad Tölz, but when he was deployed to Vietnam the second time, as a captain, he was assigned to the 1st Infantry Division at Dau Tieng as a conventional commander. During the Vietnam war there was no separate branch for the Special Forces. Special Forces soldiers had an "S" skill identifier to their job title, but there was no guarantee that they would stay in Special Forces if the needs of the Army required them to be elsewhere.

During his second deployment in 1969 he was wounded during an enemy attack on his outpost, Canada II, near Nui Ba Den. This was a major battle against the NVA that lasted throughout the night. Hundreds of RPGs were fired into the camp and one flew into his bunker, striking him in the back without exploding.[7] He recovered sufficiently enough from those wounds to continue to serve in the 1st Infantry Division until they were redeployed back to the United States in January of 1970. Though the unit returned to the United States, he did not have enough "time in country" so he was then assigned to the 1st Cavalry Division until he was able to return to the United States in July of 1970. Among his awards for his two tours of duty in Vietnam were three Bronze Star Medals, two Air Medals, a Combat Infantry Badge, a Purple Heart and two RVN Medal of Honors, second class.

---

[5] Up until his death in 2018, my father never talked about his classified missions.
[6] As of the time of the second release of this book, the United States is still fighting in Afghanistan, or the "Forever War". As of right now, 2021, we finally withdrew from Afghanistan after fighting in this war for 20 years.
[7] This wound would eventually lead to him getting 100% disability from the government.

This was the world I grew up in. This was normal. As a kid in Fort Bragg, Bad Tölz or Fort Devens my next-door neighbors and my friend's fathers were all Special Forces and all veterans of the war in Vietnam.[8]

When I was coming to the end of my tour of duty in Germany, I decided I would try out for Special Forces again. To prepare myself I went every week to the public pool in Ludwigsberg and taught myself how to swim. I did this by watching swimming at the 1988 Olympics on Armed Forces Network, recording it on a VCR and then playing it back to see what technique they were using. By doing this I was able to pass the swim test that consisted of a 50-meter swim, in uniform and boots. Since I was in F Company, 51st LRSU, I had no problem with the physical fitness test. I submitted my application for Special Forces Selection and waited.

Special Forces had gone through several changes since the end of the Vietnam War. The 3rd, 6th, and 8th Special Forces Groups were disbanded and the Army no longer focused on unconventional warfare. Like many nations throughout history, the United States in the 1970s trained to fight the last war they had won, which had been WWII. Armored warfare in the fields of Germany to hold back a Soviet invasion through the Fulda Gap was the standard training scenario. However, this all changed when President Reagan took office in 1980.

Miraculous things, to our point of view, happened to the military during that time. The Special Forces became its own branch of the service, similar to infantry or artillery. Now a Special Forces soldier, and in particular Special Forces officers, could stay in special operations without having to worry about doing time in a conventional unit to gain rank.

As the Special Forces grew it needed more men to fill their ranks. Since its creation a soldier had to be at least a sergeant to be in Special Forces, but new standards were lowered and a soldier could enlist in Special Forces and join the

---

[8] When I was in High School in Fort Devens, Massachusetts, I built a lot of military models, usually from the Tamiya brand. The only really good hobby store nearby was located in Leominster. My father was always busy doing Special Forces training, but the next-door neighbor, "Mister Miller" also liked building models, so he would take me to the hobby shop. I didn't think much of it. It was only years later that I found out "Mister Miller" was Walt Miller, the Son Tay Raider.

A-Teams right after Basic Training. These soldiers became known as "SF Babies" since they had no experience and were not as effective as the sergeants who had been in the Army for years. By the late 1980s Special Forces realized that this was a mistake and privates could no longer enlist right out of Basic Training.[9]

Special Forces also changed their selection process to get into the Special Forces School. Prior to 1988 soldiers would volunteer for Special Forces School and be sent to Fort Bragg. Many would not be able to pass the initial selection process, known as Phase I. The Army was then stuck with a soldier who was no longer assigned to a unit and had to determine what to do with him. Some soldiers volunteered for Special Forces training so that they could get out of the unit they were in. Once they arrived in SF School they would drop out and then be eligible for assignment to a unit different from the one they had left. This cost the Army money and it did not bring in the dedicated Special Forces soldier that was desired.

In 1988 a new system was devised, based upon the British SAS (Special Air Service) selection. A soldier who volunteered for Special Forces training would have to go through a selection process and if he failed, he would then be returned to his old unit. Instead of a PCS move (Permanent Change of Station), it would now be a TDY move (Temporary Duty). If the soldier passed selection, he would then be PCS'd to Fort Bragg to undertake the Special Forces Course.

Special Forces Assessment and Selection (SFAS)

My application to attend Special Forces Selection was accepted and I was given a packet of paperwork to fill out and bring with me to the course. In the packet was a recommended 12-week training schedule to get a candidate ready. I followed that training schedule to the letter, doing long runs, working out with weights and doing long rucksack marches on my off-duty time through the German countryside. I was in the best shape I had ever been in when I finally rode the civilian airliner to Fort Bragg.

---

[9] Ironically the Army has returned to the SF Babies during the war in Afghanistan. The Special Forces Groups were increased in size from three battalions to four, so more men were needed. Soldiers could enlist right into Special Forces out of Basic Training again if they were at least 21 years old. Whether or not they are effective, or as much of a liability as the SF Babies from the 1980s has yet to be determined.

The Special Forces selection course was called SFOT (Special Forces Orientation Training).[10] Those of us who were there nicknamed it the Special Form of Torture. We were the first group to go through this new selection process and most of the soldiers that had volunteered were from non-combat units, and very few were airborne qualified. If these soldiers made it through SFOT they would be sent to jump school.

The first part of SFOT was to determine if the volunteers could pass the physical fitness test and swim test. There were 220 soldiers trying out and after these two tests sixteen were returned to their units. I passed with flying colors due to all of the physical training I did while in Germany. Afterwards the class drove to Camp Mackall, west of Fort Bragg and moved into old tar paper shacks. These were temporary buildings that were built in WWII to house the paratroopers training for the Normandy invasion. These "temporary" buildings were over forty years old.[11]

The next part of the selection process was a psychological test that took about three hours. I don't think anyone failed that test. After that there was a lot of confusion about what would happen next and the rumors were rampant. The cadre did not scream and holler like Drill Sergeants, but instead just told us what to do and where to do it in a calm voice. If we didn't show up on time and in the right uniform it would be noted by the instructors.

The day after we arrived, we did a morning run. All that we were told by the instructor was to run down the road until we were told to stop. We didn't know if it was going to be three miles or twenty-three miles, so no one knew if we had to pace ourselves or sprint. The first run turned out to be three miles. During the whole time we were watched. The only standards that were told to us, again and again during the entire selection were "do the best you can."

That same day we also did a rucksack march, with the same vague standards, in the hot June sun. Soldiers were randomly picked and told to dump out our rucksacks to spot check that they had everything, and it was the proper weight of 45 pounds. If something was missing, it was noted by an instructor. There were no showers to wash off after each event, so each man had to clean himself the best he could by getting water out of a "water buffalo".[12]

---

[10] SFOT was pronounced "Soft-Tee". A few years after this they changed the name to SFAS (Special Forces Assessment and Selection).

[11] These shacks were finally torn down in 2004 and modern barracks were built where they once stood.

[12] This is a round water tank on a trailer.

Chapter 3: SPECIAL FORCES – A Storm in the Desert

No one was allowed to quit, or voluntary withdraw, until Day Four. Soldiers could be removed due to injuries or failing the swim test, but no one could just quit. When Day Four did arrive, many couldn't take the constant physical activities and just gave up. Whenever a soldier quit the loudspeakers in the camp would play "The Ballad of the Green Berets". Some candidates would wait until night, when everyone was sleeping, because they were embarrassed. However, the loudspeaker would wake us up when it blared out the song, letting us know another one couldn't take it.

On Day Four we had to do an obstacle course that had twenty-seven obstacles. Most of these were rope climbing obstacles. I was never very good at climbing a rope and I did poorly on it. I was worried it might affect whether or not I passed, but I drove on to the next mission. On Day Four they finally opened up the showers, telling us that before that time the showers had been broken. Each night we would be told to be in bed by a certain time. If a soldier wasn't in bed it was noted by the cadre, who was always watching. We were never told what time to wake up; the cadre would just tell us "We'll tell you when you need to know." The constant life of not knowing what would happen next wore some soldiers down. After we did another layout of all our gear, and right after we were supposed to go to bed, five people quit in our shack. By the morning of Day Five fifty-six people had quit and gone home.

Day Five started at 0600 with a 3½ mile run, and then we moved to the Sandhills Training Area to learn land navigation. For some, especially those soldiers from non-combat units, this was extremely hard. For those of us with some infantry background it was pretty easy. This training went on throughout the day and then into the night so that we finally returned to the shacks at 11:45. We were told that the next day the actual land navigation course would begin at 0600.

Normally if a soldier does a compass course, they would be told how many points they would need to find. Not in SFOT. We were told to go to a point on the map that we had been given and take further instructions once we got there. It was hot, real hot, and the points went on and on, never ending. We were allowed to use the roads, since so many of the soldiers had never used a compass much before. The average distance between each of the points was 2½ kilometers or a about a mile and a half. There were some dangers lurking in the woods for the unlucky. One of the candidates was bitten by a water moccasin. He chased down the snake, caught it and cut off its head so the medics would know what type of snake it was. The medics medevac him out of the camp, but he returned that afternoon because the snake had not injected any venom into his leg. The heat and the snakes were too much for some candidates and a few more quit.

I was doing pretty well, though I was physically exhausted much of the time. In some instances, I was better off for all my experiences in the Army. I finished the night land navigation course before anyone else, around 0100 and I was able to sleep in the road by the trucks until 0430. When we were driven back to Camp Mackall the truck went into a ditch "by accident" and we had to road march six miles back to camp. Back at camp we were able to sleep for an hour and a half, and then they woke us up again for another day of land navigation. A lot of the candidates' feet were blistered and sore, but I had ten years' worth of infantry on my feet, so there were no blisters or soreness. For some it was too much. The "Ballad of the Green Berets" seemed to never stop playing.

Whenever I had a moment back at the camp, I would try to wash my clothes in the outside sink. I mainly wanted to get out the salt stains left by dried out sweat, since the salt would rub your skin raw. By Day Seven the land navigation points were getting farther apart and we only had to find three points in six hours. We never knew the time period or how many points there were though. Odd things were put into the course to see how a candidate would react. At one of the compass points, I walked up to an instructor in a lounge chair, wearing sunglasses, red shorts and a T-shirt, underneath a beach umbrella, drinking from an ice-cold glass of lemonade, that had its own little umbrella. He took my paper, handed me another point and told me to move out. I laughed, but I figured it was part of the evaluation. The points were about five kilometers or a little over three miles apart. We finally finished up around midnight and were told to go to bed. We were told that we would be awakened the next day for training...everyone thought that land navigation was finally over and we could get some sleep.

First call was an hour later, at 0100. When we were awakened, we were told we would be doing another day of land navigation. A lot of the candidates just quit right then, unable to stand the pressure, the physical pain or the lack of sleep. The rest of us resumed the torture, and moved through the dark woods, looking for the points on the ground...for ten more hours. When we finally got back to the camp at noon, I fell asleep until 4:00 in the afternoon. That night they gave us ten hours of sleep and it was heavenly. They just wanted to see who would crack due to the continued land nav exercises and they achieved their goal.

Day Ten started out at 0330 with a ten-mile road march. We didn't know it was ten miles of course and were just told to "do the best you can". By the end I was feeling pretty good, though many of the candidates looked like old men or cripples as they limped around on their blistered feet. One soldier wrapped his whole foot in moleskin, making a type of moccasin, hoping it would stop the pain. It only made it worse and he ended up quitting.

Chapter 3: SPECIAL FORCES – A Storm in the Desert

When we finished the road march, we were told we had completed Stress Gate One. Each candidate was brought before an assessment board and told how they had done. Fifteen candidates were told that they had failed the first part of selection and were sent home. I wasn't one of them. The rest of the day was a vacation…we ate hot meals, washed and sewed up torn uniforms and slept as much as we wanted. It was fantastic!

During that day I bumped into Brian Jeznach. He had volunteered for Special Forces, but he was in the class before they started SFOT. He told me he had not made it through Phase I due to the heat and now he was in Echo Company, the detail company. We didn't talk for long because he had to finish the detail he was on.[13]

On Day 11 we took it easy again. The cadre must have thought the damage to the feet of most of the class was too extensive. We ate hot meals and then did some round robin classes on how to make a poncho raft, how to tie certain knots, and how to make a rope bridge.[14] This would be critical for the second phase of selection. Before we went to bed the instructors rearranged which hut everyone was in because some huts had lost so many people, they couldn't pull night guard duty.

The first half of the selection was all about the individual. Could the individual candidate put up with the physical pain and the unknown future? The second half of selection was about the team. How well could the candidate work with others? The beginning of the team phase was physical training, where the team did log drills. This consisted of making everyone hold a long log while doing sit-ups or something similar. When it came time for the run the cadre told us the standards, "do the best you can" and then told us to go. We were stopped after only 200 yards and then told to go into a classroom.

---

[13] I would continue to bump into Brian Jeznach throughout my Army career and afterwards. He continued going to the reenactments and I would bump into him there. The last time I saw him was at the 225th anniversary of the Siege of Yorktown, Virginia in 2006. He met my wife and daughters and I found out that he had divorced his first wife. He had gotten out of the Army and was in the Westminster, Maryland Fire Department. He remarried afterwards, but he was killed while I was writing this book in 2010. He had been working on his jeep and the jack collapsed, crushing him underneath.
[14] A poncho raft was constructed by taking two Army issue ponchos, snapping them together and putting all the gear inside it. It was then tied up to make it semi-waterproof. Though everything always seemed to get wet, it would allow the soldier to get all his gear across a body of water without sinking to the bottom.

Triple Canopy

We were briefed on what the future missions would be. Each team of twelve men would be doing "civic action" missions in the fake country of La Paz. After the briefing we had another easy day of sleeping, washing clothes and eating. In my diary I wrote "Morale is pretty high. There are a few worry warts that are driving the rest of us crazy, but everyone else is just blowing it off. It's all downhill from here. What could they do to you?"

Little did I know.

For the first mission I was put in charge of the team. At 0600 on Day 12 we were told to move to a point on the ground by an instructor, but before we reached it a friendly native of La Paz needed our help. We were told we had to move two "bodies" made of sandbags, on stretchers made of iron pipe, to a medic 8½ kilometers away (5 miles). The total weight of each sandbag man and stretcher was 250 pounds and each of us was still carrying the 45-pound rucksacks that we carried for the entire selection process. I came up with a sequence where two men would carry the stretcher for five minutes and then switch off, then after 15 minutes we would take a three-minute break.

As we moved across Rhine-Luzon drop zone I looked back and saw other teams moving across the sand. It looked like some of Pharaoh's slaves toiling across the desert. All the while the evaluators continued to observe and make notes. They watched who did their fair share, who went even further, and who held back and let others do the work. Our system worked well and we were able to complete the distance in four hours. We were soaked to the skin due to the heat and we were exhausted, but we were still in high spirits and we were able to eat lunch. After lunch the instructors put another candidate in charge of the team.

The next task was to road march for about 1500 meters to meet another La Paz native. When we arrived, he asked us to build him a rope bridge. This was pretty easy, due to the skills I picked up while in the Ranger Battalion. They did rope bridges in their sleep there. We finished before anyone else, so we marched another two kilometers back to camp and washed the salt out of our uniforms. After four inspections, done repeatedly because the evaluator wasn't satisfied (and I think to purposely waste our time) we finally got into the beds around 10:00 that night. Though we lay there, the instructors told us don't go to sleep. To stay awake I continued a letter I was writing to Betsy and I wrote in my diary. An hour later an evaluator came in and said we could sleep now. They had been watching through the windows to see who would sleep.

264

Chapter 3: SPECIAL FORCES – A Storm in the Desert

At 0330 we were awakened and told by the evaluator that we could not go back to sleep. Every time an instructor came in and saw someone sleeping, he would ask them their roster number. He would write it down in the notepads they all carried around. At 0600 we were told our next mission. The team would have to march 10 kilometers (6½ miles) where we linked up with another La Paz guerilla. He told us we had to deliver a jeep trailer to a second La Paz guerilla, somewhere down the road. The only bad thing was the trailer had only one wheel and we weren't allowed to touch the inside of it because it was "contaminated." If any of our team members touched the inside they would now be "dead" and we would have to carry all their gear and the "dead" person to the end.

There were poles and ropes lying on the ground around the trailer, so we made a carrying system. We moved 4 ½ kilometers down the road and right before we finished one of our team touched the trailer's inside and was now dead. We had to carry him too, but lucky for us, only for a short distance.

After we delivered the trailer to the La Paz guerilla, we marched five more kilometers down the road, stopped, and ate lunch. I saw some wild plums and blackberries growing there and I ate them too. Another native of La Paz came along and asked us to build him a bridge, so we marched to his bridge site. This was a bridge made up of wooden boards and rope, but like everything else in SFOT, there was a twist. There was a minefield on either side, so we couldn't touch it at all or we would be dead. We sat around, trying to come up with a solution, but we were too tired and exhausted to be creative. Finally, we just slapped it together. The first bridge broke in half, so we built a second one. It broke too. Finally, the third bridge stayed together, but it looked like hell. We lost two guys in the minefield, so once the bridge was finished, we had to carry them and their gear 1000 meters down the road. After washing clothes and eating dinner we were told to go to sleep at 10:00 that night. I fell asleep as soon as my head hit the mattress.

On Day 14 we were startled awake at 0200 and then told that we could not go back to sleep. Like before, all those who did had their roster number written down. Sleep was becoming a precious commodity. At 0530 we marched six kilometers down the sandy road to another La Paz native who told us that we needed to move a jeep down the road. The jeep had no tires on it and he had no jack to put on the tires that were on the ground. The team used iron bars to push up the jeep and put the tires back on. We then tied the iron bars to the front and back so that we could get everyone to help push the jeep down the sandy trail. Going uphill in the deep sand took away a lot of our energy, but when we went downhill, we had to run to keep up. We pushed and pulled that jeep over the trails for 4 ½ kilometers, where we received our next mission.

The new mission was to build an observation tower out of four telephone poles, a bunch of 2x4 boards and some rope tie downs. This was dangerous work because whenever one of the poles came crashing down you had to make sure you weren't under it. Our tower was finally built, but like most things we had constructed, it looked terrible.

When we finished the evaluator told us we had to march to another mission. The team was worn out and that day I had sweated more than all the other days of selection, but we trudged down the road for another eight kilometers. It turned out it was another psych job and we were just marching back to camp. There was no other mission. As we moved down the road, we passed a group from the 82nd Airborne and the 101st Air Assault that were doing war games against each other. They stopped and stared at us, looking at the incredible sight of these sweat and dust covered scarecrows.

Back in camp the evaluators told us we could relax, but, of course, we could not sleep yet. We sat in the dark outside, watching a huge thunderstorm in the distance. The lightning lit up the sky like fireworks and after each crash of thunder we would "ooh" and "aah" like it was the 4th of July. We were finally allowed to go to bed at 2200 again, though we knew it would only be a few hours.

Chapter 3: SPECIAL FORCES – A Storm in the Desert

Day 15 started like all the others, around 0200. We all tried to keep each other awake for three hours until we received our next mission. I kept thinking about sneaking off to a dark corner of the shack to sleep, but everyone else must have had the same idea, so it never would work.

At 0530 we marched down the road for ten kilometers and met up with the La Paz evaluator who had a 55-gallon oil drum that needed to be delivered to "the people of La Paz". To move it we had four jeep tires and some iron poles. We created a go-kart looking contraption that fell apart after we moved it a short distance. We then built a rickshaw looking contraption with all four tires on one axle, and then continued down the road for four kilometers. It didn't seem too hard as compared to the other tasks we had done.

After we dropped off the barrel, we marched eight kilometers to a lake where the evaluator told us we had to get a casualty across the lake and we also had to take all our gear with us. We were given a stretcher, some water cans and two 2x4 boards. The team made a raft out of the water cans and the rest of us made poncho rafts for our own gear. This was also an easy task and we were able to get wet while doing it. Afterwards we were trucked back to the barracks and told not to sleep. I think there must have been a consensus that we no longer cared. Everyone slowly went from sitting, to laying down, to totally unconscious in very short time. I was able to get about an hour and a half sleep before they came in, woke us all up and wrote down our numbers for violating their orders. We finally were allowed to go to sleep at 2200.

Day 16 started at 0145 when we were jolted out of bed by "Another one bites the dust" blaring over the loudspeakers. Our hut was told that we would have an equipment inspection at 0300, and no one was allowed to go back to bed. When 0300 came and went we wondered if it was part of the game. I later found out that all the instructors had fallen asleep, since they were almost as tired as we were. The inspection did happen at 0400 and then we marched six more kilometers to another La Paz guerilla linkup.

The task was fairly simple this time. We had to make a rope bridge, get all of our gear across and bring along a 55-gallon oil drum weighing 500 pounds. It wasn't that hard and we knocked it out quickly. Once we were on the other side, we were told to march down the road another two kilometers to a truck that would drive us to the next mission. This was another way of playing mind games with us. As we neared the truck, and almost touched it, the driver would spin the wheels and drive off. We would continue to march close to the truck, and then it would take off again. This continued for several miles until we were finally able to actually get on the truck.

Triple Canopy

The truck dropped us off at another La Paz guerilla who told our twelve-man team that we needed to take fifteen water cans to a village that was three miles away, and we had to do it in 45 minutes. Our team worked well together, switching out cans from time to time, because due to the odd number, someone had to carry two cans. What I didn't know was that there were two cans that were only half full and weighed less than the others. These cans were marked, so that the evaluators could see if the same person would carry the lighter can the entire way, which would hurt the team as a whole. As we moved, we saw other teams that were not working together so well and were at each other's throats. One team had a few people getting into a fight over the lighter water cans.

When we finished the water can carry, we marched back to camp and were told we could sleep if we wanted to. I slept from 0930 to 1400 in the afternoon, and only woke up to do peer reports. Peer reports were mainly so the cadre could see if anyone was not doing their share when the cadre was not around. I ended up getting some of the highest peer reports in the class. After the peer reports most of us went back to sleep until it was dinner time. After dinner we went to sleep again until 0700 the next day. Some of the candidates thought that the selection process was over, but most of us knew there was one more big event coming up.

After dinner that night we were told by a Major Velky that we would be extracted from La Paz and trucked back to Fort Bragg. Most of us had figured out that there was one more big road march, so we didn't believe anything he was telling us. We were told to go to bed at 2200, but it was a very short sleep. We were awakened at midnight and formed up in the company street while it poured down rain. Teams were put on trucks to take them back to Fort Bragg, but there were only a few trucks, so this process took awhile. They had to split the teams up to fit them all in one truck. By the time I finally got on the truck to move to Bragg it was 0200. I was wet, tired and I didn't trust anything they were telling me. So, it was no surprise to me when the truck "broke down" right at the border of Fort Bragg.

The cadre told us to get out of the trucks and to begin marching down the road. It was pouring down rain, I had my 45-pound rucksack strapped to me, and I knew that it was about 25 miles to the main post on Fort Bragg. The cadre told us that there were more trucks about eight kilometers away and they would take us all the way in. Some candidates took off at a run, wanting to finish the whole thing. I knew there would be no trucks and I began the typical pace of the infantryman, about one mile every 15 minutes.

Chapter 3: SPECIAL FORCES – A Storm in the Desert

Eight kilometers away there was another truck "broken down" and the instructor told me that there was a second bunch of trucks, only six kilometers away. I waved and kept marching. At that six kilometer point a new instructor told me that the trucks had moved and were now five kilometers away. I continued my pace, passing by some candidates that were sitting on the side of the road, psychologically broken down, and had given up. The exhaustion, physical pain in their feet and legs, the pouring down rain and the total frustration of not finding any trucks had taken a toll on their will to finish. As I marched, I passed many of the candidates, but only one had passed me.

The mind games continued all night long, but after 6½ hours of marching, and 22 miles later, the end was finally in sight. I knew it was the end because there were several trucks and the commander of the Special Warfare Center was there. I had been walking with one other candidate, a big guy from the 2nd Ranger Battalion called Eric Nason. I challenged him to run in. We both sprinted to the finish and then shook hands. The selection course was finally over.[15]

The next day most of us were brought into a briefing room and told that we had all made it. There were some candidates who were not there. Even though they had gone through the whole selection course and didn't quit, there were 29 who were told they didn't make it. We started with 239 candidates, and finished with 106. No one from our team was cut. That night Nason, a Hopi Indian we called "Arizona" and I all went out to eat and then go to the red-light district of Hay Street.[16] It was a pretty wild night, but we had a lot to celebrate.

The graduation for the selection course was not really a graduation. They gave us certificates of completion and told us our Special Forces course dates. Mine would be in October. One of the instructors told us that we had walked a total of 220 miles during the entire selection process. My dad had driven up from Jacksonville, North Carolina, where he was a Salvation Army minister. I returned with him and spent the weekend with my parents. On Monday I took my final flight back to Germany.

---

[15] I later learned that Eric Nason knew Bert Puckett and they had both been in the 2nd Ranger Battalion together. We enticed Nason into Revolutionary War reenacting and as of the [first edition] of this book he was the commander of the recreated 2nd South Carolina Regiment with CSM Bert Puckett.

[16] Hay Street today was renovated and is a nice restaurant and shopping area of Fayetteville. The strip joints, prostitutes and drug dealers of the "Fayette Nam" days are long gone.

Many of the soldiers who made it through selection took 30 days leave or were put on light duty due to all the injuries they received in SFOT. I was a little bit different. Four days after I returned to Germany, I was sent to the German Mountaineering Course in Mittenwald. For the next week I climbed the tallest mountains in Germany and then ran down the loose rocks to get to the base of the mountain, ripping apart my jungle boots from all the skree.[17] It was great and it was a fantastic get away from the Army. Every night was a party in Garmisch or Mittenwald. In the end I was awarded the German Edelweiss and then reported back to Coffey Barracks in Ludwigsberg.

Leaving Germany was a bit anti-climactic. I had to fight the chain of command of F Company every step of the way to convince them I actually was leaving and going to Special Forces. They tried to do everything they could to hold me back because they needed bodies for the big REFORGER operation. However Special Forces training trumped all other orders and I was able to get out of there and take a long 45-day PCS leave before I started Special Forces. One of the first things I got when I arrived at my parents' house was a "Dear John" letter from Betsy. She had met some college guy in Charleston and decided that we wouldn't have worked out. I started the Q-Course feeling totally down and swearing I would not mess with any woman again unless it was just for sex. I had been burned too many times and I wasn't going to get hurt again.

Special Forces Qualification Course

After in-processing at Fort Bragg and getting issued our gear we were trucked out to Camp Mackall to do Phase I of the Q-Course. Prior to SFOT the Phase I portion was where the weak soldiers were eliminated before moving on to Phase II. However, SFOT eliminated the need for that and Phase I was revamped into a course where there was no yelling or screaming, and each SF candidate focused on the training. Phase I would last for only 30 days. We started the course with 160 candidates. Prior to SFOT there would have been 300 to 500 candidates in each Phase I class.

We learned that one of the SFOT graduates had been killed in Airborne School on a jump and a few more quit jump school because of what happened to him. Before we ever went to Camp Mackall they removed seven others who failed the height/weight test. They were too fat. They didn't get thrown out of

---

[17] Skree is all the loose rock on a mountainside. You can run down this, pretty fast, and it felt like skiing. I was there with other soldiers, but I linked up with members from SEAL Team 2 for most of the course.

the course, but were not allowed to move on until they lost the weight. If they didn't lose the weight, they would then be reassigned somewhere else in the world.

One of the first training sessions we did in Phase I was the survival course. The candidates received training on fire making, signaling, finding your direction without a compass, making traps and snares, building shelters and how to slaughter animals and determine if they are healthy enough to eat. For the last class they brought in a live goat, showed the quickest way to kill it, and then after butchering it, pointed out the different parts of the goat and what could be eaten and what couldn't. Some of the candidates had never killed anything in their lives and they looked a bit squeamish.

That night we had a group camp and built hooches around a giant campfire. I found some sassafras and cooked it up so everyone could have some sassafras tea. Survival skills would not be that hard for me since I had been through the Special Forces Desert Survival course and the German Winter Survival Course. I also had a decade of reenacting experience of living in the woods using only 18th century skills. It was like camping.

For our survival exercise we had a rucksack with two ponchos, a sleeping bag, field jacket, three socks, a pack of matches, a roll of MRE toilet paper and a knife. We moved through the woods about five kilometers to where the survival area would be. We hadn't eaten since noon the day before, so everyone was pretty hungry. Along the way I picked up "lighter knot" and threw it in my ruck.[18] Some of the soldiers were wondering why was I carrying huge pieces of heavy wood in my rucksack and thought I was being foolish.

At our survival site we were given a live chicken, one onion, a cup of rice, a plastic bag to put the remains of the chicken in, and were told the rules of the exercise. We would be taken to a specific location that would be separated from other candidates. We were not allowed to leave that area, nor were we allowed to communicate with anyone else. We weren't told what would happen to anyone who broke the rules, but we assumed that they would fail Phase I.

Right after the instructor put me in my survival spot, I built a poncho hooch, gathered up a big pile of pine needles and made a bed underneath it. I then took

---

[18] Lighter knot is also known as lighter wood or fat lighter. When a pine tree dies gravity pushes the sap into the lower area, the part of the tree that will rot away last. Due to this all the wood in that area will light on fire like it was soaked in gasoline. The only bad thing about lighter knot is that you couldn't cook over it due to the high concentration of creosote.

the lighter knot from my rucksack and using dead branches from around the area I soon had a small bonfire going. I only had to use the single match due to the lighter knot being so thick with sap. Right after the wood caught on fire it began to rain, so I took my second poncho and strung it up high, over the fire pit, making it look like it was a front porch.

Dinner that night was pretty good. I dumped the rice into my canteen cup and began boiling it up. I cut up the onion and put it in the rice. While it was cooking, I killed the chicken and strung it up because I really didn't feel like having to chase it down if it got loose. My field jacket got soaked a bit, so I put it by the fire, but I didn't pay attention to it and a hole was burnt into the sleeve. I cooked the chicken on a hastily constructed spit and soon had a pretty good meal. I only ate half the chicken and was going to save the rest for breakfast.

As I was eating, I heard a rustling beside me in the darkness. I looked over and I saw one of the other candidates, soaked to skin, staring at me. He threw me a half of his uncooked chicken. I figured he couldn't get the chicken cooked due to the rain, so I held up a piece of my cooked chicken. He shook his head "no" and in a hushed voice whispered "I need coals!" I put a handful of coals in his canteen cup and he turned and vanished into the night. About an hour later I was laying in the warmth of the fire and I heard him again. He shook his canteen cup at me like a beggar on the side of the road. His fire must have gone out. I gave him two pieces of fat lighter that I didn't use, and lit one of the pieces from my fire. I also told him how to make a fire so it lasted. He didn't return again that night, so I figured he must have either made a fire, or given up.

The next morning was a beautiful day. The sun came up and I made some sassafras tea and ate the rest of the chicken. I wandered around my small area and made a "solar still" to collect water and then set up a few traps and snares. The instructor came by in the afternoon to see what I had done and he put "exceeded standards" on the grading sheet. He told me that two of the candidates quit the night before because they couldn't handle sleeping in the woods all alone in the middle of a rainstorm. I thought it was a great time. We were now down to 149 candidates after less than a week of training.

The day after the survival course we marched seven kilometers to a lake to learn about water operations. The training mainly focused on using Zodiac rubber boats that could hold 15 men. These boats were simply known as RB-15s. It was November so whenever we had to flip the boat, and do a capsize drill, we were floating in freezing water. After the training we were told that we had to carry the boats 4½ kilometers to Luzon drop zone on Camp Mackall. The boat weighed 242 pounds and we all carried 45-pound rucksacks, so it reminded us of SFOT. It warmed us up though. Once the boats were dropped off, we had to run five miles back to the same camp where we stayed during SFOT.

For the next few days, we were in the classroom learning land navigation. We were taught the basics, but we were also taught how to make maps from aerial photographs and how to read a latitude/longitude map. Being in the "rear" taking classes was great because a handful of us had developed serious colds. The only treatment was Contac cold pills and Sucrets throat lozenges. I also ended up with diarrhea for the rest of the time in Camp Mackall. This may have been due to the water there, which made a lot of us sick. The medic just kept making me drink Kaopectate until I had white water coming out every time I defecated. It didn't work, but I drove on anyway. I just took a lot of toilet paper everywhere I went.

The land navigation portion of Phase I took place near Hoffman, North Carolina. Special Forces created numerous lanes in the dense woods back there. We would be in the Hoffman training area for five days. I was never worried about land navigation since I had ten years experience under my belt, but many of the candidates, especially those from non-combat units, were pretty paranoid about failing it. If you failed the land navigation part of Phase I you would be recycled to the next Phase I class. If you cheated you would be terminated from the Q-Course.

All of us did a terrain walk for a distance of fourteen kilometers, with the instructors showing us what to look for during the upcoming land navigation tests. That night we made a huge camp near a field and we were allowed to have fires. The next day, the first day of Land Nav, was a simple test. We had to find four points in four hours. Unlike SFOT, candidates could not use the roads. If they did, they would be considered cheating. I finished pretty quickly and then lay around sleeping until everyone else finished. My sleep was interrupted with periodic bouts of running to the woods to use the bathroom, so it wasn't a great time for me. I kept it secret though, because I didn't want anyone to know about the diarrhea because I thought they would medically remove me from the course.

That night we did our first night land navigation. The candidates could not use flashlights and what we were looking for was a small white piece of wood, known as an engineer stake, stuck into the ground. It was pretty hard and I only found three out of four points, but that was considered passing. The next day I finished the day land navigation test, but I was feeling really bad. I went to the medic and he told me that I had dysentery. It was such an ancient illness, like one you would read about from the Revolutionary War. One of the candidates was as bad off as me and was sent back to Fort Bragg to recover. I kept my illness as secret as I could. I felt so weak from dehydration that I only found two out of four points that night. My team's SF instructor, SGT Gizoni, gave me some pills and it helped a little. Luckily the next day was remedial training for any candidate who didn't get at least 75% of the points in the last two days. I was good and slept most of the day.

That night we had our last and our longest land navigation test. We were trucked to a start point and began the test at 0200. My first point was six kilometers away and I had to swim across Bones Fork Creek. I panicked a little when the water was over my head. Here I was, in the middle of the night with no one around, crossing the freezing water. I said a prayer to God and put my faith in him to get me across. I made it, and though I was freezing, as long as I kept moving it wasn't too bad.

I made the first point and was given the coordinates to my next point. It was six kilometers back, going across that creek again. This went on for two more points, moving four to five miles across country, and then coming back across that creek. My final point was only 700 meters away and it seemed trivial compared to what I had been through that night. Each candidate had nine hours to finish the course and I did it in seven hours. I was told that 40% of the class had not made the standards. Out of the ten men in my team, only three completed it to standard…I was one of the three. There would only be one more retest for those who didn't make it at the very end of Phase I.

The next part of Phase I was patrolling. I was issued an M-60 machinegun and the class was taught basic patrolling techniques. I didn't think this would be a big deal since I had plenty of experience with this too. Rangers eat, sleep and breathe patrolling. I still had diarrhea, but the only thing the medic told me to do is to not eat anything at all and drink a lot of Kaopectate or water. I think the medics were stumped at what was causing it, and though I had diarrhea for over a week, I didn't look too bad, so they let me continue training.[19]

We marched to a patrol base with overloaded rucksacks that weighed more than 75 pounds. The first day and night we had classes on various subjects about patrolling or defending a perimeter. The instructors took those of us who had been in the Ranger Battalions and had us teach the classes. In our team SFC Young and I did all the teaching. The next day we taught more of the same, and for the first time I could go to the bathroom like a normal person. I was a happy camper again! We did one more day of learning how to move and fight as a

---

[19] When I was on a mission to climb the Himalayas of Pakistan in 1994, I once again had diarrhea for weeks. The medic was also stumped, but I kept going along, and just carried a lot of toilet paper, as I climbed that 28,000-foot mountain.

team and then we marched back to Camp Mackall. This portion was a "basic training" for the Special Forces candidates who had never been in combat units.

The next few days were classes about how to do ambushes or how to write an operations order, but it was pretty boring to me. I fell asleep in a few of the classes, but I couldn't help it. The final part of the tactics class was a written test that I also nodded off while I was taking it. I only missed one question though and scored a 95%. That night we were issued fifteen MREs for the six-day field problem. I jammed it into my Alice Pack and then fell asleep.

The next six days were a test in tolerance for those of us who were Rangers or Infantry. The instructors knew who needed to be taught and who had experience, so each patrol had an extremely inexperienced person in charge. They would forget many things that would get a person fired in the Ranger Battalion. To make it even more miserable it was pouring down rain the first day. The instructors must have felt some sort of pity because the next patrol leader was allowed to pick three people to help him. Our patrol leader picked SFC Young, a former Marine and me. Since we were now instructing him, we were allowed to go "admin" and we cooked up some coffee over a fire. The patrol leader asked SFC Young "what would you do if you were me?" Young told him "I would kick back here, drink coffee and take a nap." The patrol leader asked him why, and he said "well if you was me, you'd be a Ranger and you would already be good to go. You wouldn't have to be patrol leader." The sun came out that morning and raised some spirits when our clothes dried on our backs.

The mission that day was a bit better than the day before, even though the patrol leader had forgotten two people back in the patrol base who were pulling security. We finally finished the patrol after dark. It was a cold November night and we only had two meals that day. The SF instructor came to me and told me that I would be the next patrol leader. He gave me a short easy patrol because everyone had to be graded at least once. I led a recon patrol of an ambush site and got an "exceeded standards" grade. I merely did what the Ranger Battalion would have expected.

By the end of the week the team was getting better at patrolling and making fewer mistakes. There were some instructors who would throw artillery simulators if the patrol was moving too slow, because the instructors wanted to get into the patrol base and sleep that night. We really hated those sergeants. Every time a simulator would go off, we would have to run, hundreds of meters through the woods, to get away from the "artillery". We had one candidate quit because he said the patrols were "bullshit" and he didn't join SF to run through the woods.

275

When we returned, we all had a four-day weekend due to Thanksgiving, except those candidates who didn't pass the land navigation test. While we were cleaning the weapons, a process that lasted until 7:00 at night due to anal retentive armorers, those Land Nav failures retook the test. Fourteen of them failed and were recycled to the next Phase I class. In the end we finished Phase I with 120 candidates that would move on to Phase II.

When we returned back to Bragg the wives of the married candidates greeted us with cakes and sandwiches, and some pretty pornographic welcome home posters. I bumped into Brian Jeznach again and we both decided to go to Jim Collins's house for Thanksgiving. Jim Collins was the president of the recreated 2nd North Carolina Regiment and his son, Jimmy, was the 18-year-old commander of the Regiment. They both lived in Lillington; about fifteen minutes drive from Fort Bragg. Towards the end of Thanksgiving, I ended up in Jacksonville, visiting my parents and washing five loads of my dirty laundry.

The next weekend I drove down to St. Augustine to do a reenactment. Bert Puckett was there with his wife Jeannie, and his sister Betsy was there too. Betsy said she was amazed at how much she missed me and that she wanted to see me again. I didn't react at first, but that night Bert, Jeannie and I went out drinking on the town. When I returned from the camp fire Betsy was there waiting for me. We ended up in each other's arms, and I felt gloriously happy again. I was still in love with her. All things must end though and when I returned from St. Augustine Phase II began.

For the Phase II portion of the Q-Course each soldier would go learn whatever job they would do on an A-Team. I was training to be a weapons sergeant, so all of us weapons sergeants would be together for three months. Depending on the job assigned to the soldier, the training could last longer. The Special Forces Medic course lasted almost a year and they learned how to do minor surgery and dental work. During Vietnam era there was a Heavy Weapons Sergeant who learned everything about mortars, heavy machineguns and recoilless rifles, and there was a Light Weapons Sergeant who learned everything about all the major small arms used by the armies of the world. However, by the 1980s there were just Weapons Sergeants and we learned both heavy and light weapons.

All of the weapons training took place on Fort Bragg, mainly in one old WWII building near the SWC (Special Warfare Center) arms room. The SWC arms room is actually a large warehouse with a few dozen civilian workers keeping the weapon systems working. We started our training by learning about air defense missiles, such as the Redeye, Stinger and the Soviet SA-7 Grail. For practice we would take the missiles over to Pike Field and have them lock onto the aircraft taking off from Pope Air Force Base. At the end of that week, we had a test on those missiles and everyone passed. Unlike Phase I, there wasn't a threat of being recycled looming over Phase II...yet.

SWC shut down training during the last two weeks in December because of Christmas. Most everyone went home except Nason and me, so we had the barracks to ourselves. I didn't want to go home to my parents because they were in the Salvation Army. This was the busy time of the year and I wouldn't see them anyway. The one exception to this was when I went there on Christmas Day, since they wouldn't be working then.

During the holidays Nason and I went to a lot of bars, strip joints and just hung around the almost empty city of Fayetteville. It was empty due to being a military town and most of the soldiers going home on leave. A few times we went over to the Collins's house in Lillington. Nason was hooked on doing reenacting after seeing some home movies of the different battles. During that time, I also went over to the 7th Special Forces Group and talked to the Group Sergeant Major, CSM Ivan Ivanov. I wanted to get assigned there because they were the Group seeing the most action in the 1980s with the war in El Salvador and Honduras.

On New Year's Nason and I went down to Bert's dad's house in Georgetown, South Carolina. It was a surprise visit and Betsy was amazed to see me. Though we had talked through the night at St. Augustine, things had changed again and we weren't going together anymore. One of the main reasons I wanted to go was to see her brother, Bryan. He was going to the Citadel and his plan afterwards was to be commissioned an officer. I wasn't as close to Bryan, like I was to Bert, but he was still part of my adopted family and like a little brother to me. In the day we would go to a range and fire up hundreds of rounds of ammo, and then at night we went to various parties held at the houses of Betsy's friends. Betsy would drive us back since we were too drunk to do it ourselves. I wasn't happy though because I wanted to be close to her, and she wouldn't let me. On New Year's Eve Betsy and I stayed up till 4:00 in the morning talking. She kept telling me she didn't want to hurt me, but I told her it was too late. I told her I didn't think there would ever be another girl in my life.

When we returned to Fort Bragg, we began the instruction on anti-tank weapons. These were the WWII era 57 mm recoilless rifles, the 90mm and 106mm recoilless rifles, the Soviet RPG-2, RPG-7, Sagger missile and the SPG-9 recoilless rifle. Little did I know that I would be encountering all of these Soviet weapons in combat in just a few more years. At the end of that week, we were tested again and no one failed.

Triple Canopy

In our off time Nason and I would go to Rick's Lounge, the main strip joint on Hay Street.[20] Nason had a girlfriend there, a really big girl, who was a waitress. We became such regulars that when we came in the place would chant our names. Nason was "Nase" and I was "Troll" ... my nickname in the Army due to the wild hair I would have sometimes. It would look like a troll doll prize from a carnival. I also spent a lot of time at the Collins's and goofed off with Jimmy Collins, who was now a senior in High School.

During this time, I began to lose my religion a little too. I was depressed about not being with Betsy, and I wrote to my parents that I didn't want to go to church when I visited them. I felt odd about it. I also wrote to them that I had quit praying because I didn't think it made a difference anymore. I was almost 30 and I hadn't had a girlfriend or even a normal date in three years. It was a depressing time, though it should have been one of the best for me.

Towards the end of January, we began the hardest part of the heavy weapons instruction, the mortars. Mortars, unlike a recoilless rifle or an RPG, just couldn't be aimed at the enemy and fired. You had to know how much explosive force to use on each round, what the angle of the mortar tube was to hit the target and a lot of other variables to make that explosive round come down upon the enemy within a few shots. Out of all the weapons I learned in Phase II, skill with the mortar is the one that I ended up using more than any others during my time in Special Forces.

The instructor who taught mortars was an old SF veteran named Sully. His last name was Sullivan, but we all called him Sully. He had been a heavy weapons sergeant in Vietnam and now was contracted to teach us in Phase II.[21] For two weeks we learned the basic mortarman skills. We would set up the mortar, and prepare it for firing, then break it down and pack it away. We would then put new information on the mortar, for distance and elevation, and have to set it up again. We also learned how to "lay" the mortars in line using the M2 aiming circle. At the end of those weeks, we were tested and like the other tests, no one failed, though it was close for some. I received a 100% for my score. Though my personal life sucked, I was excelling at my military skills.

---

[20] Rick's Lounge was one of the most famous strip joints in Fayetteville. Today the restaurant *Huske Hardware House and Brewery* sits where Rick's Lounge used to be.
[21] Sully was one of the few people who had fired the 60mm mortar from the hip, and lived to tell about it. A round was put in the barrel; he slung the tube around his shoulder on a strap, and then pulled the trigger near the base. The mortar fired, and whirled around him, like a hula hoop, knocking him over. We saw this through a series of pictures he showed. The man knew EVERYTHING about mortars.

Chapter 3: SPECIAL FORCES – A Storm in the Desert

In February, we were taught the Fire Direction Center (FDC) skills needed to hit a target. The FDC is the soldier who would give the mortarmen the information that they would set their mortar sights on. This included a lot of mathematical calculations and trigonometry. I never thought I would use math after High School and now I was doing a crash course on advanced math. This, more than any other skill, stumped a lot of the candidates but I had a handle on it.

During this time an odd thing happened. An instructor came into the class and asked how many of us were Rangers. About twenty of us raised our hands. He then asked how many of us had been in a Ranger Battalion. This time only fourteen raised their hands. He then asked us how many of us were left-handed. No one raised their hands. He then said, "I know where fourteen of you are going" and walked out of the room. We all pretty much knew it too when he said that. The only place in the world where the left hand was frowned upon was the Middle East, and the Special Forces Group assigned to that part of the world was 5th Special Forces Group.

5th Group had been in Fort Bragg since the beginning of Special Forces, but they were about to move to Fort Campbell, Kentucky. They were going to move two battalions to Fort Campbell and leave the 3rd Battalion behind to become the 1st Battalion of the newly formed 3rd Special Forces Group. The Army would then create a new 3rd Battalion in 5th Group, but to do that they needed a lot of men. I guess the Group commander said he wanted Rangers in his new Battalion.

After a week of learning FDC procedures we then learned Forward Observer (FO) skills. The FO was a soldier with a radio, observing the enemy, and calling the FDC and telling him where the enemy was. The first rounds would usually fall short or long, so the FO would then make corrections and call it into the FDC so that the mortar rounds would impact on the target.

During this week something happened on the world stage that would have ramifications for the United States in the next century. The Soviets pulled out all of their forces from Afghanistan, ending the Afghan war for them. The cost of the war and trying to outspend the United States would eventually crush their economy. After the war in Afghanistan the Mujahedeen were waiting for some relief from the United States, to rebuild their nation from a decade of war. No relief was coming and the Taliban would rise up out of this chaos to restore order. For those of us in the Q-Course it looked like a defeat for the Soviets, but none of us knew that the Communist empire would come crashing down in just a few short years, making us the victors of the Cold War.

Triple Canopy

We went out to the range for two days and fired the mortars and recoilless rifles. Everything we had learned for the last month was put to the test with live ammo. It was not a pass or fail kind of test, so everyone was relaxed and had fun. That was the final part of the "heavy" weapons instruction, so that weekend we went out every night to Rick's Lounge or other night clubs. Rick's Lounge had started an odd competition known as midget tossing. A midget, in a special suit with handles, would be picked up and thrown as far as he could into a padded area. Nason entered the midget toss, and should have won, but the manager of Rick's gave it to someone else. When that happened Nason's waitress girlfriend jumped on the manager's back and a fight broke out. It was mainly a fight between the waitresses and the manager, but it was a great time.

When we started the "light" weapons training we had new instructors. The heavy weapons instructors were all professional, experienced soldiers or former SF. Unfortunately, two of the light weapons instructors were SF babies and they seemed to have a chip on their shoulders. They wanted to fail as many as possible. Only one instructor, SFC Johns, was squared away and wanted to make sure each of us understood the weapons. The final test of the light weapons course is something known as the pile test. We had to learn almost 300 different type of weapons, and at the end of the five weeks the instructors would randomly pick five of them, disassemble them all into one big pile, and then each student had 29 minutes to put them all together again. If you failed that test you would be recycled.

The first week we learned pistols and submachine guns. We learned on every type of weapon, some going back to WWI. Unfortunately, each night the two SF babies would give all of us a spot quiz on everything we had just learned a few minutes earlier. Needless to say, we all scored low. They would then tell our company commander that we weren't scoring high, but the company commander already knew about them and he told us not to worry about them.

The following week we learned rifles, which really weren't that hard, except for one...the BAR (Browning Automatic Rifle). This was one of the hardest weapons to put together due to all of the intricate pieces. Normally a weapon, such as the M-16, could be put together in a few minutes, but the BAR took fifteen minutes to reassemble. All of us hoped it would not be on the pile test.

Early in March we were taught machineguns. Some machineguns were pretty dangerous to take apart if you didn't know what you were doing. The compressed spring on the M1919 .30 caliber machinegun had almost taken one student's eye out when he assembled it. The M2 .50 caliber machinegun had a bolt that could go through a wall if it was not assembled correctly. To prove the point the instructors showed us the holes poked in the walls from previous

280

classes. When we went to the range to fire the machineguns, I was one of three chosen to fire the Soviet DSHK 38/46. This is the big Soviet .51 caliber machinegun. There was limited ammunition so only the top scorers on the spot tests were able to fire it.

It was pretty cold when we were at the range and the snow came down towards the end of the two days we were there. Day One on the range was mainly loading all the magazines for the weapons we would be firing on Day Two. There were thousands and thousands of rounds to be loaded and it took hours. Firing all of the weapons the next day was pretty fun though.

On the days leading up to the pile test the instructors brought in 200 different kinds of weapons and let the students pick which ones they wanted to practice taking apart and putting together. All of us were trying to guess which ones were going to be on the pile test. When we asked the recycled students who went through the test before, they were no help because there were five different weapons each time they did the pile test. One of the students who sat beside me in the class was "Nick" Nikolai. Every time he put a weapon together, he always had a few parts left on the table, and ended up with a confused look on his face.

At the end of March, we had the pile test, but only a few students could test at a time, so the rest of us waited outside the classroom. There was a fence surrounding the weapons training complex, and beside it was the Vertical Wind Tunnel for the MFF School.[22] This building housed giant fans that would blow upwards, simulating free fall out of an aircraft. While we were waiting around for the pile test Nason told me to go over the fence and check out who was there. The 2nd Ranger Battalion was training on Fort Bragg and a company of them were on the other side of the fence getting ready to do a mission. When I walked over to the fence, I saw Bert Puckett and I also noticed the sergeant stripes on his collar. I was surprised at that, since it seemed he had just gone into the Rangers a few years before. We shot the breeze for awhile, but soon it was my turn to go take the pile test and I had to tell Bert goodbye.

I wasn't worried about the test and when I took it my goal was to not just pass, but to break the record of assembling all the weapons and doing a function check on each one within 18 minutes. The weapons on my "pile test" were an M-1 Garand rifle, an M3A1 Greasegun, an M1911A1 pistol, an M-2 .50 caliber machinegun and an M1919A6 .30 caliber machinegun. I had no problems, and

---

[22] Military Free Fall, commonly known as HALO school for the acronym "High Altitude, Low Opening.

I did come close to the record, but I didn't beat it. I had assembled all of the weapons in 19 minutes.

Eight students failed the pile test, but they were allowed one more retest. Only one of those, Nason, passed the retest and the rest were recycled back to the beginning of Phase II. One of the guys who were recycled quit the Q-Course because he wanted to return to the life of the Ranger Battalion. He couldn't stand the idea of having to do Phase II all over again.

The day after the pile test, we went back out to the Hoffman training area and were taught about squad and platoon SOPs (Standard Operating Procedures). We did a night movement to a new location and the next day learned how to dig a mortar pit, and the M-60 and 90mm recoilless rifle fighting positions. As weapons sergeants we would need to know how to do this and train others once we were on an A-Team. The final part of the field problem was for the students to plan an ambush. Only one of the plans would be chosen and then all of us would follow it. The instructors chose my plan.

That night we did a movement and an ambush with blank ammunition, and then afterwards we drove back to Bragg. In my diary I wrote "Totally wasted field problem, learned nothing new." I wasn't trying to be arrogant, it's just that in 3rd Ranger Battalion we had lived and breathed tactics like the ambush, recon and raid.

Before we started Phase III, we learned what group we were going to be assigned to. Almost the entire class, including myself, was going to 5th Special Forces Group in Fort Campbell, Kentucky. I wasn't really thrilled about that since I wanted to be assigned to 7th Group. Fort Campbell seemed so far away from everyone I knew, and from reenacting with the 2nd North Carolina Regiment.[23] Nason was one of the few who was not going to 5th Group and instead was going to 10th Special Forces Group in Bad Tölz, Germany. This was mainly because Nason was fluent in German and had a working knowledge of Russian.[24]

---

[23] Many of my choices for assignment in the military were based upon whether or not I would be close to reenacting. Even my choice of a unit where I would retire, 3rd Special Forces Group in Fort Bragg, was based upon how close I would be to my main hobby of reenacting in the Carolinas.
[24] Erick Nason published a book in 2006 on his experiences in Special Forces, titled "From Desert Storm to Iraqi Freedom: One Soldier's Story". He earned his doctorate in 2014, so the huge Ranger midget tosser is now known as "Doctor Nason".

Chapter 3: SPECIAL FORCES – A Storm in the Desert

The final test for Phase II was a written test on subjects that were never taught by the instructors. As we were taking the test, we realized that much of the material was unknown to any of us. Every one of us failed the test, which angered all the students. The instructors realized they had given us an old test, and so they graded all of us on a curve, and everyone passed. We were now 18Bs, or weapons sergeants. By the end of Phase II, I felt pretty good about what I had done and since my scores on all the tests were so high, I set my goal to become the Honor Graduate from the course. Unfortunately, I became too cocky and it cost me in Phase III.

Robin Sage

The final part of the Q-Course brought all the different students together, placed in an A Team, and then conducted a field problem in the country of "The Republic of Pineland". Pineland consists of 15 counties west of Fort Bragg and covers almost a third of North Carolina. This final operation is known as Robin Sage.[25] Since some of the specialties in the Q-Course take longer to get qualified not everyone who started with us from SFOT was there. We also had another new element added into exercise, and that was the officers. Since Phase I all of the training had been done with and by NCOs. The officers had their own Phase II and we had not mixed with them.

The first part of the Phase III was classroom instruction. We learned about unconventional warfare, resistance groups, letter drops, and other areas of covert operations. We also learned how to set up message pickups, Landing Zones (LZ), Drop Zones (DZ), aerial resupply and calling in air support…especially the AC-130 Specter gunship. At the time I thought it was pretty boring, but little did I know I would be using all of this for real about a year later. The days were long and we didn't get off work until almost 8:00 each night.

After we learned all we could in the classroom we went out to Camp Mackall and practiced what we learned on the ground. The air operations were mainly

---

[25] Robin Sage was created in 1952 by Aaron Banks. It was originally known as Cherokee Trail and Gobbler's Woods because the location was in the Chattahoochee Forest in Georgia. The exercise later moved to the Uwharrie and Pisgah National Forests. Aaron Banks renamed it Robin Sage when the operation moved closer to Robbins, North Carolina. He also named it after Colonel Jerry Sage, an OSS officer friend from WWII. As of the writing of the second edition of this book (2021) Robin Sage is conducted in 26 counties. At the moment I am sitting at a field desk, in the woods near Gibson, NC. I am now a paid contractor that helps train the Special Forces students during Robin Sage, while driving my personally owned Ferret Scout Car known as "Little Alice".

done using a Porter aircraft (Pilatus Porter) that was an extremely maneuverable aircraft and could take off and land in very small places.

When we returned to Fort Bragg we went into "isolation" to plan for our mission. The isolation training simulates the way Special Forces receives a mission. Each team would go into isolation and stay there until it was their time to deploy. During isolation the teams would learn everything they could about their target, about the people in the area, about any assets they could use, and also plan the best way to E+E from the area (escape and evade) in case everything fell apart. The teams would not have any contact with the other ODAs or with anyone outside of the isolation facility (ISOFAC). At the end of isolation each team would briefback the mission to the overall commander. An example of this in "real world" was when the ODAs went into isolation prior to Afghanistan in 2001, the teams that had the best concept of the operation were the ones chosen to go in.[26] I was not worried about isolation and briefbacks since I had done many of them while I was in F Company, 51st LRSU in West Germany.

On the first day of isolation our team was created and given the designation of ODA 934. Our Team commander was a Major in the Reserves who none of us had seen before. He had done most of his Special Forces training through correspondence courses delivered in the mail. None of us were thrilled by this. In his civilian job he was public relations for UNICO oil. We also had a Lieutenant on the team that had been in the infantry, so he was partially squared away.

Before isolation our team went to the range to make sure that each of us knew the basics of what was needed to succeed in the mission. I taught the members of the Team that did not have any infantry experience on how to use the M-60 machinegun. Unfortunately, we learned that the Major didn't know anything at all. He didn't even know basic infantry skills or how to read a map. Because we had to teach him everything, we wasted a lot of time that could have been used coming up with Team SOPs. That night we stayed in GP Medium tents on the range, still playing "catch up" due to the clueless Major.

The next day we marched to the mortar range and the weapon sergeants taught the Team how to fire the mortars, how to call in an airstrike and how to use lasers to guide bombs onto the target. After the sun went down, we had some classes around our tent on basic infantry tactics. In the morning we moved

---

[26] An excellent book about the isolation and the mission into Afghanistan is *The Horse Soldiers*, by Doug Stanton. My old team in 5th Special Forces Group is featured in that book.

Chapter 3: SPECIAL FORCES – A Storm in the Desert

to the demolition range and the Engineers on the team taught how to use C-4 plastic explosives and set up firing wires. On the final day before isolation, we went to the range and zeroed our M-16s.

When we returned to Bragg, we gathered up our duffel bags and went into isolation. The LT on the team had pizzas delivered, which was excellent. The Major didn't like it, but we blew him off since we hadn't received a mission yet. In the morning an order was given to the Major, but he kept arguing with the LT about what it meant. All of us NCOs wanted to know what was in the order, but due to their bickering we were never able to read it.

The grader for our team was SFC Griffin, who gave us a packet titled "101 questions to study for the ISO test". Since both officers were pretty much useless, I spent my time coming up with the answers for the test. As I uncovered an answer, I would write it on pieces of poster board hanging around the room so everyone else would know the answers too. When it came time for the briefback I had about as much knowledge as anyone in the room. I started becoming cocky and when the Major would comment on how great it was that I was on the team, I would joke around and tell him, "We'll get along fine, just as long as you remember I'm god." [27]

I ended up maxxing out the ISO test, but we had four people fail. Though all four were able to pass the retest, there was a rule in the Q-Course that you couldn't fail test two tests in each Phase without being recycled. Due to this our team lost two men. After isolation we learned that there were a lot of ISO test failures on the other Teams. One of the teams only had one student pass, and everyone else failed. Out of the entire class only one of the 18Bs failed the ISO test. It was lucky for us that SFC Griffin had given us that study guide.

After a good night's sleep, we went through jump refresher training and rigged up to jump into Pineland. This would be a night time infiltration using a C-47 aircraft. I was torn between being thrilled about using the same aircraft that the paratroopers jumped into Normandy during WWII, and dreaded jumping into Pineland in a storm because of the threat of being injured. I was too close to the end and I didn't want to get medically recycled. Luckily it

---

[27] In the Johnny Cash song *God's Gonna Cut You Down*, he warns sinners that they cannot escape judgment. I think my blasphemy caught up to me, and I was getting too cocky. Due to this I almost failed the Q-Course.

continued to dump rain all over the drop zone, located near Ellerbe, so our jump was cancelled and we would be trucked out to Pineland.[28]

Our drop zone had been a farmer's field, so we got off the trucks and set up a secure perimeter around that field, waiting for an "auxiliary agent" of the Pineland resistance. The agent was a good-ole boy farmer and his family driving their pickup truck. Our Major walked up to the agent with his sleeping bag under one arm (because it fell off his rucksack), a case of C-rations under the other arm, carrying his M-16 rifle like a suitcase by the handle and dragging 50 feet of a 120-foot rope behind him. He told the agent that we were there to help them. The farmer just laughed and told the Major that if all of us were as sorry looking as him he might as well turn us over to the government.

The "Republic of Pineland" has been around since the 1960s. To make a realistic guerilla warfare environment the citizens of those North Carolina counties in "Pineland" are encouraged to participate. Many of these citizens are paid by the government to be the resistance against the government of Pineland. Local law enforcement also takes part in this and if any Special Forces soldier gets found by the police he will be arrested and turned over to the cadre. The people of "Pineland" have been waging a guerilla war for decades and are pretty skilled at it.[29]

Unfortunately, realistic role playing sometimes leads to deadly encounters. In 2002 two Special Forces students were captured by a highway patrolman who did not know about *Robin Sage*. The two students tried to bribe the officer with Pineland money, which resembles monopoly money. The deputy sheriff, not understanding what was going on, and not liking the looks of the men in uniform, used pepper spray on one of the soldiers. The two soldiers grabbed their M16s and took off running into the woods. The officer fired his pistol at the fleeing soldiers, killing one of the students, and wounding the other. After

---

[28] Ellerbe is located about fifteen miles west of Camp Mackall. Not long after graduating the Q-course I was able to conduct a jump from a C-47 after all, checking off a "bucket list" item.

[29] Around Fort Bragg you can see bumper stickers on a few trucks that have "Free Pineland" written on them. Only a few get this inside joke.

that incident police officers who are participating in Robin Sage wear a distinctive arm band that says "Robin Sage".[30]

Our auxiliary agent put us in the back of several family pickups and covered us with tarps. He had brought his children along and they sat on top of us in the back of the pickup, making it look like we were just some sort of cargo. The agent drove us to a church, dropped us off and disappeared into the night with all his kin. The Major had no idea where we were and had us move into the trees. Not satisfied with where we were he had us move further into the trees, and still not satisfied he had us move yet again.

One of men on our Team suggested he ought to look at his map, but sometime since we had got on the trucks, he had lost the map. I pulled out my map and plotted a course away from there. As I did this the Major quickly spun around, hit me in the eye with his M16. Blinded, I fell into a trash pit and slashed my knee with a broken bottle. I was extremely pissed off then, vowing to stay away from this officer that was a "Jonah" … a moving disaster.[31]

The Team moved about 200 yards into the woods when we heard a feeble voice in the rear yelling at us to "halt". It was the Major. He couldn't make it and told us that he had to rest. All of us were carrying 90 to 120 pounds in our rucksacks, but he had the lightest ruck because he kept throwing out anything that he thought he didn't need. He had tossed aside the 120-foot rope, but another man on our team picked it up since we figured we may need it later. The Team continued moving, but the Major would stop every few feet, wheezing for breath, whining about how he had to stop. At one point he even started crying. All of us were disgusted and knew that he didn't belong there at all, but we were stuck with him. After moving through the dark woods all night long we finally ended up at the contact point at 0400.

As the sun began to rise, we spotted the two guerillas that we were supposed to link up with. The guerillas used in Robin Sage at that time were soldiers, from the 82nd Airborne, detailed out to do this mission. They wore civilian clothing or some sort of foreign clothes and were told to act as if they were real

---

[30] The soldier who was killed was 1LT Tomeny. The soldier who was wounded was SSG Phelps. Phelps sued the officer and the Moore County Sheriff's office and received $750,000. LT Tomeny's next of kin settled out of court with Moore County.

[31] The *Jonah* goes back to the War Between the States. It is the soldier who cannot do anything right, and though he will survive most encounters, everyone around him end up dead or wounded. The term comes from Jonah of the Bible.

guerillas.[32] The two guerillas told us to follow them and proceeded to walk us around in circles in the woods. I understood why they were doing it, so that we wouldn't know where their basecamp was, but the Major continued to whine, complain and stop every few feet. Due to the Major slowing us down we didn't get to the basecamp until ten hours later.

While we were moving the Major had us stop for lunch and told the communications sergeant send a report by his radio. Unfortunately, when he did that, he was within 500 meters of the guerilla basecamp. When we arrived in the basecamp the Guerilla chief (G-Chief) was so angry about this that he shot one of the guerillas that were guiding us. This was all part of the act, but the cadre wanted to see how we would react in realistic situations.

The G-Chief was actually a Team Sergeant from 5th Group, but we didn't know it at the time. He called himself Colonel Jack Daniels. He ordered one of his men to take us back out into the woods and march us around for an hour, then bring us back and see if we could do it right. Once we were in the woods, we told the guide to move ahead without us, we would catch up. Every man on the team then told the Major that he was a major screw up and he needed to get his head out of his ass. Everything he was doing was jeopardizing the team and the mission. We also told the LT that he needed to take charge from now on.

When we returned to the camp the G-chief had us hold up outside his perimeter and then he brought each of us into his hut, one at a time, to ask us questions. He asked me what I could offer his men. I told him that I could give them weapons and tactics training. He then asked me if I could build a fire. I told him I could, and he told me to do it. This was one of those times that reenactor skills paid off. I had a fire going pretty quickly, even though it had rained all night and everything was wet. After each of us had spoken to the G-chief he allowed us into his camp.

I was tasked with setting up the security around the guerilla camp, since they had none. The G-Chief also told us that we would have to pull the security the first night to show his men how it is done. Our team pulled security, but not too well. We were exhausted from having not slept for over 24 hours and having to

---

[32] Today the guerillas are a combination of paid role players and students who had not gone through the course yet. Personally, I am not thrilled about this idea, since it gives away any surprise events during Robin Sage. Though it helps the students who are held over and waiting to go through Robin Sage, it also gives them training where they do not have to think that much about what to do next.

walk all over the woods due to the Major's incompetence. Some of us didn't make it through the night, and Nikolai fell asleep while on guard.

The next morning, we began digging fighting positions in case the camp got attacked. While the weapons sergeants did this the commo NCOs would go out every six hours, far away from the camp, and send situation reports (SITREP). The engineers of our Team built Colonel Daniels a chair, trying to build rapport with him. The medics wandered through the camp, checking on the guerillas. Some of them had diarrhea from drinking unpurified water from the creek flowing beside the camp. My job was mainly pulling security with the M-60 machinegun. The LT continued to talk to the G-Chief, trying to get him to sign a loyalty oath, and trying to get his guerillas to fill out ID cards so that he could pay all of them.

The currency in Pineland was the "Don" and the LT had 14,000 Don to use during the mission. He had to use some of the Don to pay the farmer for gas and to bribe any roadblocks that we may have come across on the way to the church. The haggling over whether or not to sign the loyalty oath went on until dark, but finally the G-Chief said he would sign one, if we all would. We figured, "why not" and signed. We also knew we wouldn't get to sleep that night until all the guerillas were paid. Unfortunately, the LT never kept a record of the money he spent and this hurt him later on. That night the SF soldiers and the guerillas both pulled guard duty, so we were able to get some sleep. Unfortunately, every time the guard duty was pulled by a guerilla, it stopped because he would just go right back to bed.

The next day I walked the perimeter and I found garbage left behind by one of the guerillas. In actuality the guerilla was SGT Jones from the 82nd Engineers. I picked up the garbage and put it by his hooch. I was chosen as the assistant patrol leader that day to go do an aerial resupply mission around noon. Just as we were about to go out on the patrol the guerillas went on strike and refused to go. None of us knew why, but it turned out it was because of me. SGT Jones was angry that I put the trash by his hooch and he told his men not to go out. The G-Chief liked the idea and let him go with it. I had to write an apology and give it to the G-Chief. When the G-Chief read it, he said that it wasn't what Jones said. I didn't understand, but the G-Chief said that SGT Jones said I had put the trash inside his sleeping bag and smeared food all over it. He also said that I had kicked Jones in the head while he was sleeping. I was amazed, and the G-Chief, and the admin cadre, SFC Griffin, told me that kicking him in the head was "fucked up".

They asked me was it true and I told them no it wasn't. I was torn now, trying to figure out if this was part of the game, or was it real. I figured they were still playing, so I formally apologized to the whole camp. SFC Griffin pulled me aside and told me that all the rapport that the team had built up had been destroyed by my stunt and that I would get a counseling statement from higher command later. I was amazed, and angry that this was not part of the game, but real. SGT Jones had been lying, for whatever reason, and he might have just stopped me from graduating the course. My head was no longer in the game, worrying about this, and so I began to make mistakes.

When we moved out on our patrol, I took out a three-man surveillance team. As we sat there, overwatching the area where we would meet one of the auxiliaries, I asked the other two guerillas what the heck was going on back there. They told me to not worry about it. They said SGT Jones had an attitude against everything, and he really didn't like white guys, anyone in Special Forces, or being tasked to do this mission. He wanted to be back at home.

As I was talking to the guerillas SFC Griffin appeared. All he observed was me talking to the other two, while I should have been watching the area. Griffin pulled me aside, again, and chewed my ass. Now I had two strikes against me. We continued with the mission, linking up with the auxiliary, paying him 1,000 Don to take us to the aerial resupply and then having him bring us back. All should have been great, but I was pretty demoralized.

The other Team missions that were happening around the same time were successful. The LT was able to get all the records on the guerillas out of the area by setting up a message pick up. A Porter aircraft flew low and slow down a cleared strip by the road and grabbed the container with a hook. The Major also had a patrol to find the cache site, but he was in the wrong place and never found it. This was bad because the cache had more Don that we needed to complete the mission.

That night there was an important meeting. The Major, LT and the Team Sergeant all went to the Area Guerilla Commander's camp. The three had to go to a meal and build rapport with the Area commander. They had to chew tobacco, eat jalapeno peppers and anything else that was put before them. Unfortunately for us, the Major promised the Area commander that the medics could cure all diseases, the 18Bs could make bullets, and the engineers could make explosives out of chicken poop. The guerilla leaders knew that it was bullshit and before the night was over the Area commander was threatening to shoot the Major if he didn't leave his camp.

The next morning, I was given a patrol where I would be the patrol leader. This was a major graded exercise and I was hoping if I did well it would make me look better to the evaluators. When it came time to have the guerillas camouflage, SGT Jones told me that he "wasn't going to put on no camouflage." I was tired of this punk, but I left it alone and let him go without camo for this mission.

We linked up with our auxiliary and he took all of us in a truck to the same DZ that I had reconned the other day. Once we were dropped off, we moved into a swamp and waited until nightfall, when the rest of the Team would arrive. When the Team Sergeant showed up, he took over from me. The Team sergeant sent me to a barn and told me to wait there, pulling security, until after the drop. He set up an "L" using railroad flares in a peach orchard behind the barn. As I was pulling security SFC Griffin came up and asked me where the team sergeant was. I told him in the peach orchard behind the barn. SFC Griffin then chewed me out because he said the "L" should have been in front of the barn. I quickly ran over and had the men holding the flares move to where they should be, just as the C-130 was coming in. The drop landed about 1,000 meters away from the target, so we had to go get it and carry it back. We put it on a truck and had the auxiliary take us to a safe house, where we unloaded it. While we were doing all this, a large crowd of civilians from Norman, a town nearby, came out to watch the show.

During the time we were in the base camp each of us had a guerilla counterpart that we had to train in our skills. We all chose our guerillas and mine was a SP/4 in the 82nd Engineers. Unfortunately, he just wanted to lie in bed and sleep when he wasn't on a mission. I would do what I could to train him about the mortars, hoping he would remember some of it.

Some good news came to us when we learned that the Major left the camp. He could not stay for the entire mission since he was only in the Reserves. I don't know if he ever got qualified and was awarded a Special Forces tab, but I hope not.[33]

SFC Griffin came to us that day and told us that we had a piece of sensitive gear missing. The team inventoried all that they had and couldn't find anything missing, so it worried us. The next day was Day 7 counseling which was the half way point of the Robin Sage exercise. The evaluators let each of us know where we stood. All of us were ripped up pretty badly by the evaluators and no one had a positive counseling. I received a counseling statement for ruining the rapport of the guerillas, talking at the surveillance position and for the aerial resupply not being on target. SFC Griffin then gave me the firing pin out of the

---

[33] We called that way of getting qualified a "paper tab" because most of the course was done on paper.

M-60. He said that this was the piece of sensitive gear that came up missing. Since I was the weapon sergeant, I got a negative counseling for that too. On top of all that I also got written up for having no security on the perimeter, having trash around the basecamp, and for the guerillas not doing weapon maintenance. SFC Griffin told me that I would be seen by higher command later on about whether I would graduate or not. I was not real thrilled by any of this and felt totally demoralized, but I drove on with the final missions.

After the counseling session the Team bought back the missing firing pin for 3,000 Don. We also decided to pull guard by ourselves. Every time we let the guerillas pull guard they would just go to sleep. The next day I was told that I couldn't go out on a mission until I had talked to a captain from higher headquarters. I waited until dark, but the captain never showed. Finally, SFC Griffin counseled me again, hitting all the same points, and telling me that if he didn't see any change, I would not graduate. He also told me that I wouldn't recycle either; I would just be terminated from the course. The next day SFC Griffin talked to the whole team, and told us that we were all doing terrible. The engineers hadn't built anything, the communications sergeants hadn't given any classes to the guerillas, and security still was almost non-existent.

Going into my next mission I felt the sword of doom hanging over my head. My next mission was to advise the guerillas and let them plan a mission. My guerilla that I was advising we called "Cadet" and was actually a PFC in the Engineers. He did a pretty good plan and operations order, but while he was giving the order the other guerillas were blowing him off and not paying attention. The mission was to move down a river and blow up a bridge that crossed it. Once we got near the bridge, I advised Cadet on what to do, and the raid went perfectly. We even surprised the aggressors on the bridge, who were in reality 82nd Airborne military police. We were on the bridge for about seven minutes, while the engineers rigged fake explosives. After the bridge "blew" we were brought back onto the bridge and critiqued by SGT Roy Tabron, a guest instructor from 5th Group. He said we did great and I was hoping that he would tell this to SFC Griffin.

That night I finished all the weapons positions, drew up range cards and sector sketches for each one and put in aiming stakes. It took me most of the night, but I was going the extra mile to prove my worth. In the morning I took Nikolai and we built the alternate and supplementary positions. That afternoon an auxiliary brought in fifteen chickens and Nason and I did a "kill" class and showed the guerillas how to skin them. We took barbecue sauce out of the MREs and made barbecue chicken and baked potatoes. The guerillas that each of us had personally advised were brought to SFC Griffin and they filled out a rapport form. SFC Griffin didn't ask the guerilla I had trained, but instead brought in SGT Jones. Needless to say, I scored really low on the rapport form, a score of 5 out of a possible 30.

The following day was the final mission of Robin Sage. This mission would be a guerilla planned and guerilla led mission, we would only be advising. Each of our Special Forces "jobs" would have a part in this mission. The mission was to blow up a microwave tower near Norman. When we left the patrol base, we took everything with us, since we would not be returning. We traveled through the forest, stopping right before dark at the Mission Support Site (MSS, which is a Special Forces patrol base). Since we could see the tower in the distance, we put out security and slept till 0200. We moved through the dark forest to the tower and attacked it at 0530. The attack wasn't the best, but it worked and we "blew up" the tower with simulated explosives.

After the mission was over, we were told to move back to the tower, where SFC Griffin told us our final grades. I was pretty nervous, but was relieved when he said that no one had failed and we would all be graduating. It felt like a great weight was lifted off my shoulders.

We trucked back to Camp Mackall and were given our final critique. I didn't do great, but I passed with a score of 79.9%. To pass a student needed to have an overall score of 70%. No one on our team got over 83%, and I learned it just wasn't me that was worried about passing. Out of the 18B students, one failed and out of all those that entered Phase III, seven had failed. I also learned that the soldiers who were going to go to Fort Campbell would be going to language school to learn Arabic first. Later that night I drove to a Laundromat to clean up all my gear and I got on a pay phone and called my dad to tell him that I passed.

The next day we turned in all our gear and had a rehearsal for the graduation. The honor graduate for our class was SFC Olmstead who came from SFOD-D, commonly known as Delta Force. The honor graduate always receives a knife during the ceremony, but during the rehearsal he was given a small pocket knife from us. When he walked up on the stage during rehearsal, we all began to hum the theme from "Mission Impossible". Even though he never said where he worked, we all knew it, so it wasn't much of a secret.

My dad came to my graduation in his dress uniform. He had to lose twenty pounds to fit in it, so I guess he knew I would make it, and had started earlier. Since he was a Special Forces officer, he was moved into the VIP section of the audience. During our graduation ceremony we had a moment of silence for Colonel Nick Rowe. Colonel Rowe was a legend in the Special Forces community. He had been captured by the Viet Cong in 1963, spent five years in captivity and then escaped from a Vietnamese POW camp by beating the guard over the head. He had written a book about his exploits called *Five Years to Freedom*.

In the 1980s he created the Special Forces SERE School (Survival, Evasion, Resistance & Escape) based upon his own experiences. When I was in the 82nd Airborne several of us were detailed to "test" the course, and I met Colonel Rowe. A few days before our graduation he had been assassinated by the communist New People's Army in the Philippines. Rowe had been assigned there as the chief of the Joint US Military Advisory Group. His body was returned to the United States and he was buried in Arlington Cemetery.

After the graduation my dad met me outside and he handed me a knife, the standard gift for such ceremonies. I gave him a Special Forces Tab, since they never had those when he had been in active duty. One bad moment came when Nason was not allowed to walk across the stage because he had worn a 75th Ranger Regiment crest on his uniform, to represent his regimental affiliation. At that time there was a fixation towards Rangers in the Special Forces community. According to US Army regulations a soldier could either wear the Ranger Tab or the Special Forces Tab, but at that time a soldier couldn't wear both.[34] When Nason wore his Ranger regimental affiliation crest, instead of a Special Forces affiliation crest, it angered the chain of command and they said he could not go across the stage. He was allowed to graduate and get the certificate, but he wouldn't be publicly recognized.

In the first week of May we started the Arabic language course. This was a new school so it didn't have the bugs worked out of it. Our instructor in the first week looked like a cartoon character. He had big teeth, huge glasses, a big nose and his body was shaped like a pear. He also had a heavy accent that very few of us could understand. Every day from 0830 until 1530 in the afternoon we would try to learn Arabic, but it was a very hard language to understand. The instructor was mainly teaching us technical terms dealing with radios and weapons. Unfortunately, whenever another instructor came in, he would go over the last thing we learned and correct the previous instructor. There were plenty of arguments between two instructors on how to say the various Arabic

---

[34] This was changed in the 1990s, so a soldier who was both SF and Ranger qualified could have two tabs. If he was assigned to an airborne unit, such as a Special Forces Group, he would wear the third Airborne Tab above his unit patch. This became known as the "Tower of Power", "Triple Canopy" or a "Triple Tabber". In the 2000s a fourth tab was created for Engineers that was known as a Sapper Tab. Soldiers cannot wear four tabs, only three.

words. I honestly believe that no one could be fluent in Arabic unless they were born there, and even then, they would argue about what to call a fork for hours on end. This is mainly because of all the different Arabic dialects that are in the Mideast. What I was learning was Egyptian Arabic, which would later get me a laugh when I spoke it to non-Egyptians.

On the weekends I went to a lot of reenactments, seeing Betsy from time to time. Nason had gone with me to many of the reenactments during Phase II, but now he was on leave getting ready to go to 10th Group in Germany. At a reenactment in Ninety-Six, South Carolina Betsy did show up with her new boyfriend, a pudgy looking guy who was a cadet at Norwich University, and he had no personality. Betsy's dad disliked him intensely and one night after having a few beers he offered me money or beer if I would just go and beat the crap out of him. I didn't take him up on it and I ignored her as best I could so I could get on with my life.

A few weeks later I went to a reenactment near Spartanburg, South Carolina to blow off some steam. After I had set up my tent, Betsy arrived with Nason. I ignored her and went about having a good time. Someone ran up to me and told me that Betsy had passed out. I knew it was her diabetes and I grabbed a bag of sugar, found out where she was and kneeled beside her. I knew where her test kit was from when we were dating, so I checked her blood-sugar and then began to feed her sugar from my fingers. I knelt down beside her, listening to her breathing and feeling her heart rate until the paramedics came. She woke up and told the medics that she wasn't going to the hospital. After they left, she started looking bad again and I told her that I was driving her to the hospital. I picked her up and carried her to my truck. As Nason and I drove her to the hospital she passed out again. I held her, whispering in her ear, telling her that she couldn't leave me yet... that I still loved her. In the hospital we were told that one of us could stay with her, so I sat by her bedside, holding her hand. I fell asleep and stayed like that until 0300. The doctor came in and told me that she could go. After she had slept off her diabetic attack, she was feeling better.

The next night, after everyone went to bed, we stayed up all night and talked. She said that she hated living at home because her dad was always on her to make her life better. She started crying, saying that she would never be able to live up to her dad's expectations. She also told me that she was only going to go to college for a year, and then she was going to leave and marry that pudgy cadet from Norwich. This shocked me, because I didn't know they were that close. She also told me that she had been pregnant with his baby and had a miscarriage a few weeks earlier.

She looked at me and told me that I was her best friend, which was ripping my heart out. In my mind I was more than a friend to her, but she never saw it. I kissed her, and she kissed me back. However, we both knew what would happen, and I didn't want to have my heart torn out again. I walked away. She told me the next day that we had never been as close as we were that night. I told her that I didn't know what to do with her and her new boyfriend. I didn't know if I should stop her, or let her go so she would be happy.

In between me worrying about Betsy and our life together, I continued to learn Arabic and was able to pick up some conversational phrases too. For two days we would learn how to teach a class on some type of military equipment in Arabic, and then we would have to recite the class back to the instructor. During the classes I figured out what to do about Betsy also. Every time we weren't going together something bad had happened to her. Whenever we were boyfriend and girlfriend, she was happy and safe. I figured this was a sign, so instead of waiting for something to happen I decided I would be proactive and make it happen. At the next reenactment she would be at I was going to ask her to marry me.

On Memorial Day weekend I went to a big reenactment up in Fort Loudon, Tennessee. This was a pretty good event and everyone had a great time. Nason was still around, using up his leave time with the Pucketts before he had to sign into his unit in Germany. Betsy finally arrived later, but because it was a pretty hectic night, I was not able to get her alone. The next day I kept trying to talk to Betsy, but there was so much happening during the event that it wasn't possible. Betsy also knew me and I think she knew something was up, so she avoided me. Finally, after it was dark, I dragged her to a quiet corner of the fort, held out a ring and asked her if she would marry me. She was shocked and didn't know what to say. I told her don't say anything yet, just think about it and let me know the next day.

Later that night I talked to Betsy's brother, Bryan, around the fire. He told me that Betsy wasn't really serious with anyone and for somebody who was supposedly engaged to that guy from Norwich, she dated an awful lot of guys down in South Carolina. The next day I tried to talk to her, but she avoided me and ended up leaving the event early. I was pretty angry and talked to Nason, who verified what Bryan had told me about Betsy. I was pretty confused about the whole situation. I didn't know if she was playing some sort of mind game, or was she just being young and naïve.

When I returned to Fort Bragg, I called Nason and asked him where he would be for his last weekend in country. He said he had a hotel at Myrtle Beach, so I went down there with SSG Stone, from the language school. We were going down there to see what the nightlife was like. It turns out that weekend was the "Beach Week" for High School seniors, and Betsy was going to be there with four of her girlfriends. She was also going to be there with her boyfriend.

A lot of people in America think of Vegas, when you hear the term "what happens here, stays here" but the wildest spot in the Carolinas is Myrtle Beach during Beach Week. It was wall to wall people, cruising at two miles an hour down the road, and underage drinking was considered normal. To top off the wild time we were having, one of the members of the 2nd South Carolina was a manager at the top strip joint in Myrtle Beach called *The Doll House*. He sent a limo out to pick us up at our hotel and when we arrived at his club, we found out MTV was shooting a video with all these beautiful girls dancing in the background. It was a great time.

The next day, at 0630, I heard a knocking at the door that had the same beat as the pounding in my head. I opened the hotel room door and Betsy was standing there. She said that she was up all night and had ended up sleeping on the beach, worrying about what to do about our relationship. I stepped outside and talked to her for an hour, but at the end she said she had to give me back the ring. She didn't want to see my face when she gave it back, so she said she would put it on my car hood once I closed the door. I held the door open for a long time, knowing that once it was closed we would never be together again. I slowly closed the door, and with it, my heart.

When I returned back to Bragg, I did something I never thought I would do, and since it was embarrassing, no one else ever learned about it. I had to get over Betsy and the only way to do that was to meet someone else. I joined a dating service called Matchmaker International. Each month the service would send me four names that matched up to my interests and then I had to meet them. I figured this way I could get on with my life and forget about Betsy.

While we were having our wild beach week, on the other side of the globe there had been a massacre in China at a place called Tiananmen Square. Over 100,000 student protesters wanted to have democratic reforms and refused to move out of the square. The Chinese government sent in soldiers, but they sympathized with the protesters and didn't fire upon them. The government then sent in soldiers who were not from the Beijing region and therefore would have no sympathy towards the students. They fired upon the protesters, killing over 3,000 of them.

A few weeks later I was at another reenactment in Wake Forest when I was told that Betsy had taken $100 from her stepmom and then ran off to Philadelphia to be with her boyfriend. It didn't really shock me though, and I swore I wouldn't have anything else to do with her. To make matters worse, the Arabic classes weren't going so well either. Most of the students were getting fed up with the instructors, who couldn't agree with each other, and constantly argued and corrected the classes that were only made up prior to the course.

We were the first language course on Fort Bragg specifically made for the newly graduated SF students. Normally students would go to the Defense Language Institute (DLI) in Monterrey, California, but to save money we were taking the course on Fort Bragg with instructors who were not to the same standard as DLI. Many of us thought it was a waste of our time and we just wanted to get on to a Team and do our jobs.

A month after I joined the dating service, I was linked up with a girl in Raleigh named Terry. I took her on a date and we hit it off. I was expecting some homely girl, but she was pretty good looking after all. After that first date we started seeing each other on a regular basis.

Towards the end of July, I drove down to Fort Benning to see Bryan Puckett graduate Airborne School. After much haggling with the chain of command down there I was able to jump with Bryan on his fourth jump.[35] Initially one of the sergeants in the S-3 shop would not let me jump with Bryan, but I went over his head to the XO of the jump school. The sergeant wasn't thrilled about it and tried to stop me, but I told him to "fuck off" and got on the C-130 anyway.

The following day I stopped off at Hobgood's grave and left a Ranger Tab on the headstone. There were flowers beside the grave, with a card that said "Happy Father's Day, Love Russie". Afterwards I drove over to Fryar Field for Bryan's last jump. I had jumped in a pair of wings the day before, and I pinned these on his chest during the ceremony. "Blood Wings" were not allowed by order of the commanding general, but I pounded them in anyway. I also pounded them into the chest of a Citadel cadet friend of his.[36]

---

[35] Students must do five jumps to become airborne qualified.
[36] Blood Wings are when the backings of the wings are left off after pinning them to the shirt. When they are pounded into the shirt, by punching them or slapping them, they leave two bloody holes. Some troopers keep their shirt with the two bloody marks as a memento. Today the "kinder and gentler" Airborne School is so strict about giving blood

Chapter 3: SPECIAL FORCES – A Storm in the Desert

Afterwards I quickly drove back to Fort Bragg, threw my dirty clothing into the laundry bag, showered and then drove up to Raleigh to spend the rest of the weekend with Terry. It had been a great week and we were both becoming closer with each weekend we spent together. I even started taking her to reenactments and letting her "dress out". However, we both knew that I would be going to Fort Campbell as soon as the language school was over, so neither of us wanted to get too serious. The good thing was that I had totally forgotten about Betsy when I was with Terry.

Towards the end of August Terry was becoming more depressed, because she knew I would be leaving. A few times we fought over trivial things, but we always made up before the night was over. When the Arabic class ended a lot of the students signed out and headed up to Fort Campbell early. I decided to take 30 days leave and spend them with Terry.[37] During those 30 days I ended up falling in love with her and she said she was in love with me too. I felt like life couldn't get any better, but I also knew this would now be a long-distance love affair. The chances of that succeeding were slim. It wasn't perfect and we fought all the time because of our uncertain future, but both of us wanted it to go on forever.

Terry was the opposite of the women I had dated in my life. She was from up north; she didn't particularly care for the military and she was extremely jealous. When we traveled to Williamsburg to take part in a reenactment, she learned that Betsy would be there. Betsy's boyfriend had thrown her out and she had returned home to her dad. Terry was spitting mad and didn't even want to be there, but I convinced her that I no longer cared about Betsy anymore. She finally decided to stay and we had a fantastic time together.

Towards the end of September, we spent our final weekend together by driving to the Outer Banks. Far out in the ocean a large hurricane was brewing, named Hurricane Hugo and it was heading straight for the Carolina coastline. While we were at the Outer Banks the waves were fantastic and we spent several days in a hotel on the beach near the Hatteras lighthouse. We talked about how serious we wanted the relationship to be. I was all for having her follow me to Fort Campbell, but she owned her own home and she had a fantastic job in Raleigh with AT&T. She also thought that all relationships were doomed from

---

wings that if anyone does it they are banned from any graduations for life. They are still given though. Some traditions will not die, whether politically correct rules try to stop them or not.

[37] Unfortunately, this cost me a good team position. The first students there were placed on the HALO team, and then the SCUBA team. I had really wanted to be on a HALO Team, but I wanted to spend the last few weeks with Terry instead.

the start and only heartache will come out of them. She told me that she didn't want to get serious because everyone she ever loved had left her. What could I tell her? I was in love with her and now I was leaving too.

When I left, I drove nine hours to Fort Campbell, with Hurricane Hugo chasing me the entire way.[38] When I arrived there was no one in 5th Group to inprocess me since there had been a Group organization week and now everyone was off. When I had been stationed in Germany the RTO on my LRRP team had been PFC Chad Klotz. I knew that SGT Klotz was stationed in Fort Campbell with 5th Group, so I called him. He had gotten married to a German girl and they were both going to head up to Chicago for the weekend. He told me there was another German girl coming along, and asked if I wanted to go with them. I told him sure and ended up heading to Chicago. Hurricane Hugo followed me like a bad omen. Klotz was going to Chicago to attend his grandfather's WWII reunion. We spent a few days there and then we drove to his sister's house in Iowa. I didn't know he was going to do this side trip to Iowa and when I woke up in his car, I found myself surrounded by miles and miles of corn.

5th Special Forces Group (A)

When I returned on Tuesday, I finally signed into 5th Group. I was almost assigned to a SCUBA team, but I told them that I didn't swim that well. I really wanted to be on the HALO team, but all the Arabic students who left early had gotten the good team positions. I was assigned to C Company in the new 3rd Battalion. There weren't many people assigned to that battalion yet, since it was so new. My team was going to be ODA 592, though I was told that the number designation might change. I was also told that the Team would not be operational until July of next year, and until that time we would just be going to schools and forming the battalion. In the beginning I was the only one on my team, so I did physical training with the SCUBA team.

Forming a unit on Fort Campbell was not an easy task. The main unit there is the 101st (Air Assault) Division. We constantly butted heads with them to get anything done. An example of this was when the Special Forces soldiers would go to get their equipment issue at CIF, they would only give us a partial issue because they said that only infantry gets a full issue. They did not know what to make of us and were not prepared to have a special operations unit on their "turf". When we ran the 101st Division Run we were put last, which was almost

---

[38] Hurricane Hugo was a Category 4 hurricane when it slammed ashore in South Carolina. At that time it was the most destructive hurricane ever recorded, killing 109 people, leaving 100,000 homeless and causing 10 billion dollars in damage.

an insult. One of the SF wives held up a sign during the run that said "5th Special Forces Group – the 101st's Bastard Stepchild".

Airborne operations were also tricky on Fort Campbell since the 101st had not done any airborne drops since the Vietnam War and they weren't set up for it. The weather was also very fickle and many of our jumps were cancelled due to this. When we planned for training on the various ranges we often had to cancel because a unit in the 101st had priority over us. It was extremely frustrating.

During this time, I continued to see Terry, but the love affair had been cooled by the nine-hour drive to Raleigh. We had long talks on the phone whenever I was the staff duty NCO, and on a few of those phone calls we would break up, and then get back together. The whole relationship was on shaky ground. In November I drove to a reenactment in Camden, taking Klotz with me. He was not getting along with his wife and needed to get away. Terry showed up at Camden, and after getting extremely drunk she began to smack me. She had done this before whenever she drank, but this time I was getting tired of it. Finally, when she smacked me, I smacked her back. She then smacked me again, and I smacked her back. I told her I could do this all night, so she stopped, but then she wanted to drive right back to Raleigh. She was drunk and I wouldn't let her, so I drove her to the hotel instead. We made up, but the future between the two of us didn't look too bright.

While I was having petty problems in my life, the world was rapidly changing. Poland voted in a non-communist government, while Czechoslovakia opened up its borders to the west. Hungary also voted in a non-communist government and then East Germany eased up on travel restrictions to the west. Finally, two years after Reagan had said "Mr. Gorbachev, open this gate! Mr. Gorbachev, tear down this wall!" the wall separating the two German countries came crashing down. This was the beginning of the end of the Soviet Union.

By December I was resigned to not seeing Terry anymore. Whenever I drove to Raleigh she was about as affectionate as a cold fish, until she drank heavily, then her personality was reversed. I couldn't put up with a rollercoaster romance like that, and the traveling back and forth was costing me most of my paycheck each month. As a diversion Klotz and I went to the 125th Anniversary of the

Triple Canopy

Battle Franklin. I had done War Between the States reenacting a few times in the past, but was never was serious about it. When I saw the recreated Battle of Franklin, I was hooked. It was huge and spectacular with almost 10,000 troops on the field. A large Revolutionary War battle may be only 1,000 troops, so this was a huge difference. I walked around the field until I found a unit that suited me. They looked the part, like soldiers, and had no tents. They slept on the ground. This style of reenacting is known as "campaigning" and it sucked me in. On that day I joined the 24th Tennessee Regiment and began researching the uniform and equipment that I needed to acquire.

Right before Christmas my ODA started receiving classes on land mines that were in Afghanistan. The three Special Forces soldiers teaching us had done the mine removal mission, codenamed "Operation Safe Passage" several times, working out of Pakistan. One of the soldiers training us had most of his hand blown apart by one of the Soviet mines, so all that was left was his palm, with no fingers. There wasn't much training going on besides this and I was getting really annoyed at the attitude amongst the older Special Forces sergeants that seemed like all they wanted to do was avoid any type of training and just treat the new battalion as a vacation home. I wanted to get out and do something, anything, but missions were nonexistent, training areas were all taken by the 101st and even the real-world threat of the "evil" Soviet empire had become less threatening.

After being in a Ranger battalion, Special Forces did not seem that special. The only place that seemed like there might be some action was down in Panama. Manuel Noriega had been threatening US soldiers down there for the past year. He had been beaten by Guillermo Endara in the national elections, but Noriega had his thugs beat up Endara and then Noriega refused to give up power. Finally, several American servicemen had been beaten up or killed, which made the possibility of war seem much closer, but this wouldn't affect 5th Group, since our area of the world was the Middle East. The 7th Special Forces Group was the ones stationed in Panama and if anything happened, they would be the ones in harm's way.

I took Christmas leave on December 19th. Right after I had signed out on leave Fort Campbell was shut down for about an hour. No one could leave or get in. I didn't know why, but eventually I was allowed to continue the drive to my parent's house in Jacksonville, North Carolina. Several things happened that night to destroy my morale even more. There was an ice storm that shut down the interstate, but I drove on anyway. I ended up slamming into some woman's car when she parked her car in the fast lane because she was too scared to drive on the ice. I tried to stop, but the ice was so bad that I just continued to skid into her car. After hitting her I quickly sped up, to get to the side of the

302

road. I was glad I did, because her car was quickly hit by two other vehicles. After I got to my parent's house I went to sleep, only to be woken up early in the morning by my mom telling me that the United States had invaded Panama. The invasion was called Operation Just Cause.

Bert was still with the 2nd Ranger Battalion and he had jumped into Rio Hato airfield that night. The 82nd Airborne had also jumped onto Torijos-Tocumen airfield after the Rangers cleared it. It was Grenada all over again and I was pissed that I was missing out on the fight. I was so angry that I swore that if Special Forces didn't get better soon, I would return to the Rangers.

After Bert got back from Panama, he wrote me a letter describing what had happened. He wrote that an hour out from the drop zone "the jumpmaster lead us in the Ranger Creed... something I'll always remember." Chris Miller, one of the crewmen on a C-130, wrote "My head was buried in the scope with one hand on the antenna knob and one hand on the joystick. Everyone was hotmike, and the noise was terrible. The Rangers were singing in the back of the plane." [39] The Rangers weren't singing, they were reciting the Ranger Creed at the top of their lungs!

Bert continued in his letter "After the Creed he told us they knew we were coming. Great, if I wasn't already scared now I was. At ten minutes we stood up. The guy behind me pulled his reserve and started panicking, so I helped him Hook up and take the bad reserve off. We got a good one passed back and I handed it to him and tried to calm him down, he said he was good to go so I turned back around and waited for the one-minute warning. At one minute this bonehead behind me started screaming again cause he couldn't get the reserve on, so I took it off and threw it toward the front of the aircraft and told him to go out the fucking Door or I was gonna shoot him. I turned back around and started hearing the rounds impact on the bird. It sounded like pop corn popping in a steel pot. Then I wanted out the door, Bad. I remember the jumpmaster screaming "follow me" and everyone rushed the door."

What Bert wrote about illustrated how no one really knows how they will react in combat until it happens. Even though soldiers are conditioned to face extreme situations, and even though they go through Ranger or Special Forces

---

[39] Operation Just Cause, A brief history of Colonel Chris Miller's experiences in Panama, extracted from:
http://webspace.webring.com/people/vu/um_1425/panama/riohato.html on 30 July 2010

training, how they will react is never known until the first shots are fired towards their location. I would find that out myself when we deployed to Saudi Arabia during Desert Storm.

Bert described "Believe me we were going a lot faster than 130 knots because my ruck lowered itself on the opening shock. I thought I was gone. I didn't have any twists and when I looked up to check canopy, I saw tracers going through it. When I looked down I saw the runway and Hit! There were tracers everywhere and I was in the middle of the taxiway. Not kool."

He ran to what he thought was the tree line so he could get off the runway and get under cover, but he wrote, "it turned out to be one tree with a bunch of Kuna grass around it. Then this truck goes past at about 40 mph with three guys firing out the back. It was weird cause the truck had yellow rotating beacons (like a tow truck) on. I emptied the magazine from my pistol at it and kept on haulin' ass. I only went about 50 or so meters and fell in a little ditch. Then I couldn't get my weapon out of my 1950.[40] So I start cutting my harness off, when this truck stops and starts backing up. Well, I could get my LAW out but I was a little late, by the time I had it in operation six or seven LAWs hit it." [41]

Bert's final message to me in his letter was "Do you still have the hots for my sister? Even after she acts like such a Bimbo sometimes?" I thought about it, and I thought that I did still care for Betsy, and now that I was probably breaking up with Terry I didn't really know if I had the "hots" for her or not.

Before I went on leave to my parents, I had been invited to a Christmas party in Versailles, Kentucky, put on by a bunch of reenactors who were with the 2nd North Carolina. I met a girl there, Jennifer, and I was trying to win her over but I didn't seem to be making much headway. I was still depressed. I wrote in my journal, "I feel like I'm wasting my life. History is happening around me at a rapid rate of speed and I'm sitting on my ass."

When I returned to Fort Campbell I had just about enough of doing nothing. The Team Sergeant, SFC Sal___, wasn't planning any training, and he wasn't able to physically keep up with the rest of the team. He made a lot of excuses, but it just disgusted the rest of us on the Team. It became too much for me and

---

[40] An M1950 weapons case, which is a canvas bag that paratroopers use to put their weapon in when they jump. The M1950 weapon case is worn on the left side of the jumper.
[41] Letter written to me from Bert Puckett in January 1999.

I told him that I wanted to transfer out of his Team so I volunteered to become the armorer for the Battalion. Sal___ refused my request. The Team became tired of his excuses and lack of leadership and went to the Company Sergeant Major asking if Sal___ could be reassigned. Every school that we had asked for, or whatever training we wanted to do, Sa___ would just tell us "No" and not give any reason. We had a meeting with the Team and I told Sal___ that he was dragging the Team down. The other Team members told Sal___ that other Teams in the Company were leaps and bounds ahead of us. He just ignored us.

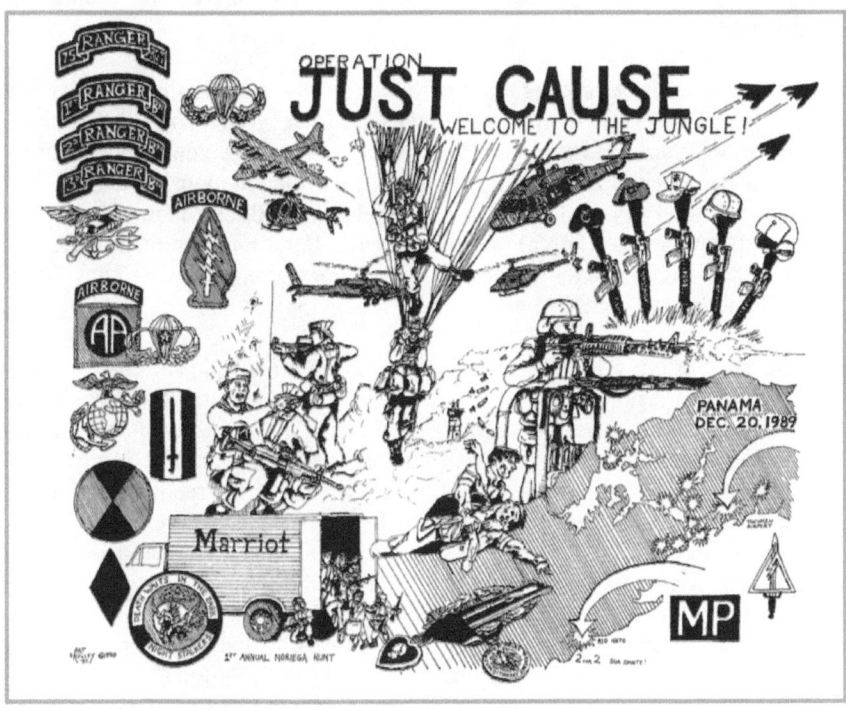

The situation came to a head on the day after I had Staff Duty NCO. SDNCO is a 24-hour duty and normally whoever has it gets the next day off, to sleep. After I got off this duty the Team junior weapons sergeant banged on my barracks door and told me that Sal___ wanted to see me at 1300, and I had to be in uniform. I really didn't want to listen to him make excuses, so I told the junior weapons sergeant to "cover for me" and I was going to try to get some sleep instead.

Later that day the Company Sergeant Major called me into his office and told me that I disobeyed an order from him, to appear at his desk in uniform. I didn't know about that, and tried to tell him, but the SGM was pretty angry. He chewed me out and he told me I was going to be removed from the Team and put in the B Team. The ODB is the company level staff. I really didn't mind because we weren't doing anything on the ODA, but now my reputation had turned into crap in that company. For the first time since I had joined the Army I considered getting out and not reenlisting. Sal___ wanted to terminate me from Special Forces, but the Sergeant Major wouldn't let him. What Sal___ did do was give me a blistering NCO evaluation report. These reports are the main thing that is used to be promoted to SFC, my next rank, so my future didn't look bright at all.

I began to consider getting a job with DEA or the US Marshalls, so that my time in Federal service would not go to waste. However, to be in either of those organizations there was a requirement to have a college education. Since I wasn't doing much of anything in 5th Group, I enrolled in night courses on-post through Austin Peay State University. Sal___ continued to try to get me removed from Special Forces by going to the Battalion Command Sergeant Major and asking him to terminate me. He had gone around the Company SGM, which really pissed the SGM off.

The only bright note was that Betsy called me and told me that she was enrolled in Charleston College. She said that she wanted to see me again and that she missed me. That same night Terry called me and said she was still in love with me and wanted to get back together. For the first time in my life, I had to choose between two women. When I went to the reenactment of the Battle of Guilford Courthouse in March, Betsy was there. We talked, drank some beers and ended up kissing by the fire. I was hooked by her again, and I knew it would probably end badly, but I didn't mind at all.

Towards the end of March, I went to my first big War Between the States reenactment, the 125th Anniversary Battle of Bentonville, North Carolina. Right before I left for the event, I was told by the SGM that I had been picked up on the E-7 list and that I would be promoted to Sergeant First Class within a few months. Luckily the bad NCOER that Sal___ wrote had not made it into the system yet. That same night Betsy called me and we talked for an hour and a half. She told me she would meet me at Bentonville because she really wanted to see me. Life was great again!

The reenactment at Bentonville was one of the largest I had ever been to. Everything was different than Revolutionary War reenacting. We marched out at 0700 and for the rest of the day we fought the entire battle. We ate food out of our haversacks, instead of cooking over an elaborate camp fire. By the end of the day, I was really tired, but it felt great. Towards the end of the day rain

306

came down on us in sheets, so the 24th Tennessee took refuge in an old farmhouse.

After it rained, I went out to see if Betsy was there yet, and found Terry walking around the camps. She had showed up, knowing that I would be there, and had been roaming from camp to camp trying to find me. I was amazed we had found each other because this was a huge event with thousands of tents. Terry took me into town to get some beer for the night, and while I was inside the store, I called Betsy's stepmom on a payphone to find out where she was. She told me she was at the event, looking for me. Terry took me back to camp, and then I told her that Betsy was there. Terry said she wanted to leave, so I let her, without talking her out of it. I felt like a total jerk, but Betsy was who I really wanted to see.

Right after I got out of Terry's car, I saw Betsy's red hair from a distance and went to her. I hugged her and asked her what her plans were for the night. She told me she didn't know, so I drove back to the same store that Terry had taken me to, and I called her dad, asking him what he wanted her to do. He told me that if it was still raining to take her to a motel.

After we arrived at the hotel we talked for a few hours, then I went and took a shower to wash off the day's worth of mud and burnt gunpowder. When I came out of the shower Betsy was in the bed. She looked incredible. That night we made love for the first time. When we were done, she began to cry, and I asked her why. She said that we would never be the same again. I told her she was right, and it was because I would love her even more. We held each other until morning, and then returned to the reenactment in time to do the surrender ceremony. The young guys in the 24th Tennessee were looking at me with something that resembled idol worship. To them they saw a Special Forces guy show up at an event, fight all day, then have two beautiful women come and pick him up, spending the night with one. They were amazed.

When I returned to Fort Campbell everything seemed to be going great, but then Betsy called me and told me she didn't want to make a permanent commitment to anything yet. I told her no problem. She then told me she had been arrested at a bar for underage drinking. Betsy wasn't 21 years old yet. She said she needed $200 for the fine, and asked me if I could send her the money. I had a feeling I might be getting suckered, but I told her I would send it as fast as I could.

After I sent her the money, she quit calling me. I tried several times, but each time the phone was either busy or no one was home. Days later I was finally able to talk to her, but she was distant, cold and said she wasn't looking for a relationship like I wanted. I was pretty angry at her. In April I was able to get a three-day pass and I drove down to see her. The whole weekend was a disaster. She acted like she didn't want to be there, and she told me that what we had was over. I felt used and betrayed. I began to think she had sex with me just so she could get money from me to pay her fine.

Later that month I got a letter from an old friend, Sue Cawood. At Guilford Courthouse she asked me if Betsy was smoking marijuana and I told her no. In the latest letter from her she told me that Betsy was doing drugs and wanted to let me know about it, so I could talk to her. I was really angry at Betsy now, since she lied to me, and I figured she had used the money I sent her to buy drugs. I wrote her an angry letter, which I knew would end our relationship. I would never trust another woman again.

That weekend the Kentucky riflemen invited me up to a small camping event held at their range. When I got there, it was snowing a little, and there were only two of the Kentucky guys in the camp. Later that night two girls showed up, one was a girl named Robin. All of us drank a little and then we fired bottle rockets at each other. Robin took my bottle rockets and ran off into the woods with them. When I went after her she teased me with them, then she came up and kissed me. We made out, standing in the middle of the woods, while the snow came down around us. It was almost magical, with the sound of snow falling all around, and little flakes falling on our faces. I took her back to my tent and we spent the night together. She told me she had a crush on me ever since she was a little girl, and when she heard I was at the camp she showed up to make me hers. Unfortunately, her father didn't want us to be together. I was ten years older than her, almost 30, and she was only 20 years old. Though we saw each other for a few more weekends, I ended up having to break it off so it wouldn't cause any problems with her family. I told her that after she graduated college, if she still wanted me, I would be around.

At the end of April my old team, ODA 592, was disbanded. At that time the Army was being reduced and there was a program called Quality Management Program (QMP) to remove any soldiers who were not meeting standards. Over fifty soldiers in 5th Group were QMP'ed and Sal___ tried to do it to me. This angered the SGM so much that he disbanded 592 and sent it to other teams. Sa____ didn't get another team, but instead was sent to Battalion headquarters, to the S-2 section. I was still on the B Team, learning how to use the new computers that we were issued.

Computers were still new to the Army, and we all were learning by fiddling with them. None of our buildings had air conditioning, so the computers would die around lunchtime each day due to the heat. Because the B Team Sergeant had been transferred, I was running the B Team until a replacement could show up. Since I had the time, I was still taking night courses at Austin Peay University and was passing all my classes with "A"s.

One major change to my life was when I was promoted to SFC I was finally authorized to live off post. I had spent my entire military life, a decade, living in the barracks. I found a furnished apartment in Clarksville, Tennessee for $330 a month and moved in. On July 2nd the 3rd Battalion was finally activated and officially became operational. Prior to that date we had been a "Provisional" battalion. The one bright note in that hot August was that I did such a good job on the B Team that the incoming SGM let me go back to an A Team. The day after I was assigned to my new team, ODA 591, Iraq invaded Kuwait. The Group was locked in and we awaited orders while we attempted to figure out where Kuwait was and why it was important.

In the 1980s Saddam Hussein's Iraq had been the only Middle Eastern country bold enough to try to stop Iran's dominance of the Gulf. Saddam wanted to be the one that would be dominate, so the eight-year Iran-Iraq War became a bloody clash of ideologies that cost Iraq 375,000 casualties and cost Saddam's country half a trillion dollars. Soon after the war ended in 1988 oil prices dropped from $20 a barrel to $13, making Americans believe that Saddam was a good ally. In Iraq the economy was not doing well and Saddam's scapegoat for his economic problems continued to be the tiny country of Kuwait. He accused the Kuwaitis of flooding the world market with oil and he stated that Kuwait had stolen Iraqi oil by slant drilling into Rumaylah oil field, which was located mainly in Iraq.

Kuwait was a country that was looked down upon by many in the Gulf as being too arrogant and wealthy. Saddam gambled that other Arab countries would not care what happened to tiny Kuwait and he hoped that Western countries would also be apathetic to a country where only four percent of the population was franchised.[42]

At 0100 on August 2nd three Iraqi Republican Guard divisions crossed the Kuwait border with 1,000 tanks. They raced to seize the Al Jahra heights overlooking Kuwait City and the Al Mutla pass leading to the city to block off any escape from the south of the city. Iraqi Special Forces helicoptered into the

---

[42] Atkinson, Rick, *Crusade: The Untold Story of the Persian Gulf War* (Boston: Houghton Mifflin) 1993, pg. 28

city seizing government buildings. The Emir of Kuwait fled to Saudi Arabia, but the Emir's brother was killed fighting the Iraqis. Four days later the fighting ended and Saddam Hussein proclaimed that Kuwait no longer existed as a country. It was now the 19th province of Iraq.[43]

We had not been officially placed on alert but all of Group was staying close to Fort Campbell, waiting for orders. We knew we would be going to the Gulf because 5th Group was the unit most familiar with the area and the language. I bought a huge Atlas of the world, just so I could study the area around Kuwait. I noticed that there were large areas of the map that were designated "neutral zones" due to a dispute with either Iraq or Saudi Arabia. Little did I know that I would soon be living in one of these neutral zones for almost half a year.

For the first time since the end of the Korean War the United Nations began to consider military intervention and had condemned Iraq. Our Battalion began to set the "war machine" in motion. All the Teams packed everything they would need to go to war in the desert, then inventoried everything else left in the Team rooms and put it in long term storage. If we deployed, other units would be using our buildings until we returned. We were all sitting on edge, waiting to go to war. We knew that this would not be a small fight, like Grenada or Panama. Iraq was touted again and again on the news as having the "5th largest army in the world". Most of us figured this war would be as big as Vietnam or Korea and a lot of us thought we would be killed. We began to get into the mindset that most of us would not be coming home.[44] I accepted it, mainly because I didn't have anything really to come home too. I told Sue Cawood that the most dangerous thing in the world was a young man without any future. She worried about my future.

My A Team began training on all the equipment that finally showed up on our inventory. For almost a year we had little to no gear being issued because we were still a new unit but this changed due to the impending war. One mission we figured we would have to be expert in was calling in air strikes, so training was increased for that skill. We learned how to use the Laser Target Designator (LTD) which would guide bombs to the target. All the teams zeroed their

---

[43] Atkinson, *Crusade* pp 52, 53
[44] Colin Powell had told Schwarzkopf that he was to "accept losses no greater than the equivalent of three companies per coalition brigade" or no more than 10,000 casualties (Atkinson, *Crusade*, pg. 113).

weapons, which consisted of M16A2 rifles and the M9 Beretta pistols. I also zeroed my M24 Sniper rifle which was brand new and had never been fired.

A week after Iraq invaded Kuwait all of 5th Group was still in Fort Campbell and had not deployed. Elements of the 82nd Airborne were the first to deploy to Saudi Arabia and the news media were calling them the "speed bumps" that were supposed to slow down Saddam Hussein's army when he invaded Saudi Arabia. We were all pretty pissed that we had not been deployed yet. One of the reasons we hadn't left was because there were so many units deploying to Saudi that there wasn't enough aircraft or room on the ships to send them all at once. We had to wait our turn.

We were placed on a countdown system, and were told we were "C-8" on August 8th. This meant we would deploy in eight days. Though I had been transferred to ODA 591, it wasn't a whole team yet and it would not be the first to deploy. We were in limbo status. We continued to square away things on the Team like wills and power of attorneys. Each day the wives of the married soldiers would stop by, bringing lunch, and to see their husbands, possibly for the last time. On a whim I bought a life insurance policy for $250,000, and made my father the beneficiary. He never knew about it though.[45]

By August 12th we were still waiting. Each day was a tedious day of doing nothing and waiting to go to war. For a second time my A Team was disbanded because we didn't have enough people, so ODA 591 ceased to exist. The team members were spread out amongst the other teams and ironically, I was re-assigned to ODA 592, my original Team that had been broken up in April, but had been recreated, with new leadership. The Team commander was CPT Mastrovito and the Team sergeant was MSG Cr___. Both weapons sergeant positions were filled, so I became the acting XO, though I obviously wouldn't outrank the Team SGT. The Team was the company's "Mobility Team". We were the only Team with the desert HUMVs, known as DMVs (Desert Mobility Vehicles) or "Dum-Vee" for short. The Team had four DMVs, that had three Special Forces soldiers assigned to each vehicle. A DMV crew was the driver, one gunner and one vehicle commander. The four DMVs had two M2 .50 caliber machineguns and a brand-new weapon that no one knew much about, the Mark 19 automatic grenade launcher. This weapon was so new that we had to send our weapons sergeants, SFC Be__ and Nikolai, to Fort Knox to learn about it from the man who created it.[46]

---

[45] At that time the most insurance the government would pay out was $100,000.
[46] This was the same SSG Nikolai that I went through the Q-Course with, and in 2020 I joined his stock market group to do day trading, but after not being so successful, Nikolai

Triple Canopy

Since Saddam had repeatedly used chemical weapons against the enemy in the Iran-Iraq War, we had to learn how to use the newest chemical warfare gear. We were told that our primary mission would be Search and Rescue (SAR) for downed pilots and our secondary mission would be Foreign Internal Defense (FID). A FID mission was when Special Forces would train or lead foreign troops to attack the enemy. Our final mission would be Direct Action (DA), where we would attack the enemy ourselves.

The "war machine" was at full speed and all around Group there were pallets of our gear, ammunition, vehicles, and equipment, ready to be loaded up. Each night a different team would pull guard on this gear. Though we were getting some equipment issued there was a lot that we were lacking. I had a lot of extra Army equipment in my personal storage locker that I brought it in for the Team members who might be missing anything. Over the years I had collected a lot of equipment and I brought in shoulder holsters, flashlights, compasses, GLINT Tape (infra-red), water cans, overalls, tow straps, chains, crowbars and magazines for both the M16 and M9s. I wrote down what I loaned out on a yellow legal pad, though I really didn't expect to get any of it back.

Outside the front gate of Fort Campbell, a surplus store called the US Cavalry Store had become media central. There were dozens of reporters and their camera crews waiting to interview anyone about where they might be going. The soldiers of the 101st Air Assault Division wore their desert uniforms proudly and talked to the reporters, while we continued to wear our green BDUs and tried to slip by them without making any comments. We had been told our new date of deployment would be August 14th, but the day before the Group commander told us we had been bumped from being one of the primary units to being on the bottom of the list. The priority went to infantry and armor units. Since everything was packed on the pallets, all we could do was wait. Each night I would either go out on the town with Klotz or stay home and wait by the phone.

Every day we didn't deploy we tried to do more training. The Team didn't have any weapon mounts for our vehicles, but we finally were able to "borrow" some from a Tennessee National Guard unit. The NG unit never knew about it though. When our team members went up to Fort Knox to train on the Mark 19, they also scrounged mounts for that weapon, water cans, gas cans and anything else that they could use. While we were waiting Captain Mastrovito

---

removed me from his group, and has never talked to me again. I don't know why and to this day I have wondered what made him turn his back on a buddy in such a way as this.

married his girlfriend. They initially were going to wait until Christmas, but he correctly figured we would be in a war zone by then.

The Team conducted night training over the next couple of nights, learning to use the new PVS-7 night vision goggles, and how to drive with them in blackout mode. We also learned how to do combat formations in the vehicles by driving across Sukchon Drop Zone. One night we almost wrecked one of the DMVs when we were practicing a fast-moving wedge formation. As the wedge turned the outer DMV had to go extremely fast to keep up, this almost made the vehicle flip. Though it was a scary moment, we laughed about it and called this new formation the "flying wedge".

Unfortunately, there was dissent in the Team. MSG Cr___ was treating all of the men like privates and belittling them constantly. The Team met secretly and talked about what we should do about him. Many of the team members wanted to go to the Company SGM and have Cr___ removed. I told them that I had to sit this one out since I had already been part of a team that wanted to get rid of their sergeant and I didn't want to get wrapped up in another fight. I didn't want to have a reputation as the guy who is always trying to remove the leadership. The Team decided to give Cr___ one more chance and see if he would change.

While I was waiting to deploy Terry called me from North Carolina. I was surprised, but she said if I was still there the next weekend, she would come to Fort Campbell and see me. She said she still loved me. I was no longer seeing Betsy and I figured I might be dead in a few months, so I told her go ahead and come to Clarksville and spend the weekend with me. Unfortunately, she wasn't able to make it and promised she would come up the next weekend.

Before the weekend we took the DMVs to a shop on the post and had them do a major overhaul for combat. We put armor plating on the floor so that if we ran over mines, it would not do as much damage.[47] We had canvas doors and a canvas cover made to replace the hardback hatch on the vehicle. This was to cut down on the heat that we knew would be over there. We installed a mine sweeper device for the lead vehicle and made individual bags for all of the NBC gear. That weekend we were given four days off, though no one went anywhere except to their homes, to stay by the phone.

When I visited Klotz's home his German wife Petra would get angry if any mention of the war came up. So, we avoided any discussion about it when we

---

[47] The idea of "up armored" HUMVs would not happen for another decade. Our DMVs had no armor except a few Kevlar reinforced areas.

were around her, but it was the only thing on our minds. I told Klotz that the latest word put out by the Group commander was that we would deploy on August 27th and were going to some place called King Khalid Airfield in Saudi Arabia. Over that four-day weekend I washed and waxed my truck heavily, since I figured I would not see it again for awhile. I also bought the best tobacco I could find, figuring I could use it to negotiate with Bedouin tribesmen.

The next week we filled our company's empty teams with ODAs from 3rd Group. One of them became our new ODA 591. Our Team didn't have an XO, but we were assigned 1SG Bishop.[48] We nicknamed him "LT Bishop" since he was in the XO slot. I had been the vehicle commander on DMV C-11 (DMV 3 on our team), but with the 1SG coming in I became the gunner of that vehicle and became the assistant intelligence sergeant on the Team.

To prepare for the mission I read up on anything dealing with desert warfare. I read T.E. Lawrence's *Seven Pillars of Wisdom,*[49] *Popski's Private Army* and *The Desert My Dwelling Place* about the British Long Range Desert Group in WWII.[50] None of us had ever been to Saudi Arabia and we were just guessing about what we would need to know. After reading the book about the British in North Africa I started teaching myself celestial navigation. I went to a local boating store and bought a sextant and then went about trying to find current nautical almanacs for the mathematical calculations needed to figure out where to locate yourself in the world.

Since all the Teams were just sitting around, waiting to deploy, there was a lot of free time. This was a curse whenever the higher chain of command is near. The Group commander and CSM would constantly be in our Team rooms, and worrying about whether or not we had haircuts, or whether our moustache was too long. Our Group Sergeant Major was CSM Simms, but his nickname became Simba, though never to his face. He would come into a Team room and tell the Team SGT that his men were all "violators" of the moustache policy. He told our team that our LCE was not according to SOP. There was no SOP and each Team made up their own SOPs, so we were confused. We listened to him though, because he was the Group CSM. He annoyed us to no end and we wondered why he was worrying about having a haircut if we were about to go to war. We all began to pray to go to war just so we could get away from CSM Simms.

---

[48] Bishop would later be my battalion CSM when I was assigned to 3rd Group in 1997.
[49] T.E. Lawrence is also known as "Lawrence of Arabia".
[50] The LRDG was the predecessor to the British SAS.

Chapter 3: SPECIAL FORCES – A Storm in the Desert

Each day was also an exercise in trying to keep the 101st Air Assault Division away from our gear. One day they tried to take our Mark 19s, because they said that their MPs were supposed to be issued them. The Group commander told them they couldn't have them and they would have to find their own equipment.

One of the reasons that we thought we weren't being deployed first was due to a rumor that General Schwarzkopf hated Special Forces because he was a conventional commander and didn't understand the capabilities that we could offer. Some of the older SF soldiers said that his dislike of Special Forces may have started back when Schwarzkopf was tasked with setting up 7th Infantry Division. During *Operation Celtic Cross* the Special Forces and Ranger Regiment were tasked with being the aggressor force against the newly created 7th ID. The special operations forces hit the logistics areas so hard that they had to stop *Celtic Cross* for 24 hours just to regroup.

At that time General Schwarzkopf wanted to see who was defeating his new unit and paid a visit. One SF soldier wrote to me that "He showed up to meet the team in the woods and all the clowns including him had on their K pots and full perfectly clean and matching LBE. If I recall we all looked like refugees from Yutzputz."[51]

Another soldier wrote "Somebody knew what an idiot he was cause when I took him to the B Team, they had this phony wall rigged in a barn on sliders and took Normy in....looked like a real barn that it was...then One Ball Willy (another SF soldier) wheels the phony wall back and there are two commo guys tapping out phony code and a bunch of maps with arrows all over them strung up.... like freakin' Hogan's Heroes... Normy got all happy again like 'oh yeah' that's super sneaky shit.... I had to go outside cause I couldn't maintain, laughed so hard my guts hurt." [52]

Another soldier wrote "The Schwarzkopf/SOF disdain was part of the mindset that many conventional officers had towards both SF and Rangers. Rangers were slightly more tolerated because most conventional officers could at least understand some of the basic concepts of employment. SF was considered a 'dead end' for career officers. There has been a LOT of that history in the Army going back to the 50's and before. Mostly with SF in the 50/60's... Growing up in Mech/Armor made most of Ranger/SF concepts incredibly foreign for most officers growing up in that environment. Also, the community was pretty small till the Regiment stood up in '84, when SF became MOS's, not just skill identifiers. 1st Group was reactivated along with 3rd Group. In the

---

[51] K pot = Kevlar helmet. Special Forces only wore helmets on jumps.
[52] Retrieved from http://groups.yahoo.com/group/3-5SFG on August 5, 2010

same time frame USASOC stood up as an Army MACOM. All of this happened within a very relatively short time. All of a sudden SOF had a voice at the main table that they had not had before. General Lindsay stood up USSOCOM, but General Downing moved its chair further up the table so to speak. General Schwarzkopf in turn was the 800 lb guerrilla in theatre during Desert Storm. As was related to me at the time; he thought that SOF used a lot of resources on a few people with small payoff. He wanted CENTCOM to have absolute control over SOF in his theatre. To say the SOCCENT had a short leash was an understatement. That is why General Downing had his turf and General Schwarzkopf had his. That came down from NSA to CENTCOM. That all changed though when the SCUDs started flying. Scud hunting became a major part of the game for some time. That brought in a lot of SOF units to prominence that otherwise might have remained much deeper in the shadows."[53] Whatever the truth was, there was no love lost between the Special Operations community and General Schwarzkopf.

By the last weekend of August, we still had not deployed. I spent my time at home renting a VCR and some movies, and then re-recording the tapes to keep for myself. One night a few of us drove over to Nashville and went to a strip joint called *The Classic Cat* to see a stripper known as "Suzie Boobies" who had a 62-inch chest. Afterwards, as we drove back, we were pulled over by a police officer, who thought we might have been driving drunk, but he let us go once he found out we were about to go to war.

Friends and relatives would call me at my apartment at all hours asking me when I was going. I couldn't tell them, mainly because I didn't know. My dad told me that he had volunteered to return to active duty, but he didn't tell my mom about it.[54] Terry didn't come up that weekend, since she had to work. She promised she would come up over Labor Day weekend, if we were still there. However, since she hadn't come, I was pretty angry by the time Labor Day arrived.

We finally got our first briefing about the situation in Middle East twenty-four days after Saddam invaded Kuwait. The Battalion commander, COL Shaw, told us that the SEALs had been trying to do the FID mission, but they weren't doing a very good job. They had no vehicles, and in the Middle East a vehicle was needed to do just about anything. The SEALs asked the Navy commander to send Special Forces over to take over that job. Our Battalion would not be doing FID, but instead would be assigned to King Fahd airfield to do Special Reconnaissance (SR). SR was extreme long-range reconnaissance behind the

---

[53] Retrieved from http://forum.armyranger.com on August 9, 2010.
[54] He was never called back to service.

enemy lines. We were told that we would leave by August 30th, and due to that we were placed on a one-hour recall alert.

The 1st Battalion began to deploy to Saudi that same day. We received a class on what to do if we were captured by the enemy, either during this initial "peacetime" phase, or after the bullets started flying. The female support troops of Group were told that they may not be deployed, since Saudi Arabia was making a stink about having female soldiers over there. We did the final packing of our personal equipment, known as palletizing, and we packed the vehicles to deploy. We then went home and waited. Watching CNN was almost comical. They reported that "Special Forces Green Berets" were training Kuwaiti resistance and striking at the Iraqis behind the lines, however we didn't have anyone over there yet. The news media always gets the story wrong.

I sat around, renting movies, and doing movie marathons, wishing I had someone to spend my last days in America with. We were told we may have a three-year commitment to this war, with a minimum of 13 months in the combat zone. It was lonely in the apartment and all I had to look forward to was going to war.

Klotz deployed with the headquarters section, leaving me to fend for myself for entertainment. Due to this I would go out late at night, cruising the riverfront or other areas where folks gathered, seeing if I could find any girl to talk to and maybe bring back to the apartment. It felt like it was high school again, cruising around until three in the morning. We had been told by the chain of command that we had to stay in our houses until 6:00 p.m. each day, but then we could go out as long as we were able to get back within one hour. I returned my cable box to the cable television company and then had to rely on rented videos for something to watch. August 30th came and went and we still didn't deploy, but the new date for our deployment was designated for Saturday, September 7th.

Right after Saddam Hussein had captured Kuwait there was a mad rush to get troops into Saudi Arabia so that it could be defended in case Iraq decided to invade. Though historians today think that Saddam did not plan to invade Saudi Arabia, the United States never knew this and by the end of August General Norman Schwarzkopf had assembled seven Army brigades, three aircraft carrier battle groups, fourteen fighter squadrons and 17,000 Marines. In addition, another squadron of Vietnam era B-52 bombers were located on the island of Diego Garcia in the Indian Ocean, ready to be used when the bombing campaign started.

317

Saturday, September 7th, arrived and there was no news at all. No aircraft were coming into Fort Campbell and no new date was assigned to us. We were in total limbo. I would have paid for my own ticket to the war zone if it was an option. This was going to be a large war, and all the warriors wanted to be part of it. One rumor floating around Group was that the Group commander personally paid for a civilian airliner to get us over there, but it had been commandeered by the 101st.

That night Nikolai and I went out to a bar called *Texas East*, where I did meet an incredibly sexy blonde, Margie, who danced with me all night. We ended the night with passionate kisses in the parking lot and she swore she would see me there the next night. She was true to her word and the next night we danced again and she almost took me home with her. At the last minute she said she wanted to meet me when she hadn't been drinking as much, so I agreed to take her out to dinner the following night. Unfortunately, every time I called her the next day her sister would hang up on me and call me a stalker. I was pretty angry at all women by that time. I was really getting annoyed at being told one thing by a woman I was interested in, and then being lied to.

To get over being dumped by Margie I headed over to a redneck bar called *The Golden Jukebox*. It had a dance floor, pool tables and played a combination of country music and Southern rock. Standing by one of the pool tables was a girl that looked like a wholesome "girl next door". Her name was Shawna. I talked with her awhile and she told me she was there with her brother, who was in the 101st. We struck it up great, until her brother said she had to take him back to the barracks.

After she left there was a bar fight between two girls that ended up throwing punches and ashtrays at each other. It was a hoot! I ended up holding a bloody cloth against one girl's head that had been beaned by one of the flying ashtrays. As I was holding the cloth, Shawna walked up to me and smiled. I asked her where her brother was, and she said he took off, with his truck. I asked her how she was getting home, and she said "I'm going home with you!" After that she commenced to drinking hard, and after she had loosened up, she told me what she did for a living. She was a 20-year-old stripper at a club called the *Cat West*. It was a franchise of *The Classic Cat* in Nashville.

When the bar began to close at 0200, I tried to take her out of there, but some of her drunken friends stopped us. They didn't know me and didn't trust me and said they were going to take her home. I didn't put up much resistance, but as soon as their attention was diverted by another bar fight, I slipped out with Shawna, poured her in the truck and drove her to my apartment.

When we got to the apartment, she was so drunk that she couldn't even stand up. I carried her to the bed, dropped her on the sheets and then went into the bathroom. By the time I came out she was passed out. No matter what I did to try to wake her she would not respond. I figured it was a lost cause and crawled in beside her. Around 0600 she was talking in her sleep and sounding pretty pissed off, so I moved to the couch in the other room and fell asleep. At 0900 I woke up with her standing over me, naked. She was shivering so I pulled aside the covers and she slid in beside me. We made love on the couch, and then moved to the bedroom. I spent the rest of the day in my apartment with her, only going out once to get some lunch.

As the day went on, she told me about herself. She told me her nickname that the police gave her was "wild thing" and that she had been an informant with them on a handful of drug cases. She also told me she was married to an MP in the 101st. I felt odd about this, since it was the first time I had committed adultery, but since I would most likely be dead soon, I shrugged it off. I figured we would have fun with each other while our time lasted.

She said she had married the first time when she was 14 and had two kids with her first husband. Her body did not look like she had any kids at all though. She remarried again when she was 18 and then married the third husband a year later. She also told me she had a warrant out for her arrest. She was definitely a wild thing. We spent the night together and then next day I drove her to the fairgrounds and dropped her off with another soldier that she called her "adopted father". He was going to take her to her alcoholic rehabilitation class. For me, it was a pretty good weekend.

Two days later I was still waiting for word that we were going to deploy. There was no news at all and we were all tired of just sitting around doing nothing. Shawna called me and asked me if I could come and get her at the fairgrounds. I didn't want to have a relationship with her in case we were told we were leaving, so I told her no. She continued to ask if I could come and get her, and she said she really wanted to see me again. Since it was 10:00 at night on a Tuesday I relented and told her that I would come and get her. When I picked her up, she had a suitcase and told me she wanted to stay with me. I told her OK, but I laid down the rules first. The biggest rule was that she would have to leave when I finally deployed. She agreed, and she moved in with me into my apartment.

Triple Canopy

Back in Group we continued to get ready to deploy. I was hurting in several places due to all the shots we were getting. Some of the shots were for protection against biological weapons that might be used against us, such as anthrax. The Group commander and CSM had both deployed, but when they got to Germany, they were bumped off their aircraft for other commanders who were considered more crucial to the mission.

While I waited for some sort of news Shawna was there to take my mind off of the upcoming war. I began to think she was a nymphomaniac, because all she wanted to do was have sex. Wake up to sex; after breakfast have sex, turn off the news of what was going on in the Gulf to have sex. It was constant. She wasn't smart enough to strike up a decent conversation, so I think she made up for it by making sure she didn't have to talk.

The next day I was pretty worn out and she asked me if she could take my truck to get a soda at the corner store. I told her no problem, but she had to be back soon because I was on alert and I needed the truck to get to base if we were recalled. She drove off at 10:00 that night and didn't return. She had stolen my truck. I should have seen it coming, but I was enjoying the constant workouts with her.

I called the police and told them that my truck had been stolen. After asking me several questions they decided that this was a domestic disturbance and said I would have to wait and see if she came back before morning. Finally, around 0400 she came back with my truck. The truck I drove was a customized Chevy that had all types of ground effects kits and other cosmetic features that made it look more like a sports car than a truck. The front end was off and lying inside the pickup. She told me a story about how she had hydroplaned on the road and the front end came off. She also told me that the police had stopped her, had found my pistol in the truck and had taken all the ammo, but they left the pistol. I went out to the truck and found the pistol on the front seat, empty, and the seats were soaked in beer, or urine, I couldn't tell which. Almost all the fuses were blown so nothing electrical was working right.

I didn't mince words. Before she came back, I had already packed her suitcase and handed it to her, then told her to leave. It had been thundering and lightning as I said all this, then when I opened the door, the rain began pouring down. I felt no sympathy at all as she walked away in the rain. I called the police again to tell them that I had my truck back, but I needed a report for my insurance company. The insurance would later assess the damage as $550. This was a lot of money in 1990. I decided I would get the truck repaired when I returned, if I returned, from this war. I also told the police about the bullets that were missing, and I didn't know if the police had them taken from her, of if she had shot someone with them.

Our Battalion finally got word that our aircraft had come in and we would be deploying soon. Unfortunately, the C-5a transport had its engine catch fire when it landed and they had to replace the engine first. I finally received a call from Group telling everyone that they needed to be at the company at 10:00 in the morning on September 15th.

That same night, before I began to shut everything down, Shawna called me again asking if I would take her back. I laughed, told her never, and hung up. I then called the phone company, cancelling my phone service until I returned. I unplugged the refrigerator and defrosted it though I would continue to pay the bill for my apartment. I had to make sure that nothing would spoil in the refrigerator if the electricity went out for any reason. I kept paying for the apartment because it was where all my stuff was kept. I was worried a little about Shawna breaking in and stealing things, but anything of real value I had put in a rental storage unit. I took my truck over to Klotz's house, put baking soda in the floor to stop any damage done by the beer/piss from Shawna, and then I took a taxi to the company. As long as the C-5a would not catch on fire again, we were finally on our way to the war.

On to War

Our first stop on our long trip was in Westover, Massachusetts to refuel and change aircrews. The Red Cross and the Air Force wives had set up a departure station there and everything was free. There was junk food, cokes, books, ice cream, writing paper, personal hygiene items and newspapers. We watched videos on a VCR in the lobby and then ate breakfast in their mess hall when it opened.

Our next leg of the journey took us to Torrejon, Spain, where the aircrews changed out again. There was a similar departure station there too, but now we had to pay for everything. There were hundreds of drawings and cartoons on the walls from all the units that had passed through there. I drew up a cartoon of Bart Simpson in a DMV, saying "Liberate the Oppressed Dudes!"

We found out that the first C-5a from our company, that had left two days before, was still in Spain because their engine had broken down. While we were there our company commander, Captain Star___, was promoted to Major and Master Sergeant Lloyd of ODA 595 was promoted to a captain, since their team didn't have an officer. As we flew to Saudi Arabia, we wore BDUs without any

Triple Canopy

patches so no one would know who we were, or that we were headed to the desert. None of us wore any berets either and had left them home. Once we would arrive in Saudi, we would change into the desert uniform, nicknamed 'Chocolate Chips". Our "Chocolate Chips" also didn't have any patches.

We arrived in King Khalid Military City (KKMC) around noon on September 17th. The only Americans on that base were 5th Special Forces Group and Task Force 160th Aviation. When I took the first step off the plane, I knew we were in a foreign land. Stepping off the ramp was like stepping into an oven. The heat almost knocked me over and the exotic smell let the senses know that we weren't in Kansas anymore. It was a combination of dust, spices and animal dung.

We were issued 5.56mm and 9mm ammunition and then were told to have magazines in our weapons, but not to chamber a round. There was no security in the base that we could see and there were only a few Air Defense Artillery pieces scattered around the airfield. We were told that the Saudis didn't want us to put up any sandbags or barbed wire, because they didn't like the way it looked. Though it wasn't very tactical, we didn't mind at all because we ended up in a building with four floors, which looked like a modern apartment complex.

Our entire battalion was located in the one building that was air conditioned. The center of the building was open so that I could see all four floors and the lobby. Each team had one floor, but on that floor were our beds, stoves, showers and washers. It was pretty good and I thought that the conditions here were better than our barracks back at Fort Campbell. The showers were a lot different than what we were used to, there was no "sit down" toilets and the shower doubled as the toilet. There was a hole in the floor at the back of the shower, which was the toilet. There was also no toilet paper, since Saudis did not use it. They would use their left hand, and then wash it when they were done. We had known about this and brought along our own toilet paper, but there was no way

322

to flush it down. So, we soon started doing what the Saudis did and rinsed away the waste as we took a shower.

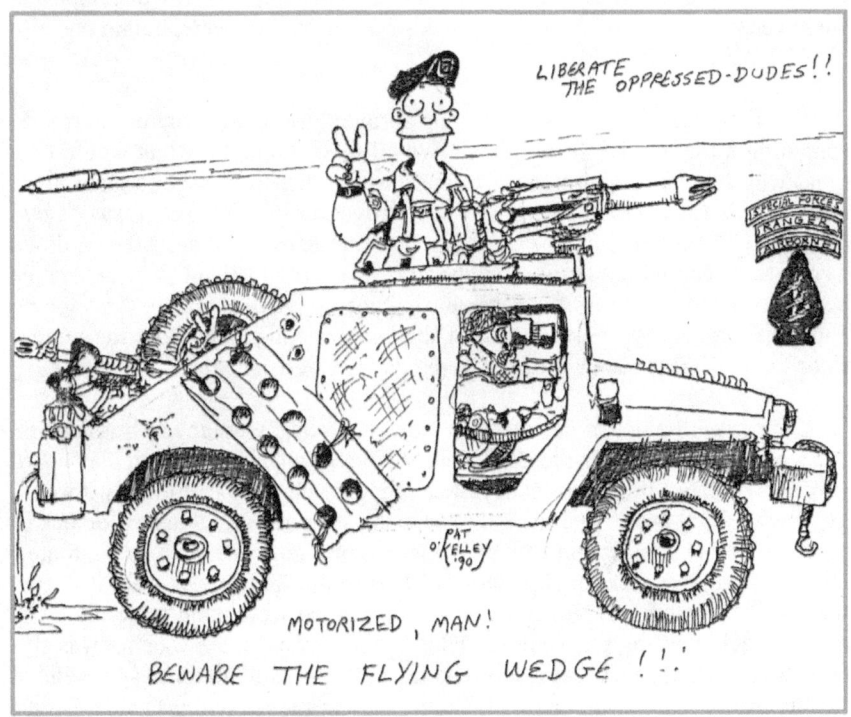

When we arrived, we had no mission. There were Syrians, Kuwaitis and Egyptian units at KKMC, but most of those countries had their military up on the border with Iraq and Kuwait. They were there to slow down Saddam Hussein's army if they decided to push on to Saudi Arabia while the Americans built up their forces. Our battalion commander told us that the Teams that did not have any vehicles might end up with Toyota Landcruisers for transportation. My team already had the DMVs, so it was not a problem. There was a shopping center nearby, that everyone called the "Wal-Mart". We ate MREs for breakfast and dinner, while lunch was fruit, Pepsi and pita bread. I really liked the pita bread since it was baked fresh each day. I didn't see a lot of Cokes in the Middle East and I learned that the Arabs thought that the Jews drank Coca-Cola, so they only drank Pepsi. I figured whatever marketing executive in Pepsi pulled off that one did a major coup in the cola wars.

There were mosques and minarets all around KKMC and the call to prayer could be heard every four hours starting at 0700 and ending at 2100. The call to prayer was on an incredibly loud speaker system and there was no way anyone could miss it. For our own opsec we didn't wear anything that would tell who we were.[55] We covered the bumper numbers on the vehicles with duct tape and we never saluted any officers. We knew who our officers were, but no one else did.

The first night we were there a "bug sprayer" truck drove by our barracks, spreading a fog around the building. We all joked about how that would be a great way for the Iraqis to sneak in and gas us, then we all got serious for a minute and moved closer to our M17 protective masks. Our Team was on two different C-5a aircraft, and the second half of our team didn't get there until two days later because their engine caught on fire.[56] When MSG Cr___ arrived he began to make our lives miserable by having us move things around for no apparent reason. We moved a lot of the heavier equipment from the bottom floor to our 4th floor, so by the end of the day we were smoked.

There wasn't much to do each day, since we weren't assigned a mission yet. We would drive around in the DMVs, exploring the city, or we would play board games.[57] A lot of time was spent just catching up on sleep. At night we would go over to "Wal-Mart" and get a dinner meal, such as fried chicken, kabobs or gyros. We had a television and VCR that we brought with us from the States, but the TV didn't pick up any stations because the frequencies were different than ours. Someone forgot to pack videotapes, so we couldn't use the VCR either. Each morning we would do PT pretty early, when the weather was still cool. Luckily one of our engineers, KC Dreller, was a linguist who spoke fluent Arabic, Russian, Persian, German, Bulgarian and French and was able to start making contacts with the Kuwaitis on the post. Due to his ability to speak the language we were able to find their shopping area in what was known as the "5000 area". While we were there, Captain Mastrovito picked up the last map of Saudi Arabia in the store for 100 Riyals, or $25.

By the end of the first week, we had moved our DMVs from the motor pool area to the parking lot in front of our barracks. We also strung up barbed wire around the buildings and had a roof guard overwatching all of it. The stories that we had been told when we first got there, about the Saudis not wanting us to put up barbed wire, had been bullshit rumors. Since the chain of command

---

[55] OPSEC = operational security or keeping information from anyone who didn't need to know.
[56] C-5a aircraft were notorious for breaking down and begun being replaced by the C-17 in 1993.
[57] Computer games were not that common yet and the few there were out there didn't have a lot of graphics.

didn't seem to think there was much of a threat we were ordered to turn in our weapons and ammo so they could be secured in one central location. This really angered us on the Team because we were trying to put ourselves in a war time mentality, and the rear echelon chain of command wasn't thinking that way yet. The rumors were that the Iraqis had moved chemical units to the border and that they had test fired a SCUD-B missile with a chemical warhead. We ignored most rumors, but we never knew if they were true or not.

We finally received our first mission, to teach the 4th Saudi Brigade FDC procedures. This FID mission was supposed to last four weeks, but we all figured we would be at war by that time. After we received the mission, our Team zeroed their rifles and was able to train on their .50 calibers for the first time. Nikolai and others drove down to Dhahran to learn how to do close air support missions with the A-10s down there. I was supposed to do a flight along the Iraqi border, but it got cancelled due to VIPs wanting to see what the border was like. We never did find out who the VIPs were. We all bought Saudi shemaghs, the traditional headdress, in case we had to E+E during the mission, but none of us wore it like the Saudis and we would have stood out easily. Some of the Teams painted their weapons, ponchos and LCEs with tan spray paint. We didn't, since we were a mounted Team and figured a big vehicle would be noticed before the weapon was.

We were issued bromide pills that we weren't supposed to take until the war actually started. We were told that these pills would supplement the atropine injectors that we would use if we came in contact with a chemical nerve agent. There were a few of the soldiers who took them early and their side effect was "erratic behavior". During this time, we also had numerous accidental discharges (AD). This was because when each soldier left a building they were supposed to lock and load their weapon with a live round. Then when they went back into a building, they were supposed to clear the weapon and then aim it into a barrel full of sand and pull the trigger to make sure that it was clear. This constant loading and unloading was annoying and many forgot about it, firing the weapon into the barrel when they cleared it. I just kept mine unloaded the entire time, since the chance of the enemy hitting us was pretty slim. The ADs got so bad that the Battalion commander said that anyone who had an AD would get Article 15 punishment.

Though mail was free, we weren't getting any. The rumor was that the Air Force wasn't taking the mail out, or bringing it in. Supposedly there was 35 tons of mail sitting in Heathrow, England, but there was no aircraft to deliver it. The mail was free though, since Saudi Arabia had been declared an imminent danger zone by Congress. We were now collecting the additional $110 a month for combat pay. I continued to write down what we were doing in my diary, but

I would only do it when I was in the shower/toilet. I didn't think the idea of a diary would go over too well in a unit that prided itself on secrecy.

One night we had a scare when all the chemical detectors went off in the building. MSG Cr___ came running in with a panicked look on his face, yelling "GAS!" and we all put on our masks. The SCUBA Team went one step further and ripped open their chemical suits and put them on. These suits weren't supposed to be open until the war started because they only lasted a few weeks. After that the SCUBA Team had to get new ones. It turned out that it had been a false alarm that was set off by the bug spraying truck.

The rumors floating around were that Saddam Hussein had a nuclear bomb now, and that there were twenty-three divisions in Kuwait waiting to cross the border. The Iraqis also had placed huge minefields along the border that were 50 kilometers long, one kilometer wide and had a tank ditch running the length of it. If this was true, we would be outnumbered about ten to one. We were also told that the French Foreign Legion and the British SAS would be coming to KKMC.

Our Senior Medic was nicknamed "Poppa Doc", but we didn't have a junior medic yet. Papa Doc told us that the Army women in Dhahran were having sex with just about everyone down there. He said that some did it for money, while others were doing it to get pregnant and sent home. Supposedly one of the women was caught with $900 on her from prostitution. We were also told of the Kuwaiti resistance fighting in Iraq. The name for the resistance was "Sabbah" which meant "morning". A lot of this talk of nuclear bombs, and chemical weapons was spooking some of the guys. I really didn't care one way or another and figured there wasn't much I could do about it so why worry.

Nikolai came back from Dhahran and told us about how the 1st Battalion was living in King Fahd airfield in an underground parking garage, nicknamed The Batcave. They had arrived in country on September 5th. Initially the 1st Battalion was going to do the FID mission, mainly to be advisors to the Arab units on the ground. Our 3rd Battalion was originally slated to do the Special Reconnaissance missions, but Delta Force wanted to have this piece of the action. The rumor was that 1st Battalion tried to train the Saudis, but somehow offended them and was thrown out of the Saudi Army (SA) camp. They ended up with the reconnaissance mission, and the 3rd Battalion then ended up with the FID mission. Nikolai said that those living in the Batcave never knew if it was daytime or night time, and since there wasn't anything to do there, no one bothered to go to the surface. The 101st Air Assault and the 82nd Airborne were both at King Fahd, so we figured we had it easy where we were. Nikolai told us that the 101st would only let SSGs and higher have any ammunition, since eleven of their men had been shot accidentally since they arrived in country. One of the 101st privates swore that terrorists stole his pistol and shot him in the foot. A little morale boost came when Nikolai told us that the grooming standards were relaxed a bit. We could now have a little longer hair and we could have a bigger moustache. This was mainly to build rapport with the Arabs. They all sported moustaches and they thought that anyone who didn't have one was not a soldier. So, we all grew big bushy "Magnum PI" moustaches for the duration of the war.

On September 26th the Battalion finally moved out on our first FID mission. At 0500 we headed north out in a convoy and drove to Hafr el Batin, one of the last cities in the northern part of Saudi Arabia. We really had no idea where we were going, but we followed the vehicle in front of us. Driving down the road was like being in a hot air cooker, with 100-degree wind whipping around us. We were glad that we had removed the doors from the DMVs and left them back at the barracks in KKMC. The terrain looked like a vast, white pool table. No trees, no vegetation and flat as far as the eye could see. I wondered how we would ever be able to hide in this desert from Saddam's armor divisions.

After driving for two hours, we finally passed through Hafr el Batin and came out the other side, where our Team was dropped off after we met the Saudi guide. We followed the guide's armed Toyota Landcruiser through the desert, down a road that had been plowed through the sand. We drove for twenty-five kilometers until we came to the 4th Saudi Armored Brigade. Their camp was huge, with white and tan tents spread out for miles in every direction. Interestingly we didn't see any armor vehicles in this armored brigade.

MAJ Star____, CPT Mastrovito, MSG Cr___ and K.C. Dreller went into the Saudi commander's tent. Dreller, who knew all the languages, became the translator for them. The whole experience began to resemble the Robin Sage exercise we had done in the Q-Course. MAJ Star____ made several cultural mistakes, such as drinking tea before the Saudi commander did, but it didn't anger the Saudis. Star____ told the commander that we were there to teach FDC procedures to his men. The Saudi general said his men didn't need FDC and that they were already trained on that, but he wanted them to learn what to do in the case of a chemical attack. MAJ Star____ kept trying to push the FDC on him, since this is what we had been training on for the last two weeks, but he only angered the SA General.[58] The General, who was also a sheik, told Star____ that they had just been issued the M17 mask from the United States

---

[58] SA = Saudi Army.

and his men didn't know anything about them, not even how to put them together.

While these negotiations went on, we tried using our limited Arabic with the SA military police that were escorting us. They invited us into their tent, which was covered in rugs and pillows. The tent had rugs sewn into the walls and there was a refrigerator and two air conditioners set into the sides. It turns out that this was the tent that they set up for any reporters that might show up. The MPs brought us some tea on a silver tray that was served in what looked like a shot glass. The tea was extremely hot, and really sweet. I liked it and so did most of the other Special Forces soldiers. As we waited the Saudis continued to bring in tea, about every ten minutes. Some of the tea tasted like peppermint, and their coffee was the most horrible thing I had tasted over there so far. It tasted like hot ground peppercorns. There were no chairs, so we lounged on the floor or leaned on pillows, while eating dates and drinking tea.

When it was lunch time, they brought in a huge plate that was three feet across and laid it on the floor. There were no forks or eating utensils and each man grabbed the food with their right hand. The left hand was never used, since that is the one used to clean your backside. Though most Americans would think the average Saudi to be unclean, they washed their hands, feet, and face five times a day before they prayed. The lunch was rice, lamb and cut up vegetables. There were also bowls of watermelons, apples, oranges and pomegranates. After we ate other bowls were passed around to wash our hands. When we saw the Saudis all lay down and take a nap or "siesta", we figured it was the thing to do, so we slept until around 1600. The Arabs always slept through the heat of the day if at all possible.

The junior Engineer on the Team, Darren Crowder, was also the driver of the vehicle I was in. Both of us were told to go back to Hafr el Batin with the Saudi MPs and guide the rest of the company to the camp. Our two guides knew no English at all, but we were able to get them to understand us with our ragged Arabic. While we waited in Hafr el Batin we talked to them and watched the war machines drive past. There were no American vehicles this far north, so all that passed by us were coalition vehicles. We drank Pepsis, while watching the Egyptian tanks roll by. The Egyptians were still using old Soviet vehicles, just like the Iraqis, which I later learned, would make recognition hard in the middle of a fight. Around the time of noon prayers, we learned that there was going to be an execution in Hafr el Batin.

A Pakistani worker had murdered his girlfriend. The condemned man was brought out to the center of town, and appeared to be either drugged or in shock. He was crying when he knelt down on a sheet of plastic that had been spread out beforehand. The executioner carried a large sword, almost a stereotype of the one that would be in the Aladdin cartoon. After giving a short statement the executioner cocked his leg back, and touched the prisoner on the back of the head. I don't know if this was supposed to let the Pakistani know that he was about to die, but it served the purpose. The condemned raised up a bit, ready for the blow. The executioner then swung the blade, taking off the Pakistani's head in one blow. The blood shot out from his neck, like from a garden hose, and then slowed. The head rolled to a stop, only to have another man, I was told it was a doctor, grab the head and put it in a sack. The executioner then wiped the sword clean with a cloth, and then licked the blade with his tongue. Needless to say, crimes, such as murder, are few and far between in Saudi Arabia.

SGM Beuckman, our company sergeant major, finally arrived with the rest of the company, and after much confusion we led them to the camp site in the desert. We arrived after dark and unloaded all the cots behind the sheik commander's tent. There were five teams in the open desert on cots that night and when the temperature dropped down to 65 degrees it felt like it was freezing. However, everyone was so tired that they decided to wait until morning to set up the ARFAB tents.[59]

The next day we began training the Saudi units with our NBC equipment. MAJ Star____ put one big team together consisting of five 54Bs (Chemical weapons specialist) and three assistants. The Team sergeants tried to get Star____ to change the configuration to one 54B with each Team, so we could train more of the Saudis, but he wouldn't budge from his original decision. He didn't seem to trust the Teams when they were not in his sight.

Star____ constantly wanted situation reports about where we were. When one of the radios had a broken handmike, he wouldn't believe the Team leader and called him a liar. I wasn't impressed by him and he seemed to me like a conventional unit commander placed in an unconventional situation. Since we weren't tasked to train anyone, our Team went over to a Saudi artillery unit, drank tea and tried to build rapport. We asked the artillery unit commander, another sheik, what he needed in the way of training. He wanted to have a command exercise, which we weren't able to do since none of us had an artillery background. We passed on the information to headquarters to see if we could get an artillery officer up there to help them out.

In the afternoon our Team dropped off one of the NBC training teams and then we drove to Hafr el Batin to get ice. MAJ Star____ didn't want us to go

---

[59] ARFAB stands for Army Fabricated Tent. It was a canvas tent that could be set up over aluminum poles in a very short time.

there, but we finally convinced him that the ice was needed. While we were in Hafr el Batin, Dreller talked to the Saudi soldiers and learned that there was an enemy propaganda radio show that had been nicknamed "Baghdad Betty". She had told the Saudis that we were all apes and that our female soldiers were running around naked in Dhahran.

When we returned to the camp, we were told that General Schwarzkopf would be arriving with five Blackhawk helicopters, so Star_____ made everyone put on all their gear, to include helmets and wait in the sun during the "siesta" time. It was the first time I had heard about a general named Schwarzkopf, but he never did show up. We felt pretty stupid standing there in the heat of the sun, waiting for no reason.

The next day Star_____ "grounded" our Team. He didn't like the idea of the Team being able to leave whenever they wanted, so he took away one of our DMVs and held it hostage in the camp. Unfortunately, it was my DMV, so I just sat in the tent all day with the other two guys, doing nothing. CPT Mastrovito was also grounded and was ordered to stay in the camp. I ended up getting diarrhea from all the flies on everything, so the only time I left the tent was to use the bathroom. I had pretty low morale because on top of the bullshit we had to put up from Star_____, I was an extra weapons sergeant that had no real job.

The next day our DMV Team was tasked with being the delivery truck for the company. We drove back to the FOB (Forward Operating Base) and dropped off our spare tire. It didn't have a wheel, so we picked up one with a wheel. We also picked up a mechanic and a bottle of battery acid to put in the camp generator. Unfortunately, the bottle busted on the way back and we had to douse the rear end of the DMV with water so it wouldn't eat through the vehicle. When we returned to the camp, we learned that Star_____ was going to give two of our DMVs to ODA 596 to do a CAS (close air support) mission near the border. We were pretty pissed off since we had spent hundreds of hours getting the vehicles ready and training on how to use them, and now Star_____ was giving them away.

The Team approached SGM Beuckman and told him we wanted to talk to the Battalion CSM about our problems with MAJ Star_____. The SGM had Star_____ talk to the Team instead. We offered him several solutions to his problem, but he would not change his mind. CPT Mastrovito also told him that two DMVs are tactically unsound, and you needed at least three to pull security and operations. Finally, Star_____ told us that if we could get a HUMV, he would let us keep our DMVs. No problem! We immediately drove back to KKMC to drop off laundry, pick up mail, and steal two trucks.

330

Chapter 3: SPECIAL FORCES – A Storm in the Desert

Our life of crime was stopped before it began when we learned that two 2½ ton trucks and a HUMV were on their way to the company. The rumors we heard back at KKMC were that there had been a drive-by shooting against the 101st Division in Dhahran and a couple of the 101st soldiers were missing. The other rumor was that the Iraqis were going to attack that night because it was Mohammed's birthday. Every time we were back at KKMC we heard wild rumors, but none ever proved to be true.

I got back to the company at 0400, but the rest of the Team had drove to the border of Kuwait to link up with the forward element of the Saudi 4th Brigade. When Cr___, Dreller and Nikolai arrived they all sat around for about four hours waiting to see the Saudis. This was because an American US Army Reserve general was being entertained by them. After it got dark the fat general walked over to Cr___ and told him he needed to shave and that we should change our uniforms every day. K.C. told him that he would change with Nikolai and Nikolai could change with Cr___.

After awhile we began to call each other by our nicknames. K.C.'s nickname was "Ghengeez", after Genghis Khan. The Saudis had called him that for some reason. Nikolai was just called "Rat" because sometimes he looked like one. Gentz, our senior engineer, was called "Inspector Gadget", because he liked to take everything apart and try to improve it. Blake, the Intelligence Sergeant, was called "Beast" because he looked like a big slab of muscle, with some eyes. CPT Mastrovito was called "Captain Nick" by the Saudis. He also had a code name of "Miser". Crowder was called "Crow Dog" and Cunningham was simply called "Doc". I was called "Troll" due to my hair.

At the beginning of October, the conventional artillery officer arrived that we had requested to help out our Saudi artillery unit. He was stumped by some of their equipment, since it was French armor. The Saudis used GCTs and AMX-10s. During the time that we had been training the Saudis an Egyptian Division had slowly been moving into our area. Every time we did our runs to Hafr el Batin, there were more and more Egyptian tents and vehicles. As we drove along the manmade road to our camp, I estimated that there were 25 kilometers of Egyptian tents.

Our Team finally began to teach some classes to the Saudis. Gadget taught NBC classes, Papa Doc checked out the Saudi medical gear and Be__, Rat and I taught CAS. Mastrovito, Cr___ and K.C. drank tea and continued to build rapport with the commander. They ended up drinking some tea that they thought had a narcotic in it, because it had their head swimming.

Triple Canopy

Soon the Syrians moved into our area, having just come from Lebanon. They looked a lot like us and some were fair haired. We didn't trust them that much since they were a possible enemy of the United States. During one of our trips to Hafr el Batin we acquired a can of black paint and I painted our Team logo on the side of the DMVs. The Team decided our logo would be the Flying Wedge. There were other DMV Teams that had their own logos. The Team with the Foreign Legion had a sexy French girl on theirs, with the words "French Kiss". The 1st Battalion DMVs had a WWII Afrika Korps design, but instead of the swastika on the palm tree, they replaced it with the SF crossed arrows. One Team had the comic character Spaceman Spiff, from Calvin and Hobbes. I also drew the symbols of each person who was in the vehicle, such as a Rat on Nikolai's turret ring. I drew an Orion constellation, since my code name was Orion for use on the radio.[60]

Everyone on the Team kept getting diarrhea due to the flies. To combat this, I would wash my face, hands and hair every night. I took a "whore bath" every other day out of a folding cloth sink and I would wash my clothes every fourth day. The US Army didn't have desert-colored boots, so we decided not to polish our jungle boots until they began to look the right color. As we drove around K.C. and I would wear the Saudi shemaghs to keep the dust out of our faces.[61]

The rest of the Team wore green "drive on rags" (triangular bandage) and goggles. Our goggles were all bought from civilian stores before we left because we never had goggles issued to us. We didn't get military issue goggles until December. My own goggles were workshop goggles from Sears. I filled the holes in the side of the goggles with glue to keep out the sand. We never wore our LCEs and kept them in the DMVs with the rucksacks. All we wore were the M-9 pistols on our belts and our M17A1 protective mask. I also wore my Randall knife on a sheath on my belt.

---

[60] Our code names were all five letter words, with the first letter being the same as the last name.
[61] In my journal I called them Khafiyas, but it turns out that the Khafiya was the female scarf.

Different Team DMV Markings

On October 5th I saw the border of Kuwait for the first time when the Team drove up to the Saudi covering force located there. We linked up with ODA 596, who told us that they had just found twenty dead sheep that had been killed for no apparent reason. They were investigating it to see if it might be some sort of chemical agent. All along the border between the two countries there was a large dirt berm that was ten to fifteen feet high. This had been built to slow down smugglers and drug traffickers. ODA 596 also told us about an Iraqi BMP that had come over to the Saudi side when it had become disoriented. The Saudi border patrol fed the Iraqis, gave them water, and then sent them home. On our return drive back to our camp we saw miles and miles of Egyptian armor, and anti-aircraft weapons. We also saw the French Army for the first time. Overhead I saw two jets cruising high in the stratosphere. I didn't know if it was ours, or theirs. They were the first aircraft I had seen over the skies of Saudi Arabia.

We returned to KKMC to change the oil in the DMVs and learned that we were only one of two companies on the border. Everyone else in Group was still back at the rear so whenever we walked into the messhall everyone stopped what they were doing and stared at us. We were covered in dust, had the improvised goggles on and all were wearing "drive-on" rags. Some captain came up to us and told us we had to take off the rags. We did, and then commenced to eating everything we could. For weeks we had been eating the same food as the Saudis, which were basically goat and rice or chicken and rice. In the messhall there were cheeseburgers and real cold Pepsis. Not the warm Pepsis covered in flies, but cold ones, and there were no flies anywhere on the food. We felt like we had died and gone to heaven. Everyone in the messhall just watched us like we were from another planet.

While we were in the FOB, we had to be a taxi service for the B-Team since they didn't have any vehicles. They didn't leave their tents much, so they had not learned a word of Arabic. We all became the translators for them. CPT Mastrovito pulled radio watch with me, when we learned we had the detail, right after we had both driven six hours to get there. MSG Cr___ knew about this duty, but didn't tell the CPT. Mastrovito was getting pretty angry with Cr___ because he never let anyone know what was going on. It was some sort of power play, where Cr___ would have information and therefore others would have to rely on him. On a good note, our morale rose when we found out that there was a direct phone line in one of the tents. The crown prince of Saudi Arabia had it installed for us and he was paying the bill. I didn't use it much because I didn't have anyone back home to talk to. I called my mom and dad a few times, but that was about it.

In the second week of October, we were coming back from a supply run in Hafr el Batin when Crowder, in the lead vehicle, got lost. It was pretty easy to get lost because there were so many new tents and units showing up in our desert. Crowder was driving pretty fast down the sand road and ended up driving through the Egyptian perimeter. All the rest of us followed in our DMVs. I looked out the door and saw an M-60 tank go zipping by, followed by a dozen running Egyptian soldiers all pointing their AK-47s at us. I told Crowder to stop so we could explain that we were lost. We were soon surrounded by angry Egyptians waving their AKs and bayonets in our faces. I silently jacked a round into my M-9, figuring that if one of them began shooting I would return fire. The Egyptian guards led us to their commander, poking some of us in the back with their bayonets. Rat got stabbed and was ready to have it out right then. It was a pretty intense moment, but luckily, we had K.C. with us and he was able to explain to the commander that we were with the Saudis. We were escorted back to our DMVs, but after that we all swore we would not let Crowder take the lead again.

On October 9th our mission with the 4th Brigade was over and we returned to KKMC. Our Team was debriefed by our Battalion commander. We told him about our encounter with the Egyptians and that we were always running out of water, Pepsis (when there was no water) and food. He was surprised at this because Star____ never reported any of this to him. He also learned that we never knew the evacuation signal or the evac plan from his operations order. Star____ had never told any of us and we didn't even know there had been a signal.

Chapter 3: SPECIAL FORCES – A Storm in the Desert

While we were back at KKMC Papa Doc was able to acquire some medical supplies for the Saudis and we helped liberate fifteen cases of Pepsis from the messhall loading dock. Life in the rear was pretty boring compared to what we had been doing. Our Team had the mission of making a sandbagged walkway for the B Team, which served no purpose except for cosmetic appearance. We did get issued a new piece of equipment that was pretty amazing to us. It was called a NAVSTAR and it was about the size of a hardback book. It weighed five pounds, but it would change the way we would navigate from now on. The NAVSTAR was able to tell us where we were, to within a few meters, by gathering the signal of three satellites. This was the very first GPS and it amazed us. The only drawback is that it took a long time to find three satellites to triangulate our position. It also didn't work well inside a DMV, so we duct taped it to the hood of the DMV so our driver could read the display through the front window.

Our Team continued to go up to the Iraqi border to teach classes to the Saudis, but they were only short trips and each night we would return to the B-Team camp and our ARFAB tent. The tents were dug into the ground and then sandbagged, just in case the Iraqis fired missiles at KKMC. There were trenches to each tent, but it was almost impossible to dig in the sand. It was easy to dig for about three feet, and then the soil turned into something that resembled concrete. The only way to dig further was with pick axes or a backhoe. I would see this again when we had to clear Iraqi trenches in Kuwait. They never went down more than a few feet.

We still got mail out in the desert and my dad sent me a Newsweek magazine that had Raquel Welch, at 50 years old on the cover. We put her on our tent wall and pretty much worshipped it. It was the only girly picture we had. I thought this was funny, because I think my dad did the same thing with Raquel Welch in Vietnam. We were also issued some additional ammo, such as claymore mines, M203 grenades, more .50 caliber ammunition and the newer anti-tank weapon called an AT-4. It was a one-shot weapon, like the LAW rocket, but it could penetrate more armor. The Team also picked up our junior medic, straight out of the Q-Course. He naturally became known as "Baby Doc" while Cunningham became known as "Papa Doc". Even though he was only a few years younger than us, he seemed like a kid out of basic training.

On October 11th our Team was sent back to KKMC to learn of our next mission. We would be doing the SR mission and would have a SOTA team (Signals intelligence Operational Tasking Authority) attached to us. The SOTA team was military intelligence (MI) and would process whatever information we found and report it back to the rear. We were also told of the first shots fired into Iraq by US forces...by accident. A Warrant Officer from our 2nd Battalion

335

had leaned against a locked and loaded .50 caliber and pushed the trigger, sending rounds into Iraq. No one was hit, but it was an embarrassing situation. We were told he had been relieved and sent home. Due to that our Team decided we wouldn't go around locked and loaded.

We continued to improve our tent in the FOB whenever we had free time. We built a wall in the front and back out of sandbags, built shelves along the walls, and put-up ponchos in between the roof and our cots to try to keep the intense heat out. We also put up a volleyball net outside our tent and would play against the Saudis in the evening. Mail came regularly now and with each batch of mail something new had appeared. These were letters addressed to "any service member". The letters we received were usually from kids all across the United States. Since we didn't have much to do, we answered all of them. We never got any letters from girls in their 20s or 30s though, just kids and folks over 50 years old. We suspected that the perfumed letters, or ones with girly writing, were taken out by the REMFs long before they arrived to us.[62]

The bullshit in the FOB continued to grow with each day. Star____ made us the assistant instructors to a sergeant, while the sergeant taught Kuwaitis on how to use their gas masks. Our Team's Captain, MSG and the four SFCs were not thrilled about having a SGT order us around. Star____ also took back our NAVSTAR, making us the only Team without one. By mid-October our Team captain had just about enough of MAJ Star____'s petty bullshit. Captain Nick wrote a letter to the Group commander trying to see if he could get our Team detached from the company. So far, we were the only Team to have accomplished a FID mission, while the rest of the Teams had not left the camp yet. Star____ also kept putting us on guard duty at KKMC, so we no longer could do a regular mission. Our off time was spent playing volleyball and picking up little rocks to gravel our tent floor with them.

When we weren't on guard at KKMC we sat around in the air-conditioned barracks answering the "Any Service Member" letters from the kids, and staring at our lone Raquel Welch poster. Crowder came up with the idea of writing to our favorite female celebrities, to see if we could get them to write back, send a picture, or do a USO trip over to Saudi. Charlie Blake wrote to Markie Post, an actress on a TV show called *Night Court*. She did write back, sent some pictures and a belt buckle with a gavel on it. She also sent a cassette tape with all of the cast wishing us luck. Crowder wrote to some super model that I had never heard

---

[62] REMF is an age-old acronym that means "Rear Echelon Mother Fuckers" and pertains to those who never see the front line, but are the first to brag about how dangerous their job is and how heroic they were. In the later wars in Iraq and Afghanistan they became known as "Fobbits" because they never left the FOB.

336

of, Cyndi Crawford.[63] She wrote back, sending the team a photo of her, sort of naked, and autographing the picture in gold nail polish.

I wrote to an actress named Lolita Davidovitch; a movie star who had been in a movie called *Blaze*. She had red hair and I liked her. I never heard back from her though. We also wrote to various businesses, trying to use some of the Desert Shield celebrity status to work for us. Crowder wrote to the president of the Moon Pie Corporation, and he sent us back a case of Moon Pies each month. I wrote to a DJ called Slats, who worked for a rock and roll station in Nashville. He sent me cassette tapes of current songs.

We also heard about one of the most repeated stories in Desert Storm. In our version of the story a 101st soldier received a videotape from his wife. Initially in the videotape she filmed the kids and she talked to the camera, but then it clicked over to her having sex with some guy, and then back to her looking at the camera, saying "I want a divorce".[64]

In mid-October we were training the Kuwaiti Army on how to use German gas masks. None of us had ever seen them before so we were teaching and learning at the same time. The Kuwaitis used variety of foreign equipment; their armored vehicles were British Chieftain tanks and Soviet BMPs. While were teaching one company they showed us a gold-plated AK-47 given to them by Saddam Hussein. They said they were going to take it back to Kuwait and shove it up his ass. The Kuwaiti armor commander told us how they had destroyed seventy Iraqi tanks when they retreated out of Kuwait. We didn't know if we should believe him or not. Every Kuwaiti we had met so far told us of how they all made valiant stands in Kuwait, so it was pretty doubtful. The primary instructors teaching the Kuwaitis were the 54Bs; however, they didn't know how to teach a class and didn't speak any Arabic at all. When their classes were over, we would return later that evening without the 54Bs and retaught the Kuwaitis.

It was during this time that we experienced our first sandstorm. It wasn't like in the movies, where the wind is blowing like a hurricane force gale. Instead, the sandstorm came in slowly, without any wind at all. Everything

---

[63] Cindy Crawford was just beginning to be known then and I had never heard of her before.
[64] After the war I heard this same story from just about everyone who served in the Gulf. The Navy swore it was a sailor who got the videotape. The Marines swore it was one of them, and Anthony Swofford, author of *Jarhead* wrote that he saw the tape. This story was the most famous rumor during the war and it turned out to be not true. Versions of this story went back as early as Grenada.

turned an orange color. It was extremely quiet and hot and humid in there, like we were covered with a wool blanket. As we drove, we were able to see about a mile, but no more. Normally we could see about ten miles across the flat desert surface.

The Air Force came out to the FOB to teach us how to use the radios and how to call in CAS missions. We kept getting classes on this because this would be our primary mission with the Arabic units that we would be attached to when the war began. During the class we learned that MAJ Star___ had the IPs (Initial Point). These were places on the ground where the Air Force would meet the Teams to call in airstrikes when the time came. He was supposed to have put that information out to the Teams, but he never did. The following day Be__ and Nickolai were sent down to Dhahran to learn how to call in missions with the AC-130 Spectre gunships.

We did finally receive our new mission. We would be replacing ODA 596 on the border doing the SR (Special Reconnaissance) mission. The SR missions were all located in buildings along the border of Iraq or Kuwait. The first mission, SR002, had consisted of a combination of ODA 562 and 505 and was commanded by Captain Ken Takasaki. They were combined with the Saudi Special forces, commanded by Captain Fahd, a Saudi prince. Prince Fahd had graduated both the US Army Special Forces and Ranger schools. Their mission was to patrol from the Saudi border town of Ruqi out to about 60 kilometers.[65]

To assist in this mission, we received a new piece of equipment that we hadn't used before, the PAQ-4 infrared laser for the M16 rifle. The device was a tube that attached to the side of the rifle's handguard and it would project a blinking laser that could only be seen with the night vision goggles.

Once we learned we were going up on the border I bought a short-wave radio at the "Wal-Mart" and we were able to listen to the BBC to find out what was happening in the world. The news media was predicting that the attack against Iraq would most likely happen in November or December. Chief Gaglione, of 595, came in one night and told us that his wife told him that the media reported that Saddam Hussein was going to withdraw from Kuwait. They reported that Saddam had a dream in which Mohammed told him he would be a hero if he pulled out before America attacked. I didn't believe it. Our Team drove out to the FOB, struck our tent and then packed it away. The FOB would be moving somewhere else in the desert and we wouldn't need a tent on the border.

---

[65] Nigel Cawthorne, *Warrior Elite: 31 Heroic Special-Ops Missions from the Raid on Son Tay to the Killing of Osama bin Laden* (Ulysses Press) 2011, pg. 118

Chapter 3: SPECIAL FORCES – A Storm in the Desert

To get into shape again I ran up and down the four flights of stairs in the barracks at KKMC. Just sitting around doing nothing was not helping us to get ready for a fight. B Company set up a fast rope on the fourth floor and they practiced sliding down it to the ground floor. The temperature also cooled off to about 90 degrees in the daytime.

I was annoyed that all the good-looking girl's "Any Service Member" letters never got to us. All we ever got were letters written by children. So, I decided to start my own program that I called "Any Sorority Member". However, I needed a large group of young girls to write to. Crowder had dated a girl from the University of Tennessee who had belonged to a sorority called AOΠ. He told me that the sorority was all over colleges in the South. So, I started "Any Sorority Member" and wrote to the AOΠ sororities in various Southern colleges. It worked and I started getting dozens of letters each week, and many had pictures! I also began to plan on what I would do if I made it out alive. I decided I would go skiing in someplace like Vail, or Aspen. To motivate me to survive I put I put up pictures of skiing and snow around my bed.

At the end of October CPT Mastrovito had enough of all the bullshit and was threatening to resign his commission. Between MSG Cr___ playing mind games and Star____ treating us all like we were children, he was extremely frustrated. We had a midnight meeting to discuss what we were going to do as a Team. Some of the Team members didn't want to get rid of Cr___, because if he was relieved Charlie Blake would be in charge. A few members didn't think Charlie had enough experience. Finally, we decided to talk to Battalion CSM Griffith about both Cr___ and Star____.

Griffith told us that the solution for Cr___ may come in a few days, and left the question unanswered. He also told us that the Battalion commander would talk to Star____. When our company SGM found out about it he came in and chewed us out for going over his head, however we thought he was part of the problem, so we had skipped him in the chain of command. Cr___ had not been there when this went down, and had gone to King Fahd airfield to have some dental work done. When he returned, he found out about the mutiny and then went out of his way to discredit everyone who spoke out against him.

The French Foreign Legion showed up at KKMC and we went over to check them out. They gave us beers, which Americans weren't allowed to have at all. We also traded for some of their berets or pins. When they visited our barracks, we showed them how to use the PAQ-4s and some of our other gear. Gaglione came in and talked to them in French. He had been in Vietnam with the 173rd

Airborne and learned how to speak it there.[66] The Legionnaires told us that they also thought we would go into Kuwait sometime in November.

At the beginning of November, we trained up for our SR mission. We drove around with our NODs (night observation devices), used the PAQ-4 and trained on the heavy weapons we carried on the DMVs. We were told that up on the border three French soldiers had been captured by the Iraqis, who had camouflaged themselves as Bedouins.[67] We were also told that the Saudi border patrol had been in several firefights with the Iraqis. The Iraqi air force supposedly would fly over the Saudi border to see if anyone would fire upon them. Hafr el Batin was now off limits to US soldiers due to possible terrorist threats against them.

We were told that the new rules of engagement stated we were allowed to defend ourselves, but we could not do any offensive actions. Our SR mission was classified as SR002. We still didn't have enough ammunition, and when we trained with the Mark 19, we could only fire 35 rounds. We also still didn't have any radios for use in between the DMVs.

At the beginning of November, we went into a semi-isolation and planned for the SR mission. Our twelve-man Team would be spread out in three OPs along the border. I would be in OP 3 on the far right of the border on Kuwait. Also at that OP was MSG Cr___, Nikolai, Pappa Doc and two SOTA Team members, O'Brien and Novotny. The SOTA Team were radio intercept specialists. Our other two OPs were on the border of Iraq, while a control OP would be located at a small border town called Ruqi. Each OP would have one medic and one commo man. Be__ was the "medic" for us, though he was only combat lifesaver qualified. Due to what I learned in my time in the LRSU unit I was designated the commo man. Each OP would have two DMVs, one with a .50 caliber machinegun and one with a Mark 19 grenade launcher. The additional DMVs were the SOTA vehicles. We finally got the radios we needed for the mission that included the PRC-90, PRC-3 SATCOM, PRC-70, the DMDG (the new encryption device) and each OP had a NAVSTAR.

---

[66] CW3 Michael Gaglione had an interesting career in the military. He had jumped into Vietnam with the 173rd Airborne in the only large-scale airborne operation of the war in 1967. He had not been Special Forces then, but came back into the military and joined Special Forces as an enlisted soldier in the HALO Team. Right before Desert Storm began, he became a Warrant Officer. He would continue his military career and retire to Montana, where he continues to "ride bikes with big boobed women with thin waists sitting on the back".

[67] As usual, none of this turned out to be true.

Chapter 3: SPECIAL FORCES – A Storm in the Desert

We would not be alone at the OPs. We would be there with a combination of Saudi Border Patrol (known as the Selaah Haduud), civilian workers and the Saudi Special Forces (Kuwat el Qassa). Each post had about thirty men assigned to it. The Saudi SF had plenty of weapons, to include a 60mm mortar. One of the things we would have to be on the lookout for, was the Iraqi deserters, who crossed over constantly. The SR mission was the only real-world mission the US Army was doing at that time because everyone else was training for the invasion into Kuwait or moving into position to do the invasion.

The trip to the border took four and a half hours. The border posts were 120 miles from KKMC, and 250 miles to the nearest conventional US Army unit. We were less than fifty miles from the main Iraqi divisions. We were the "speed bumps" who were supposed to slow down Saddam's army with air strikes until the coalition forces could move forward to stop the Iraqis. The post that I was assigned to was Al-Amarah, but we called it OP 3. The post was located fifteen miles east of Ruqi. It resembled an old French Foreign Legion fort out of the movies, like *Beau Jeste*. Though the dirt berm was right in front of the OP, the actual border was five miles in front of it, so the Iraqis were far enough away that I had to use the binoculars to see what they were doing. The fort had thirty Saudi Border Patrol and I learned they were actually part of the Saudi Coast Guard. The Saudi Special Forces at the post were driving Toyota Landcruisers with Belgian made M-2 .50 caliber machineguns and German MG-3 machineguns.[68]

The post had a kitchen, flush toilets, a shower and our room had an air conditioner in it. The air conditioner only worked when the power was on, and that cut off all the time. The blackouts were scheduled so we had power from 0900 to 1400 and then again from 1900 to 2230. Our job was to patrol the border at night. We weren't supposed to let the Iraqis know there were Americans at the post so we only patrolled with our one DMV.

It was cold whenever we patrolled so I would wear my gortex jacket and my wool watch cap. Each night we would patrol the border and then link up with the Border Patrol at the end of our route. This was a control measure, because if either one of us didn't show up, we would know something happened to them. The border patrol would make a fire out of little bushes they could find, and cook tea (which they called "shai"). Sometimes they would roast a goat and we would sit around in the dark with them eating meat. They would pass interesting parts of the goats to us, like the testicles and then watch our reaction as we ate them. They usually laughed uproariously when we discovered what we had

---

[68] The Belgian .50 caliber had a selector lever to go from automatic to single fire, unlike ours that just fired automatic.

341

eaten. They also passed around sour milk that came in a carton. It came already soured and I figured that they must have acquired a taste for it when they had been Bedouins.

On our first patrol we all were wearing the PVS-7s when Nikolai ran the DMV into a tank ditch. I was in the Mark 19 gun turret and I slammed forward so hard that it knocked me out for a few minutes. I ended up with a slashed hand, a bump on my forehead and a slash on my eyelid from the PVS-7s.[69] After the accident we decided to drive with the lights on. The Border Patrol would drive around with their brights on, and we learned that the 2nd Battalion SR patrols to the east also drove with lights on, so we decided it would be safe for us.

The patrol would end around 0400 and then we would get some sleep. The SOTA guys would listen to the radio each day, while we slept until 10:00 or so. Usually, I slept until it was too hot because the air conditioner had quit running, or because there were too many flies crawling on my face. After we ate breakfast and drank our shai, we would try to entertain ourselves until the night patrol started. Entertainment consisted of listening to BBC on my radio, listening to the seven pirated cassette tapes I bought in the Saudi store, reading books, writing letters, playing volleyball or catching up on more sleep. We did get mail every day, which raised morale. The OP circuit rider would drive each day, about a 60-mile route. The circuit riders were two counter intelligence guys, because sometimes they would also be able to pick up an Iraqi defector.

The food wasn't bad and usually consisted of goat and rice. Once every few weeks a Saudi truck would deliver a couple of dozen goats and put them in a pen beside the OP. The cook from Pakistan would slaughter one goat a day and then cook it up for everyone in the OP. The only thing I couldn't stand was a spice we called the "green goat spice" that was put on the food. I think it was made of crushed up cloves. The Saudis loved it, but it made the food taste terrible. We would alternate our patrols by either doing one long patrol with four guys in a single DMV, or we would break it down into two three-man patrols. The closest we ever came to an Iraqi OP was the one that was about 3,000 meters north of the dirt berm.

---

[69] This was one of many concussions that I would have in the military. Research done in the 2020s showed that Traumatic Brain Injury (TBI) contributes to early dementia. I often worry about this when I cannot remember someone's name or the plot of a TV show that I had seen. My father would die of complications due to dementia in 2018, so it is something that does worry me.

Chapter 3: SPECIAL FORCES – A Storm in the Desert

On my off time I would go up to the roof of the OP and map out what I saw in the distance on the back of a Pepsi carton. I used the laser range finder to figure out the distances. I was able to see three Iraqi OPs on the horizon from our OP. They had small armored vehicles, but no tanks. We figured the tanks must be hidden over the horizon. The closest Iraqi OP to ours was 6,000 meters away, and the farthest point that I could see was thirteen kilometers from my rooftop.

Beside our OP was a Mercedes Benz that a Kuwaiti refugee had left parked there, and then returned on foot back to Kuwait. The Saudis were watching the car for him. I never had to worry about the Saudis stealing anything and they were a very honest people. There was a three-legged dog standing guard on the Mercedes each day, but it would never come to us. I left food and water by the car every day for the dog. My "Any Sorority Member" program paid off and I began to get packages back from the college girls. One who wrote me frequently was Patricia, who had blond hair, green eyes, and described herself as 38-26-24. I also wrote to a bunch of people from Cupertino, California. I wrote so much that the newspaper printed my letters and the mayor of the town made me an honorary citizen.

Since our OP was the closest to Ruqi we would get VIPs all the time. They were usually various generals who wanted to see the border. Initially we were told we would be on the border for 30 days and then rotate out with ODA 595, but by mid-November we were told there were not going to be any rotations and we were there until the war started, or peace broke out. The nights got colder, down to 40 degrees, and I wore my polypro long underwear, gloves and our new desert night camo parka. Supposedly the night camo would break up our outline if someone with NVDs looked our way.

War began to look like it was imminent. We were told that all the units in Germany, except a handful of support units, were coming to Saudi. There were two armored divisions already in Saudi and the 1st Infantry Division from Fort Riley, Kansas was coming. When we drove around each night, we would use division level call signs to talk to our OP, so that it would fool the Iraqis who might be listening. I used "Big Duke 6" as my call sign. Usually anything with "6" in the call sign was a Brigade commander. For our actual radio communication back to Dhahran we used the SATCOM that couldn't be intercepted by the Iraqis.

On November 14th Operation Imminent Thunder started. The Air Force was going to fly the trace (the border) and see what the Iraqis would do. I think they were trying to find the Iraqi AAD (anti-aircraft defense) radar sites. We were told the Marines were also conducting a practice landing on the Saudi coast.

343

Supposedly if the exercise turned hot, then we would be going to war. We monitored the radios more than normal that night, and remained fully alert. We had finally mastered driving with the PVS-7s, so no lights were on that might give us away that night.

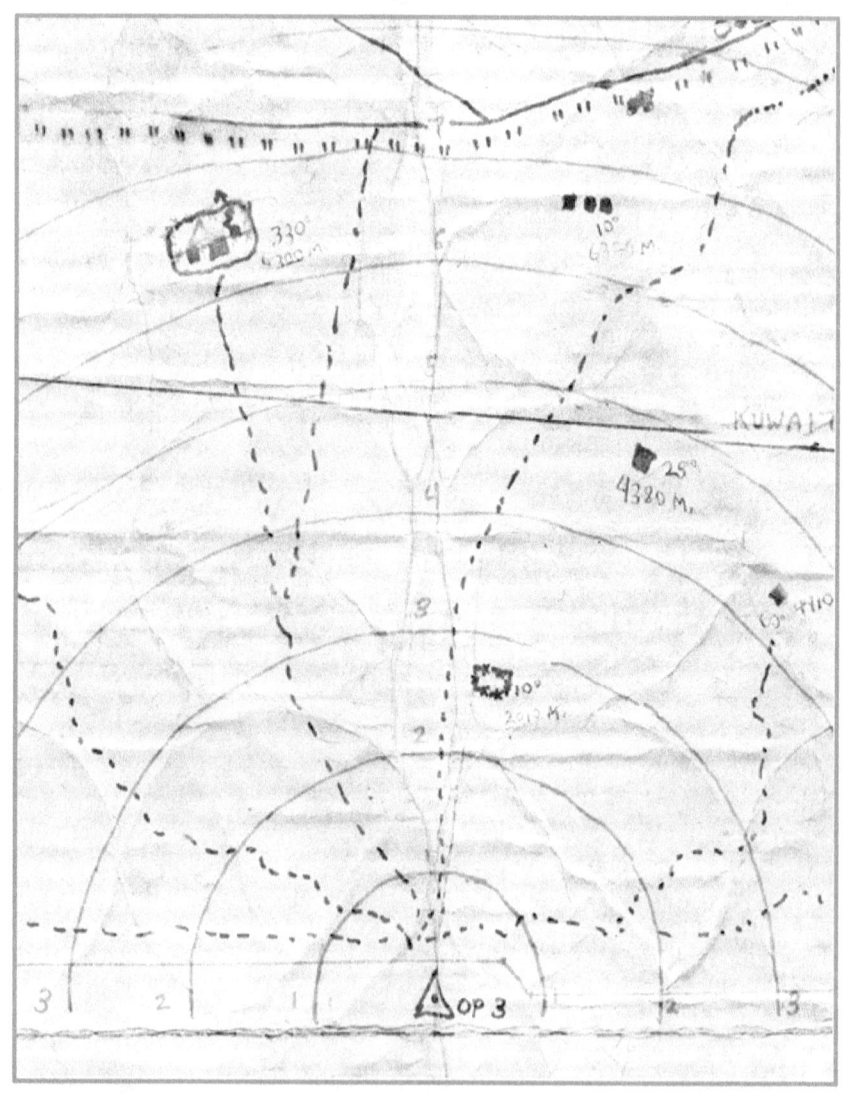

A part of the range card I drew.

Chapter 3: SPECIAL FORCES – A Storm in the Desert

Imminent Thunder was supposed to last three days, but it ended the next day. When we came in that morning, our Border Patrol was being relieved by another group of Border Patrol. They had been on the border for three months and now were going to be able to see their families. As they drove off, they fired their G3s in the air, scaring the hell out of all of us!

As a result of Imminent Thunder, the closest Iraqi post brought in two truckloads of soldiers and began digging trenches around their OP. I could see them putting in what looked like a minefield through the binoculars, but I wasn't sure. The Saudis told me that the Iraqis were using Kuwaitis to dig the trenches and lay in the mines. To get a better look at what they were doing I would use my sniper rifle scope, which had a better magnification. Other than the scope and the binos we had nothing else. For an early warning post, we had terrible optics. Later that day we learned that Alpha Company had rolled a DMV because they were driving too fast, and had killed a French soldier who was in the turret. The SF driver and another French soldier were sent to the hospital.

I continued to get as much information about the enemy as I could, such as watching the Iraqis I observed strengthen their position. However, every time I tried to report what I had seen, Cr___ would not let me. He just wanted to send a message that said "we can't see the enemy; we need better optics" so that he could get a telescope up there. I was pretty angry about that, because this was real world intelligence that might save some lives. The SOTA guys were able to send whatever they wanted, since Cr___ was not in their command, so I slipped information to them to send back on their radio.

Cr___ was terrified that the war would start soon and he wouldn't let any of us talk about it. We would get copies of *Newsweek* or *Time* and there would articles by reporters about how the war probably wouldn't happen. Cr___ would use these reporters' opinions as his basis for us not taking the war seriously. When Bob Kerry, a Democrat Senator, wrote an article about how he didn't think the war would happen at all, Cr___ told us that it was the final word. How could we argue with a Medal of Honor winner? [70]

When the OP began to run low on food, I was able to take the SOTA guys and do a food run to Ruqi. The SF guys had plenty of food, mainly MREs that we never touched, but the Saudis were down to just a few chickens.

The SR CP in Ruqi was only about 500 meters from the Iraqi border. Ten kilometers south of Ruqi was a small town called Asher Kilo, which literally

---

[70]This incident should show that the media can affect how our soldiers would react to the situation they are in and can do considerable damage to the morale of soldiers in the field.

345

translated into "10 kilometers". While in Ruqi I was able to call my dad on the phone in the headquarters. I wasn't able to talk long, but that didn't matter. The SF guys in the SR CP had tried to make a satellite dish from Pringle cans, and they swore that it worked. They also had a pet kangaroo rat (called a gerboa) that would scurry about the room and go from one Pringle can to the next to hide.

The weather at night was turning colder and I wore all my cold weather gear and two pair of socks while we patrolled. When we patrolled, we now worked in two different patrols, or what the Border Patrol called a "durea". The Border Patrol would do the first durea and we would do the second with the Saudi Special Forces. When we met up at the shai fire we would have tea, drink habib (a really sweet, hot milk) and eat bread and cheese. The Saudis would bake the bread by mixing up the flour in a bowl and then pouring it right on the coals. It was pretty good. They told me that sitting around the fire and eating with each other was a religious experience and part of Islam.

On November 20th something odd happened at our OP. As we prepared to start the patrol around 2200, all the Saudi Border Patrol ran into the building, hollering out orders, grabbed their weapons, then ran back out again, screaming and hollering. They got in their vehicles and tore out of there, heading to the east. All of us grabbed our gear, and went to the roof, armed to the teeth, ready to repel any invaders. We searched with our night vision goggles for Nickolai and Cr___, who were on a patrol. They had a radio, but it was broken and Cr___ had not sent in a request for a new radio yet, so we couldn't talk to them to tell them something may be coming their way.

We could see the Border Patrol high beams in the distance, about thirteen kilometers away, racing back and forth across the desert. It wasn't until morning that we found out what happened. It turned out that the Border Patrol from OP 2, east of us, did some donuts in the sand, and the Border Patrol from our OP thought they were being attacked and raced out to save them. On their return the Border Patrol shot the dog that I had been feeding for the last month. The Arabs didn't really care for dogs and thought they were unclean. It really pissed me off, but I had to hold my anger in check so it didn't screw up any rapport we had with the Saudis.

The next day I was awakened by two generals, five colonels and a captain from the 1st Cavalry Division. They told us that they were allowed to be up on the border, but they had no escorts. Cr___ didn't know if they were official or just tourists, so he drove over to Ruqi to figure it out. They turned out to be tourists, just seeing what they could on the border.

346

Chapter 3: SPECIAL FORCES – A Storm in the Desert

The SOTA team had hundreds of blank cassettes to record messages they intercepted from either the Iraqis or from the Kuwaiti resistance. They gave me a few blank cassettes and I started sending recorded letters home and to the various "Any Service Member" people that wrote to me on a regular basis. My dad had done this in Vietnam and I still keep the tapes he sent me. I figured if I ever had any children, they may want to listen to what I did during the war.

Thanksgiving 1990

Thanksgiving was uneventful and the meal was the usual goat and rice. During the meal a civilian vehicle had drove to our OP from Kuwait and the passenger got out and talked to our Border Patrol commander. The commander told me that he was from the Kuwaiti resistance. I tried to talk to him but Cr___ told me I couldn't and he had to order me to stay away from the Kuwaiti. I was amazed at how he could blow off the chance of talking to someone who had been in Kuwait, but Cr___ was becoming more and more scared as the possibility of war loomed closer. I guess he thought that if we all ignored it, it would go away.

Since it was Thanksgiving the whole chain of command showed up to visit our OP. Poppa Doc told CPT Mastrovito that Cr___ would not let us send out any information and then he told the Deputy Group commander about the Kuwaiti resistance member that we were ordered not to talk to. SGM Beuckmann pulled Cr___ aside and chewed his ass. After that, we were allowed to send any information we thought was valuable on to Dhahran.

We were told that AT&T was letting all the servicemen in the Gulf have free phone calls until November 25th. Cr___ and Doc went to Ruqi first to make their calls, but they couldn't get through at all. Imagine a half a million soldiers, sailors, airmen and Marines all trying to call at the same time. We also learned that President Bush was in country, visiting the troops on Thanksgiving. We didn't think we would see anyone up there, but the Group commander did show up, bringing our mail. I got a package from my mom with some cans of chili in it. Due to that mail run our Thanksgiving lunch was fresh pita bread and chili. We broke open the MREs and found all the ham slices. For Thanksgiving dinner, we had ham slices, glazed in honey and pineapple chunks (from another care package) with macaroni and cheese.

That night we learned that OP5, the OP west of Ruqi, had been going out of control. Gadget had gotten bored and taken apart the turret on the DMV, and now it didn't work anymore. Cr___ told us that Baby Doc wanted to beat his ass but Charlie Blake wouldn't let him. Then on patrol Charlie thought he was in Iraq and started to freak out so much that Baby Doc was ready to give him a sedative to calm him down. Captain Nick was playing around with Baby Doc and yanked out his Gerber knife, slicing a huge chunk of meat out of his own hand.

I was having my own problems in our OP. Cr___ would play power games and not let any of us know who was going on patrol each night until right before we left. Then he would harass anyone who wasn't fast enough. Whenever the CP circuit rider would come through, he would bring the Intelligence Summaries (INSUM). Every person in the OP was supposed to read the INSUM, which gave the latest intelligence on what was happening and what to expect. Cr___ decided he would be the only one to read them, and he wouldn't let us near them. This angered all of us, since he was now withholding information that might get us killed. I was so angry that I began to consider terminating my Special Forces status just to get away from leaders like Cr___ who might get me killed.

On November 26th we got a visit from Prince Fahd and Major Phalen from our Battalion. Phalen told us that VII Corps had moved up behind us, about sixty miles to the south. He also told us that when the war started, we would be left in place to conduct air strikes on any targets of opportunity.

Saddam Hussein called up an additional 250,000 soldiers since the coalition forces now equaled his army. Phalen said that we would be increasing the OPs so that they would extend west of the Neutral Zone area. Right now, there were only six OPs on the border, and our Team manned three of them.[71] That night I wrote my dad and told him about Cr___'s stupidity and told him that if Cr___ gets me killed, do anything he can to destroy him.

On November 30th we had several generals visiting OPs 5 and 6, the far western OPs. Poppa Doc had gone into Ruqi for supplies and he saw thirty Iraqi T-62 tanks going across Wadi el Batin. The wadi was a huge dried out river bed that was a mile across at points. This was key terrain in a flat desert because either side could hide in that wadi. While we were waiting for the generals to visit us, a vehicle arrived with four majors inside. They were tourists from the 101st so we told them to get out of our area. An Iraqi truck had driven up to OP 2 to check it out, but the Border Patrol chased them back into Kuwait. It seemed that both sides had their tourists.

---

[71] OPs 1 & 2 were manned by 2nd Battalion, while OPs 3, 5 & 6 were manned by our ODA 592. OP 4 was the Ruqi SR CP.

Chapter 3: SPECIAL FORCES – A Storm in the Desert

The Team would listen to my radio during every bit of free time that we had, trying to see if UN Resolution 678 had passed. Cr___ came in while we were huddled around the radio and told us to quit listening to BBC, that the UN Resolution had not been passed. We knew he was lying and continued to monitor the situation. Later that day we learned that it did pass, by 12 to 2, with China abstaining. Only Cuba and Yemen voted against it. The UN Resolution gave Saddam Hussein until January 15th to get out of Kuwait. After that the use of force was authorized by the coalition forces.

In the same broadcast we learned that Margaret Thatcher had resigned as Prime Minister of England. Initially I thought it was because she had no support from her party for the war, but it was due to another matter. That night I finally got some great news. I would be leaving OP 3 and going over to OP 5. Another SOTA team member was coming up to OP 3 and that would give our OP seven American soldiers. Meanwhile OP 5 only had five Americans. They also didn't have any weapons sergeants over there. Captain Nick wanted me to go to OP 5 because they were having a bit of cabin fever over there and wanted me to try to settle them down. Cr___ tried to take away my M24 sniper rifle right as I was beginning to leave, but after much argument, he allowed me to keep it. It was just another final petty power play before I left. Mastrovito told me that he was pretty pissed off at all Cr___ had done and he was going to try to get the Battalion to relieve him from being a Team sergeant before the war began.

OP 5, Al Awja, was the total opposite of OP 3 in temperament and in location. The buildings in OP 3 were corrugated steel houses, like a Quonset hut. There was one shower for the whole OP and it was located outside. There was one toilet, also located outside. Beside the toilet was one sink, exposed on the outside of the building. The cook at OP 5 was from Bangladesh and spoke horrible Arabic and no English at all. The room that the SF soldiers lived in was shared by three Saudi Special Forces. There was wood paneling on three walls of the room, with the ever-present giant photograph of a woodland stream on one wall. Most of the Saudis had one of these peaceful stream photos on their wall, or in some cases, snow covered mountains. I think that this was their impression of what heaven looked like.

The Iraqis were much closer to the OP here. Our OP was four kilometers from the berm, but the berm was only 500 meters from the Iraqi OP, all within small arms range. We had two DMVs with three Americans to each vehicle. We only patrolled with the two American DMVs and didn't patrol with the Saudis. Our route was patrolling 12 kilometers one way, and then patrolling 12 kilometers in the other direction, for a total of about 25 kilometers each night. Each night the patrol would start at 2200 and end around 0400. We did not do any shai fires, unless we linked up with the Saudi SF.

349

Triple Canopy

The six Americans at OP 5 were myself, as the weapons sergeant, Gentz as the engineer, Baby Doc Westover as the medic, KC Dreller as the communications sergeant, Casper Wallace as junior communications, and Charlie Blake as the intelligence sergeant. We had no SOTA Team members at Al Awja and everyone there liked being isolated from the bullshit drama happening at OP 3.

I was glad to learn that OP 5 had an E+E plan (Escape and Evade), that we had not known about at OP 3. This would be used if the shit hit the fan and we were cut off or over run by Iraqis. The E&E link up point for OP 5 and 6 was a dry lake bed located south of our location.

One night, in the beginning of December, we saw Iraqis firing flares into the sky. We drove to our western limit and linked up with CPT Mastrovito and the guys from OP 6. Soon afterwards the Saudi SF linked up with us and we had a shai party. They fixed Foul Beans (pronounced "fool" and tasted like refried beans), Shaksika (scrambled eggs) and they had an adoption party for me, the newest member.

I was adopted into the Amrii tribe. Arabic doesn't have a "P" sound so the Arabs use a "B" sound with anything that has a P in it. For example, they called Pepsi, "Babsii". They also called me Bat and after the adoption I was now Bat Amrii. They said I was a sheik of the Empty Quarter and my subjects were the Jerboas.[72]

Each week we would try to cook something "American" for the Saudis. I was the head cook for our group, since I had some knowledge of such things. I picked up some ground up goat meat in Ruqi, and we had Goat burgers and French fries. The Border Patrol were mainly Bedouins, who were offered a job working for the government, working in the same place they had wandered around in all their lives. They had never eaten a French fry before. They loved it and after that meal they wanted Mas, the Bangladeshi cook, to make French fries with every meal. I was amazed at how calloused their hands were from living in the desert all their lives. While Mas was making fries, I saw one of the Border Patrol pick up a French fry from the sizzling oil with his bare hands. It didn't faze him at all. Once I also saw one of them light his cigarette by picking up a coal from the fire, in between two fingers, and touching it to his cigarette.

On December 4th Charlie and I decided to get a closer look at the Iraqi station on the other side of the berm. Before I showed up Captain Nick had gone across

---

[72] The Empty Quarter, or Rub al Khali, is the southern-most desert in Saudi Arabia, a vast desert that no one lives in.

the border, with his Saudi counterparts, and had tea with the Iraqis. He didn't talk while he was there and the Iraqis didn't know he was an American. They thought he was one of the Saudis. We took a Saudi Landrover to camouflage who we were, and Charlie took pictures of the Iraqi OP. I sketched what I could see as we laid on the berm. The entire time the Iraqis were watching us through giant binoculars and were pointing their machineguns in our direction.

I continued to get letters from the different sorority girls and I passed off a few of them to the other guys in OP 3. I didn't want to hog all the attention. As the war, and Christmas, was getting closer, we began to receive more and more mail and packages. We made a Christmas AT-4, instead of a tree. We took decorations and cards and taped them all over the anti-tank rocket. I got a small nativity scene from the Collins, and I put them in my pocket. I joked about how I had Jesus in my pocket for luck. I put a stained-glass sticker on the side of the DMV rear window that my Mom had sent me.

We were issued Saudi uniforms, so that we would blend in better. They used the same "chocolate chip" camouflage we did, but their field jackets were all tan. We also got Saudi desert boots, which our army didn't have yet. Dhahran delivered a telescope to our OP, which helped a lot for trying to identify what the Iraqis were doing. We also started our own disinformation program, to make the Iraqis wonder what we were doing. Since we were always being watched, we did things like take a cardboard box, all painted black, out to the berm, and leave it there. Or we would stop the vehicle, dig a hole, and then cover it up, though nothing would be inside the hole. One night we were patrolling when we bumped into a herd of camels heading towards Iraq. We knew they were the King's camels, so we got out of the DMVs and herded them back to the south, away from the Berm.[73]

On December 10th we had great news… Cr___ was being relieved! Word had gotten back about how bad the conditions were on the border due to him, and Poppa Doc had told the chain of command that when the war started, he was going to sedate Cr___ so that he didn't get anyone killed. Our new Team sergeant, Al____, came from the HALO team. His nickname was "Gator". We learned from him that Cr___ would be leaving Special Forces and he would be reclassified as 12B engineer.

---

[73] We were told that all the camels that were loose along the border were "the King's Camels".

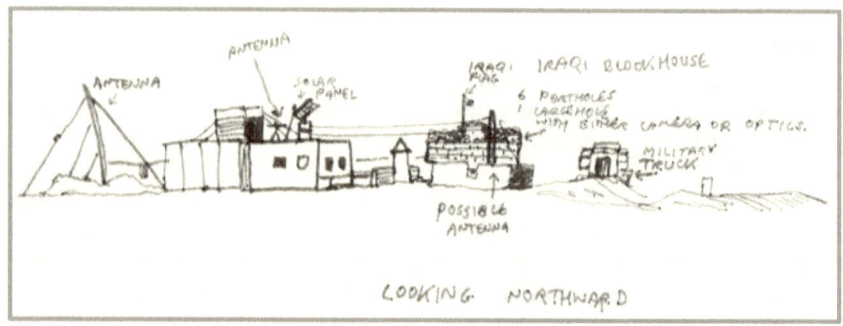

Sketch I made of the Iraqi OP.

Fifteen Iraqi deserters came over to OP 6 on December 13th. One of them was a sergeant major. OP 2 had several Iraqis come over and one of them was a colonel. I figured that the Iraqi army must be having some serious problems if these high-ranking soldiers were deserting to the Saudis. Each night the Iraqi OPs along the border would fire flares and light up the desert with search lights, looking for the deserters. We were told that the reason there were so many deserters was because Saddam had given sealed orders to the commanders on the border, that weren't supposed to be opened until a certain date, but some of the commanders opened them up anyway, and the orders said "Attack the Americans on December 30th." The Iraqis on the front were not the best quality troops, or were considered "expendable" by the Iraqis, like Kurdish soldiers, so they decided to desert instead.

We had a problem when the Saudi SF officer, Lieutenant Ibrahim, went home on furlough, leaving Sergeant Sayyid in charge. Sayyid was pretty immature and we didn't get along with him as well as we did Ibrahim. That night we decided to do the patrol with just one vehicle, like they did at OP 3, but when Sayyid found out, he thought we did it to make him look bad. He drove over to OP 6, where the Saudi SF commander was, and told him that we weren't even patrolling anymore. The Saudi SF commander sent him back and told him to quit being an ass. After that we didn't trust him, like we did the other Saudi SF guys.

When Ibrahim returned, I made them goat cheese pizza and the Saudis made us Sambuka, a lamb, egg and spinach egg roll. I had a catalog from a military supply store called *The Cav Store* and when the Saudis saw all the high-speed gear, they wanted to place an order. The Saudis in the Special Forces were all highly connected and were extremely wealthy, so the idea of spending hundreds of dollars was no big thing to them. That night I sent off an order for $400 worth of gear. I wasn't sure it would get to us before the war began, but the Saudis said that if it was meant to be, it would happen.

The Muslims had a saying, Insh-Allah, which means it is the will of Allah. Whenever we said something, like "the war may begin tomorrow", they would reply, Insh-Allah. We soon picked up the saying, and used it in common speech with each other. The other phrase we would use all the time was "Momken, Mesh Momken", which meant, maybe yes, or maybe no.

As the war loomed ever closer, we would snap at each other for little things. I almost got in a fight with Baby Doc on whether or not Vanilla Ice was a good singer. Baby Doc thought that Vanilla Ice would be around for decades and I told him he wouldn't last more than his one hit wonder. This went on until we were close to beating the hell out of each other. On another occasion Gentz wanted to take apart the turret on the DMV again. It never rotated the right way because of all the weight placed on the turret. Gentz had put several boxes of ammo up there and a dozen smokes and grenades. After Gentz took it apart the first time it never worked right. When he decided he would take it apart again, Baby Doc blew up and was ready to beat his ass. Gentz also lost his LCE, with his pistol, when it fell off the turret during a patrol, however he never told any of us until we returned. We had to drive the route again, but saw nothing. When the sun came up Gentz went out with Sayyid and finally found it, but now Charlie wanted to smack him around. This continued each day until Gator came over and stayed with us for a few days. Due to the constant incidents of stupid things happening I called us the C.L.O.W.N.s, the Clandestine Liberators of Wimpy Nations.[74]

---

[74] To this day I have old Special Forces veterans who remember the CLOWNs in Desert Storm.

In mid-December each of the OPs were told to rotate their people into KKMC for R+R. I didn't want to go, but it was mandatory. I wanted to stay on the border, just in case the war started. Since we had to go, Casper and I took one of the DMVs so that it could be worked on back in the rear. As I got closer to KKMC there were miles and miles of brown ARFAB tents that were being set up for the soldiers coming in from VII Corps in Germany. There was also a MASH hospital, one of the last ones in the Army.

I was told that there were nurses there and they had all night card games if I wanted to go on over. 5th Group was located in the center of the city and they had put a chain link fence around the Group area. The team members wore the uniform of the units they were advising. The Team advising the French were in French uniforms and also had the "high and tight" haircut like the French.

While I was in the rear, I had to get my first haircut in two months. I let one of the cooks cut my hair and he gave me a high and tight, though I didn't want it that short. I found out that the rapport that Special Forces had built up with the Saudis in KKMC over the months was being shot to hell by the regular Army soldiers, who didn't care about the customs or traditions of the area.

Chapter 3: SPECIAL FORCES – A Storm in the Desert

My R+R was pretty terrible. All I wanted to do was get back to our OP. I spent my time doing maintenance on the DMV. I was not a mechanic, but I learned pretty fast how to keep the DMV up. We replaced the glow plugs and put in a new speedometer. Cr___ was still there, in the rear, harassing us. He made me take a "tape test" to see if I was overweight and so he could try to punish me on the "change of rater" NCOER. I avoided him whenever possible.

When I finally returned to Al Awja two days later, I felt happy. There were rumors that we were going to be pulled back from the border and placed in the Kuwaiti "Liberation" and "Martyr" Brigades that would be assigned to retake Kuwait City. Major Phelan had been able to get 3rd Battalion the mission of advising the Kuwait Army when the ground war began. That same night thirty Iraqis deserted, some dressed in what we thought were women's clothes. After we had questioned them, they admitted to us that they had been given the mission of crossing over into Saudi Arabia and sabotaging the coalition forces.

On December 20th seventy-one Kuwait refugees showed up at Al Ulaymah (OP 5). Some of them said that they had paid $12,000 to be smuggled across the border to safety to get away from the war. That day we learned that our communications net was less than desired. Several vehicles filled with officers arrived at our OP and told us that we were going to be talking to three different locations when the war started, with two different types of radios. We would talk on the SATCOM to Dhahran, but we would also talk to Ruqi with the FM radio. As far as we knew only our SATCOM worked. The FM radio probably worked, but Ruqi was too far away. We were told that on the first day of the war there would be 2,000 aircraft flights each day, bombing the Iraqis into submission.

That afternoon K.C. and I drove up to the berm right before sundown, and watched the Iraqis. We stood on top of the Berm and noted that it took them about fifteen minutes to notice we were there. One of the Iraqis made a "come here" gesture, but I returned it and waved for him to come to me. I put four Pepsis on the berm, and put a chemlite beside it. We then backed off a kilometer to see what would happen. Four Iraqis walked to the berm and then argued back and forth what to do. Finally, I saw the chemlite vanish when they picked up the Pepsis. It was ironic that in a matter of weeks we would be trying to kill these same men.

Two days after the refugees showed up, we were told they were now being sent back to Iraq. The Saudi Minister of the Interior said he couldn't identify their nationality, so he didn't know if they were Iraqi or Kuwaiti. Many of the refugees had said they were Kuwaiti Army deserters, but none of the Kuwaitis recognized them. I didn't like this turn of events and after learning from the earlier deserters, telling us that they had been sent across to sabotage, we were worried about our security. No matter who they were, they now saw what we had at OP 5 & 6 and they also knew how few soldiers we had there. When it was time for the refugees to cross back over into Iraq, an Iraqi military vehicle arrived and picked them all up. I figured it had been a setup, since the Iraqis were all prepared to pick them up.

Each day there was a countdown to "K-Day" or the invasion of Kuwait. The Saudi newspapers started this trend by putting the "K-Day" in red letters on the front page each day. Sayyid had been transferred to Tabuk to do instructor duty and the new top sergeant at OP 5 was Khalid. Khalid and I got along fine. On K-25 (December 22nd) both Khalid and I went up to the berm with some other Saudi SF soldiers. We watched the Iraqis digging into their OP, while four Iraqi machineguns were aimed at us. Khalid kept yelling at them, in Arabic, "25 more days!" They continued to dig, ignoring the taunts.

On the next day, a Sunday, we were visited by the Group commander, COL Kraus and the Group Sergeant Major, CSM Simms. With them was the Deputy Commander of ARCENT and another general. They wanted to take a look at the wadi and the OPs to the west of the wadi. We escorted them down into the wadi and took them along the border. The only thing CSM Simms said to me was that I needed to trim my moustache. I thought that was pretty funny, since we were living with limited water, limited facilities, and only had the rear mirrors on the DMV to shave by. I figured if my moustache was too long it might cause us to lose the war.

That night OP-5 got our first uniformed Iraqi deserter. He was wearing a green camouflage uniform and a black beret. K.C. interrogated him and told us that the Iraqi deserted because he hadn't eaten in two days. Prior to that all he had eaten were two pieces of bread a day.

He told us that their trenches were only waist deep, because the ground was so hard. This was true, since the Saudi desert was like concrete two feet down and it took heavy equipment to make trenches. The Iraqi told us that each soldier only had ten rounds a man. We also learned that there was a battery of BM-21s within striking distance, about twelve kilometers away. They would soon be one of our biggest threats.

Triple Canopy

We lost Baby Doc when he went down to get the mail from KKMC. All the other OPs had their men return, but not ours. Finally, on Christmas Eve, he returned, with lots of packages and mail. Prince Fahd came to our OP and gave us all Christmas presents. We all drew numbers to see who got what. I got a brass Saudi tea set with Saudi markings on the cups. We decided to make Christmas Eve special, so we all planned to meet to the rear of all three OPs, at a place we nicknamed "The Haunted House". It was an old mud hut, a Border station, partially built into the wall of the wadi. Each team was supposed to bring a dish to eat. Our OP brought macaroni and cheese that came in with the care packages. As we cooked the macaroni, I played Christmas music on my cassette player. Baby Doc called it "suicide music" because it made you want to blow your brains out.

All of us had dark humor that was coming out more often. Charlie got a brass urn from Prince Fahd, and we told him that would be great to put his ashes in when he was burned alive in the DMV. He told me my tea set could do the same thing. Someone had sent us an inflatable Santa Claus, and a lot of the boxes from the States came with Styrofoam packing peanuts. I put the Santa behind the .50 caliber and then we drove down the berm as fast as we could, throwing packing peanuts out the top of the turret, so it looked like it was snowing. I figured the Iraqi report to higher headquarters that night must have sounded weird. "A fat man in a red suit was driving down the berm as fast as he could!"

At midnight the whole Team arrived at the Haunted House. This was an SF only thing, but we brought LT Ibrahim. One of the other OPs brought some chickens and we roasted them over a fire. Gator wanted to be tactical and put ponchos over the windows to keep the light out, but we all started choking due to the smoke. We took them down so we wouldn't die of smoke asphyxiation. The party lasted till about 0400, then our DMVs drove to Ruqi so we could call back home. I wasn't able to get my mom and dad, but since they ran the Salvation Army in Jacksonville, NC, I figured it was their busy time of the year. I was able to talk to the Jim Collins though.

At 0900 on Christmas morning the Battalion commander and CSM woke us up, with a cheery "Merry Christmas!" They did not bring mail due to some foul up with the vehicles in Dhahran. They said 600 tons of mail was just sitting there, waiting to be delivered. We were pretty tired and went back to bed as soon as they left. Whenever I slept, I had on my uniform and wore my shoulder holster. I would sleep on my left side, with my hand on my M-9 pistol, ready to pull it out if anyone busted into the room.

At 1300 we woke up and drove into Ruqi for Christmas dinner. This was much better than Thanksgiving, when we didn't get anything at all. Trucks had brought food in Mermite cans and they had gone all out on the meal. There was turkey, ham and roast beef! We sat in the parking lot at the headquarters and ate until we were stuffed. We then divided up the left overs and took them back to the OPs. Each of us was given a case of Snickers bars for dessert. It was fantastic.

The far western OPs had their Christmas dinner parachuted to them by C-130 aircraft. It was cold and overcast, making everything seem gray, so it was almost a white Christmas for us. White sand and white sky. I tried calling home again, but no one answered. Instead, I called Sue Cawood and talked to her for about 20 minutes. She was the person I was closest to during the war, and she was also the only girl friend that I never had romantic notions for.

That night we had our first Red Dragon report. We were told that whenever a SCUD missile launched, whoever saw it would get on the radio and say "Red Dragon". All the other radio traffic was supposed to stop since this was the highest priority. Everyone who saw the missile launch would then shoot an azimuth from their location and report it, which would triangulate the launch site. The Air Force would then quickly fly to the area, and take out the missile launcher. That day we saw four SCUDs launching into the sky, but no one knew where they landed. We were told that back at KKMC the soldiers went a little nuts, running around in a panic trying to get into shelter and putting on their MOPP suits (Mission oriented protective posture).

The day after the Christmas feast, Gator wanted us to start getting ready for when we would be pulled off the border and train the Kuwaitis. However, we also had to do patrols and be the "speed bump" against Saddam's armor, so it caused some heartburn with the team members. We were issued a new piece of equipment with the next mail call; a small plastic box called the KL-43. None of us had trained on it or seen it before. It was a new encryption device for radios. It was pretty neat, a small box, about six inches long and four inches wide, as compared to the DMDG, which was the size of a laptop computer. We were also told that we would only be transmitting to one location now, Riyadh. That same day we noticed two AMX-30s parked a mile behind our OP.[75] We drove back to check them out and found out it was a Saudi Tank company. So now we had a little backup in case the Iraqis came over the border.

Towards the end of December, we were visited by MAJ Richardson, and his driver SFC Sa____. I hadn't seen Sa____ since he had been relieved as the Team Sergeant of ODA 592. Since that time, he had been fired from two other jobs. MAJ Richardson was the combined operations commander and he told us the options for our future missions. The first option was that we could go back to the Kuwaiti battalions, advising them and doing CAS missions as we retook Kuwait City. Another option was that we could go back to Group and do CSAR

---

[75] The AMX 30 is the main battle tank of France in the 1960s and 70s, and was also used by many other nations, to include Saudi Arabia. It was comparable to the American M60A1 tank or the Soviet T-55.

Triple Canopy

missions for downed pilots. However, Richardson was pushing for a new mission. He wanted us to go to Tabuk, the Saudi Ranger School, and have us train to do direct action missions with our Saudi SF counterparts. Richardson said it would be like Project Omega in Vietnam. We nodded our heads, but I had never heard of Omega. I was interested because it sounded pretty high speed. I was also looking forward to it because if we did the "Omega" mission, ODA 592 would be the most experienced team in the war, having done FID, SR and then DA.

Richardson told us that he wanted to pull us out on January 1st, but Schwarzkopf wanted us to stay on the border until the first shots of the war were fired. Schwarzkopf had said that when the ground war began the Marines would fight their way to Kuwait City, and then halt, letting the Kuwaitis take the city. The Marines would form a cordon around the city and no one except Kuwaitis and American Special Forces would be allowed inside. Before MAJ Richardson arrived, we thought we had observed several F-15s flying low over our OP. Richardson told us that they were Iraqi MIGs locating future targets. We also heard a rumor that the Rangers were in Saudi Arabia, but we didn't know for sure.[76]

On New Year's six Saudi 155mm Self Propelled artillery pieces joined the armor company to our rear giving us added security from any possible border raids by Saddam Hussein. The artillery turned out to be the same group of Saudis we taught back in October. We drank some shai with them and then worked it out so that one of their vehicles would come up to the OP and we could use their radar system.

Some mail arrived when Baby Doc returned from his R+R in KKMC. In one package he had gotten a bottle of Jack Daniels from his brother and two porno magazines! We were amazed. Since we had arrived in country there were constant reminders that there was to be no alcohol or pornography at all, because it might endanger the rapport we built up with the Saudis. The lack of alcohol turned out to be a major factor in some soldier's lives. We heard rumors of soldiers being sent back to the United States because they found out they were alcoholics and had begun to have withdrawals symptoms. We also heard stories of soldiers who had been busted making stills in their camps. In the same mail

---

[76] Only one company of Rangers served in Desert Storm. This was another case where Schwarzkopf seemed to have a dislike of Special Operations forces. Prior to deploying the Rangers, General Schwarzkopf and General Downing (75th Ranger Regiment commander) argued about how the Rangers were to be used. Schwarzkopf wanted the Rangers to be his personal bodyguards. Downing told him no, that is not how Rangers were supposed to be used. In the end Schwarzkopf didn't use them at all. The only company that did deploy to Desert Storm did so mainly because of Secretary of Defense Dick Cheney's insistence that he wanted the SCUD missiles to be located and destroyed. The Rangers deployed with DELTA force and worked with them behind Iraqi lines.

360

call, I got a Sports Illustrated swimsuit calendar from Bert for Christmas. It went right up on the wall beside my bed. The Saudi SF guys in the room were highly interested and wanted to know if all American women looked like that. I told them yes… yes, all American girls looked just like those swimsuit models.

Sayidd returned to us from vacation and brought a couple of bottles of honey. The Saudis made us "ariika" that night, which was semi-cooked, doughy bread, dipped in honey and butter. Honey was a prized possession over there because it was so rare. They also served us "henini", which was a doughy bread, covered in brown sugar that resembled funnel cake at a fair.

Keeping the camp clean and antiseptic was almost impossible at the OP. The Saudis would take all their trash and throw it in a "dump" right beside the OP. After awhile all this accumulated trash was a breeding ground for flies and maggots. The biggest problem was the remains of the goats that we ate each day. Our Bangladeshi cook would slaughter a goat, and put the carcass in a wheelbarrow. He would then roll the wheelbarrow back to the kitchen, skin and clean the goat, and put the non-edible parts in the wheelbarrow. When he was finished, he would take the refuse out to the dump and throw it on the pile. This led to a stinky, fly covered mess. Due to this I would go out to the dump each week, pour diesel on the remains and light them on fire.

I decided that I would do something special for New Years. For the last few weeks, I began piling the trash into a mound, trying to get it as tall as possible. I put all the Christmas boxes in the pile, along with bottles of gasoline, oil and old batteries. I planned to light it on New Years at midnight so it would be our "fireworks display". I figured the Iraqis would get pretty excited seeing the huge fire on the Saudi side of the border. In one of my Christmas cards sent to me by my WWII reenactment group there had been a black and white picture of all the guys in their British WWII gear.[77] To confuse the Iraqis even more, I threw the picture over the border so that the wind would blow it towards the Iraqi OP. I figured they could try to analyze that picture for intelligence on what unit was in front of them.

On New Year's Eve I was not on a patrol, since I had radio watch. Every night one of us stayed back to monitor the radio in case something happened. Exactly at midnight I set the dump pile on fire and the flames shot thirty feet into the air. The guys on patrol told me they could see the fire from where they were, miles out in the desert, and it looked great.

The New Year brought about a few changes. The Saudi civilians in Ruqi realized that they were the only town on the border and they would receive the most artillery and rocket fire. As each shop sold out of their goods, they locked the doors and headed south to Riyadh to wait out the war. K.C. was told to take

---

[77] My WWII reenactment unit was the 21st Independent Parachute Company, a British paratrooper unit.

R+R in KKMC, but he refused to go. He threatened to terminate Special Forces instead of going back. None of us wanted to go to the rear, and preferred to be up on the border away from the bullshit back there. K.C. was ordered to go and I was sent with him to make sure it happened.

On the trip to KKMC we passed convoys of American M2 Bradley vehicles and APCs from the 1st Armored Division heading out to their jumping off points. In Hafr el Batin a flatbed from one convoy had taken the corner too fast and had tipped over, spilling out the APCs on the back and spreading them across the road. K.C. and I looked like Bedouin gypsies with our part-Saudi, part-American uniforms, but we stopped to interpret for the American MPs who were trying to talk to the Saudi MPs. A 2½ ton truck full of French Senegalese troops were also involved in the wreck, so K.C. was speaking Arabic, French and English all at the same time. The American MPs were probably trying to figure out who the heck we were, and why we were at that intersection. We looked like heavily armed Pirates of Penzance to them.

Back at KKMC Bravo Company had moved out of our air-conditioned barracks and moved into a tent city with their Kuwaiti counterparts. So far, they hadn't done much of anything except lift weights, get lots of sleep and trained their Kuwaitis. They all began wearing "Free Kuwait" buttons that we hadn't seen yet. I encountered one Kuwaiti soldier pulling guard on our roof. I asked him what he was doing and he said he was on SCUD watch. I thought this funny, since if you could see a SCUD coming, you only had a few seconds before it detonated. There were all sorts of wild rumors floating around back in the rear. One night we were told Iraqi Special Forces had infiltrated past the OPs and were on the way to KKMC, then on another night someone told us the Iraqis had captured an entire OP. We laughed at these rumors and didn't take them seriously.

We were told that one of the OPs had a soldier blinded by an F-16 firing a laser at them. Due to this we were issued a new type of sunglasses, known as BLIPs that were supposed to stop that from happening. However, we never wore them.

Before I left OP 5 to go to KKMC we all wrote our final letters home. These were the equivalent of an unofficial will to our next of kin or friends, to be opened if we were killed. I wrote to my dad and put my "just in case" letter in there. I told him to not open it unless I was killed in the fighting. I had these letters, plus boxes full of stuff that I wouldn't need in the war, and mailed them home at the post office in KKMC.

While I was at the post office there was a girl from the 3rd Infantry Division. She smiled and flirted with me and I just stared with my mouth wide open. I felt like an idiot. I was dressed in my mixed uniform, which made me look like I was a pirate. I had a shemagh around my neck, a desert patrol cap (which our army didn't have yet), with my Sears work goggles on the top of the cap. I also

Chapter 3: SPECIAL FORCES – A Storm in the Desert

wore my sheepskin vest that Ibrahim had given to me for Christmas, so I looked more cowboy than pirate. The whole "bad boy" look must have attracted her.

There were thousands of American troops in and around KKMC and when we went to our "Wal-Mart" it was filled to capacity with American soldiers. I picked up some trays of T-rations and coffee from SGT Nance, the messhall sergeant and we bought a large Saudi flag from the "5000 area". When I walked into the shop, looking like Rambo the goat herder, the shopkeeper asked me if I was Pakistani. I figured it was a compliment that I was able to speak Arabic well enough to be thought of as something besides being American. It may have also been an insult, since I did look like a Pakistani street sweeper.

Before I left, I took down my poster of skiers in Vail. I wanted those to motivate me on what I would do after the war. On the back of one of the posters there was taped an article about Kuwaitis being trained by Special Forces. It was an article that I had cut out of a magazine and put it on our bulletin board while I was with Cr___ at OP 3. The article had come up missing, but I never could find out what happened to it. There was the article, taped on the back of the ski poster, and it had been ripped in half. Written across it in Cr___'s handwriting were the words "Here is your article O'Kelley." I figured Cr___ thought I wouldn't see it until after the war was over. Cr___ was still in the rear harassing us, but we ignored him. He tried to put K.C. on guard duty, but we just blew him off. We were on R+R and the orders from the Group commander were that if you were on R+R you pulled no guard or details.

When we returned to our home at OP 5, we learned that a lieutenant had been going to each OP, asking them what they wanted to do after this mission. Did we want to go with the Kuwaitis or stay with our Saudi SF counterparts. Out of all the OPs our Team was the only one that wanted to stay with the Saudis. We had built up some strong rapport with them and had made friends with a few of them. Gator wanted to go with the Kuwaiti battalions, but he had not been up there as long as we had and didn't get along with his Saudis as well as we did.

To try to learn more Arabic, each night I would study my ARAMCO Arabic-English dictionary.[78] I would try to learn fifty words a night, two from each letter of the alphabet. This paid off on the long run and I was able to communicate a bit better due to this.

On January 6th or K-10, Charlie, Baby Doc, Ibrahim and I went to Hafr el Batin to do some final shopping before the war began. Ruqi had shut down so we had to go there to get our supplies. I bought a tea pot, a frying pan, some food and tea. When the call to prayer began wailing from the minarets, we saw the "religious police", the "Mutaween", ordering stores to close and herding

---

[78] ARAMCO = Arabian American Oil Company.

363

people to go to the mosque.[79] We must have not looked like Americans because the Mutaween carrying a stick came towards us, but was quickly shooed away by LT Ibrahim.

While we were there, drinking Pepsi and eating the local food, we watched the war machine continue to drive past. Miles and miles of tanks, Bradleys and other military vehicles moved through the town and then headed west. While we sat at the intersection a black female truck driver came over, wondering who we, the desert gypsies, were. When she found out that Ibrahim was Muslim, she tried to convert him to Christianity. She was extremely religious but she had no clue about Islam. She thought Islam, Mohammed and Muslim were three different religions. Ibrahim took it in stride, but it could've gotten ugly. We told him that since he was allowed more than one wife, he should hook up with her. He waved her off, not even wanting to deal with that.

While we were there, we tried to get either Iraqi Dinar or Saudi Riyal from the local bank, but they were out. We also tried to buy some gold, which we could use in case we had to E+E. While we were looking at the gold a regular Army colonel walked up to Baby Doc and asked him where his helmet was. Baby Doc told him "Fort Campbell" and we all laughed. The colonel walked away, muttering to himself.

When we returned to our OP that night, we saw an Iraqi truck driving right beside the berm with the lights on. We thought he might have been dropping off troops. We parked the DMV while Charlie and Baby Doc went up on the berm to see what was going on. I locked and loaded the .50 caliber, putting the TVS-5 night vision sight's cross hairs on the truck. At that point I figured I was the first American to lock and load on an Iraqi. I could see the driver in the crosshairs. The driver continued driving west down the berm, so we followed with our lights off. The Iraqi truck drove to the Iraqi OP, then headed north into the desert. We figured he must have been lost and when he bumped into the berm, he realized he was really messed up. I wondered what he would have thought if he knew that there was death, in the shape of a 700-grain 50-caliber round, staring at him in the distant dark desert.

The next day we drove to Ruqi for Poppa Doc's promotion ceremony. He had been promoted to SFC. He didn't have any orders, so since I was the Team's

---

[79] The Mutaween "have the power to arrest unrelated males and females caught socializing, anyone engaged in homosexual behavior or prostitution; to enforce Islamic dress-codes, and store closures during the prayer time. They enforce Muslim dietary laws, prohibit the consumption or sale of alcoholic beverages and pork, and seize banned consumer products and media regarded as anti-Islamic (such as CDs/DVDs of various Western musical groups, television shows and film which has insults on the Islamic law or Islam itself). Additionally, they actively prevent the practice or proselytizing of other religions within Saudi Arabia, where they are banned." BBC News, http://news.bbc.co.uk/2/hi/middle_east/2399885.stm

Chapter 3: SPECIAL FORCES – A Storm in the Desert

S-1, I hand wrote the orders, signing it for the Secretary of the Army, John Stone.

Gator wanted me to be the primary CAS guy on the Team, so he planned on sending me back to the FOB for additional training. While in Ruqi I bumped into LT Taitt, who had been the XO in F Company, 51st LRSU when I was in Germany. Though he was now in 5th Group, he told me that F Company was at KKMC. The world was getting smaller.

Since the war was getting closer, our OP had stopped linking up with the Saudis or the Border Patrol to do the midnight shai parties. We all thought the idea of sitting around a fire, this close to the enemy, was not real smart. When Captain Nick told us we ought to do it so the Saudis wouldn't be insulted, we still refused. Finally, he ordered our OP to go to a shai party. It turned out that three quarters of the team was there, within one mile of the Iraqi border. Though all the guys in our patrol went, I refused to go near the fire and stayed with the vehicle. I continued to watch the border with my NODs on, pulling security. Captain Nick brought some goat out to me and asked me if I was going to the fire. I told him that I felt better sitting in the turret pulling security.

While there were rumors of Saddam getting ready to pull out of Kuwait, another mission opened up in our part of the world. There was a coup in Somalia, and our Battalion's A Company and our Company's HALO Team were alerted, ready to go and evacuate any Americans. They were ready to take off from Saudi Arabia when the mission was cancelled. This was just the beginning of our problems with Somalia, which would culminate in the fighting described in the movie *Blackhawk Down*.

The night before I went to the FOB for CAS training, I was on radio watch. The team that patrolled would go out and return around 0400, but the radio watch would stay up until dawn. When the team returned Gentz wanted all the lights out. We never turned the lights out because the radio watch needed to have them on. Though the lights had been on every night since we had been there, Gentz was angry and ripped the wires out of the wall. I exploded! I went over to his bed and tossed him out of it, and we almost went to blows over this, until Charlie Blake got between us and stopped it.

Captain Nick also began accusing me of odd things that he heard other folks saying about me. He accused me of not fixing the DMV, drinking a bottle of Jack Daniels and taking money from the Saudis. None of this was true and it really pissed me off. I told him that if he couldn't trust me, then I didn't need to be on his Team. We were about to go to war and he didn't need someone there that was a risk. I later found out that Nickolai at OP 3 was making these stories up and Mastrovito believed them. Gator took care of that mess and cleared my name. I think the stress of going to war was getting to us.

We were also exhausted, we patrolled all the time, and when we weren't patrolling, we were pulling radio watch. There was very little sleep each day, and the little bit that we did get was interrupted by the worries of a possible enemy attack. I tried to do some PT each day, mainly running around the OP in a circle, to relieve stress and continue to stay in shape.

That night when I had been on radio watch, the border was pretty busy. I counted forty-seven lights on the Iraqi side of the border, then artillery flares began to burst behind us from the Saudi artillery. An AC-130 gunship flew slowly up and down the border, turning on its bright searchlight when it saw something suspicious. Out there on patrol Charlie and Ibrahim found a new Iraqi bunker complex in the wadi, and I reported it. Finally, at the end of the night, there were reports of Iraqi helicopters defecting and flying into Saudi Arabia.[80]

When I was dropped off in Ruqi to pick up a vehicle to go to the FOB, I learned that they had not been getting any support for the last few days. No mail, no resupply and no information. We were on our own with six days left until the January 15th deadline. Back in the rear C Company was living a WWI existence. They were living in trenches and bunkers, while ODA 311 was living in a tent with a VCR and television. ODA 311 was assigned as the company's CAS Team and they were training up the other team members. Those of us who showed up for training didn't get to do any hands-on training, and instead we just watched and listened to 311 as they had some A-10s simulate attacks on the friendly tanks all around the desert.

I returned to the OP with Gator and he put out the latest information. Five deserters had crossed over at Ruqi and one of them was an Iraqi captain. He told Gator that Saddam said he was going to attack on January 11th, the next day, at 0400. Due to this we increased our security. That night odd things did happen along the border. For the first time there were no lights anywhere on the Iraqi side of the border. Meanwhile on our side there were still the headlights of the Border Patrol vehicles and the shai fires that dotted the desert.

When we geared up to go on patrol, I would put on a Benny Goodman cassette tape to get into the mood. Each night I played Goodman's *American Patrol* as we dressed up in our cold weather gear. All of us would be dancing around to the music as we dressed. I figured we could have dropped into 1944 and fit right in. It was really cold each night now, with the temperatures dropping down into the 40s. We replaced the doors and rear hatch back on our DMVs since the canvas ones we had improvised were not keeping the cold out.

The next night during patrol we saw an Iraqi vehicle, cruising within 200 meters of the border with the lights off. This time the Iraqi side of the border was alive with lights, unlike the night before. Some of the lights going off on

---

[80] I found out after the war that this was actually a covert operation done by the ISA, where Americans, in Iraqi uniforms and flying in Iraqi helicopters, flew into Iraq to do an aerial reconnaissance, and then flew back in as "deserters.".

the Iraqi side were quick, like a strobe light. I never did find out what they were for. Later, after sitting underneath bursting Iraqi anti-aircraft rounds, I realized they must have been the Iraqis testing their ADA. The Border Patrol told us that they had seen an aircraft go down in Iraq, but we couldn't verify it.

Since we didn't have any more support from the rear, we had to send back one of our vehicles to KKMC to pick up supplies and mail. Captain Nick suggested to the 1st Cavalry Division that they send up a Ground Surveillance Radar (GSR) team to one of our OPs, and augment them with the Cav's LRRP unit. This would also help us with pulling security and the radio watch. We had to monitor three radios while on radio watch. The SATCOM talked to the combined cell in Riyadh, the PRC-70 talked to KKMC and the Saudi Jaguar radio talked to Ruqi.

The Saudi artillery unit built a bunker near us for their forward observers. I figured if shit hit the fan this was a location we could run to. Five days before the war was supposed to begin, we finally came up with our E+E plan. It was late, but better late than never. The plan was to move south and work our way through the Saudi lines. Captain Nick also had put some of his ammo and supplies in a cache to the south of their OP.[81]

On January 11th the Iraqis became more emboldened by doing their own recon of the border. At Mastrovito's OP 6 an Iraqi MI-8 HIP helicopter buzzed his OP and then flew back around like it was about to do a gun run on the post. Mastrovito's men manned the guns on their DMVs, and the helicopter flew past, heading south into Saudi Arabia. It flew east for awhile, heading towards our OP. Our adrenaline was rushing since light skinned vehicles versus armed aircraft didn't seem like it would be to our advantage. [82] We faced it down in a tense "Mexican standoff" with our .50 calibers and Mark 19s on our DMVs versus its rockets and cannons. I figure it thought that it might take out one of our vehicles, but the others would turn it into a flaming metal scrap pile, so it reconsidered and flew back into Iraqi airspace.

The next day we received a radio message that our alert status had risen from "medium" to "high". None of us knew that this meant. Up until that time we had been given notifications of DEFCONs or OPCONs, but never a "high". We didn't know if this meant that hostilities were imminent or not. For our OP we broke open the case of hand grenades and put them on our LCEs. Prior to this we didn't think we needed them and we didn't want grenades rolling around the barracks getting in the way and possibly going off. Our Saudi SF received a Belgian .50 caliber for their vehicle and fifteen cases of armor piercing

---

[81] We never did recover that cache. For all I know, that ammunition and supplies is still buried in that desert.

[82] The idea of an armored HUMV would not be thought of for another decade.

ammunition. We didn't have any armor piercing ammunition so we traded them some boxes of ball ammunition for some AP.

As I watched through binoculars the Iraqi OP near us received a water trailer. At the same time Captain Nick and Charlie went back to the Saudi armor unit behind us to coordinate for a passage of lines, if and when we had to retreat through them. A message was sent to all the OPs to find an alternate position to withdraw to when the bombing began. We chose to go to Wadi el Batin, since you could hide a tank division in there without being detected from the surface. We were hoping the Iraqis weren't thinking the same thing and targeting the wadi. We packed up everything we didn't need, but wanted to keep, in small bags and then destroyed everything else. I fixed everyone a big pot of spaghetti from a care package that my mom sent to me, since it was better to eat it than destroy it. During that night six more deserters came into Ruqi, and were sent to the rear.

There was no set SOP on how to set up our gear, so we all wore the gear the way we preferred. K.C., the driver of our vehicle, wore a thigh holster with a Tanto knife attached beside the holster. He also wore a serrated knife that looked like an old bayonet, on his right hip, and an M-9 bayonet on his left hip. He wore a Leatherman pocketknife beside the M-9 bayonet. He wore his Kevlar flak jacket over all of it, with an LCE draped over the flak jacket. He also carried a claymore bag full of M-16 magazines slung over his shoulder. On his head he wore a black watch cap and a black Palestinian kafiiya around his neck. K.C. was our "hippie", and believed in peace and love…but he'd also blow you away. He had a habit of adopting whatever culture he was in. So, he was part communist, part Muslim, and a little French.[83]

I was the only one on the Team who had seen combat, so my gear was a little more lightweight. I didn't wear a flak jacket, but kept it on the back of my seat in case I needed it. The Kevlar flak jacket would stop pistol bullets and some shrapnel, but it wouldn't stop a high velocity rifle bullet, so I figured "why bother". My LCE was setup with four ammo pouches with grenades on the sides, a first aid pouch, and an M-9 bayonet with a 9mm magazine in the pouch on the bayonet. I had two 1 Quart canteens on each side. In my buttpack was a Saudi flag, an American flag, a VS-17 panel and a Confederate flag.

The Confederate flag was given to me by some girls back in the early 80s and I had carried it to Grenada, and everywhere else I deployed since then. All these flags were to be used for recognition symbols, because what Iraqi would have a Confederate flag? The butt pack was a canvas one that my brother had used, and in the clear plastic window for an ID tag it said "SGT Sean O'Kelley, USMC, STA PLT" in his handwriting.

---

[83] After KC left the army I think he became a chef in Lake Havasu, AZ.

Chapter 3: SPECIAL FORCES – A Storm in the Desert

I wore my M-9 Beretta pistol in a Bianchi Ranger shoulder holster and it was loaded with 9mm Hydroshock rounds. I also had an ammo pouch filled with small things like a signal mirror, 550 cord, mini-mag flashlight, weapons oil, shaving brush (to spread the oil), a heat tab stove, camouflage stick, a pair of pliers, a cleaning rag and a small piece of VS-17 panel.[84] I carried a soft PAQ-4 pouch to hold my 35mm camera, an extra roll of film and four SLAP rounds. SLAP rounds were sabot launched armor piercing rounds. They fired from a 7.62mm weapon and were mainly used by attack helicopters. I was able to get some to use with my M24 sniper rifle. The sniper rifle was in a soft case, slung from the ceiling of the DMV, where I could pull it out quickly if I needed it. On my LCE suspenders was my compass in a pouch, a flashlight with an infra-red lens, a strobe light with an infra-red cap, and a Blackie Collins knife under the strobe light with the handle upside down so I could draw it quickly if I needed it. I mainly used it to open MREs. I also had 100 feet of 550 cord wrapped around the back of the pistol belt of the LCE. Finally, I carried my Randall Attack Knife on my right hip, on my pants belt and a Swiss Army knife on my left hip.

Every night after our alert status went to "high" I slept wearing the shoulder holster, belt knives and the Saudi desert boots. On the night of January 13th, two days until K-Day, we learned from the BBC that the House and the Senate of Congress passed the motion to allow the use of force against the Iraqis. I was amazed by that since I thought they had already done those months earlier.[85]

Two days before K-day we learned from deserters who came in that the Iraqis were going to conduct a surprise attack at 0300, which would consist of a missile barrage followed by a tank assault. There was also a report of "Black Bedouins" who would actually be Iraqis conducting attacks behind the lines. We had sent Baby Doc to KKMC again to get mail and supplies and when he returned, he told us that everyone back there now had to wear their helmet, mask, weapon, LCE and carry their MOPP suit wherever they went. I was glad we weren't back there. Baby Doc said that Major Paxton told him that as soon as the war starts, we would be pulled back and attached to 1st Battalion to do DA missions.

While Baby Doc was briefing us, we received a message on the DMDG. We decrypted it and learned that no Special Forces could be on the road after the sun went down. We were also told to stay away from KKMC and Hafr el Batin, because they were main targets for terrorist raids. Ibrahim said that he was told that the terrorists would be using car bombs and helicopter suicide bombs. To

---

[84] A VS-17 panel was a large waterproof cloth that was Orange on one side and hot pink on the other. It was used for signaling.

[85] The Congressional vote happened on January 12th, with the House of Representatives authorizing the use of force by a vote of 250 to 183. The Senate passed the resolution by 52 to 47.

further complicate things we also received a message on the Jaguar radio from the Saudi AMX-10 that told us they had detected a soldier on their radar, heading our way. The Border Patrol went out and picked the deserter up and brought him to us.

The deserter was Kurdish and had come from an ADA site. He walked over twenty kilometers to get to our position. K.C. and Ibrahim interrogated him for three hours, and then we sent him back to the rear. Afterwards Baby Doc gave us our final two gamma globin shots, which were literally a pain in the ass. We got a message over the radio to start taking our bromide pills. This was a key indicator, because we were told that we would not take the pills until the war started.

Casper, Baby Doc and I went out on patrol that night, looking for any indication of the 0300 armor attack. It rained the entire time we were out and it was extremely cold. The desert was turning into sandy mud. While we were on patrol Charlie and the rest of our guys went into Ruqi to drop off the intel report from our Kurdish deserter. They learned that the OP at Ruqi was packing up and moving to the rear. They told us that all the OPs were ordered to evacuate their posts and pull far enough back that they could observe their OPs. Charlie found out where our patrol was in the wadi and linked up with us. He had taken all of their stuff out of the OP that they wanted to keep, and they told us we should do the same.

When we arrived at the OP it was in chaos. The Border Patrol was scrambling to get out of there and Ibrahim and his Saudi SF were packing everything up. We quickly grabbed our things and then headed back out on patrol. By 0400 we couldn't see anything due to the rain, so we moved to our new location, which was a bunch of goat pens about a mile behind our OP. The 1st Cavalry had a GSR Team and a radio intercept team waiting for us there. So far, we had a pretty exhausting day, so we let the 1st Cav guys pull security. We quickly fell asleep sitting up in the DMV seats. No one wanted to lie down in the mud in sleeping bags. The sleep was uneasy, and any noise or movement woke me up.

The next day the war was supposed to start. It was January 15th…K-Day. We were ordered to return to Ruqi to get our "blood chits" and a belt of gold coins to use in case we had to E+E. The blood chit was like the ones that were used by the Flying Tigers in WWII. Each of us had our own chit with a serial number on the four corners. No two numbers were alike. My number was 103828. The chit had an American flag on it and written in Arabic, Farsi, Kurdish and Turkish were the words: "I am an American and do not speak your language. I will not harm you! I bear no malice towards your people. My friend, please provide me food, water, shelter, clothing and necessary medical attention. Also please provide safe passage to the nearest friendly forces of any country supporting the Americans and their allies. You will be rewarded for assisting me when you present this number and my name to American authorities." If someone helped us, we would tear off a corner of the chit so that the government would know who was missing by the serial number.

Baby Doc was issued a medic bag that had something we called the "E+E drug", that was a version of speed that would keep us awake for several days.[86] Grant Winkler, a medic on ODA 581, wrote that the medic bag he was given had "5 boxes Morphine syrettes, 20 Diazepam injectors, 5 bottles Ketamine, 5 bottles Xylocaine, 1 bottle Demerol, 2 bottles Anthrax vaccine, 1 bottle Botulism vaccine, 2 boxes Amyl Nitrate inhalers, and 500 tabs methamphetamine."[87]

Though we were given the drugs, we never used them. Some Teams did take them as they crossed the border during the Ground War. J.T. Wilson on ODA 595 wrote "the speed was crap; I was given two packets. I was told by Joe Garrido that one was for a limited time frame for keeping you alert the other was for a longer duration. On the night we crossed the berm, around 0100, I swallowed one of the short duration pills...I fell asleep. I never bothered with the long duration; I think they were placebos."[88]

When it was time to return to our new positions, I convinced Charlie to move into the wadi. It would have been suicidal to stay at the goat pens that we were at the night before, since there was no cover or concealment from enemy tanks or aircraft. We drove down into the wadi, and then moved onto one of the hills in the middle of the wadi. There were mounds of dirt and holes all over the hill, so we could hide all our vehicles there. We nicknamed the site "Castle Clownskull" after the *He-Man* cartoon's Castle Greyskull.

When the sun went down, we were all tense and no one slept. We were waiting for something, anything, to happen. When it was dark, we saw a single jet fly into Iraq, but it was pretty high. I also saw some flares and explosions in Kuwait, but they were so far away we didn't hear the sound. We figured it was friendly, nervous, fire. Charlie was ready to emplace our GSR team, but this was cut short when we received a message that all the Teams should return to their OPs.

We didn't know why they had us leave the OPs the previous night. Chief Torbert, in Ruqi, had some theories. He thought that we had been a diversion, moving around to the front of the Iraqis, while the VII Corps moved into their attack positions farther to the west. Another theory was that it had been raining so hard that we may not have been able to get air support, so the chain of command had us move out of the OPs just in case there was an attack by the Iraqis. Whatever it was, the pullout of the Team from Ruqi caused some anger

---

[86] The drug was dextroamphetamine.
[87] Personal email from Grant Winkler received on 17 August 2010.
[88] Personal email from John Wilson received on 17 August 2010.

with their Saudi counterparts. They called them cowards for not staying to defend the OPs and our guys called them stupid for staying to defend an office building. It seemed like we were the only OP that still had any rapport with their counterparts.

While we were in Ruqi we picked up one more group of American soldiers. A LRSU Team in a HUMV had moved up to augment our forces, so now our OP had seventeen Americans crammed into our small barracks. The LRSU soldiers were amazed at how we were living. They thought we would be in bunkers and trenches, instead of in a paneled room with beds.

The next day, K-Day, we augmented our patrol with the LRRPs to give our guys a break. On this patrol there were two of us in a DMV and one LRRP. Our other DMV had one SF and two LRRPs, and they stayed near the GSR position in the wadi. The six LRRPs were pretty young, commanded by a SSG who was from Australia. They were glad to be up on the border near the enemy, and attached to Special Forces, instead of back with their own unit and having to put up with petty bullshit.

On the night of January 15th, I was the lone SF soldier in the wadi with two LRRPs. At first it seemed that the war would begin because I saw a string of American aircraft circling behind our position, and lights were burning all across the Iraqi border. When all the planes flew to the east, the lights in Iraq went off. Right at that moment the GSRs radar broke, and we were blind. To make matters worse I couldn't get anyone on the radio because we had lost commo. I didn't know if we had been hit with some sort of EMP weapon that jammed radar and radio, but it was a pretty intense night.[89] Nothing happened though and we returned to the OP at 0500.

Major Richardson arrived in Ruqi, and he told us that he would be there until the war started. He brought another LRRP team up, that would be assigned to Captain Nick at OP 6. Our GSR guys were able to get another radar sent up to Ruqi so they wouldn't be blind anymore. MAJ Richardson was mainly there to make sure that the OPs got the word to extract when the order finally came down. Captain Nick was given a new E+E plan and had to go to Hafr el Batin to coordinate with the 101st Division that had moved into the town. They surrounded the town and dug into it like it was Bastogne in WWII.

We now had two choices for our withdrawal…we could try to E+E through the friendly units to our rear, or we could go west and try to avoid the armor battle that would be happening around us. If we went to the rear, we would have to try to work our way through lines full of trigger happy, inexperienced privates of the 101st. We decided if we had to E+E we would head west, and if we did that we would keep going until we got to Israel.

---

[89] EMP = electromagnetic pulse.

Chapter 3: SPECIAL FORCES – A Storm in the Desert

After the sun went down on January 16th, I was on radio watch at the OP, while the DMVs went out on patrol. I got a call on the Jaguar radio from the Saudi AMX-10 that there were tanks moving around in the wadi. This had to be Iraqis, because there were no other tanks up with us. I called our DMVs to verify if this was true, but they couldn't see anything yet. I then got a radio message with the code word "ZONAL". This was the order for us to extract out of our OPs and move to the alternate positions. I told Ibrahim that we were ordered to leave, but he told me that his men were going to stay at the OP. One of our DMVs came by the OP and picked me up, then we went to the "Castle" in the wadi.

When all of us had gathered at the Castle another message came in, telling us we needed to go to MOPP level 2 (put on the NBC suits) because Desert Shield had just become Desert Storm. The order said hostilities would begin at 0300 or 2400 ZULU time. This was the first time any of us heard of an operation called "Desert Storm", and we wondered if it was like Operation Imminent Thunder two months prior. We didn't put on the MOPP suits because we only had two of them, and there was no resupply for us. The other OPs did break open their MOPP suits and wore them for several weeks.

Lieutenant Ibrahim and his men arrived at the Castle and told us that he also had been ordered by his commander to leave the OP. That night we had twenty-three soldiers at the Castle and five vehicles, but the hill was large enough to hide them all. Most of us didn't want to get any sleep and stayed up waiting to see if the war would begin at 0300. I went to the highest part of the hill, a mound where I had put two LRRPs on an OP position. It was cold, with the wind cutting through us, but we didn't care. None of us talked, just looked out across the desert for any sign of movement. My world was green, and grainy, through the lens of the PVS-7s. The night was clear, the sky was full of stars, and we waited.

At 0300 the world was still quiet. No movement at all, and no sign of a bombardment. I glanced off to the distance and I saw the sky light up, like it was a thunderstorm seen from across the ocean. There were no sounds, since the explosions were far away. No one else saw the flashes of light, because I was the only one with NVGs on. I called down to the rest of the Team to get up to my position, "NOW!" Something was happening! I told them that I think I was seeing Baghdad getting bombed.[90]

In an instant the world changed from a quiet desert night to a continuous sound and light show. The Iraqi anti-aircraft guns at the end of the wadi opened up, firing into the air at targets that either weren't there, or we couldn't see them. The big 23mm cannons sounded like they were right next to us. Farther back

---

[90] Baghdad was 300 miles from our position, so I was most likely seeing Basra being bombed.

surface to air missiles roared into the sky and exploded in a fireworks display over our heads. The closest missiles and ADA could be seen without the NVGs, but since I had the night vision goggles on, the entire sky looked like it was alive with a thousand balls of light, all racing to the sky, searching for something to kill.

Over the Berm

The war began at 0237 when four modified Apache helicopters flew north into Iraq and destroyed the radar sites that guided missiles into attacking aircraft. Hellfire missiles destroyed generators and command vans, while smaller Hydra rockets sent thousands of small finned darts through wires, electronics and men. The bodies of the Iraqi crewmen were hit again and again, ripping them to shreds as the radar site burned brightly in the dark night. After four minutes of unleashing death upon the Iraqis, the Apache helicopters flew off into the desert.

A second attack group had taken out another radar station twenty miles away, creating a corridor through which aircraft would be able to pass through, unmolested by enemy anti-aircraft weapons. Special Forces soldiers followed the attack helicopters in Chinook helicopters and set up eleven reflector beacons that would be used to mark the corridor.[91]

I looked around our little knob on the hill and all seventeen of the Americans were there. The little hilltop was small, so each man was holding onto the other so they wouldn't fall off. K.C. began to sing, in a quiet voice, "all we are saying, is give peace a chance". Soon the others on the hill picked up the song and holding onto each other rocked back and forth. I figured this was a scene that would be hard to describe to anyone later. Here we were, the closest Americans on the ground to the enemy, probably within a few miles of the anti-aircraft guns and Iraqi positions, and we were singing "give peace a chance."

I realized that we were all easy targets and broke up the mob, telling them to spread out. As we did one of the LRRPs pointed up and said "Holy shit... look up!" I looked up with my night vision goggles and saw hundreds of aircraft, in a line, heading north towards Kuwait. The aircraft heading into Kuwait had their running lights on, but cut them off when they reached the border. There was a second line of aircraft beside the first, heading south, coming back from their bomb runs, and turning their lights on.

Kuwait City got hammered next, and we could see the huge explosions with the naked eye. The sky all along the border erupted with anti-aircraft guns and

---

[91] Atkinson, *Crusade*, pp 32-33

missiles. The tracers from the ADA would snake back and forth across the sky, like a kid squirting a garden hose. As the closer explosions went off the guys around the OP would say "oooh" and "aaah" like it was a 4th of July fireworks display. One of the guys nicknamed that night "The Lightshow" but I called it the "Footprint of Allah". It was exciting and terrifying to watch. Oil pipelines and refineries went off in gigantic flame balls, spraying burning oil into the sky. Kuwait City burned, and even though it was 75 miles away we could see the sky bright from the flames. Ibrahim had come up the hill to join us and said "Oh my God" when he saw all the destruction. That was strong language coming from him, because he never took God's name in vain.

I looked up and saw one of our own aircraft, pretty low, go screaming down the wadi. It looked like it had sparks or flames coming out of the engine. I continued to watch to see if the pilot ejected, but it continued flying to the south. As I watched the aircraft disappear in the darkness, I saw explosions in Saudi Arabia, behind us. Ruqi kept their lights on throughout most of the night, but at 0530 the lights went out. This may have been from Iraqi artillery that was shelling the town. The Iraqis shelled Ruqi more than any other place, because they knew where it was on the map.

Right before sunrise one of the OPs to the east of us was shelled by Iraqi artillery. The Special Forces there requested artillery or aircraft to take out the enemy guns. Two and a half hours later they finally got an airstrike on the Iraqis in front of their position. Close air support for us that night was a low priority and other targets were more important. Between 0300 and 0600 I counted six separate flights of aircraft going northward. Each flight had hundreds of aircraft in them. At sunrise I finally decided to get some sleep, though I wanted to keep watching. I knew I would need some sleep in the future, so the light show would have to wait.

In the 20th century there had only been one other war where the outcome was decided in a single battle, and in a single day. This had been the 1967 Arab-Israeli War. Now Desert Storm achieved that same distinction. The bombing missions in that single night had crippled the Iraqi air defenses so much that their ability to mount a coordinated assault was now gone. Iraq was destined to lose.

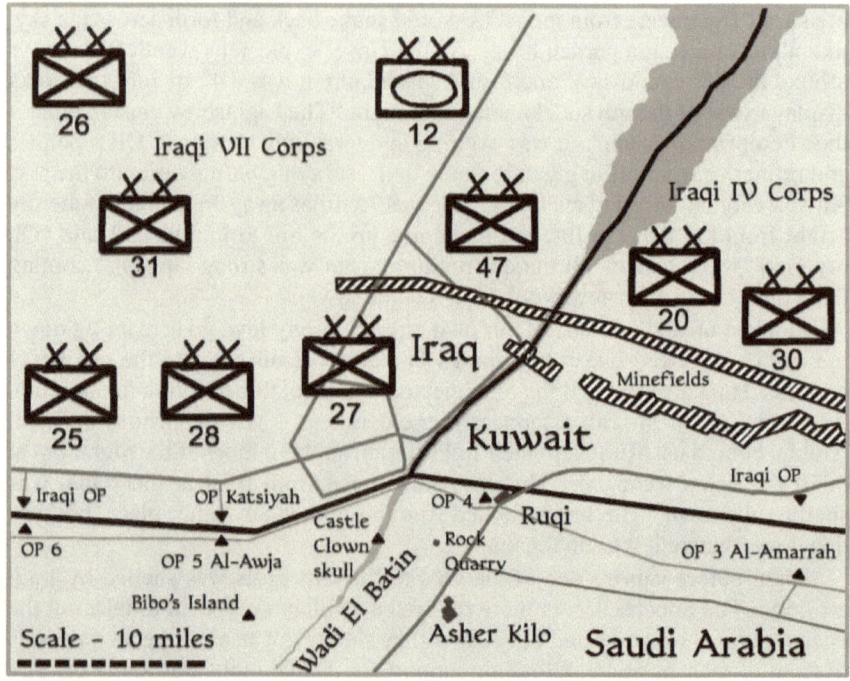

At 10:00 I got back on the berm, but stayed low in case the Iraqis were looking for us. I had my radio beside me, tuned into BBC, so I could hear what was happening in the war. The reporter said that the Republican Guards had been destroyed and so were the SCUD missiles and chemical plants. It wasn't until after the air war that I learned this was far from the truth, but reporters always get the story wrong when it is first reported.

A dark cloud spread across the border, from horizon to horizon, consisting of the smoke from all the burning vehicles, oil wells and towns. Just across the border from our position was a giant fire from a burning refinery. The jets roared overhead, but I couldn't see them because they were too far up.

We spent that day at the Castle, not going anywhere and just observing what went on around us. We put orange VS-17 panels on the hoods of our vehicles so that we wouldn't be mistaken for Iraqis. When we did get some sleep, we did it beside the vehicles, but not in them. I dug a hole beside my vehicle, so if the DMVs got hit by artillery, I would be protected a little from flying shrapnel. The hole that I slept in was not a fighting position, but looked more like a grave. The digging was easy since it was sand, but I could only dig about knee deep due to the concrete-like rock after that.

376

Chapter 3: SPECIAL FORCES – A Storm in the Desert

The oil well in front of us continued to burn into the night. No one was around to put it out. Other burning oil wells dotted the horizon. The flames were a bright red color. There was very little bombing until around 2200, when the Air Force went back to work. I could see strikes around the oil well fires, and figured they were bombing Iraqi positions. Only one anti-aircraft gun fired, which meant a lot of them had been taken out or they had moved. They might have been given orders to quit firing up in the sky unless there was a definite target. The A-10s roamed the border, looking for targets of opportunity. We gave them the nickname of the "Sharks", but their call signs were Nail 21 and Nail 22. Nail 21 was later shot down by a lucky "golden BB".

Sometime after midnight I saw a large missile race up into the sky, but unlike the anti-aircraft missiles, this one didn't explode, and it continued to go into the upper atmosphere. This had to be a SCUD, so I got on the radio and called in a Red Dragon, and gave the compass reading to where it launched. Later we heard from BBC that seven SCUD missiles had been launched at Israel and one had been launched at Dhahran. The one I reported was the one that headed to Dhahran.[92]

Right after the SCUD was launched three batteries of 120mm rockets slammed into the Kuwaiti desert to the east of us. I figured the Iraqis were searching for the Special Forces OPs over there. The missiles had hit Ruqi and caused some minor damage to some of the buildings. When the forty missiles exploded it sounded like rolling thunder and the ground shook. As the air war went on, I got pretty good at figuring out if the missiles were going to hit near us or not. When the missiles launched, I could see all forty in the air, racing into the sky. At their arc I could then figure out if they were going to hit near us, or go miles to the rear. It was tricky, but I had to keep an eye on them until they peaked.

The next day, January 18th, four DMVs from 2nd Battalion linked up with us so that we could escort them out to OP 7 to the west. Charlie took them out there, where he was planning on handing them off to Captain Nick. Unfortunately, Charlie wasn't able to get communication with OP 6. Charlie

---

[92] The first SCUD missiles launched on January 18th at 0300. This missile was destroyed by a Patriot missile battery in Dhahran. There is some dispute as to what the Patriot missile shot down, since no wreckage was found. However I did see a missile launch from the Iraqi desert. Eight missiles flew into Israel, forcing the United States to change the priorities in the air strikes against Iraqi targets. Israel had to stay out of the war, or the coalition with the other Arab countries would fall apart. To ensure that Israel would not strike back at the SCUDs, they would be the primary target for US bombers.

saw Mastrovito's DMV on the horizon and headed towards it with the other vehicles.

Dave Be__ was in the gunner's turret of Mastrovito's vehicle and when he looked up, and saw five vehicles approaching in the distance, he smacked the driver, Darren Crowder, in the head. He began yelling "There's Tanks! Iraqi Tanks! Go, GO!" Crowder tried to look where he was pointing but Be__ kept smacking him in the head and telling him to go. Both of the DMVs took off, racing to the south.

When Charlie got to where they had been he found their radios and intercept equipment, still running. He collected the radios up and put them in his vehicle. When he came back to us, he gave the 1st Cav guys the radios to use in their vehicles.

The two DMVs from OP 6 continued to run southward until they reached friendly lines. It wasn't until the next day that they returned to resume their observation of the border. Since they weren't able to link up with OP 6, the 2nd Battalion DMVs returned with Charlie and stayed with us that night.

After Charlie had returned there was another BM-21 rocket attack, this one near our position. Though it was miles away, it shook the ground and made us all take cover. We were still not used to being bombarded by the enemy, but that would change as the war went on. Gentz came running up to the observation point and told me that we had ten minutes to get off the Castle. Someone on the radio told us that thirty Iraqi helicopters were coming to attack our position. We all did a mad rush to get to our vehicles, but we were able to grab all our gear. We didn't want to leave anything behind, like Captain Nick's team did. We drove out of the wadi and raced across country to Asher Kilo, the town south of Ruqi.

When Gentz had received the message to abandon our position we were also told that we were supposed to link up with other Americans at Asher Kilo. None of us knew what was going on and we figured we were going to be pulled off the border.

When night came, we drove up to the road and watched the "light show". There was more bombing on the Iraqi side of the border, and we saw another jet returning, this time on fire. The pilot didn't eject and it continued to the south like the one I saw the night before. The Iraqis conducted harassing fire all night long, and artillery rounds and missiles would explode here and there, with no discernible pattern. All the Saudi Special Forces drove up to the road where we

378

were and LT Ibrahim and his soldiers told us goodbye. They then drove south and we never saw them again. An American vehicle did rendezvous with us and said that we were not pulling back, but we had to stay in place, and continue to observe. I never did find out who was in the vehicle.

We chose to head into Ruqi. I say "we" because a lot of the decisions we made were decided by consensus. Charlie was in charge, but sometimes he deferred to me. Ruqi, though it had been shelled repeatedly, seemed like a good spot to lie up for the night. There was a gravel pit there with a lot of vehicles where we could hide. While Charlie reconned into the area, I saw the largest explosion in my life.

I had on my PVS-7s and I was looking towards Ruqi, when a giant ball of flame began to grow, and grow, and then break apart into little balls of light. The dome of flame was huge, and looked like a nuclear bomb had gone off. There was still no sound, but I could see the shock wave coming our way. I yelled down to K.C., "Hold on!" just as the shock wave and noise hit us. The whole DMV rocked to the side, rising off the ground on two wheels, and then slammed down again, jarring everyone in the vehicle. We all thought it might have been a nuke.

Charlie had linked up with Gator at the gravel pit during his recon. Gator had the rest of OP 3 up on the berm since there was nowhere to hide where they were located. He told us that we would be losing our LRRPs and Cav guys in the morning. Even though there had been the largest explosion I had ever seen, I was also tired, so I ignored trying to figure out what the explosion was. I was able to get two hours sleep in the pit that night since I was pretty exhausted from the day's activities. There were more 120mm rocket strikes that happened around us throughout the night, but I slept through them.

That morning we said farewell to the LRRPs. They wanted to stay with us, but they were ordered to return to their unit. We still had to get the 2nd Battalion DMVs to OP 7, and after eating an MRE breakfast we headed west, hoping to link up with Captain Nick's DMVs. We decided that if he wasn't there, we would continue on, another ten miles, to the location of OP 7. None of us had been to that OP, but we knew about where it was supposed to be located.

Luckily, we found the lost DMVs of OP 5 about five miles back from the berm. We linked up with them, and Dave Be__ told us how they were chased back to our C Company by a platoon of Iraqi T-72s. We then told them that it was only us bringing the 2nd Battalion DMVs to them, not Iraqi T-72s. At first, they didn't believe it, but eventually they realized it was true. They were pretty embarrassed about it. Charlie gave them back their antennas and radio that they had left behind. Someone on our team called them the "Kevlar Kuwaitis", since the Kuwaitis had also run from the tanks. I warned them about the rocket strikes all along the border, especially around Ruqi. We handed over the OP 7 team and returned to the Castle in the wadi. It was still the most defendable ground in the area.

That night, as I sat on radio watch, I heard Be__ trying to call in an airstrike on the BM-21 launchers I told him about. I didn't know why he was trying to do it, since the launchers were mobile and moved each time they fired. He continued to try to get communications with aircraft, but they weren't receiving him. He called me up, asking if I would call in the aircraft and direct them to his position. I told him that the targets he was trying to attack were long gone and he didn't know what was to the front of his position. There might be deserters out there, or another SF element doing a mission out there. He kept trying to call in the aircraft until I finally told him to abort the mission. It was too dangerous to attack a target you couldn't see, and didn't have any information about. I figured I might have spared some Iraqis that night who was trying to escape from Saddam's army.

It rained hard that night and each of us decided to sleep in the vehicle. If we slept in our holes, we would have been sleeping in pools of water. The visibility was nonexistent and we couldn't see more than a few meters around the vehicle. It would have been a great time for the enemy to strike, if they knew where we were. Ruqi got hit with rockets again, and Iraq was bombed. I could see flashes through the rain, and hear the rumble of the explosions in the distance. I didn't see any anti-aircraft tracers at all and I figured the A-10 sharks must have destroyed them.

On the morning of January 19th, we got a message to head south and link up with our C Company and pick up more ammunition. We were told that there were 40 cases of Mark 19 ammunition, and three cases of AT-4 rockets. Our vehicles were already so full of ammo, food, water and fuel that we hardly had any room to move. C Company was also 40 kilometers to the south of our position. This made us pretty angry. We had no support at all from the rear and they wanted us to leave the border unsecure while we drove back to get our own supplies. To compound our feeling of isolation the Saudi armor unit to our rear sent a radio message that they were leaving also. So, the only thing covering the western front was four Special Forces OPs. We packed up our vehicles and drove into Ruqi, to find Major Richardson and ask him about this resupply mission.

Chapter 3: SPECIAL FORCES – A Storm in the Desert

As we were heading into Ruqi Iraqi artillery rounds were dropping around the town. The visibility was almost nonexistent due to a fog that had rolled in. We drove into the actual town of Ruqi, and not just the construction supply area. We could hear the rounds impacting, and feel the shock waves, but we couldn't see the explosions. When each round detonated there would be a flash of light in the fog, and the sound of shrapnel zipping by our vehicles and slamming into walls.

There was no one in Ruqi, so we drove to a quarry nearby that we thought the SR OP would be located. One of the SF support guys there jumped out, and stopped us from going any further. He told us that he had almost shot us, until he finally recognized the DMV shape in the fog. They were all pretty jumpy after three days of being shelled around the clock. Richardson told us that we could blow off "those idiots in the rear" and don't worry about getting more ammunition. He also told us that we needed to "hunker down and cease movement". Every time there was movement near the border, Iraqi radar would detect it and send rockets into that area. This may have explained why the artillery rounds increased when we drove into town.

We chose to drive on the road to get out of town quicker, instead of driving across country and we headed south so the radar would not know which direction we were headed to. As we drove out of town, we saw a civilian Toyota SUV driving toward us on the other side of the road. We stopped our DMV and made them pull over. The people inside the vehicle turned out to be a male and female cameraman from CBS News. They asked us some questions about what it was like on the border and had we seen any action. They also asked us what unit we were with, but we didn't let them know. They did give us some Ramen noodles and some other food, since our supplies were running low. We told them that they should not go any further north because the Iraqis would fire rockets at them. They asked us what we would do if they did, and we told them they were fools if they tried. I told them that we would watch them burn from a distance. They wanted to know if they could film us, and I told them no.

The reporters filled us in on some of the news from the day before. They told us that six more SCUD missiles had hit Israel the night before, but there were no casualties so far. They also told us that sixty-four Israeli F-16 fighters had flown into Saudi Arabia, and that they were going to paint out the cross of David symbol and put on U.S. markings. The Russians had supposedly supplied us with all the frequencies to the Iraqi SAMs, so we could jam the missiles once they took off.[93]

When we asked what was happening back home, the reporters told us that 84% of the American public was behind the war. When I asked about the huge, nuclear type bomb that detonated near us, they told me that it was a Fuel Air Explosive (FAE) and was the largest bomb they were dropping on the Iraqis.[94]

As we were ready to drive away another vehicle showed up, but this time it was a French reporter from Sygma, and he talked to us. As K.C. was speaking to him in French another vehicle drove towards us, but this time it was an Egyptian and a Border Patrol major. I told the Border Patrol major that he needed to seal off the road leading to Ruqi, so all these civilians wouldn't end up in Iraq and get themselves killed.[95] We told the reporters to be careful and don't go into Ruqi either, because it was the most bombed town in Saudi Arabia, and then we drove towards the south to get away from that traffic jam.

A mile down the road we saw a tour bus on the side of the road, with a bunch of the Saudi Border Patrol standing beside it. They told us that the bus was there to bring back all the deserters that were expected to come across the border that day. Captain Nick's OP had picked up eleven more deserters that day. We stopped and drank some tea with the Border Patrol soldiers then continued on to the Castle. Throughout the day we saw several deserters, but we stayed hidden and let them wander by, without giving away our position. Charlie and I agreed that we didn't have the facilities to take any prisoners.

There was no bombing that day due to the fog. Targets were well concealed and there was worry amongst the Air Force of accidental bombings of friendly targets.[96] I called in four pre-planned airstrikes on Iraqi OPs that we were

---

[93] These rumors turned out to be not true, but we thought the reporters might have gotten it right at the time.

[94] We weren't the only ones who thought it looked like a nuclear bomb going off. When another FAE detonated a British SAS team did send in a report about a nuclear blast.

[95] Around that time three CBS reporters did continue across the border and were captured by the Iraqis and held prisoner.

[96] On the first night of the war there was great argument about whether to attack from high altitude or low altitude with the air strikes. By striking from a high altitude, above 15,000 feet, there was no risk from small arms fire or anti-aircraft artillery; however there was a threat from SAM missiles and Iraqi MIGs. By striking from low altitude there was an element of surprise and there was no threat from SAM missiles, however small arms would be the largest threat. Each air wing came up with their own plans about how to attack. On January 17th four A-6 Intruders from the USS Saratoga flew to bomb a SCUD support site at H-6 in the western Iraqi desert. The aircraft flew as low as possible,

keeping an eye on. The difference between a preplanned strike and a regular air strike is that the preplanned strike is one on a fixed location and the Air Force can do an analysis if it is critical enough to destroy. I reasoned that these OPs were the ones calling in the rocket attacks on Ruqi. The air strike cleared through and I was told to expect the attacks on the next day when the fog lifted.

While I kept an eye on the four Iraqi OPs, Charlie had the rest of the team unload the DMVs and take off anything that wasn't absolutely necessary. We buried all the unneeded items in the desert, and would come back for it later, if we had time. We also spread out all the VS-17 panels on the ground behind the vehicles so we wouldn't be mistaken for the Iraqi OPs when the airstrike came in.

On Sunday, January 20th, the quiet night was shattered when Ruqi was jolted awake by another rocket attack. To our west the OP 6 building was hit by several mortar rounds. Captain Nick had been five miles from the building, but after that attack they moved south twenty more miles. Over on the eastern sector OP 2 had been hit with a FROG 7 missile, but there were no injuries. There was still a dense fog hovering near the ground and I think the Iraqis were using it to their advantage, knowing that our aircraft couldn't see them. Due to the fog my preplanned airstrikes were postponed. I was told by the FACSOP that we would be getting helicopter gunships when the fog lifted.[97]

We learned the fate of MSG Cr___ when he was seen working the resupply mission for the OPs. He radioed us and told us that he would meet us on the road intersection leading to Ruqi. He said he would deliver us ammo and take back any equipment we didn't need. Charlie drove over to Al Awja OP 5 to get the bags we had left there when we evacuated. He also refueled from the diesel tank at the OP.

When Charlie's DMV returned, a GMC truck followed them and drove right to our position. We didn't know what to expect and I thought it might be some sort of Iraqi suicide bomber. We took cover and locked and loaded on the

---

sending decoy drones ahead to give the impression of more aircraft. As the aircraft dropped their bombs, one A6 was shot down by a SAM missile and another had taken a missile into the engine and fuselage, but was able to limp back to Saudi Arabia after jettisoning its bombs. The two pilots who ejected from the A6 that was shot down, were captured by the Iraqis a short time later. Due to this an order was sent out that all pilots would bomb at high altitude, foregoing any element of surprise, since there didn't seem to be any. This made it extremely hard to find the SCUD missiles that were both mobile and camouflaged.

[97]FACSOP = Forward Air Controller Senior Officer Present. His callsign was *Wyme 3.*

vehicle. K.C. told the vehicle to stop and all the passengers to get out of the car. When the passengers got out Baby Doc rushed towards them with his pistol drawn. The situation looked like it might get out of hand, so we all rose up and showed ourselves to the vehicle, pointing weapons at them. We searched the passengers and found out that they were unarmed. K.C. questioned them and learned that they were the Moxhabbarat, or the Saudi version of the CIA. We pointed them to Ruqi and told them to drive south. I chewed out Baby Doc and told him to not run out into a dangerous situation like that again.

After they left a Border Patrol vehicle drove up to our position. They had followed the tracks left by the Moxhabbarat vehicle. The Border Patrol was from OP 5 and were happy to see us. They were driving GM pickups and had a dozen or so men in the back, all waving their G-3 rifles in the air. Within minutes a second Border Patrol vehicle, from OP 6, drove to our "hidden" location. There was a lot of handshaking and kissing and firing weapons in the air. We figured our security was shot, so we told them they had to leave because we were trying to hide. They told us they wanted to stay and fight with us. I couldn't get rid of them, so we told them to stay off the border, because the Iraqis were attacking all along it. They all hooped, and yelled "Allah Akhbar" and then raced to the border like a bunch of drunken Vikings. We were amazed at the lack of any fear in these guys.

Before the Border Patrol left, they gave us a baby goat for food. Crowder put him on a leash, but we didn't know what to do with it. We had MREs and we were not going to stop to slaughter and cook up a goat. It didn't help any that the goat looked like it belonged in a petting zoo and not on a plate. In the end Crowder let it go in the desert, hoping it would find some way to stay alive.

While I was on the radio MAJ Star____ came on and asked us when we were coming to get the ammunition, forty kilometers to the rear. I made some excuse, but didn't tell him that we weren't coming for it at all. I didn't want to leave our location, since I had that helicopter strike coming in. I called the FACSOP to confirm the strike and he told me that it would be coming when the fog lifted.

That night the helicopters did come in, but they went to Gator's position to strike the Iraqis in front of OP 3. They fired up the OP and knocked down the radio antenna there. I don't know how many Iraqis were killed in the attack. I resubmitted my airstrike, but the weather was still pretty terrible. It had been raining and foggy almost every day since the war began. It was also extremely cold. One morning the front windshield on the DMV cracked due to the freezing conditions and the concussions from distant explosions. So much for the idea of a hot desert war.

Chapter 3: SPECIAL FORCES – A Storm in the Desert

Charlie took our bags that had been at Al Awja to the Ruqi quarry so that Cr___ could take them back to KKMC. Richardson told Charlie that we could begin patrolling again to try to stop any infiltrators. He also told us that there would be no more CAS missions for the OPs on the border, since all the additional aircraft were now tasked with finding SCUD missiles.[98] Richardson told us that we would be on our side of Ruqi for five days, and then we would go over to OP 3 for ten days and operate as a whole team. After that we would fall back to the rear and train the Kuwaitis to take back Kuwait City.

Dave Be___ called in A-10s on the Iraqi OP across from OP 6, and as the sharks fired their main guns, Be___ saw the of Iraqis run out of the building and into the desert. We could hear him on the radio yelling "Get them; there they go, GET THEM!" Captain Nick returned to the Iraqi OP with a Border Patrol major, to see if he could get the Iraqis in the building to surrender. When they were unsuccessful, he brought up the DMVs and fired 40mm grenades and .50 caliber rounds into the building. The building was made of stone, so the grenades didn't have much effect. The .50 caliber rounds tore through the masonry though. The SOTA vehicle also fired on the building with their M-60 machinegun, but to no avail. The Iraqis wouldn't surrender, so Captain Nick drove south, away from the Iraqi's view. It did work a little though, because that night eleven of those Iraqis gave up.

When the Iraqi chain of command learned of this, they sent several tanks to level the Iraqi OP. The tanks fired several times into the building until it collapsed, they then drove over the trenches in case anyone was hiding in them. Before they left, they fired several rounds into OP 6, blowing large holes in the rock and knocking down one of the walls.

That night Charlie, Baby Doc and Gentz went on a joint patrol with the Border Patrol, while my vehicle remained behind on radio watch. Since there was no more air support for us, we decided to take out the Iraqi OPs by ourselves. I devised a plan where I would destroy the solar panel outside the Iraqi OP with the sniper rifle. That panel gave power to their radios. We would then try to get the Iraqis to surrender.

---

[98] To keep Israel out of the war the United States devoted dedicated bomber squadrons, the SAS and DELTA Force to try to find the missiles. There were also plans to have the 82nd Airborne jump into the Iraqi western desert on suspected SCUD locations, and bombing the western half of Iraq with thousands of aircraft with something that was "not carpet bombing, but close". Those plans were rejected by Schwarzkopf due to the prospect of heavy losses from the Iraqi civilian population.

385

Triple Canopy

K.C. said he wanted to cross into Iraq because there were puppies over there. When he said this, we all looked at him like he was nuts, but he insisted. There were a bunch of puppies over there. He had seen them before the war began. He wanted a puppy. At the time it didn't seem too crazy, so I told him that he could do it, but only after we had neutralized the OP. That night rockets fell in and around our position in the wadi, but none were close enough to cause us any alarm.

I had planned to attack the OP at 0530, right before the sun came up, but the sky opened up and the rain came down in sheets. It rained so hard that it would have been impossible to navigate and we might end up bumping into an Iraqi tank, so we waited until 0645. We no longer had the dark to hide behind, but I figured the miserable weather would lend us an element of surprise. Before we left the team put on the Kevlar flak jackets and some of the men put on helmets. This was the first time we had worn them in the war. The concept of coming under fire was a lot more possible that morning.

As we drove up to a pipeline we stopped at a set of "dragon's teeth" obstacle located where the pipeline crossed into Iraq.[99] We noticed an Iraqi jeep parked there. I had all the vehicles stop about a kilometer away to observe the suspicious vehicle. There was no movement so Baby Doc and I crossed into Iraq to check out the vehicle. There was a suitcase on the front seat that I thought might be a bomb, so I told Baby Doc to back off. We could return to destroy it after we hit the Iraqi OP.

We drove west along the berm until we were parallel with the Iraqi OP. Everyone in my vehicle got out and laid on the berm facing the enemy OP. The other vehicle would cover us with their .50 caliber. I told Gentz to not fire unless I called him on the radio or he saw targets. Baby Doc, K.C. and I were the sniper team. Casper would also cover us with the Mark 19 on our DMV. My plan was to fire ten sniper rounds, knocking out radios, antennas and optics. K.C. would then call to them to surrender. If they didn't, we would then level the OP with the .50 caliber and Mark 19 fire.

The whole mission was not authorized. The rules of engagement stated that we could only fight defensively and not offensively, because the ground war was not supposed to start yet. However, Captain Nick's attack on OP 6 gave me the idea, and we really wanted to get rid of any Iraqi FOs that could call in

---

[99] Dragon's Teeth were pyramid shaped concrete blocks that were meant to stop a tank, but could be easily blown out of the way or bulldozed.

rockets on our position. We justified it because the daily rocket attacks against us would make this "defensive fire".

I looked through my rifle scope, but didn't see any movement; the Iraqi flag was still flying from their flag pole though. I aimed at the solar panel on the Iraqi radio building 500 meters away, fired, and spun it around. Right after I had taken the shot, Gentz opened up with his .50 caliber, riddling the building. Charlie got out of the car and jumped in his shit, telling him to cease fire! I figured if there was anyone in the building, they were laying low now. I aimed at the big "ship to shore" binoculars on the roof of the OP and fired, knocking them off their mount. I told K.C. to see if he could get them to surrender. He got up on one knee, yelling at them in Arabic to give up. I don't know if they could hear him or not, since the building was 500 meters away.

I still didn't see any response from the building so I radioed to Charlie to "level it". Gentz and Wally both opened up with their big guns. Wally dropped a string of Mark 19 rounds right into the radio shack and then Gentz sent the heavy .50 caliber bullets through the brick walls. Baby Doc got up on one knee firing his M16, when he saw some movement in the trenches around the OP. I could hear a dog howling in pain and then saw it run out of the trench and into the desert. K.C. fired his 9mm pistol and I laughed at him, asking him what he was doing? The OP was way out of range for a pistol. He just laughed and said "I'm making some noise, man!" He then got up and ran towards the Iraqi trenches, yelling back "cover me!"

K.C. sprinted across the ground, as Gentz and Wally were still firing up the main building. He had his M-16 rifle in front of him, and jumped into one of the knee-deep trenches. He grabbed up a bunch of the puppies, and then started running back. As he ran the puppies were biting him and he would drop one every few feet. Each puppy he dropped would roll to a halt and then it would run back to the safety of the trench. When K.C. made it over the berm he only had one puppy left. We named the puppy Katzia, after the Iraqi Op we liberated him from.

I got on the radio and told Charlie it was time to go, and we both drove east, using the berm for cover. I never saw any Iraqis at the OP, so we don't know if we had terminated the FO position or not. When we got to the abandoned Iraqi jeep, I told the team to destroy it too. Each man fired a couple of magazines at the jeep. Though it was abandoned, we had too much ammunition and unloading it into the enemy vehicle would lighten our load. Wally fired a Mark 19 round at the vehicle and it impacted on the hood, tearing a two-foot hole in the metal.

Triple Canopy

Our offensive patrol continued back to Al Awja, so we could refill our gas tanks. As we approached the friendly OP, I saw some Iraqis looking around the corner. I told Charlie to come around from the other side and hit them from the rear. When Charlie began to move in their direction the Iraqis dropped their weapons, put their hands in the air and surrendered. K.C. yelled at them in Arabic to lay on their stomachs, and then we searched them.

They were carrying an AKMS and a Browning High Power pistol that was engraved with something saying that the pistol was a gift from Saddam Hussein. I gave K.C. the folding stock AK to use as a vehicle weapon. K.C. and I searched Al Awja and found the place trashed, either by Iraqi patrols or deserters. We took the prisoners, tied them up, and drove them back to the Haunted House.

When we arrived at the House the Iraqi prisoners were shivering from the cold. Since it was still raining K.C. decided to interrogate them there, instead of taking them to Ruqi. The Border Patrol showed up, having followed our tracks in the mud. They were wondering what was happening since they had heard all the gunfire that morning. Once they found out we were OK, they built a fire in the house.

The Iraqis told us that they were from an artillery unit in Iraq and that our air strikes had been ineffective. They let K.C. know where their artillery unit was located. After K.C. had gathered enough information, we drove the prisoners to Asher Kilo. The Iraqis wanted to know who we were. I told K.C. to tell them we were Israeli agents. We all laughed, but the Iraqis looked worried.

When we arrived at Asher Kilo, Major Richardson was there and wanted to know what all the firing was about. We didn't tell him the whole truth. We did tell him that we had found the Iraqi car with the suitcase on the front seat, thought it might be a car bomb, and then hit it with the Mark 19. We left out the part about attacking the Iraqi OP Katsiyah.

After dropping off the prisoners we traveled back to the shot up Iraqi vehicle and inspected it. Inside I found a bayonet, which I kept, a Soviet grenade and a magazine pouch with seven magazines for the AKMS, which I gave to K.C. We found some food, two cans of gasoline and some tools. We still needed fuel for our vehicles, so we drove back to Al-Awja to get diesel and top off the DMVs. The way we would get diesel was by siphoning it out of 55-gallon drums, using a garden hose. While we were filling up the DMV tanks Baby Doc found an Iraqi compass in the sand. It was probably from the Iraqis we captured.

388

Chapter 3: SPECIAL FORCES – A Storm in the Desert

We returned to the Castle for the night, trying to see through the rain that kept coming down in sheets. The wadi was turning into a big mud hole. I did see three HUMVs going north on the desert floor of the wadi. When I called it in, I learned that it was a Special Forces major and a recon element for the 1st Cavalry. They were going to be moving into our area soon.

General Schwarzkopf had ordered that no Special Forces units do "cross border" missions until the Ground War began. There were a few exceptions to this though. Early in the Air War, on January 23rd, four Special Forces soldiers conducted a mission to cut a fiber optic cable in the desert. Though the team found several cables, and cut them, they could not find the fiber optic cable. The team put 800 pounds of explosives in the hole that they had dug and detonated it, hoping that the cable would be severed. After the huge explosion lit up the night sky, they boarded their helicopters and flew back to Saudi Arabia.[100]

On another mission several six-man detachments from 3rd and 5th Special Forces Group flew into Iraq and landed by helicopter on the routes that VII Corps and XVIII Airborne Corps would use during the Ground War. Schwarzkopf had been told by CIA analysts that the terrain would not be able to handle armored vehicle traffic. Not wanting to rely on an analyst sitting in a room in room 7,000 miles away, he sent in Special Forces teams. The teams performed penetrometer tests on the soil to determine if it could handle the weight of an armored assault without the tanks sinking into the sand. They also filmed the terrain with low-level light lenses which proved to be the most valuable data collected on the mission.

On January 23rd (K+8) we were ordered to return to Asher Kilo and meet our Battalion commander. We had to answer for why we fired into Iraq. I thought it was odd that pilots could fire up the enemy, but we couldn't. Captain Nick also had to come to Asher Kilo and explain why he had an accident with his SOTA vehicle. We didn't lie to our colonel and told him the whole story. He told us it was no big deal and don't worry about it. He did tell us to try not

---

[100]Extracted from *US Special Operations Forces in Desert Storm 1990-1991*, on January 3rd, 2012, located at
http://www.specialoperations.com/Operations/Desert_Storm/Operations1.htm

to assault into Iraq again, unless we had to. Afterwards he chewed out Captain Mastrovito more for the accident with the DMV.

It wasn't much of an accident. The SOTA vehicle was following too close and when Crowder hit the brakes, the SOTA HUMV smashed into their rear. The damage wasn't bad and it looked like cool battle damage. The door wouldn't shut right anymore, so they put on a steel hasp and held it closed by using a Master lock. While we were there, we learned that LT Ibrahim and his Saudi SF returned, and now were a quick reaction force that would stay in Asher Kilo.

After our debriefing, we drove into Ruqi to the old SR OP and grabbed some MREs and water. We had to be quick because if we lingered too long the Iraqis would pick us up on their radar and bombard the town with rockets again. While we were there, we attempted to use the pay phone to call home, but the lines were all dead.

We learned from the SR OP that OP 7 had been attacked and that some Border Patrol had been wounded, but no one had heard from them since the attack. We also learned that some SCUD missiles had hit Riyadh the night before. The damage to Ruqi from all the artillery was minimal. There was one giant crater from some sort of bomb, and the concrete water tower had shrapnel and bullet holes covering it. The windows around the post office were all blown out from the concussions. Iraqi gunnery did not impress us much. It didn't impress the SR OP guys much either. They said that they hardly ever took cover anymore when they saw the rockets coming towards them.

We were getting no support from our battalion and had not received any food or water since the war began. We used what Charlie referred to as the "Mad Max Resupply System". We would get food from wherever we could find it, mainly the old OPs or from the abandoned shops in Asher Kilo. We would get diesel from the 55-gallon drums outside the OPs or from the gas station in Asher Kilo. We would either siphon it out with a garden hose, or drop a bucket down into the huge storage tank at the gas station. We would pull the bucket up with a rope and pour it into our gas tank.

Captain Nick told us that we had to move from Castle Clownskull and get closer to his location. This was because one of the times that they ran to the rear they had broken their satellite antenna. We were able to talk to them with the FM radio, but only if we were closer. If they had any information, we would relay it to the rear by using our SATCOM. We drove back to the Castle and picked up our cached supplies, then moved to the abandoned Saudi AMX-10 position that had been located behind the Al Awja OP. This was a sandbagged position that we could park both DMVs behind and be partially hidden. Charlie nicknamed it "Bibo's Island", since it sounded like something from "Mad Max". We cleaned weapons, changed clothes, and fried some potatoes we had foraged from one of the Ruqi shops on my small portable "Peak 1" stove. The sun came out, which made this day seem almost perfect.

Ambush at Al-Ulaymah

Before it got dark, we drove to Al-Ulaymah/OP 6 to link up with Captain Nick's DMVs and to check out the damage done by the Iraqi tanks. At first, we didn't see where the OP had gone to, but once we got close, we could see that the entire Iraqi OP was flattened, and had been run over several times by the Iraqi tanks. While we were checking out the OP there was a distant popping noise and I saw two puffs of smoke on the horizon. We all stood there, waiting to see where the artillery would hit, when two huge balls of flame burst to our left and right. The concussion nearly knocked me down and the noise was deafening. Mastrovito's DMVs took off to the rear and safety, but our two DMVs stayed where they were, while K.C and I shot azimuths to the puffs of smoke and waited to see if they would fire on us again. If they did, we would know exactly where they were and would either call in an airstrike, or we were going to attack them with our two vehicles. We remembered that the colonel told us to only fight defensively, but if they are firing first, then we would be defending ourselves.

There were no repeat rounds. The Iraqis must have fired, and then moved so they wouldn't be found. We both estimated that it had to have been 122mm artillery that fired at us. I think that we were getting used getting shot at by artillery because neither K.C. nor I budged when the artillery hit. We drove back for two miles until we found Captain Nick's DMV. K.C. set up the SATCOM and we called in the attack. In response we were told that there were 100 tanks attacking Hamatiyat/OP 1 and that the Iraqis had taken Az-Zarah/OP 7.[101] There was no word from the Special Forces team at OP 7, so we were given a mission to go and assess the situation and try and find them.

The captain gave a quick operations order and then we headed out to find a Special Forces Team that may have been taken out by hundreds of Iraqi tanks. There were only twelve of us. At the time I thought that this was the mission that would kill us all, but we didn't slow down. We drove towards Az-Zarah in

---

[101] The report about 100 tanks was not true and it may have only been a few, like the attack on Al-Ulaymah.

a diamond formation, rolling across the muddy desert at 55 miles per hour. It was almost sundown so we were heading into the setting sun. As the cold wind whipped through the open windows, I thought that this was one of the most exciting times in my life, and it was probably my last.

Things became really intense when we saw some helicopters heading our way from Iraq. They had been attacking something on the horizon and now turned their attention to us. All our turret gunners spun their weapons and aimed at the approaching helicopters. We didn't slow down, but all of us held our breath, wondering if they were enemy attack helicopters. We didn't know if they were our own helicopters, but we figured they would probably fire on us anyway, thinking we were Iraqi. The last vehicle, the SOTA HUMV, didn't even have a weapon in the turret, and instead had an M-60 pointed out the passenger door. We all breathed again when we recognized the shape of a AH-1 Cobra attack helicopter. It was ours, but they still continued to bear down on us.

Due to the great chance that we would be killed by our own aircraft each of our vehicles had a flag tied to the antenna to try to identify us as Americans. One vehicle had an American flag, one a Confederate Flag, one had a Saudi flag and one had a Pirate skull and crossbones. We figured no Iraqi vehicle would fly those flags. We also had VS-17 panels taped to the hood of the DMVs. The Cobra helicopter continued coming at us, looking as if they would fire at any moment. Finally, one of the Cobras buzzed overhead, only a few feet above us. The pilots looked down on us and gave the peace sign. We all breathed a sigh of relief and then focused on the suicidal mission at hand.[102]

After the sun went down, we switched to our night vision goggles. Only Captain Nick had been to the OP, so we didn't know what we were looking for. In the green grainy light of the goggles, I saw a dark shadow on the horizon that grew as we drove closer. The building was a large, two story, brick or stone building. It hadn't rained since we started the mission, so the sky had cleared up a little, giving us enough ambient light to make out details in the night vision goggles. All four vehicles moved slowly forward, creeping quietly, at about five miles per hour. We were waiting for someone to fire on us, and were ready to react to anything. When we got within 300 meters of the OP we stopped and K.C. got out of the DMV and shouted towards the building in Arabic, telling

---

[102] During the fighting at Khafgii a few days later about half of the American deaths were due to friendly fire. All those deaths occurred at night, by crews that were tired, stressed and did not have any way to tell friend from foe in the dark. So our flags would not be able to help us once the sun went down.

them to surrender. He also shouted in English, just in case the Special Forces soldiers were holed up in the building. There was no response.

The double doors to the OP were wide open, and the shadows could hide any number of Iraqis with automatic weapons, waiting to spring a trap on us. There was some radio chatter between the different vehicles, and finally Captain Nick said he would take one other person and go and clear the building. Whoever went with him would be going into a possible ambush, a forlorn hope, and may not make it out alive. Mastrovito came to me and asked me if I would go with him. I didn't ask why... I knew already. I was the best shot on the team and he knew I didn't panic when the bullets started flying. At the time I was not real thrilled with the idea, but later I realized that this was a great honor.

We both walked towards the OP and I saw a flash of light in the distance. I had Mastrovito hold up, and using sign language I pointed to where I saw the light. Mastrovito told me that it was the Iraqi OP, that was within small arms range of us. Both of us figured that clearing the building with just two people was too much for us to handle, and we backed off. It had been an extremely intense moment and I was glad I didn't have to go into that dark and foreboding OP.

All of our vehicles backed off a few miles, and then got on the SATCOM to tell them we hadn't seen anything. All of a sudden, the horizon lit up with multiple BM-21 rocket launchers firing missiles. I recognized the arc of the missile and realized that it would hit where we had been located earlier in the day.

The silence was shattered when Dave started yelling "INCOMING!!" I instinctively crouched down and then looked around to see what he was yelling about. He was pointing to the BM-21 rockets, arcing off to the east and yelled "INCOMING!" again. I told him to shut up, and then calmly said "Dave, you don't yell incoming, unless it is coming in! You scared the hell out of me! Those missiles won't land within 10 miles of us!"

Right when I finished chewing him out the missiles hit, way off to the east at our old OP. Embarrassed, Dave hunkered down in the turret of his vehicle and was quiet for the rest of the night.

Right after the BM-21 strike we saw three, really close, SCUD missile launches. They were almost on top of us, but I knew it was an optical illusion, and was most likely twenty miles away. Captain Nick got on the radio and called in a RED DRAGON. The radio traffic increased with all the other OPs

Triple Canopy

calling in the same RED DRAGON, and everyone giving their azimuths to the missile launches. Supposedly there was a flight of F-16s in the air at all times that would go after the launchers when we called in that RED DRAGON code word. We later learned that those missiles slammed into Israel, killing three people. The radio was now occupied with the SCUD attack, so we couldn't call to ask what our next course of action was. We didn't want to stay there, and be in the open in a part of the desert we weren't familiar with when morning came, so we drove back to our old area, splitting off from Captain Nick's DMVs when we got to his sector.

Our DMVs got back to Bibo's Island at 0200, where I promptly dug my hole/grave and fell asleep. When I woke up the sun was shining on my face. Sunlight... in the desert... it felt great! Wally had been on radio watch and told us that there were vehicles moving to our rear, and he could hear the sounds of tanks throughout the night. It was most likely the 1st Cavalry moving into place. Though we couldn't see them, we could also hear helicopters flying around behind us and the A-10 sharks firing up targets in Iraq. When the A-10 fired it sounded as if God was belching. We also heard some sporadic small arms fire coming from the wadi. Both vehicles needed to be fueled up, so we decided one DMV would go at a time, while the other DMV kept watch from the AMX-10 position. We were able to watch OP Al-Awja from the sandbagged position.

Later in the day the Border Patrol came to us and told of an Iraqi vehicle that was driving up and down the border. All of us decided that we would ambush that vehicle and see if we could get any information from the survivors. While we waited for the other DMV to get back, I modified K.C.s captured AKMS. I took off the folding stock so he could wear it around his neck while he drove.

You may wonder why we wanted an AKMS for a "vehicle weapon". The issue M-16A2 rifles issued at that time were about a yard long, and they didn't fire fully automatic, they only fired a three-round burst. The AKMS, with the stock removed, was only 25 inches long. It also fired full auto, which would come in handy if we were trying to escape an ambush. The plan was to stick the weapon out the window, aim at the enemy, and fire full automatic. Though it would not be accurate at all, the hailstorm of bullets passing near the enemy may make them duck and allow us to get away.

After we had fueled up, we drove out to Captain Nick's position, to see if he needed to send any messages to the rear. We weren't able to find them, so we reconned the Iraqi OPs along the border. We didn't see any movement there either. Since the tank attacks a few days earlier the border didn't have a lot of movement. We noticed that the flag from Al-Katsiyah was gone, so I assumed

394

the Iraqis had evacuated that position after we had destroyed their radio site.[103] Overhead I saw two jets chasing each other, the one to the rear firing missiles at the first one. The missiles didn't hit, but I hoped that it was our guys shooting the missiles. Other aircraft were flying overhead, and firing missiles into Kuwait. Smoke from the explosions spread across the far horizon.[104]

As we watched the border, we saw the Iraqi vehicle that the Border Patrol had told us about. It was an Iraqi truck, similar to a US 2½ ton truck. We immediately pursued the truck, putting the pedal to the metal, going as fast as we could. We kept the berm between us and the truck, trying to sneak up on it. When we got within 200 meters of the truck the driver saw us and did a sharp turn to the north. Gentz fired his .50 caliber at it, but it was too far away to hit. We had lost the race to a slow-moving truck. I figured we were slower because of all the weight we had, due to ammunition, gas and water, and it was also because we were racing in the mud, while the truck was on a road.

When Be__ heard our gunfire, he drove over to find us and then he guided us to their new location. He told us that they moved every night. Their new location was in our sector, and about six miles from the berm. The Border Patrol were with them and were cooking up hot tea and goat. We joked with him about how it was great to get back to the "rear" and get hot chow again. While we rested, I drew a new cartoon on my DMV. It was a cartoon of a Mexican Bandit,

---

[103] It turned out that someone on Mastrovito's team had crossed into Iraq and had taken the flag.

[104] What I might have seen was a very controversial pair of kills made by Royal Saudi Air Force pilot Captain Iyad Salah al-Shamrani, flying an F15 from the No. 13 Squadron. Two Iraqi Mirage F1EQs and two MIG-23s were heading towards coalition naval vessels in the Persian Gulf, when it was decided to let the Saudis get the kills, though there were two F-14 Navy Tomcats standing by in case the two Saudi F-15s did not get them. United States Air Force (USAF) AWACs aircraft vectored Salah to the Iraqi Mirage F1s. However, Salah was having problems trying to find the enemy jets and intercept them. The two MIG-23s noticed they were being tracked and flew away, but the Mirage aircraft continued on to the Gulf to sink any ships they could find. The Iraqi Mirages were within striking distance of their anti-ship Exocet missiles, the same type of missile that had sunk the destroyer *HMS Sheffield* in the Falkland Islands war, just nine years earlier. Finally, Salah was able to get behind the Iraqi jets, talked step by step by the American AWACs plane, and fired off two AIM-9P Sidewinder missiles. Selah told reporters later, "They started breaking in front of me [trying to get away], but it was too late. It was my day." The two Iraqi Mirages were destroyed, giving the Saudis the only two victories that were not American in the war. Extracted from The Aviation Geek Club, The Controversial Kills Scored by Saudi F15s During Desert Storm, by Dario Leone, on April 23, 2021, located at: https://theaviationgeekclub.com/controversial-kills-scored-saudi-f-15s-operation-desert-storm/

a comic version of Pancho Villa, and beneath it I wrote "Border Banditos". Captain Nick saw it and asked if I could draw one on their vehicle. I suggested "Band on the Run" for his vehicle, but he wasn't amused.

We returned to the "Island" for the night, and were serenaded by the 7th Cavalry moving around to our rear. They were probably over fifteen miles away. We could hear the engine noises and the clanking of the treads, but we couldn't see them. Over in Iraq there were numerous flares and large explosions. This may have been from the AC-130 Spectre gunship that patrolled the border during that time. As I fell asleep, I saw the oil wells blow up again, sending plumes of flames shooting high in the sky.

The primary bombing targets for the night of January 23rd were the Iraqi aircraft bunkers. Since the Iraqi Air Force had reduced their missions to just a trickle, planners in the Pentagon thought that they were planning to do a massive strike against Tel Aviv and Riyadh. Though such an attack would prove to be suicidal, it may change public opinion in the same way that the Viet Cong attacks did during the Tet Offensive in 1968. The best way to destroy an air force is on the ground, and waves of coalition bombers concentrated on the reinforced bunkers holding the Iraqi MIGs. Their weapon of choice was the "Paveway" GBU-27, a 2,000-pound laser guided bomb. That night's mission was so successful that Saddam Hussein ordered his remaining aircraft to fly to Iran, where they would sit out the war. To make sure the coalition forces did not do bomb strikes in their country, Iran gave assurances to Washington that the aircraft would not be used during the war.[105]

At 2300 I saw two large missiles launch, and they looked like they were coming our way. I woke everyone up who was sleeping in our vehicle, and told them to get ready to take cover. The missiles peaked, and continued flying to the south. It was hard to tell what might be a SCUD missile, going to Dhahran or Riyadh, and what was a FROG 7, coming straight for us, until the missile reached a certain altitude. The two SCUDs headed to KKMC, but didn't cause any damage because the Patriot missile batteries intercepted them. I could see the explosions when the missiles hit from our location. I called in a RED DRAGON and we were told later that the Air Force took out the launchers.

---

[105] Atkinson, *Crusade* pp 157-158

Sleep for us was constantly interrupted by either artillery or missiles, and we all slept very lightly, if at all.

During the war the largest PSYOPs mission in military history was conducted trying to convince the Iraqis to surrender before the Ground War began. Starting on January 12th millions of leaflets were dropped on Iraqi positions. They were delivered by rolling them off the ramp of MC130 aircraft, or dropping them in "bullshit bombs" from B-52s, F/A-18s or F-16s. The bombs would detonate above the Iraqi trenches and shower them with the leaflets. The "intimidation" leaflet would warn Iraqis of a B-52 strike or the dropping of a 15,000 pound "daisy cutter" bomb. After the massive bombs dropped, more leaflets would follow telling the survivors they now had a second chance and needed to surrender. Other leaflets illustrated the effects of war on the Iraqi's family back home, or they would show a welcoming American soldier giving food and protection.

Leaflets dropped on Iraqi positions.

Triple Canopy

A local smuggler was hired to drop twelve thousand bottles off the Kuwaiti shore. Inside the bottles were leaflets showing Marines hitting the beach while riding a giant wave. Fifty thousand cassette tapes were also smuggled into Kuwait, playing popular Arab music interspersed with anti-Saddam messages. Though skeptics thought the leaflets were not effective at the time, after the war 70% of the Iraqis that surrendered said the leaflets had been a factor in their decision to surrender.[106]

On January 24th both our DMVs drove over to Captain Nick's position, picked him up, and then headed into Ruqi. Captain Nick told us that they had gone into their crushed Iraqi OP and found some leaflets, a helmet, three AK magazines and some Saddam posters. The reason we were going into Ruqi was to get the "fills" for the KY-57 encryption device and we needed the most current codes. Captain Nick got a new satellite dish, to replace the one that had been damaged on one of their runs to the rear. The SOTA Team got mail, but none of us did. We had ripped off chunks of cardboard and made our own postcards to home, and sent them out. In a combat zone soldiers didn't need stamps and just had to write "free" in the corner where a stamp would go.

There was still no food or water resupply and hadn't been any since the war began. Major Richardson was there, but he had been gone the last two days doing a CSAR mission to rescue a downed pilot in Iraq. He told us that the newest information about what mission we would have next is that we would be a DA team or go to Hafr el Batin and be on a combined Saudi/American reaction force. He called it a MIKE Force, after the teams used in Vietnam. We were told we wouldn't be going to train the Kuwaitis or back to our own company, which suited us fine.

While I was there, I talked to Nikolai and he told us about life over on OP-3. Nikolai said they lived on the berm, and had dug into it for protection. Every fifty meters there was an Egyptian OP. They had picked up a busload of Iraqi deserters the night before. Five of them were Iraqi Rangers who were captured by the Border Patrol beside Al Awja. When we were busy trying to rescue the missing Special Forces, the Iraqi Rangers tried to cross our sector. That Special Forces team was still missing and no one knew what had happened to them.

Before we left Ruqi we picked up a truck with Saudi Special Forces soldiers that would be working with us. MAJ Star____ had come to Ruqi in a Toyota truck with two guys from ODA 311. We were to escort them to the Saudi recon APC that was near our position. We were escorting them so the 7th Cavalry

---

[106] Atkinson, *Crusade* pp 336-337

Chapter 3: SPECIAL FORCES – A Storm in the Desert

wouldn't fire them up, thinking they were Iraqis in their Toyota. When we drove to where the Saudi recon vehicle had been, we didn't find anything. We drove even farther back and found a Bradley fighting vehicle from the 2nd Cavalry. They were surprised to see us, and didn't know that there were any friendlies in front of them. Surprised by that information I linked up with their company commander and told him all about the OPs to his front, and about the Saudi Border Patrol. I told him that if he wanted, he could drive up to the border with us and check out the locations, but he declined, saying he had to have permission first.

We left the Americans and continued down to the berm with the Saudi SF, watching for anything suspicious. It was quiet up there on the berm, with no signs of life. Over in Captain Nick's area two Cobras flew over his position and landed nearby. Mastrovito tried to get them to attack to Iraqi OPs in the distance, but the Cobra pilots told him that they could only attack if they were shot at.

That night I saw a new weapon used against the Iraqis. High up in the sky I watched cruise missiles being launched from B-52 bombers. When they launched there were long streaks of yellow flame coming from the B-52 wings and then the cruise missiles took off. On BBC we heard that the coalition forces had captured an island in the Gulf. The name of the island sounded like "Haruuf" island. It started raining again in the morning and it was so cold that everyone slept in the DMV that night.

When an Iraqi minelayer tried to emplace mines near Qarah Island, it was sunk by the guided missile frigate USS *Curts*. Helicopters were dispatched to rescue the Iraqi crewmen of the minelayer, but Iraqi forces on the island fired upon the helicopters. The OH-58 Kiowas returned fire, as the *Curts* maneuvered near the island, firing her 76mm guns. After six hours the island surrendered and SEALs from the destroyer USS *Leftwich* landed on the island and raised the Kuwaiti flag. This was the first part of Kuwait to be liberated and the first Iraqi soldiers to be captured in battle. Sixty-seven Iraqis were captured in the assault.[107]

---

[107] Extracted from http://es.rice.edu/projects/Poli378/Gulf/gwtxt_ch7.html *The Gulf War, Maritime Campaign*, on January 6th, 2012.

Triple Canopy

On January 25th we all agreed to go into the Iraqi OPs that were opposite our OP, to see what we could find. This was another situation where we weren't authorized to go into Iraq yet, but after hearing about how CPT Nick went into his Iraqi OP, we figured we would too. The first OP we went to was a small one, which was just a shack out in the desert. There was no movement at all, so K.C and I got out of the DMV, crossed over the berm and walked into the hut. Wally stayed with the vehicle, on the gun, pulling security. Inside there was a safe that I was able to crack, with some ledgers inside. K.C. read them and said that they were from August, so they wouldn't have much intelligence value. We snatched a chemical decontamination kit that we would later turn over to the S-2. There were about a dozen pictures and posters of Saddam around the walls and a green Iraqi beret. Outside there was a trailer with an unassembled radio tower in it.

We left that small OP and drove over to larger Al Ulaymah/OP 6. The place was trashed and looked like a dump. There were old batteries and MREs all over the floor. We figured the Iraqis who were across the border had eaten all of our abandoned MREs when they occupied the building. There were big holes in the walls from when the Iraqi tanks attacked the OP.

I wanted to see the where the artillery rounds impacted that almost got K.C. and me two days earlier, so I walked over to the shell holes from the Iraqi artillery strike, and picked up some pieces of shrapnel. I gave everyone on the team a piece of the jagged steel. We picked up some food, batteries, coffee and maps from our abandoned OP and then returned to the "Island".

The fighting along the border that day was almost nonexistent, which we figured was due to the bad weather. We spent most of the day writing letters home on pieces of paper from the ledgers we had found in the Iraqi OP. Supposedly there was going to be a mail run the next day, and we wanted to send some notice to our people back home that we were still alive. One of the Saudi SF cooks, that was the spitting image of Louis Gossett, Jr., cooked us a hot lunch of rice with butter. It tasted great!

It rained all afternoon, but finally quit after the sun went down. That night we had a front row seat to another massive SAM missile launch aimed at the B-52s soaring overhead. We got word on the SATCOM that the Iraqi ADA would be "weapons tight" (turned off), for 30 minutes, so that their MIGs would be able to come through. Due to this, we had to prepare for an attack by enemy aircraft. We figured we were "hidden" as well as possible inside the Island. No aircraft appeared though. I wondered when the last time any American soldier had tried to hide from enemy aircraft. Vietnam? Korea?

Chapter 3: SPECIAL FORCES – A Storm in the Desert

We reported on two SCUDs launched at Israel and five that were launched into Saudi Arabia. One of the SCUDs that hit Saudi Arabia killed one person. The Marines, way in the eastern sector, fired several rounds at Al Jaber Air Base during an artillery raid. We could see the flashes from their large guns across the horizon. The Egyptians, not to be left out, fired their artillery, and it landed near Gator's position on the berm. As usual, the friendlies were more of a danger to us than the enemy.

In the morning we drove to Ruqi, anticipating a mail call, but all that was delivered from the rear was food, water and diesel. It was great about the resupply, but we really wanted to get our mail and hear of news from home. The supplies had been dropped off on the side of the road and no one stayed with them. We figured that whoever delivered the mail, most likely MSG Cr___, didn't want to spend any additional time near the front and all the possible danger, so he just threw out the boxes and took off running to the rear.

When we returned to the "Island", we saw some American Bradleys on the horizon. It was the first time they had been close enough to be seen by us. The American Army was slowly creeping towards us. Charlie's DMV was out on a border recon, so I took my DMV back to the Bradleys to let them know they had friendlies to their front. I was a little worried about getting fired up by our own vehicles, and approached them in such a way that it would appear to be non-threatening. I also wore as much US Army gear as I could, so I didn't look like a desert pirate.

We drove our vehicle right up to the Bradley and surprised the crew anyway. I had the Bradley commander take me to their 1SG and FO, so I could show them where all the friendly OPs were on the map. They were surprised that anyone was out there and had been told they were all alone, and they were the closest American unit to the Iraqis. They said that their men had seen us moving back and forth in their gunsites, and had identified us as T-72s, BRDMs and SA-6 or SA-8 missiles. The 1SG said that he might come up to our position that night, but they were ordered to not go any further north than 29 degrees and 5 minutes. That was the "north line" for them, which was located about where Al Awja was.

After Charlie returned to the "Island", we saw an AH-1 helicopter circling nearby. Charlie had found an entire case of leaflets that had landed in Saudi Arabia and he put it on the ground, trying to make a signal to not shoot. The Cobra kept lining up on us, like it was going to do a gun run. We all began to wave our arms, trying to get the helicopter to realize that we were friendly, but it was pretty far away. I wanted to get something orange, so I grabbed a six pack of orange sodas that we had "liberated" from the store in Asher Kilo, and held it over my head. I climbed onto the tallest part of the Island and gave a peace sign with my other hand. It was a tense moment, and I was ready to be blown away by the attack helicopter's minigun. The Cobra flew in closer, hovered within 100 meters of me, and then flew slowly by. I signaled to him to land, and he brought the chopper down beside us. When the helicopter landed the leaflets blew all around us like snow in a snowglobe.

When the pilot cracked open the window, I told him who we were. He, like everyone else so far, was surprised that we were there and had not been told about any Special Forces in the area. I showed the pilot the map and pointed out all the friendly locations, and then pointed out suspected enemy targets. He told us "Good luck" and then flew on, moving to the east.

K+12, THE CAVALRY ARRIVES

The next day, Sunday, January 27th, we were listening to BBC, as a misting rain fell around the vehicle. The announcer told us that it was Superbowl Sunday. We didn't even know who was playing, and didn't really care. We had our own competition to deal with.

Chapter 3: SPECIAL FORCES – A Storm in the Desert

We got a radio message that told us that the night before, Captain Nick's guys went into the destroyed Iraqi OP, across from OP-6, and found ten bunkers around the OP. They picked up an RPK machinegun and an RPG and called us asking if we wanted to assist in the salvaging. We drove over to the OP, and went through the bunkers, many that had collapsed. We found another RPK, two AKMs, a 60mm mortar and another RPG rocket launcher. We also found crates of ammunition for all these weapons. When some artillery blew up, miles away on the horizon, it spooked Captain Nick's guys and they wanted to back off, so we withdrew with them.

We finally got our mail on that Super Bowl Sunday! A second truck came up from the rear and met us on the road. There were letters from all of the different people I wrote to, and there were cans of food from my mom. We didn't have a food shortage anymore, since we got the MREs, but anything besides MREs was always great. When we got the mail, we learned of another Special Forces team that had gone deep into Iraq, and had been compromised. They were outnumbered, but were able to fight their way out of there and all of them make it back alive.

We also learned that the "missing" Special Forces soldiers over on OP-7 had finally been found. Their radio had quit working and their extra batteries had not worked either. They decided to leave and drive to the rear to get another radio and batteries. No one knew this and assumed they had been captured, until that team finally showed back up in the rear.

It was too cold to do much of anything where we were located, except watch the horizon. Later in the day I did escort the 2nd Cav FO Bradley vehicle to our location to look at some of the targets we had been watching. The main target we wanted to get rid of was an Iraqi OP in the wadi with over 40 Iraqis entrenched around it.

After the sun went down, I listened to OP-1 get ambushed while I was on radio watch. When they drove up to their old OP a squad of Iraqi Special Forces fired on them with RPG rockets and AK-47s. The Special Forces DMVs were hit several times with small arms fire, so they returned fire with their .50 caliber machineguns. The heavy bullets tore through the walls of the Saudi OP and drove off the Iraqis. None of the Special Forces soldiers had been hurt.

Pat Ballogg, the Team Sergeant of ODA 553, later wrote of the fight at OP-1, "We had a few alternate positions we'd take nightly for obvious reason. Half

403

of the team reconned forward in a Hummer with one Saudi Land cruiser and the Saudi Special Forces, and came under fire enroute to the OP. We then moved forward with our vehicle and put .50 cal fire on the Iraqi's....they beat feet to a vehicle hidden behind the burm and moved out. The funny thing was the Saudi LT had been all Muslim touchy feely about his "brothers" across the way, and had gotten out of his vehicle to approach them, when they opened up on our element (he thought they'd come to surrender as had others) ...I got out of the vehicle to police him up and he was screaming "They shot at ME! they shot at ME!" over and over again ... after that he was all "we gotta kill these mad men."[108]

That night I was able to kill a SCUD missile. At 2130 I was sitting in the driver's seat, doing radio watch, and trying to stay dry. On the far horizon I saw the bright flame of a missile launch, and it headed up into the stratosphere. I quickly shot an azimuth with my compass and then called in a RED DRAGON. Other stations in the area called their azimuths in too. Within five minutes, jet aircraft came screaming overhead and flew through very heavy anti-aircraft of the SAM missile launches. The site where the SCUD launch had come from was decimated with multiple explosions. There were secondary explosions for minutes afterwards. The pilots told us on the radio that they had got the SCUD.[109] After that it was quiet on the Iraqi front. For the rest of the night, it was so cold that my two canteens, draped on the outside of my vehicle, froze solid.

On January 28th our OP decided to go check out the Iraqi OP that we had attacked the week before during the "Great Puppy Raid". We watched the OP Al-Katsiyah from the berm and didn't detect any movement at all. None of the antennas had been repaired and the trenches looked like they had been abandoned. We left Gentz and Wally on the guns to overwatch us, and the rest of my Team crossed over into Iraq to see what was in the building.

The place was abandoned and there wasn't much of anything there, just a few RPG rounds that we took with us. I checked out the damage that we had done during our attack. The Mark 19 had blown the roof off of the radio shack and had knocked a hole in the base of the main building. The M16s peppered the walls with holes. The .50 caliber had missed most of the building, but where

---

[108] Email message from Pat Ballogg on 7 January 2011.
[109] After the war we learned that many of the SCUD missiles that were "destroyed" were not SCUDs, so I do not know whether this was really a SCUD missile or not.

it had hit, there were chunks torn out of the wall. We took the RPG rounds over to Captain Nick, since he could use them with his captured RPG.

After we left Captain Nick's position we decided to go up to their OP on our own. Every time we had gone up there with them, they quickly left when any artillery landed nearby. Due to this, the OP was never fully searched. We decided to do our own looting, to see if we could find some good intelligence or weapons. It was good that we had done this, because we found a huge stash of weapons and ammunition in the trenches and put them in our DMVs. There were cases of AK ammunition, 60mm mortar rounds, grenades, RPG rockets, gas masks and chest pouches full of loaded AK magazines.

I walked along the trench line, looking into the crushed bunkers, and found the barrel of an AK-47 sticking out of the sand. The other guys had missed it because it was in a collapsed bunker, with a crushed roof over it. When I yanked on the weapon it wouldn't come out of the sand. I dug down farther, into the sand, to loosen it up and there was a hand holding the AK. I jerked it back and it came free from the dead Iraqi's hand. This would be my "driver weapon". I figured the dead Iraqi had been caught in the bunker when the Iraqi tank spun around on top of it, or he had been killed when CPT Nick's Team had fired up the OP a few days earlier. We dropped off the RPG and mortar rounds at Captain Nick's location and then headed back to the "Island". Since Nickolai had been a mortarman before coming to SF, and since his Team had found a 60mm mortar in an Iraqi trench, it made sense to give them all the mortar ammunition.

When we returned to the "Island", I cleaned up the AKM. It took me awhile to clean it and I had to use gasoline and steel wool to get it functional. I also cut off the wooden stock with my Swiss Army knife saw blade and then smoothed it out with the file. I know that a weapon without a stock had no real way to aim it, but I was using this as a "break contact" weapon, and mainly wanted it to stick out the window and go full automatic if we were ambushed. Our DMVs were already full of ammunition, food and diesel, so we cached all of this newly found ammunition at the "Island".

One of our goals was to get that Iraqi truck that resupplied each enemy OP. The Iraqi truck still moved down the road paralleling the berm, but only at night. What we needed was a mine that we could put on the road it the Iraqi truck used. I decided to go back to the American Cav unit that was near our position, and ask to borrow an anti-tank mine. We drove our DMV back to one of the Bradley vehicles, and started bartering with the crew. I told them I needed a mine, just one, and we had some stuff to trade for it. They asked me what I had, so we threw some berets, gas masks and spent T-62 tank rounds onto the ground.

Triple Canopy

Our appearance was bizarre, to say the least. I stood there, wearing my fadewa, black ski mask, rolled up to make me look like a Pakistan street sweeper, my kafiiya, goat fur vest and the cut down AK-47. We were covered in several weeks worth of dirt and grime. Our DMV was painted with "Border Banditos" and there was a Saudi flag flying from the antenna. Katzia, our puppy, looked out the window while he sat in K.C.'s lap. The Bradley crew just stared at us like we were from Mars.

Their platoon leader came over and I explained the situation to him. The lieutenant told us that he couldn't give us any mines unless the company commander authorized it. I asked him where his company commander was, but he had left to do a reconnaissance. Realizing I would get nowhere with the regular army, we drove off, back into the cold desert. I always wondered what those tankers thought about that encounter with the odd-looking soldiers who wanted a single anti-tank mine.

In the afternoon we heard firing down in the wadi and drove there to see what was going on. We discovered two new Iraqi positions with about a platoon of men in each location. I had K.C. park the DMV on a hill in the wadi, where we could keep an eye on them, and then I tried to call in an airstrike using the SATCOM radio. Unfortunately, the SATCOM didn't want to work that afternoon, so we had to let the opportunity go.

As we patrolled back along the border, I saw a single Iraqi soldier running away from the berm and heading north back into Iraq. I told K.C. to stop the vehicle, and as he skidded to a halt, I grabbed the M-24 sniper rifle mounted to the inside roof of the DMV and headed to the berm. The Iraqi was over 800 meters away and the wind was blowing steady from left to right. The Iraqi continued running until he reached a building beside the pipeline that carried oil into Saudi Arabia from Iraq.

As he stood there, trying to catch his breath, I decided to take the impossible shot. I really didn't expect to hit him, but it would scare the heck out of him and let him know that there were hunters on the border. I led him by about two feet, to compensate for the strong wind, and fired. The distance was great enough that I was able to bring the scope back onto the target before the bullet hit. The Iraqi was leaning up against a small building that had a solar panel on the roof. The bullet shattered the solar panel, spraying glass all over him. He hit the dirt, so fast that I thought I had hit him, but then he jumped up and took off running into the building. I told K.C. to yell at him and tell him to surrender, but there was no movement. He either couldn't hear us, or he didn't trust us. I didn't blame him for not wanting to expose himself.

Chapter 3: SPECIAL FORCES – A Storm in the Desert

That night there was a massive BM-21 barrage over on OP 7. I could hear the SF Team over there calling in an airstrike to take out the launchers, but they were having no luck. All aircraft were devoted to SCUD hunting out in the eastern desert. Charlie was over in the OP 7 area on a patrol, so we didn't know if he was safe or not.

Later that night we saw flashes of light and heard gunfire on the border near us. This also had us worrying about Charlie. The next morning, we learned that the gunfire we heard had been Captain Nick getting ambushed. He had gone out with one of the SOTA guys, Gary Rodeffer, riding with the Saudi Special Forces in their Toyota.[110] As they approached the berm, Captain Nick thought he saw someone and told the Saudis to stop so he could check it out. He thought it might be another group of deserters coming over to Saudi. He and Rodeffer guy got out and walked toward the berm. Right then an Iraqi rose up and fired an RPG at the Toyota. The RPG flew straight at them, but it exploded on a small mound of sand in front of the truck, showering the vehicle with fragments and shattering the Toyota's front windshield.

AMBUSH AT AL-ULAYMAH JAN. 27, 1991

---

[110] Facebook post from Nick Mastrovito on 27 April 2021.

Triple Canopy

The Saudi SF soldier on the back of the truck tried to fire his .50 caliber machinegun, but it jammed. Rodeffer fired his M16 and then ran back to the vehicle. Unfortunately, he had on his NVGs and didn't see Captain Nick, and he slammed right into him, sending the SOTA soldier sprawling in the sand. Captain Nick fired a long burst from his captured RPK at the Iraqis on the berm. He was wearing his fadewa, and Rodeffer described him as looking like an old west gunslinger wearing a duster.[111]

One of the other Saudi SF soldiers had fired a few rounds from his Steyr AUG, and then tried to get back into the truck, holding onto the open door as it began to back up. Captain Nick jumped into the back of the truck as it began to move away from the berm, and then they drove off as fast as they could. The Iraqis behind the berm all opened up, sending tracers flying high over their head as they raced to safety.

It turns out that the Iraqis knew that there were Special Forces patrolling the berm, and they wanted us badly. We were later told that there was a price on our heads, and any Iraqi that could capture one of us would get a 10,000 Iraqi dinar reward.[112]

Earlier that day my DMV was on patrol along the berm, watching for any signs of Iraqis trying to lay in an ambush before night came. The pipeline had been a pretty active area for enemy movement. As we approached the pipeline, we saw about a dozen Iraqis take off running to the pipeline building. We sped towards the berm, wanting to get a shot at them, but by the time we got there they were safely behind the building. This was the same building where I had shot at the lone Iraqi two days before. We weren't allowed to pursue the Iraqi soldiers, since the ground war had not started yet. Even if we had wanted to, we couldn't because the berm was in the way.

K.C. and I got out of the vehicle to see what they had been up to. They were part of an ambush that hadn't been executed because we spotted them. We found an RPG, four rounds with the safety caps taken off of them, and a cleaning rod to extract any missiles that would misfire. We figured that the Iraqis had seen us coming in the daylight, and decided to do a long range shot with the RPG. When the soldier pulled the trigger, it just fizzed and didn't go off. He then had to remove the old round and put in a new one, but it was stuck. I think that was when they realized that we were getting too close, and they took off

---

[111] The fadewa was a heavy wool coat that looked like a bathrobe. We had been given them to us by the Saudis for protection against the cold.

[112] About $10,000 US dollars.

408

running to the pipeline house for cover. So, we had an angel on our shoulder for that patrol.

We headed back to the east towards the wadi, when the monotony of the patrol was shattered by a huge explosion in front of the berm. This was the biggest explosion I had seen in the daylight, and it was only about 400 meters away. It rocked our vehicle. The other vehicle with us sped to the rear, but we stopped our vehicle and watched the horizon to locate what was shooting at us.

K.C. hit the gas and raced towards the smoking crater in the berm, telling me "It's like lightning... it never hits in the same place twice!" I looked at him like he was crazy, and bent down, putting on my helmet and flak jacket. This was one of the few times that I put that protective gear on. After I had the gear on, I bent over so that the glass from the DMV's windshield wouldn't cut my face up when the incoming artillery hit around us. As I sat there bent over, K.C. asked me what I was doing. I told him that I was trying to duck down so that the shrapnel wouldn't tear me up. He laughed, and said, "If it hits near us, it won't matter anyway." I realized he was right, sat up, and waited calmly to be blown to smithereens.

Luckily for us there was no second attack. I never did find out what was fired at us, and I have always suspected that it was a bomb from one of our own aircraft, mistaking us for an Iraqi vehicle. Since our vehicle stopped, and none of us ran out of the vehicle in a panic, the pilot might have thought he killed us with the concussion.[113]

We continued on our patrol and drove down into the wadi to see if the Iraqi positions were still there. The Iraqis hadn't left, and now had been reinforced. There was now a battalion of Iraqi soldiers at the wadi OP, and there were supply trucks coming and going. This target was too good to pass up, and we drove to Ruqi to try to find the 2nd Cav LNO (Liaison Officer) to see if they could fire up the wadi with their artillery. He wasn't there, and the team at Ruqi told us that there might be some artillery available in a few days. All the artillery and aircraft had been diverted to a higher priority mission that we didn't know about. The team at Ruqi said that they had the same problem. A single T-62 Iraqi tank would show up, fire a round into Ruqi, and then disappear below the ridgeline. It was beginning to annoy them.

---

[113] Many people watching the war on CNN must have thought that the United States were mainly using laser and TV guided bombs. However out of the 227,000 munitions dropped on the Iraqis, less than 4,000 were "smart" bombs. The rest were the same type of bombs dropped since WWI, what are known as "gravity" or "dumb" bombs.

Triple Canopy

We drove back to the "Island" and stopped the patrols for the night. There was too much Iraqi activity to our front and too much friendly activity to our rear. We figured we were in the no-man's land and the chance of us getting hit by one side, or the other, was getting to be a likely thing.

As we watched the horizon an extremely loud roar exploded from behind us. It was our own multiple rocket launcher vehicles, the M270 MRLS, and it fired all twelve of its 227mm missiles into the wadi where the Iraqis were. Someone had finally listened to us and answered our pre-planned strike! The wadi erupted with a violent rumbling noise that sounded like God's was rolling bowling balls down the wadi. This was the sound of over 500 exploding submunitions, from each rocket, bursting among the Iraqi positions. Who knows how many Iraqis we had killed with that strike.

The next morning, we heard the rumbling of the approaching Bradley vehicles. They moved up to our sandbagged island and then stopped when they came on line with us. There were now Bradley fighting vehicles spread out across the horizon. The U.S. Army had finally caught up to us on January 29th, or K+14. We made coffee with a small stove I had carried through two wars and shot the breeze with the Bradley crew closest to us. They were amazed, like everyone always was, that we were out here. They thought they were the only ones at the front.

As I drank my coffee and ate a cold MRE I saw another dogfight overhead. Two jets chased each other, making serpentine vapor trails in the early morning sky. One launched a missile at the other, but I didn't see it hit. The one being chased punched out flares every few seconds, marking his trail in the sky. They continued on until I lost sight of them.[114]

---

[114] There was only one Iraqi aircraft shot down on January 29th. Captain David Rose, of the 58th Tactical Fighter Squadron, flying his F15C, shot down a MIG-23 with an AIM-7 missile as the MIG tried to escape into Iran. This was probably not the dogfight that I had watched. During the war the Iraqi Air Force lost 39 aircraft, fourteen of them in air-to-air combat. It was reported that no coalition aircraft were shot down by Iraqi aircraft, however Iraq claimed that two pilots, Zuhair Daywood of the 84th Squadron and Jameel Sayhood of the 9th Squadron, shot down one aircraft each. Daywood supposedly shot down CPT Scott Speicher F/A-18C on January 17th, with an R-40 missile. Speicher's remains were not found until August 2009 after the second Gulf War. Sayhood was reported to have shot down Royal Air Force pilots Gary Lennox and Adrian Weeks in their Panavia Tornado on January 19th but official records show that those pilots died on January 22nd. Sayhood was shot down that same night. There were no "aces" in Desert Storm. An ace is a pilot who shoots down five aircraft. The highest number of kills were Thomas Dietz and Robert Hehemann, of the 53rd Tactical Fighter Squadron, flying F15C

410

Chapter 3: SPECIAL FORCES – A Storm in the Desert

That day we didn't do any patrols, since the Bradley parked on either side of us could cover the whole border with their 25mm chainguns. We spent the day sleeping, cleaning weapons, and entertaining a question or two from curious soldiers from the 2nd Cav. We enjoyed being able to let most of our guard down, since sleeping between two Bradley vehicles, a few hundred yards on either side of us, meant we were the least important target.

That night all hell broke loose along the border. In retrospect I think Saddam was trying to start the ground war on his terms, and did a general push along the avenues of approach on the border. The biggest fight was over on OP-1's sector at a town called Khafgii. Fifty Iraqi tanks crossed the border and took on US Marines and Saudi Armor. That night Iraqi tanks also crossed the border and entered Ruqi, firing their main guns and machineguns as they approached. The Special Forces in the town evacuated their positions and moved back to the gravel pit. One of the SF soldiers fired an AT-4 at an Iraqi tank, but the round slammed into the building beside it and set it on fire. The withdrawing Team got on the SATCOM and called in an immediate airstrike onto the Iraqi armor. The A-10 sharks turned their noses to Ruqi and searched for their prey.

Further to the east an AC-130 Spectre gunship, *Spirit 03*, lingered over the border too long and was shot down by an Iraqi surface to air missile. The gunship had its wing torn off and it spiraled into the gulf, killing all fourteen on board.

Gator and his team that were dug into the berm, saw the flashes of light and heard the explosions in Ruqi. The Egyptian commander nearby also saw the fighting and had his Soviet tanks move to Ruqi to engage the Iraqi Soviet armor. Gator had heard the radio exchange between the SR CP and the A-10s and realized that the pilots would not know who was enemy and who was friendly since both sides were using the same Soviet tanks.

Eagles, who both had three kills. The United States has not had an ace since the Vietnam War. Cesar Rodriguez, of the 33rd Tactical Fighter Wing, shot down two Iraqi MIGs. The first was the MIG-29 of Jameel Sayhood. The second was a MIG-23. Rodriguez shot down a third enemy aircraft during the war in Kosovo, in 1999. He shot down a Yugoslav MIG-29, which gave him three victories. Though he was not an official ace, Rodriguez is sometimes called America's last ace, due to having the most victories.

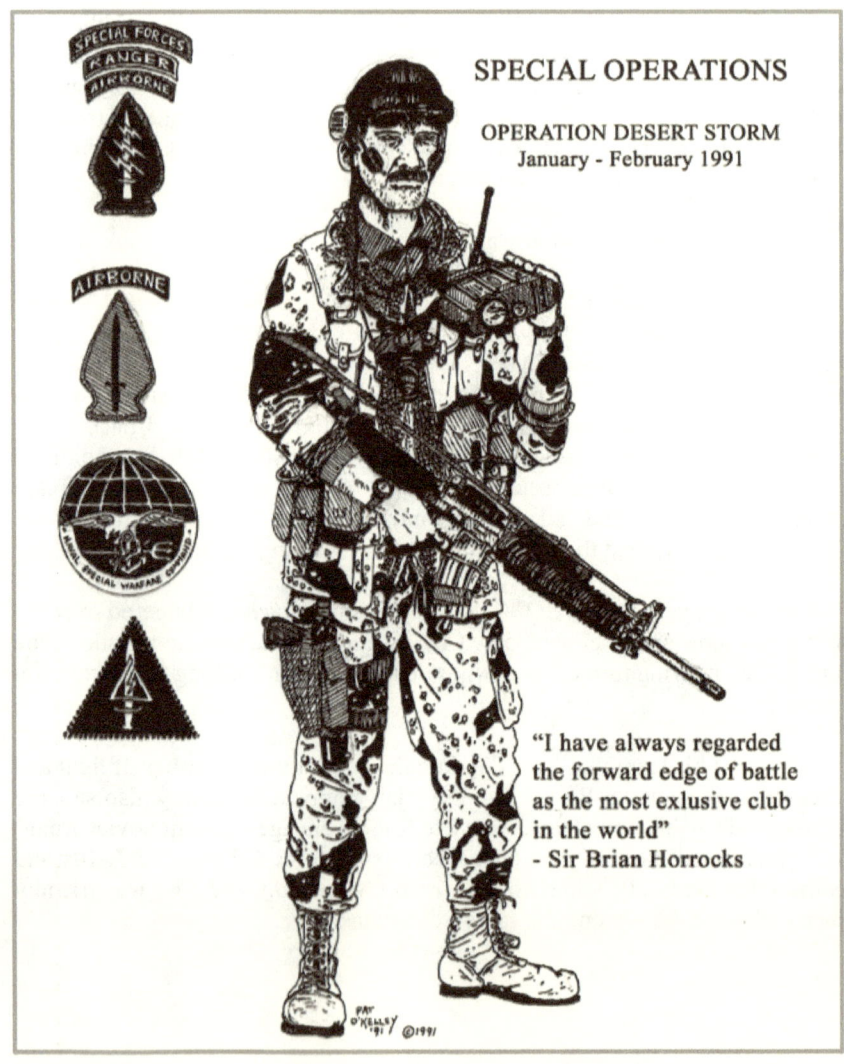

SPECIAL OPERATIONS

OPERATION DESERT STORM
January - February 1991

"I have always regarded
the forward edge of battle
as the most exlusive club
in the world"
- Sir Brian Horrocks

Gator hitched a ride on the Egyptian commander's tank, and rode it into battle, holding up his flashing strobe light to identify the friendly tanks. The Team in Ruqi, hearing Gator on the radio desperately warning off the A-10s, called the pilots and had them abort the mission. There were too many things that could go wrong. The Iraqis may have realized that an armor column was heading towards them, or they may have only attacked halfheartedly, because the enemy tanks moved out of Ruqi and headed down the road, back into the darkness of Iraq.

Chapter 3: SPECIAL FORCES – A Storm in the Desert

To the east, near Al Jabo (OP-1), a Saudi National Guard battalion observed Iraqi tanks crossing the border into Saudi Arabia, but they had their turrets turned to the rear. The Saudis thought that this might have meant they were surrendering. The Saudis attempted to negotiate with the Iraqis, and convince them to surrender, but after two hours of talking, a firefight erupted and the Iraqis retreated back into Kuwait.[115]

At the same time, at Al Jabo a split detachment from ODA 553, commanded by Captain Dan Kepper, had to abandon their OP when they came under intense firepower as the Iraqis stormed their position. The Team Sergeant, MSG Pat Ballogg, wrote that it was the "Iraqi Division across the way that came over nightly (looking for food I'm pretty sure)". Ballogg wrote that they had many exchanges of gunfire with the Iraqis, before that night, due to the Iraqis firing artillery at them. Ballogg wrote that the Iraqis occupied the post "30 minutes after we exfiled." No equipment was lost, but Ballogg's vehicles had been damaged by machinegun rounds. Ballogg wrote "you know, in one side out the other...was a good experience for my young, never been fired at guys...actually squared away a young cocky medic that was on loan from another ODA...all of a sudden he was in 'the shut up and listen mode." [116]

During that chaotic night we received a message that we would be heading back to the rear, and the SR mission was now completed, since the regular Army had finally caught up to us. The next morning, we drove over to the K-10 (Asher Kilo) gas station, where I met up with the 2nd Cavalry commander. I filled him in on what was to his front and what opposition he could expect. We were not too thrilled to find out that we were going back to our own company and that none of Major Richardson's plans for us had gone into effect.

We drove south, to Hafr el Batin, stopping there to get something to eat and make phone calls home. I was able to get my dad on the phone and he told me about the fighting in Khafgii that he was watching on CNN. I told him about our fighting in Ruqi and the artillery raid. While we spoke hundreds of American tanks and Bradley fighting vehicles were moving through Hafr el Batin and heading out to the west.

Those vehicles were from the VII Corps, moving into position for the "left hook" movement into Iraq once the Ground War started. They were attempting

---

[115]   Atkinson, *Crusade* pg. 208
[116]   This may have been the same action as mentioned on January 28th. Email message from Pat Ballogg on 2 January 2011.

to move in secrecy, though their movement was noticed by everyone who drove on the Tap Line Road.[117] The movement was historical in that over a quarter of a million troops and 73,000 vehicles moved halfway across Saudi Arabia, without the Iraqis detecting the movement. Lieutenant General John Yeosock was the commander of this movement, but by mid February he could not handle the stress and the constant tirades from General Schwarzkopf, so he was medically evacuated to Germany where they removed his gall bladder.

The new commander that assumed control of this juggernaut was General Calvin Waller, and he leapt to the chance. He knew that moving the XVIII Corps would bog down the already congested Tap Line Road, so he came up with the innovative plan of practicing the "left hook" maneuver they would do in Iraq, in the desert of Saudi Arabia, as they moved to their attack positions. A force larger than the 3rd Army commanded by Patton in WWII rumbled across the desert, fifty miles wide. One officer commented that "it's like being in the middle of the Spanish Armada." [118]

To confuse any Iraqis watching from the border a grand façade was being constructed. A PSYOPs platoon drove along the border, creating smoke from fictitious tank engines and broadcasting tape recordings over loudspeakers at night, of tanks and trucks repositioning. Engineers bulldozed passageways through the border berm for tanks that would never pass through. These same Engineers built bunkers within sight of the border, that had generator powered lights emitting from them. Inflatable fuel bladders, HUMVs and helicopters were situated in the desert, along with more elaborate decoys that emitted the heat signature of an M1A1 tank. Computers broadcast coded messages to and from dozens of false divisions, while messages in the clear, in Egyptian Arabic, told of American units wanting to know where the water point could be found.[119]

As the war drove on by, we ordered a pizza from the Hotel Fao and had lunch while watching the spectacle. The hotel had several reporters staying there, and seeing that we were covered in dust and wearing some odd clothing, some of the reporters wanted to interview us. We looked like we had been living in the desert for weeks, which we had, but we turned them down. We did let one sexy French woman reporter sit with us, but all she wanted to do was argue about the morality of the war. A reporter from *Soldier of Fortune* magazine sat

---

[117] This was the two lane road running along the border of Kuwait and Iraq and Saudi Arabia, heading west to Jordan.
[118]Atkinson, *Crusade* pp 309-310
[119] Atkinson, *Crusade* pg. 332

with us, drinking coffee, and asking general questions about what we had seen up there. We didn't mind talking to him since he had been a soldier once.

While all this was going on Captain Nick and Gator drove out to the SFOB (Special Forces Operating Base) to argue our case against being sent back to our company. They were able to get us a better assignment than what we expected. We would be assigned to a Kuwaiti battalion and be their CAS team. We would be doing split ops; one half of the team would be going into the NAC (Northern Area Command) and the other half would be in EAC (Eastern Area Command). We would be attached to a different battalion, and a different B Team commanded by CPT Davis and 1SG Bishop.

While we were in Hafr el Batin we were able to pick up Armed Forces Radio and learn about the fighting in Khafgii. Though it was a bloody fight, it looked like it was all one sided, with the Iraqis getting badly mauled. Twelve Marines had been killed in the fighting, but most of them had been killed by friendly fire when an A-10 had fired a Maverick missile on their Marine LAV-25. Thirty of the Iraqis had been killed and 466 were taken prisoner. Those who surrendered were marched to a prison compound with grins on their faces and waving to the cameras. Nineteen Saudi and Qatari soldiers had been killed and 36 were wounded. There was a total of 25 Americans killed in the two days of fighting, and most of the killed in American and coalition forces were from friendly fire. We were the deadliest force on the battlefield, even to ourselves.

Friendly fire casualties in Desert Storm were higher than any other conflict that the United States had fought in. During WWII there were about 15,000 killed and wounded due to friendly fire, or less than 2% of the 774,000 casualties. During the Vietnam War the number rose to 3%, or 1,326 casualties. However, in Desert Storm 24% of the casualties were caused by their own forces. One theory on why this number was so high was because in previous wars soldiers accumulated battle experience over time, which prevented them from firing on their own troops. Desert Storm, in contrast, was a fast "blitzkrieg" that lasted only six weeks and that combat experience didn't happen.[120]

---

[120] After the war a major effort was made to make sure that soldiers on the ground and pilots in the sky would be able to tell friend from foe. These anti-fratricide measures included wearing infrared patches on uniforms, and having thermal signaling devices put on vehicles. The most significant development had been the use of satellite tracking for

Triple Canopy

After the battle of Khafgii a new confidence arose in the coalition forces. Prior to this, the news media estimated that 10,000 coalition forces would be killed in the first few days of the invasion against the "world's fifth largest army." Khafgii proved that the Iraqi army, though brave, could not do a coordinated attack, and due to the lack of airpower, could be easily destroyed if they go on the offensive. The Iraqis also proved that if they were hit hard enough, they would quickly surrender. The coalition Arab armies, who had been planned as follow-on forces, now pressured Schwarzkopf to let them have a more direct role in breaching the Kuwaiti border. Schwarzkopf agreed, since it allowed him to free up thousands of Marines who could be concentrated at more significant targets, such as the Kuwait International airport and the high ground near the Bay of Kuwait.

When we neared KKMC we passed by hundreds of empty tents that had become a ghost town. We were moving pretty fast along the highway, when a Military Police HUMV chased us down. Our vehicles were looking pretty battered, with one of the DMVs missing a rear panel, and another had some bullet holes in it. We had a variety of weapons on board and a mix of Saudi and American equipment. The MP got out of his vehicle and approached ours, looking it over like it was from another world. When he came to our door, I asked him what he wanted. He told me that our weapons were loaded. I told him "No shit! You know there is a war out there with people wanting to kill you." He told me that no one was allowed to go into KKMC with loaded weapons and he also told us we needed to slow down. It was too much for us and we all busted out laughing. We drove off, with the lone MP staring after us, wondering who we were.

When we drove back into KKMC we had a bit of a culture shock. We had been living on the border, with Iraqis searching for us, and enduring artillery barrages from day to day. However back at KKMC there wasn't the same sense of danger and urgency that we had lived through while in the wadi. When we drove through the gate, we cleared our weapons, and then we were told by the gate guard we needed to have our Kevlar helmets on at all times. We laughed at this because we had been the closest soldiers to the Iraqis for weeks, wearing only our wool caps on our heads, but now, hundreds of miles to the rear, we had to wear helmets. It took us awhile to find our helmets, but we finally put them

---

vehicles and for people, allowing commanders to see exactly where their soldiers are at all times.

416

on. They felt strange and uncomfortable. We were also told that we had to wear our LCE and mask everywhere we went.

Back in our building not much had changed, though there weren't many of the A Teams there. The teams were all out with their Arab counterparts getting ready for the Ground War. There was a television in the messhall that was turned to the CNN channel at all times, and we watched the war happening without us. We also were able to get AFN (American Forces Network), and listen to a play-by-play action of the battle of Khafgii, like it was a football game.[121] Two days earlier we were waiting to be ambushed by the Iraqis, and now it had become the least of our worries.

The following day Gator briefed us on our mission. We were to be assigned to the NAC, with a newly organized Kuwaiti brigade. We were going to be with ODA 584, the HALO Team of B Company.[122] I was going to be the head advisor for the Kuwaiti battalion commander of the 57th Battalion of the Khulud Brigade.[123] My team would consist of Gentz, Wally, Baby Doc and Rat.

The Kuwaitis we would be advising were brand new, right off the street civilians, and had no training or equipment. We only had a matter of weeks to train them to take back Kuwait City, which was expected to be fierce house to house combat. The Kuwaitis would be issued M16A2s, RPGs and German MG-3 machineguns for weapons. I was going to take over the job as advisor from SFC O'Dell, who I had gone through the Q-Course with. Later that day I bumped into Chad Klotz, and he told me he would be going into Kuwait City too, as part of a radio relay station. Desert Storm continued to be a military reunion for me.

We began training the Kuwaitis by sitting down with the Kuwaiti commander and determining what he needed done. They had a soldier that had

---

[121] AFN was created in WWII as an American counterpart to the BBC for the American forces in England preparing for the Normandy Invasion. It is entirely run by the US Armed Forces, with the DJs, the reporters and the station engineers being all military. AFN also has a television network in countries around the world where US forces are stationed.

[122] MAJ Phelan, the commander, once described B Company's personality when he said "I could call a formation right now for PT and this is what you would see: 585 would be out there in their beach jams, Oakleys and sun screen, '84 would be in MOPP 10 with rucks and weapons, '83 would be in muscle t-shirts and weightlifting belts, '82 would be in Army gray PT uniform dress right dressed and 581 would show up in bathrobes, shower shoes and coffee cups in hand.".

[123] Khulud meant life, forever living, or immortality. We would be going into battle with the Immortals.

been in the Army before they were driven out Kuwait, but most had never served a day in the military. One night after training the Kuwaitis, a few of us went to the "Wal-Mart" to get something to eat. When we walked in, I saw two beautiful girls coming out to their HUMV. Crowder and I both followed them to their HUMV and talked to them, finding out they were living in a tent city outside of KKMC and were from the 57th Casualty Collection. Later that night we decided to go and find the girls, and decided to take our puppy, Katzia, to break the ice. One of the guys also had a bottle of Everclear that had been sent to him from the States in the mail, camouflaged as a shampoo bottle.[124] Several of us roamed around the desert, going from tent city to tent city, but never found the girls or the 57th CC. There were over 200 support units spread out over 50 kilometers in the desert around KKMC.

We were able to find one casualty unit, basically a morgue, and Crowder was able to get each of us a body bag from them. It was kind grim, but we thought that each of us carrying around our own bodybag was funny as hell.

The "every day" living in the rear echelon still felt odd. We wore more combat equipment and gear to go eat in the messhall than we ever did when we were fighting the Iraqis. Our Team also had a bit of notoriety back in the rear. Everyone had heard of us, the C.L.O.W.N.s, and of some of the things we did while on the border. Unfortunately, the chain of command also heard about our exploits and found out we had captured weapons in our DMVs. We tried to hide them so that we could use them when we went into Kuwait, but we were ordered to turn them into the arms room. We were told that the weapons would be given back to us when the ground war started, but none of us believed it.

Our days consisted of training the Kuwaitis in our battalion on weapons maintenance, chemical warfare and basic infantry tactics. Each day we would eat breakfast in our own messhall, and then we would go over to the 57th Battalion barracks in the "5000 area" to wait for the Kuwaitis to form up. Some days it took longer than others because the Kuwaitis wanted to sleep in, due to being up most of the night.[125] Training would then be delayed so that they could drink tea with us and lounge around, talking about what was going on in the war. Eventually, around noon, we would be able to start training, but this was broken up by prayers ever few hours, or by the weather. It no longer rained,

---

[124] Everclear is 190 proof (95%) grain alcohol, the strongest alcohol that is still able to be legally consumed.
[125] The average Arab male was used to taking it easy during the day and then be up for most of the night socializing when the temperature was more tolerable.

like it did while we were in the wadi, but sandstorms would come in shutting down our training ranges due to unsafe conditions.

These were not the regular army Kuwaitis that the other A Teams were training, but civilians, who did not have the warrior mindset yet. It frustrated us to no end when they did not take the training seriously. Whenever I would tell them that their life depended on what we were teaching them, they would not believe it. They thought that if their cause was the right one, Allah would protect them from the bullets. They also did not believe in learning to fight by using cover, and instead wanted to run out in the open, spraying their M16A2s on three round burst. When I told them that shooting like that would not work, they would not believe me, and instead referred to various Hollywood movies where such tactics did work. On one occasion one of the Kuwaitis told me that our tactics wouldn't work anyway, because we had lost Vietnam using those tactics. I was so frustrated that I refused to wear the "Free Kuwait" buttons or the Kuwaiti lapel pins like the other team members did.

After training was cancelled due to a sandstorm, we saw a sign for the 47th CSH on our drive back.[126] Crowder and I were the only ones in the vehicle and we thought it might be the unit we had been looking for. We drove over there and found one of the women that we met before, SGT Anna _____ and we introduced Katzia to her. The puppy went everywhere we did, and roamed around the inside of the DMV like he owned it. It turned out that we had gotten the unit number wrong. It wasn't the 57th CC but the 47th CSH. After some flirting Anna invited us back that night and told us to bring the puppy for the other girls to see.

That afternoon the sandstorm still raged on, so instead of training outside we taught the Kuwaiti leaders about company level tactics in their trailer. They were slowly picking up on the way that soldiers moved in the different formations, but we were going about it too slowly in my opinion. I was not sure the Kuwaitis would be ready in time to go into Kuwait City, that was supposed to happen in the next two to four weeks.

Later that night Crowder and I took Katzia, a six pack of near-beer, and a canteen full of Everclear over to the 47th CSH. Crowder had taken a yellow bit of cloth and made a bow around Katsiyah's neck, to make him look even cuter. The puppy had become the bait for our trap. We parked our DMV and then walked around the camp looking for Anna's tent. We had no idea what tent they were in so we asked around, and then heard girly noises coming out of one of

---

[126] Pronounced "Cash" which stood for Combat Army Surgical Hospital.

the tents. I told Crowder to slip the puppy underneath the tent flap and wait a few seconds. He pushed Katzia under the tent and we immediately heard the sounds of "ah, look how cute!" and "ooooh!" The trap was set and ready to spring, so both of us walked into the tent "looking" for our puppy.

I didn't recognize Anna, mainly because she was now wearing shorts, a T-shirt, and her hair was down. I felt like Sir Galahad in the Castle Anthrax, from the movie *Monty Python and the Holy Grail*. There were girls everywhere, eleven of them in various stages of undress. There was Kelly, a short brunette, Quincy, a pert little blonde, and Stefanie, a red head who we had met before at the store. Most were wearing skin-tight long underwear or shorts. They all fell in love with Katzia. When we offered up the Everclear the party began. They offered us "near weed", which were clove cigarettes that were legal, but smelled like marijuana. The girls then started talking about how long it had been since they had sex, and what their favorite positions were. Crowder and I just stared, listened and tried not to drool on their cots.

After a few more drinks four of them got up and starting dancing right in front of us. We just watched, not moving, utterly amazed... hoping that no sudden movements would make them vanish. One of the girls asked if we were bored and I squeaked "no". They kept trying to get us to dance, but I didn't really want to stand up right at that moment. One of the girls grabbed me, and made me stand up, then three of them started rubbing all over me, doing a *Dirty Dancing* move. Anna kept trying to undress me by unbuttoning my Chocolate Chips. I kept trying to get dressed by refastening the buttons and she kept undressing me. Finally, she stepped out of the tent making a "come here" gesture with her finger and I quickly followed her.

When I went outside, we talked for a few minutes, continuing to walk away from the tent and towards the motor pool. When we got to the parked vehicles she reached up, grabbed my face and kissed me. After some heated minutes, she pulled away and told me that she couldn't do this, she was married and had a kid. However, she then pulled me towards her again and we continued kissing. She then stopped again. This kept going on for several minutes, until I found an empty HUMV and we both got inside.

I was amazed we were able to do anything in the vehicle, but both of us were able to figure it out. Afterwards she began to cry, saying she shouldn't have done that. Crowder walked over with Kelly and Quincy. He was kissing Quincy, but Kelly was on his arm looking jealous. Anna walked back with Quincy and I pulled Kelly over and kissed her. I told Crowder we had to go, and Kelly made me promise we would come back, and bring the puppy. I promised, and Crowder and I walked to our DMV.

Chapter 3: SPECIAL FORCES – A Storm in the Desert

When we got in the DMV, we did a quick search and discovered that Crowder's flak jacket was gone. I found the SGT of the guard and told him we had to find it. He asked his guards where it was, but none of them had seen it. He told us to go to the previous SGT of the guard's tent. When we went to the previous SGT of the guard, I quickly made up a story, and told him that we had to find the flak jacket because there was a sensitive piece of equipment in the pocket of the flak jacket, and we couldn't leave until we found it. Crowder was trying to be the "good cop" and told him that we really needed it back, but I played the "bad cop" and told the SGT that if we didn't get that piece of equipment back the NAC commander would turn the camp upside down until it was found. He asked what it was and I told him that it was extremely sensitive and it was part of the OPPLAN for Special Forces during the Ground War.

The SGT was angry when I accused him of taking it, and he stormed out of the tent. He walked about halfway to the current SGT of the guard's tent, but then came back, and told us to wait in his tent. Two minutes later he returned with the flak jacket, and admitted that one of his guards had taken it. He asked me what the sensitive gear was and I told him it was two cyanide pills that we were ordered to use if we were captured. I told him these pills were sewn into the fabric of the vest. I also told him that our DMV was extremely dangerous and it was rigged to explode if certain parts were hampered with, so his men should be warned to stay away from it when we came back.

Crowder was trying his best not to bust out laughing at the bullshit I was making up, and the SGT told me that if it was that dangerous, we shouldn't be carrying it around. I told him that we had been in country since September and the only people who stole from us up to that point was his troops. I told him that we never had to lock up anything when we were with the Saudis. We both drove away from camp, waiting until we were out of earshot before we started laughing out loud. I was amazed that the SGT had bought that line of bull.

The next day we trained the Kuwaitis again, but I was extremely tired after the previous night. We had stayed out with the girls until 0200, but neither one of us were going to tell anybody on the Team about the girls. It was our cache of horny women and we were going to keep it to ourselves. The Team knew that we had found some women, but it was driving them crazy trying to figure out where we had the cache.

I didn't know if Anna would speak to me again or not, but we figured we would try to go back again. If all else failed we were ready to give them Katzia as an offering. The other guys on the team were getting tired of cleaning up the puppy poop from inside the DMV and around the barracks area and wanted the puppy gone.

421

Triple Canopy

The next day we marched five miles to a MOUT city with our Kuwaitis, and all of them completed the march.[127] We trained them on how to go around doors and windows and move down the roads of a city. Most of the Kuwaitis were getting the idea of fighting in a built-up place like Kuwait City. The 2nd Company commander was a former commando and his company was the best and was extremely motivated. Unfortunately, the other companies just wanted to sit around and watch.

We weren't the leaders, just the advisors, so the Kuwaiti leaders had to get their men to do the training. Most of the time it was an uphill battle. When we marched back to the 5000 area, we had the men get into platoon wedge formations. This worked for about two miles, but then they disintegrated and turned into a mob. By the time they finished the march they were spread out across the desert over a mile.

That night Crowder and I decided to go back to the 47th CSH. We picked up two quarts of ice cream and two six-packs of near-beer, and drove to the front gate of the girl's camp. Unfortunately, the guard would not let us in and he told us that there were no visitors allowed in the camp after dark. I figure that our sexual adventure the night before must have made some of the men in their camp angry. We asked to see the SGT of the guard, but the guard told us "No, you have to leave." Underestimating the strength of our libidos, we decided to infiltrate the camp.

We drove down the road and then turned off our lights, putting on our NVDs. Both of us got out of the car and slowly worked our way to the camp, stopping every few minutes to listen for any movement. We wore helmets, just in case someone began firing at us. We were within 100 meters of the camp when we bumped into the only OP we had seen so far. I tried to get Crowder to slowly back away from the area, since the OP was only 25 meters away. The still of the night was shattered when Crowder accidently kicked a rock and it clicked across the ground. A female voice yelled "HALT!" Neither one of us moved, but after a few minutes we slowly retreated back to our DMV. The female continued yelling "Halt! STOP!" every few minutes. As we sat in the DMV, we saw other vehicles out in the desert, zipping around, looking for us with their lights on. We slowly drove between them, using our night vision goggles, and finally got back on the road.

We quickly drove to a nearby Saudi guardshack a few miles down the road and went inside. The Saudis greeted us like we were long, lost brothers. We

---

[127] MOUT stands for Military Operations in Urbanized Terrain, and a MOUT city is a training area that looks like a city neighborhood.

sat down on the rugs and they brought out tea. We gave them the ice cream and they were delighted. After spending enough time there for a legitimate alibi, we headed back to our barracks. It wasn't the way we wanted it to end, but the night had been an adventure. We were lucky we weren't shot by our own soldiers.

The next day, February 10th, our Kuwaitis went to the range for the first time to fire live rounds. After trying to get the Kuwaitis to zero their M16A2s, we discovered that none of them could shoot at all, and the chance of them hitting the small, printed, silhouette target that they used to zero their weapons would be an act of God. I told the team to go back to basics and teach them sight picture and shooting techniques again. We also taught them how to fire an RPG and the AK rifle. In the end we could not hold them to the same standards as soldiers in the US Army. If they were able to hit a standard piece of paper at 50 yards, they were considered qualified.

One of the problems was that we stapled the zero target, which is a small, printed black silhouette that was only about four inches high, to a large plastic silhouette that was the size of a man. Instead of trying to hit the small, printed silhouette, they believed that if they could hit the large plastic silhouette it was stapled to, then they were good.

At the end of the day Crowder and I stopped off at the 47th CSH. We discovered that their rules on visitors had changed due to our visit. The word was that an entire SF Team had gotten the girls drunk and had sex with them. We were allowed in and I found their commanding officer, a colonel, in their messhall and talked to him. I wanted him to relax the rules for us so that we could visit the girls. It turns out that he had been Special Forces at one time, and after talking to him he told us he didn't care if we visited.

We wanted to get in on the good side with all the officers, so they wouldn't shut us down, so we let them handle the AKs and RPGs in the back of our DMV that we had used to train the Kuwaitis, and let them take pictures holding them. We also let the female lieutenants play with Katsiyah, because we never knew when we might want to "visit" them too. Our rapport building mission with the officers worked, and they said that anytime we showed up, just let them know and they would let us in.

I found Anna and told the girls that anytime they wanted to come to our building in town they would be able to use the hot showers and hang out. I told them we also had a stove, hot food, a TV and VCR. They were falling over themselves to be able to do it, but they had to get clearance from their chain of command. I told her that I would be back the next day for an answer. When I returned to our barracks, I taught my advisors what we would be teaching the next day, then passed out. I was exhausted from three days of chasing women, getting little to no sleep and then training the Kuwaitis.

The next day when we showed up to train the Kuwaitis; we learned that that they were on "holiday". I couldn't figure out why, it was just some sort of holy day for them, so training was cancelled. Due to the break, I was able to sit down with my battalion commander and teach him and some of his officers battalion level tactics. Unfortunately, only one company commander, the former commando, showed up for the training. A few lieutenants showed up, but not many. Baby Doc was having a hard time training the medics too, and only three out of eight medics showed up.

We were all getting pissed at the Kuwaiti's attitude, because the war was going to start in a week or two and we needed every day we had left to try to turn them into soldiers. I told the battalion commander we needed to train, and the more training they had, the fewer of his men would die. He told me that they would train after lunch, and he would make sure the whole battalion would be out there.

My advisor team showed up after the noon break (1200-1400), and waited. We were still waiting an hour later when only one company, out of the eight, showed up for the training. Only one lieutenant showed up with that company. An hour later another company showed up, but with no officers. Thirty minutes after that a bunch of officers showed up, but it wasn't the same officers that I had taught in the morning, so I had to reteach them. Due to this no real training happened until after 1600.

Everything they did was half-assed and with no real effort or motivation. If they had gone against the Iraqis, they would have been slaughtered. The Kuwait's excuse was that they didn't need to fight as hard as the Americans, because the Iraqis would give up, not wanting to shoot their Muslim brothers. Some others said that all they needed was a gun and ammunition; they didn't need any training because the Iraqis would run from them in a panic. I figured they were in for a rude awakening and hoped they wouldn't get any of our guys killed.

I finally went to CPT McDonald, the Brigade commander advisor, and our own commanding officer, and told him of our situation. He said he would get with the Kuwaiti Brigade commander and try to get him to motivate his officers. I talked to my Kuwait battalion commander, and he asked, "What could we do?"

He said that he couldn't fire the officers, and punishment didn't work. They would just quit and no one wanted to take their place. He said they all were civilians and thought like civilians. I was getting tired of hearing that as an excuse for everything, and told him that he needed to replace officers who would not come to training, with those who would. He ended up only moving one company commander, the 1st Company, to a staff position and replacing him with a more competent officer.

By the second week in February the air war had expanded to include artillery fire from the Battleships *Missouri* and *Wisconsin* out in the Gulf. Every day they fired 16-inch shells onto Iraqi positions. Initially Schwarzkopf was concerned that the battleships would use up all their ammunition and told the Navy to "quit wasting their ammunition hitting the same area over and over again", but Admiral Waller told him "The battleships are about to be decommissioned. I can take all these sixteen-inch shells and shoot them at the enemy, or I can take them home and put them in a museum." Schwarzkopf laughed and told him to "Have at it." The shelling continued for the rest of the war, with over a thousand shells falling on the Kuwaiti shoreline.[128]

The VII Corps had been successful in fooling the Iraqis into thinking the main attack would come from the Kuwait border, much to the annoyance of the Marine force that was actually going to attack from Kuwait. The Marines then conducted their own PSYOPs to make the Iraqis think they would be attacking from the "bootheel" of Kuwait, and not from where they would actually be attacking. The Marines also set up decoys along the border, and played tapes of tanks repositioning at night. To make sure the deception was complete the tapes they played were of the Marine's M60 tanks and not of the Army's M1A1 Abrams tanks.

The unit in charge of this deception was known as Task Force Troy. TF Troy also conducted drive by shootings along the berm, launching TOW missiles or firing cannons toward Iraqi positions. TF Troy had several air strikes using napalm or fuel air explosives dropped on the Iraqi trenches in Kuwait. The Marines even took a page out of the terrorist's playbook and created two truck bombs, packed with 3,400 pounds of C-4 plastic explosive. The Marines drove toward the Iraqi positions, using artillery and mortar fire as cover, then jammed the accelerators. After the Marines rolled clear of the speeding trucks, the vehicles continued on towards the Iraqi trenches. One vehicle rolled into a

---

[128] Atkinson, *Crusade* pg. 262

ditch, but the second traveled to the edge of the enemy trenches before exploding. The detonation was huge, creating massive fireballs, but it inflicted little, if any damage.[129]

Crowder and I went out to the 47th CSH to visit the girls, and found Anna and Stephanie. Anna was the unit's mail clerk, so I gave her a ride to pick up their mail. When I saw the condition of the mail depot, I finally knew why our mail sometimes never showed up. All the mail was in large bins, out in the open air, and people milled around, trying to find their bins. The bins were never covered, even when it rained. There were huge piles of water bottles spilled out around the postal depot. They had originally been stacked up in their cardboard cases, but the rain had dissolved the cardboard boxes and hundreds of thousands of bottles had spilled out onto the ground. While we were there Crowder and I mailed some of our captured items home, before they were declared illegal to ship. We figured we were the only ones in the country to have war trophies so we took advantage of this "grace period" to mail them home.

Training with the Kuwaitis was still an exercise in incompetence. On Friday they wouldn't train, because that was their holy day. The next day they marched out to the MOUT city, and did a decent battalion level march, but then the sandstorm hit us and all the Kuwaitis wanted to go back to the barracks and not train. I told them that they couldn't go home if there was a sandstorm when we went to war. They told me that they didn't need to train anymore because now they knew everything. I told them that if they wanted to leave, they could, but none of my advisors would go with them. We would stay at the training site.

Most of the companies decided to stay, except for Rat's company. I pulled Rat's officers aside and gave them a speech about how there were no bad soldiers, only bad leaders and if we went to war and any of their men died, it would be on their conscience. Gentz's company was next and said they didn't want to train either, and Gentz chewed them out. One of the Kuwaitis told Gentz that there should be ten American soldiers around each Kuwaiti soldier, so that the Kuwaitis would live and there wouldn't just be women and children in Kuwait after the war.

I guess all of our speeches and ass chewings worked, because Kuwaiti soldiers became embarrassed and trained harder than they ever had before. That night Colonel Shaw, the 3rd Battalion commander, came out to brief us on the

---

[129] Atkinson, *Crusade* pp 334-335

latest developments up on the front. He told us that they had found fifteen Iraqis in the other Kuwaiti brigades, so we needed to be careful and keep an eye on our men. He also told us that the war would start in ten days or less. The start of the war was being called "G-Day" for the beginning of the Ground War. We were ordered to paint an upside down "V" on all the DMVs. It was a recognition symbol for the Americans when we began the Ground War.

Each day we were becoming more and more frustrated with the Kuwaitis we were training. The Kuwaitis didn't want to train anymore and the officers didn't want to take charge and make them train. So, every day we would go out, try to convince them to do the training we had planned, and then end up not doing it, or doing a very modified version of it. Each night the Iraqis would launch SCUDs into Saudi Arabia or Israel, since it had become the main weapon they could use against us. We joked about how ineffective it was and it was just a missile with a hand grenade charge for a warhead. The SCUD was basically the same weapon as the V2 rocket used by the Germans in WWII. It used most of its weight carrying rocket fuel, so the warhead did not have that much of an explosive punch.

Each night, as someone on the border reported a launch, the air raid sirens would go off. It was like being in a WWII movie. When the siren sounded the soldiers around KKMC would quickly put on their masks and flak jackets and then run into bunkers to ride out the SCUD attack. The roar of the launching Patriot missiles seeking out the incoming SCUDs could be felt from inside the bunkers. Everyone went down in the bunkers, but not us.

Whenever the air raid siren went off, we would go up on the roof or look out the windows, to see if we could spot the SCUD strike. I would try to record it on the cassette player. If we saw a strike, we would drive out to the impact site quickly, before anyone else got there, and take pieces off of the missile. We would then use the pieces for barter, and trade it as souvenirs to supply units to get what we needed for the Ground War.

The Air War continued to take its toll on the Iraqi military, though anti-SCUD missions were still the priority to try to keep Israel out of the war. One air wing of F-111 bombers out of Taif, Saudi Arabia had discovered a novel way to destroy the tanks and other vehicles that were dug into fortified positions and camouflaged against the high-flying aircraft. The F-111's used their Pave Tack infrared targeting system to identify the warmer tanks from the colder desert floor. The Iraqi tankers would often keep their engines running for

Triple Canopy

warmth or to make sure their engines didn't quit working. The tanks would show up as tiny white boxes on the F-111 targeting system. The pilots would easily guide a GBU-12 five-hundred-pound bomb into the tank box with a laser, destroying the vehicle in a fiery explosion.

After learning how easy this was, the squadron commander launched forty-four F-111s on a mass tank killing raid on February 7th, killing over a hundred tanks in one night, and every subsequent night afterwards. The process was so easy that it became known as "tank plinking". Some of the pilots even complained that the missions were mundane and boring, but each night the bombings decreased the numbers of tanks that the allies would face when they finally crossed the "Saddam Line" into Kuwait.[130]

The United States has repeatedly stated that foreign leaders are not targeted during a war since this would be considered assassination. At the very beginning of Desert Shield the Air Force Chief of Staff, General Michael Dugan, was relieved of his command when he stated that Saddam Hussein was "the focus of our efforts." Though Saddam was not an official target, the command posts where Saddam would coordinate the war did become targets. Throughout the bombing campaign Baghdad was hit each night, targeting communications buildings, supply lines and bunkers where Saddam might be staying.

This all ended on February 13th when a pair of F-117 Stealth fighters dropped two GBU-27 bombs on the Al Firdos bunker. Over 200 civilians were incinerated when they had been taking shelter inside the bunker from the nightly bombings. The bodies were fused together due to the heat and charred corpses were laid out for the news media to film.

This was the first time in the war that Iraq had allowed the media to film without censorship. The effect back home was negligible. Only thirteen percent of the American public believed that the United States should go to greater efforts to avoid hitting civilian areas in Iraq. However, due to the possibility of negative sentiment around the world, Colin Powell would now personally approve all strikes against targets in Baghdad. The Air War would now focus on softening up the Iraqi military prior to the Ground War and keep away from targets in the city. Life for the Iraqi soldier would become a living hell, either from being torn apart by bits of steel, having bones turned into dust and organs liquefied by the concussion of a bomb blast, or just from the fear of being killed at any moment.

---

[130] Atkinson, *Crusade* pp 263-265

428

By mid-February our Company gained another team. ODA 593 had been training in Pakistan, and had finally been able to get a flight to Saudi. Though we were at war, there were Special Forces Teams around the world doing their normal peacetime missions, and each one was trying to get to Saudi Arabia.

On Valentine's Day we tried to get our Kuwaitis to train, but half of the officers and men didn't even show up. Gentz was pretty pissed off and went into their trailers, kicking them out of their beds. Even after that, they still wouldn't show up to training until after lunch. Later I told the 52nd Battalion commander about his men not wanting to train, and he in turn told the Kuwaiti Brigade commander. The Brigade commander had a formation and told the men that if anyone didn't want to be there, and train with the Americans, they needed to pack their bags and get out. While he was giving the speech the air raid sirens went off. The Iraqis fired two SCUDs at Hafr el Batin, but they were intercepted by Patriot missiles deployed up there.

On the way back to our barracks we stopped off at the 47th CAS. They were all dressed in full gear and their tents had come down. They were moving out to the front for the Ground War. I said my goodbyes to Anna, and then found out that Stephanie would be staying behind with eleven other girls and going to the 5th MASH. The 5th MASH was located right beside our Kuwaiti Battalion's trailers!

Things were looking up!

That night K.C. and Crowder went over to an Air Force Valentine's Day dance. We had an Air Force TACP (Tactical Air Control Party) with us now, and he was able to get them in. When they returned, I jumped in the DMV and we all went over to the 5th MASH to see our girls. They had more lenient rules than the 47th CAS, so I was able to talk them into to coming over to our barracks to take hot showers. Stephanie was excited about taking a shower and hanging out with all of us. I forgot about our one shower that had dozens of laminated Penthouse and Playboy pictures taped up to the wall. We tried to keep the girls out of that shower, but they soon discovered it. They took it in stride and giggled about the "porn shower". We stayed up till 0230 with the girls and I had a great Valentine's Day! The next day was yet another Kuwaiti holy day, so we were able to sleep in after the girl's visit. We needed it.

On Valentine's Day Captains TB Bennet and "Chewie" Bakke were soaring above the Iraqi desert in their F-15E Strike Eagle, on SCUD patrol when the AWACs aircraft guided them to several Mi-24 HIND gunships that were closing in on Special Forces teams operating in the desert. However, as Bakke got closer to the target, the missile would not lock onto the HIND and instead went "intermittent" due to the targeted HIND accelerating too fast. Bakke switched to "ground attack" mode, and saw the HINDs rotors using his LANTIRN pod.[131]

Bakke released a laser-guided 2,000-pound GBU-10 bomb, aimed at the HIND that was just taking off. The heavy bomb hit the HIND, went right through the rotors, smashed through the cockpit and pilot, and then the delay fuse exploded as it exited the HIND and hit the munitions hanging from the helicopter. The Iraqi helicopter blew up in a massive ball of fire, and the other HINDS quickly left the area in a panic, not knowing what destroyed their fellow gunship. The Special Forces team that was being hunted by the HINDs sent Bennett and Bakke a "Thank you" message from their radio. Bennett and Bakke had the only "air to air" kill of an enemy aircraft by an F-15E in the war.[132]

The next day Crowder was able to find some more girls at the 44th EVAC when they went over to the "Wal-Mart". We didn't know where their camp was, so we stopped off at the mortuary to get directions. For some reason the mortuary guys knew where everyone was located. We linked up with the new girls and set up a cookout with them.

While we were in line at the PX to get some supplies, CNN began broadcasting that Saddam Hussein was pulling out of Kuwait. Everyone in the line cheered and the Kuwaitis began dancing around. None of the Special Forces guys reacted to the news because we didn't believe it. Sure enough, the story turned out to be false. Saddam said he would pull out of Kuwait if the air attacks stopped, if Israel got out of the West Bank and Golan Heights, and all the non-Arab forces left Saudi Arabia. This would never happen, so the bombing campaign continued.

---

[131] LANTIRN was the ground targeting system used by the F-15s.
[132] Extracted April 20, 2021, *This F-15E scored an air-to-air kill by dropping a bomb on an Iraqi helicopter*, by Blake Stillwell, posted on *We are the Mighty*, https://www.wearethemighty.com/popular/this-f-15e-scored-an-air-to-air-kill/

Chapter 3: SPECIAL FORCES – A Storm in the Desert

The Soviet Union told President Bush that Iraq was willing to go to Moscow to talk about a possible withdrawal, and that the United States should not "conduct any massive ground operations" until the talks were over. George Bush told them in clear language, "No way, Jose!"

Bush called the announcement that we had heard while waiting at the PX, a "cruel hoax, dashing the hopes of the people in Iraq and, indeed, around the world." President Bush in response called on the Iraqi people to end the war before it destroyed them. He stated "There's another way for the bloodshed to stop, and that is for the Iraqi military and the Iraqi people to take matters into their own hands, to force Saddam Hussein, the dictator, to step aside." [133]

That night the 1st Infantry Division cut twenty lanes in the dirt berm separating Saudi Arabia and Iraq. The berm became known as Phase Line Vermont. As artillery pounded Iraqi positions on the horizon, Task Force Iron moved three miles into Iraq and set up a twenty-mile-wide screen for the impending Ground War. This became known as Phase Line Minnesota.

The 2nd Armored Division reported that it had engaged an Iraqi vehicle with a TOW missile. Lieutenant Colonel Ralph Hayles commanded the Apache helicopter battalion in the division and decided to personally lead the night mission against these enemy soldiers. By doing this Hayles was violating an order from the Division commander, that stated commanders were to avoid direct participation in the fighting. Hayles and another Apache helicopter identified possible targets 4,000 meters away. After talking to the 2nd Armored Division on the ground, he determined that the targets must be Iraqi. What he didn't know was that his helicopter was not flying in the direction he thought and instead he had identified a Bradley and an M113 armored vehicle.

After flying closer he decided to take the shot and fired at the US vehicles with two Hellfire missiles. Both missiles slammed into the vehicles, making one erupt into a ball of flame. As the other Apaches fired on the vehicles with 30mm cannons, a voice told them to stop firing, they were friendlies! The next day Hayles was relieved of command, and Task Force Iron was ordered to return to Saudi Arabia, leaving behind two smoking vehicles, one US soldier killed and six more wounded. [134]

---

[133] Atkinson, *Crusade* pp 302-303.
[134] Atkinson, *Crusade* pp 317-320

K.C. was able to get some expensive meat from one of his sources, and kept it marinating in some secret sauce all day long. One of the reasons we were going over to the 44th EVAC was to find a home for Katzia. We figured that there was no way we would be able to take the puppy with us into combat. The girls at the 44th EVAC were not like the other girls we had been hanging around, but were "nice" girls, who seemed like the girls next door.

The cookout was done on a 55-gallon drum outside the tent of the girls. While K.C. was cooking the steaks there was a loud explosion near the camp. The rest of us were inside the tent and we all hit the dirt, then ran outside to find the source of the noise. The camp was running all around in masks, yelling out that there was incoming artillery hitting the camp. We laughed, because it wasn't loud enough to be artillery. I told them to mellow out and quit panicking. Eventually things got calmed down. While this was all going on K.C. continued to cook the steaks, not wanting to leave this rare treat unsecured. The steaks were fantastic and dining with the girls made it even better.

We finally returned to our barracks at 11:00 that night and passed out in our beds. Around 0300 the sirens woke us up, but instead of reacting to it, we just rolled over and went back to sleep. Katzia was still with us because we didn't want to give him up just yet.

Around sunset of February 17th Captain Scott Thomas's F-16 was shot down sixty miles into Iraq. Some of the Special Forces Teams located at KKMC had been given the mission of C-SAR, or Combat Search and Rescue. Three MH-60 Blackhawk helicopters from the 3rd Battalion, 160th SOAR (Special Operations Aviation Regiment) took off loaded with Special Forces soldiers from our 2nd Battalion. The CSAR team was led by CWO Dettrick and consisted of men from both ODA 541 and 566.[135]

---

[135] One of the men on this mission, Rick Darr, died in 2003 in a car accident. Some of the Special Forces veterans thought that it was due to something he had picked up during Desert Storm. Grant Winkler wrote in an email "He'd developed a weird blood anomaly where oxygen wasn't clinging to his red blood cells and had been to Walter Reed a few times while he was at Seton Hall. LTC Davis of 2nd Bn had the same problem. He'd gotten to a point where one Heineken would make him pass out or climbing a step would make him dizzy... He asked me to keep it quiet ... Doc... suspected Ric got into something in the sand, maybe a chemical weapon."

This team was the "Special Project Team" of 5ᵗʰ Special Forces Group that was nicknamed the FMG, or Fulton Moore Gang after the Team Sergeant.[136] The men were only armed with their M16s and some AT-4 rocket launchers. After two hours the pilot, CWO Thomas Montgomery, located the downed pilot, but he was surrounded by enemy vehicles searching for him. Captain Thomas had turned on his infrared beacon to guide the Blackhawk in on his position. As Montgomery flew his helicopter directly at the downed pilot an Iraqi missile locked onto him. Montgomery ducked behind a sand dune and requested support from a nearby AWACs aircraft, and an F-16 was sent to deal with the Iraqi vehicles. The Blackhawk was able to land, rescue the pilot, and speed off, chased by Iraqi tracers and missiles. 5ᵗʰ Group had conducted the only night time rescue of a pilot during Desert Storm.[137]

The next day I had my Kuwaiti Battalion conduct an attack on the MOUT city. I gave them the operations order the day before, but when they heard the false news about Saddam leaving Kuwait, they no longer wanted to train anymore. When they finally did the attack on the city it was a huge chaotic blunder. They eventually ended up shooting at each other by mistake. I was hoping that the Ground War would not start for another month or two so that I might be able to train the Kuwaitis into not killing each other by accident.

That night we brought the girls over again to eat at our barracks. They took more hot showers, the big attraction of our place, and I took them back around 10:00 that night. After I dropped the girls off Stephanie and I stayed in the DMV, kissing for another half hour, but no more than that. She was sure that she would be leaving tomorrow, or the next day, for the front, and didn't want to get involved with anything right then.

Out in the Gulf the Navy had begun the preparation for Operation Slash, the assault on Faylaka Island. This was a diversion to make the Iraqis think there would be an amphibious landing by the Marines. The *USS Tripoli* led seven ships, to include three minesweepers, on a mine clearing mission. The *Tripoli* was classified as an amphibious assault ship, but resembled a small aircraft carrier. A little after 0430 on February 17ᵗʰ the *Tripoli* struck a LUGM-145

---

[136] Email from Richard Toellner on January 5ᵗʰ, 2012. The Special Projects Team was a highly classified detachment in Special Forces.
[137] Extracted January 5ᵗʰ, 2012, from *A History of the 160th Special Operations Aviation Regiment (Airborne)* by Ronald Dolan, at http://helpingsoar.com/history160.htm

contact mine tethered fifteen feet below the surface. The mine blew a thirty-foot hole in the ship, vaporizing hundreds of gallons of gray paint stored in the hull locker. The ship was engulfed in a gray cloud as it took on water. Many sailors thought they had been struck by a chemical mine and donned their gas masks, only to collapse from the fumes. The masks would not stop the vapors. Though damaged the crew of the *Tripoli* were able to shore up the bulkheads with wooden beams and pump out enough of the water that the ship would not sink. After twenty hours of this heroic work the ship was declared to be stable again and resumed combat operations.[138]

Within hours of the *Tripoli* striking a mine, the *USS Princeton*, a newly commissioned missile cruiser, struck an acoustic Italian mine. The detonation sent shock waves rippling under the hull of the Princeton, and detonated other mines in the area, leaving the ship without power for her *missile* systems and dead in the water. The ship was out of the war and had to be towed to Bahrain for repairs.[139] Though the Navy had suffered some setbacks, the job of clearing the mines continued without any more detonations. Soon the battleships Missouri and Wisconsin were within range of Kuwait City and Faylaka Island.

On February 18th we took the Kuwaitis to a range that we had created. We would have used the Saudi ranges, but the Saudis didn't like the Kuwaitis and every time we had a range scheduled, they would cancel it or send one of their units there instead. There was no "Muslim brotherhood" that we could notice. So, we decided to make our own range in the desert, that the Saudis couldn't stop us from using. I had the Kuwaitis do a squad live fire, and then had them do a platoon live fire. The officers wanted to micromanage the squads and tell them what to do, but I told them to back off. Those officers would not be there during the actual fighting.

Afterwards I had them do a battalion movement, but without any live ammunition. We had to stop training early when an F-16 came over us, smoke trailing from its engine, and it jettisoned two 2,000-pound bombs near us. The bombs didn't explode, but a Huey helicopter landed shortly afterwards and the EOD guys went to the bomb to detonate them. We had to leave the area for safety reasons.

Each night I would try to answer all the letters that were sent to me from my "Any Service member" people. Most of my letters came from Cupertino,

[138] Atkinson, *Crusade* pp 322-324.
[139] Ibid., 327-328.

California or San Jose. When I wasn't with the girls from the 5ᵗʰ MASH, I would spend my nights answering all those letters. We also listened to BBC, AFN and to Radio Moscow for news. Radio Moscow broadcasted that Saddam Hussein had been wounded in an assassination attempt, but neither BBC nor CNN mentioned the story. We heard of the two Americans being killed and six wounded by our own Apache helicopters during a border fight.

February 19ᵗʰ was not a good day for me or my advisors. When I went to training there were only a few officers around. CPT McDonald wanted to know where they were at all times, so I tried to find the Battalion commander. He wasn't anywhere near the training area. The officers that were there told me that they were not going to train that day and they were on "holiday."

I drove to the Kuwaiti battalion commander's apartment in KKMC, getting more pissed as the day dragged on, and found him and his officers lying around, drinking tea and playing cards. He told me that he had given the men the day off, since they would be out all night on Wednesday doing a nighttime field problem. I told him "OK" and left, but I was pretty pissed off at the whole situation. I drove back to the training area and pulled my advisors off of the few Kuwaitis who had showed up. I told the Kuwaitis that American's were dying out there, and all they wanted to do was drink tea.

That night CPT McDonald had a meeting with all the advisors. He seemed pretty angry too, but there was not much we could do about it. He told us that we would try a new method. We would no longer train the Kuwaitis, and would only be a strict advisory role. We only had five days left, so in those five days the Kuwaitis would run the show. They had become too reliant on us being there to take over if their officers didn't show up.

We were told the Ground War would begin on February 23ʳᵈ, and the Marines would lead the attack into Kuwait.[140] That night I drove over to see

---

[140] There were 17,000 Marines floating in the Gulf, waiting to attack the Kuwaiti coast. The Marine landings were code named "Desert Saber" and were supposed to have started as early as February 2ⁿᵈ. The problem with the Marine assault is that it would take too long to put into operation. Rick Atkinson in *Crusade* tells of the problems facing the attack, "Before the Marines could be put ashore, minesweepers would have to clear a channel from the middle of the Persian Gulf to the Kuwaiti coast. But to protect the sweeper force of boats and helicopters, Iraqi missile batteries and shore guns had to be obliterated first...the latter task would take a week. The minesweeping would take another eighteen days. Finally, three to five additional days of naval gunfire would be necessary to suppress Iraqi gunners who could pick off Marines landing along the Ash Shuaybah coast. All told, the timetable called for at least twenty-eight days of preparation before the first Marine set foot on the Kuwaiti beach."

Stephanie, where we kissed for awhile, and then held each other as we watched *Little Mermaid* on a television in her tent.

Spending time with Stephanie must have helped me some, and our anger at the Kuwaitis that day must have worked because the next day, when the Kuwaitis attacked MOUT city again, they showed some motivation we hadn't seen before. We were the aggressors and they slammed through wooden doors with their bodies to get to us. The attack actually made sense and didn't turn into a fratricidal chaos. I told the Kuwaiti battalion commander that they had done 100% better than before and acted like real soldiers. He then gave a speech to his men about how well they were doing. When he finished, I told the Kuwaitis that I never wore any "Free Kuwait" buttons or pins because I didn't know if they were going to actually be able to do it, but I told them that I would wear a "Free Kuwait" pin now because they showed me they were soldiers now. They all cheered and began shouting my name, like a scene from *Lawrence of Arabia*. Dozens of them came up and gave me different pins to wear, so that I looked like a waitress at Ruby Tuesdays.

I really didn't think they were trained to the level they should be, but since they would be going to war in a few days I had to do something to raise their morale and make them feel like they would be able to fight their way into the city. I spent that night with Stephanie, and towards morning she told me that she was married. So far, my luck in finding single women was 0 for two. I really didn't expect to see any of the girls again after the war, but I was hoping.

On February 20th Apache gunships from the 101st Division attacked a bunker complex at Thaqb al Hajj, forty miles north of the Iraqi border. The gunships and A-10 aircraft attacked the Iraqis for four hours, inflicting little damage. Afterwards a three-man PSYOPs team landed and broadcast surrender messages over a loudspeaker. Immediately white flags appeared out of the trenches and 435 Iraqis from the 45th Division surrendered to a handful of Americans. The Iraqis were packed into Chinook helicopters and sent south. Schwarzkopf was thrilled by the mass surrender, but he quickly became angered when photos of the prisoners were published, showing Americans wearing the 101st Division patches guarding them. Schwarzkopf feared that this would alert the Iraqis to his planned "left hook".[141]

---

[141] Atkinson, *Crusade* pp 337-338.

Two days before the Ground War was supposed to begin my Kuwaiti battalion went to the "known distance" range and fired at the 100- and 200-meter targets. Unfortunately, they didn't hit too many of the targets. They next learned how to clear trenches and how to fire the German MG3 machinegun. There were two of these in the whole battalion, and none of the Kuwaitis knew how to operate them, so I gave them classes on the machinegun and let each man fire five rounds. While they were on the range my advisors fired our own .50 caliber machineguns, and Rat zeroed his M203 grenade launcher. I also test fired my captured AK-47 that I carried during the Air War. We kept our captured weapons in the Kuwaiti arms room, so that Group wouldn't confiscate them.

Since the Kuwaitis were now running their own training, they were not competent enough to be able to get all their companies through the live fire range and they weren't able to do a night fire. Initially they didn't want to do a night time defense, but I insisted that they had to do it. After a dinner of goat and rice they moved into their trenches for the night. We were going to be the special effects to their make-believe trench warfare. As soon as all the Kuwaitis had settled down, I had Rat fire the Mark 19 down the valley in front of the defensive positions, like it was an artillery barrage. This made all the Kuwaitis duck and take cover, since many had never heard an explosion that close before. Throughout the night we fired tracers out of the AK-47s, the MG-3 and the .50 calibers, to light up the sky around their positions. Gentz set off TNT charges with our new M122 remote detonators. All we had to do was program a number in the transmitter, hit the button and one of the detonators would go off. We kept the Kuwaitis up all night long.

In between the explosions, we probed the lines, seeing if they were alert. The Kuwaitis didn't have any blanks, so they would yell "bang-bang-bang" if they saw us. Around 0300 I fired six CS gas rounds from the M203 into their trenches, gassing them all. This made them put on their American issue gas masks. Right before the sun came up, we did a "mad minute" and fired everything we had. This woke them all up and we could hear them yelling "bang-bang-bang" every few seconds. Though this type of training would be outlawed on any US facility, no one got hurt and the Kuwaitis loved it. We all had breakfast and tea with them, and they thanked us for doing the "show". When we got back to the barracks we slept until 5:00 that afternoon, only waking when five Patriot missiles roared skyward to knock out a SCUD missile heading our direction.

That same night the 1st Cavalry Division was ordered to make a feint up the Wadi el Batin, where our "Castle Clownskull" had been. This was merely supposed to have been a deception to make the Iraqis think that this would be the invasion route for the Ground War. It was the most logical choice for an invasion and was similar to how the Germans expected the Invasion of France to come through at Calais and not Normandy.

The 1st Battalion of the 5th Cavalry pushed forward for six miles before they detected the enemy. When the soldiers fired upon the Iraqis, seven of the enemy immediately surrendered. As the Cavalry troopers began to collect the prisoners all hell broke loose. Artillery, mortar and anti-tank fire rained down upon the American armored vehicles from Iraqi positions that were camouflaged from aerial reconnaissance. As two tank companies rushed forward to engage the enemy a Vulcan anti-aircraft gun was struck by an Iraqi tank round, killing the gunner. One of the Bradley Fighting Vehicles was hit several times by mortar rounds, eviscerating two crewmen and spraying the inside of the vehicle with blood and bone fragments. The Americans returned fire with their tanks and Bradleys.

For over an hour the fight continued, artillery and A-10s added to the destruction of the Iraqi bunkers, dropping bombs and firing their main guns. Finally, the order to withdraw was given, but an M1A1 tank was destroyed by running over an anti-tank mine. The crew survived, but the tank was abandoned in the Iraqi desert. Three soldiers had been killed and nine wounded in what many thought was a foolish daytime attack. On the contrary, the feint exposed four Iraqi divisions dug in north of the wadi and it led the Iraqis to believe that this would be the main avenue of the Ground War.[142]

That night I was going to see Stephanie, but two PVS-7s had come up missing. We searched until it got dark, and then finally found the night vision goggles in Baby Doc's gear. He ended up with guard duty every night that our Team had it, and he would continue to have it until the war began. CPT McDonald gave us a copy of the Ground War operations order, and our brigade was not included in it. It didn't look like we were going to be included in the fight. I didn't blame the planners; our guys were not ready. We also learned about 5th Group's only casualty of the war. Two support guys were playing cards at King Fahd airfield, when one of them pulled out his pistol and jokingly

---

[142] Atkinson, *Crusade* pp 332-333.

told the other that he had better not be cheating at cards. The pistol went off, sending a bullet through the head of SGT Leonard Russ, killing him instantly.

To add insult to our perceived injury, that night our vehicle lost our .50 caliber machinegun and had to give it to ODA 311. They were going to be with the Kuwaiti brigade that would be breaching the border berm. We weren't too happy about being with a Kuwaiti unit that was going to go in last because they weren't trained enough.

Prior to the start of the Ground War there was skirmishing across the front as the units that were going to cross the minefields began their breeching operations early. During the week before the assault Marine aircraft dropped napalm and fuel air explosives on Iraqi artillery and infantry positions. F117s Stealth fighters attacked the pumping stations that were needed to fuel the "fire trenches" in front of the Iraqi positions.

The Marine commander, General Walt Boomer, was able to get permission to infiltrate 2,000 Marines into Kuwait to mark the lanes through the cleared minefields. These Marines cleared the mines using the WWII technique of probing with their bayonets or other probing sticks to find the mines. By sunrise of February 23rd the Marines had moved eight miles north of the border, hiding in shell craters during the daylight. To the east Marines had moved twelve miles, where Iraqi artillery wounded two of them. Fourteen Marine snipers were sent ahead to cover two Iraqi deserters, who marked a lane through an anti-tank minefield with chemlights. Afterwards an entire battalion of Marines was pushed through to secure the breach only to discover another minefield 800 yards away.[143]

To the west of the Marines, Saudi Arabian engineers cleared lanes through the mines located in front of their attack position, with the assistance of US Army Special Forces soldiers that were attached to them. The soldiers were able to clear six lanes without being detected by the exhausted Iraqi soldiers covering the minefields. Special Forces soldiers attached to the Egyptians were able to infiltrate the minefields and find the pumping stations that would pour oil into the "fire trenches" that Saddam threatened the coalition with. Instead of

---

[143] Atkinson, *Crusade* pp 367-368.

alerting the Iraqis with large explosions, the Americans simply turned off the pumps and then disabled them.[144]

That night my DMV died as I was going to see Stephanie. It needed a new generator. Klotz was going to go with me, because he and his wife Petra were having problems, and she was talking about leaving him. I figured a visit to the nurse's tent would cheer him up. We decided to leave the DMV in the parking lot and worry about it in the morning. Instead, we took a CUCV pickup truck that Klotz was able to sign for.[145]

As we drove over to the 5th MASH there was another SCUD attack. The Patriots rushed skyward with a booming noise, and took out the incoming missiles. Every time the Patriots took out a SCUD, Armed Forces Network would play *Another one bites the dust* by Queen. One of the SCUDs landed a kilometer from our Kuwaiti's barracks, without exploding. When we drove over to check it out a Saudi guard at the perimeter fence gave me a piece of it that looked like an old oil pan from a 1950's Chevy.

On the night of February 23rd, the Special Operations forces went to work. Up until that night Schwarzkopf did not allow the Special Forces to use their talents that they had been trained for. Schwarzkopf did not trust the Special Forces enough to let them out of "sight" of Saudi Arabia. Special Forces did not have this problem when the United States first went into Afghanistan in October of 2001. Special Forces were allowed to use all their skills and by just deploying less than 500 men they were able to defeat the Taliban and capture Kandahar.

However, on February 23rd, the special operations forces were finally unleashed. SEAL teams infiltrated by rubber boats to the Kuwaiti coast, and then swam the last 500 yards to the beaches. They set up buoys that were used to mark amphibious landing zones, and planted over 100 pounds of explosives. When they were detonated the SEALS peppered the beaches with machinegun

---

[144] Extracted from *US Special Operations Forces in Desert Storm 1990-1991*, on January 3rd, 2012, located at:
http://www.specialoperations.com/Operations/Desert_Storm/Operations1.htm
[145] After the war Klotz ended up divorcing his wife and joining Special Forces. He was wounded during the war in Bosnia in 1995. He got out of the Army and became a Sky Marshall.

fire, while calling in air strikes along the coast. This diversion kept the Iraqis on edge, though many had been on edge for weeks with all the shelling from the battleships.

At the same time ten Special Forces teams from the 5th and 3rd Special Forces Groups were inserted into the Iraqi desert by helicopter. The teams from 5th Group landed north of the Euphrates River to keep an eye on any movement toward, or retreat from, Kuwait. The teams from 3rd Group were placed in the path of the VII Corps attack to keep an eye out for the Iraqi Republican Guard. These teams had spent weeks training for this mission at King Fahd airfield, experimenting the best method to create a "hide site" to monitor the Iraqi avenues of approach. A hide sight was simply a hole in the ground, where four to six soldiers would live for several weeks until the war passed them over, or they were extracted. The goal was to not be seen.

Some of these missions were called off before they began. Two of the Teams were told to abort due to enemy movement at their infiltration site. One team was supposed to land in an area that looked like it was covered with rocks, in the high-altitude photographs used to recon the landing zone. Those "rocks" turned out to be Bedouin tents, so that mission was also aborted. Another team landed and heard Arabic voices all around them and were immediately extracted. Another team landed, moved to their hide positions, only to discover that they were now under water. They had to be extracted back to Saudi Arabia. This left four teams on the ground that were able to move to their hide positions and dig in before sunrise.

The conventional army was also active that night. At 0110 the 2nd Cavalry artillery fired into Iraq, along with howitzers and MLRS (multiple-launch rocket system). Each MLRS vehicle fired twelve rockets that would burst above the target. Each rocket would then shower the area with 644 "hand grenades" capable of cutting through two inches of steel. One 12-rocket salvo could cover an area with almost 8,000 explosives. The MLRS was called by some soldiers the "grid square removal service".[146]

Officers watching the bombardment checked off the known Iraqi bunkers and defenses on their lists as they were destroyed. The artillery also fired Copperhead rounds (laser guided artillery) that destroyed six T-55 command tanks. At the end of the bombardment two white phosphorous rounds burst in the area, letting all know it was time to move forward. The 2nd Cavalry vehicles moved into Iraq as *The Ride of the Valkyries* blared through speakers. They

---

[146] A grid square on a map was one kilometer or about 1,000 yards.

stopped fifteen miles later at Phase Line Bud, where they were supposed to wait for thirty-six hours for the VII Corps main attack.[147]

The next day Gator was pissed at me for not fixing the DMV before I went over to see the girls. I told him that I had no idea how to fix a generator, so it wouldn't have mattered. I noticed that we were on edge and pissed all the time because we were being left out of the war. We received the OP Order and read that our Khulud Brigade would stay in KKMC until Phase III of the Ground War. Phase III was the battle up to Kuwait City. CPT McDonald didn't like it, so he drove north to the border to try to do some politicking on our behalf. My Kuwaiti battalion commander told me that they were going to do night training from now on, so our schedule for training would be from 1600 to 2400 each night.

I was going to go over to see the girls at the 5th MASH, but Crowder instead brought over some of the girls from the 44th EVAC to our place. We cooked hamburgers and tacos for them that night, as we waited for some word from CPT McDonald. CPT Nick came in and told us that the politicking must have worked because we were going to move out at 10:00 the next morning and head north to the border. Unfortunately, we had no mission. We were just going to head that direction and hope a mission opened up for us. I asked the Captain if I could have an hour to say goodbye to Stephanie at the 5th MASH, and he told me "Go".

It was a pretty emotional farewell, and a lot of the girls were crying when they learned we were leaving. They were also pissed because they also wanted to go into the fight and didn't want to be left behind. When I returned to the barracks, I learned that we weren't leaving after all, but instead might move out in two days. Everything was hinging on what would happen during the Ground War. The girls from the 44th EVAC were still there eating tacos. Darren had fallen in love with one of the girls, Melinda, who came from New Mexico.

While we ate tacos and bean dip the SCUD missiles attacked. When we heard the air raid sirens, we all ran to the roof, holding on to each other, while the rain came down. We watched the explosions in the sky like it was a fireworks display. While we watched someone came on the roof and told us

---

[147] Atkinson, *Crusade* pp 373-374.

442

that the Ground War had started. It had just turned Sunday, February 24th... D-Day had started 19 hours earlier.[148]

At 0530 the 6th Marines moved across the line of departure and into Kuwait. Though Iraqi artillery rounds rained down, they soon realized that the Iraqis had no forward observers, and the rounds were landing at random. The Marines returned fire with over 140 artillery pieces, pummeling the Iraqi artillery into silence. Within thirty minutes the Marines had moved twelve miles to the first minefield. Instead of sending men out to clear the lanes, the Marines would use the MK-154 MICLIC (mine clearing line charge, nicknamed mick-lick). The MK-154 would launch a two-inch cable, carrying a ton of explosives, across the minefield. The explosion would detonate the mines, creating a path. After the charge detonated M60 tanks fitted with plows would push any remaining mines off to the side or detonate them. Though the MK-154 looked good in training, it did not work as planned during the attack. The charges would not detonate the way it was supposed to, so a Marine had to leave the vehicle and detonate them the old-fashioned way, by using a fuze and a blasting cap.

At 0635 an M-93 FOX chemical vehicle detected mustard gas and sarin nerve agent. All across the front soldiers and Marines donned their heavy, cumbersome MOPP gear to stop any chemical exposure. Luckily the war would happen during rainy, cold weather, or else there would have been hundreds, if not thousands, of heat casualties due to wearing the heavy suit. The alarm from the FOX vehicle turned out to be a false one, and no chemical mines were used during Desert Storm. Hussein knew that if he used chemical weapons the United States, Britain and many others would continue the attack into Iraq, not stopping at the Kuwaiti border.

The mines took a toll on many of the mine clearing tanks, taking them out of commission. Eleven other armored vehicles, to include nine tanks, were also

---

[148] After Vietnam the focus of the American military was to stop a Soviet invasion of Western Europe. The new doctrine was created by General DuPuy in 1976 and was mainly a defensive war to slow down the Soviet advance. The idea of fighting a war with just defensive measures did not appeal to many commanders so in 1982 the doctrine changed to allow commanders to strike deep into the Soviet rear lines, using aircraft, artillery and Special Forces. This became known as the AirLand doctrine. The Marines admired this strategy and adopted much of it for their own manuals in 1989. Though AirLand would have had several drawbacks in Europe, it was the best strategy the Americans could have had for desert warfare. Iraq and Kuwait gave the American military a perfect killing field.

taken out of action by the mines. Fourteen men were wounded, but none mortally as intermittent Iraqi artillery detonated around them and small arms fire pinged off the armor. Even though many of the MK154 charges would not detonate or worse, became entangled in nearby vehicles, the Marines were still able to bully their way across the minefield.

Fifteen miles to the south the 1st Marines breached their minefield, also discovering that the MK154 charges would not detonate unless they were manually primed. By 1030 all mine lanes had been cleared with ten Marines wounded by mortar fire. One Marine stepped on a mine and would die two days later. As a convoy of trucks and HUMVs moved into the second mine belt they were bombarded by Iraqi rockets. For some reason the Marine tanks thought the vehicles were Iraqis and began firing upon the convoy. The Marines, seeking cover, abandoned their vehicles and ran into the minefield. Lance Corporal Porter was killed when a tank round smashed through his truck window and entered his chest. As usual, the American army was the most dangerous element on the battlefield, even against itself.

At 0727 over 100 helicopters from the 101st Division took off from Saudi Arabia and crossed into Iraq, many flying as low as ten feet off the ground. Their objective was set up a forward logistics base, Objective Cobra, for the XVIII Airborne Corps attack, ninety miles inside Iraq near As Salman. After a thirty-minute flight soldiers from the 327th Infantry seized the objective, while F-16s and A-10s attacked nearby Iraqi positions. Chinooks carrying 105mm howitzers soon landed, and the artillery added to the pummeling of the Iraqi defenders. Soon three hundred Iraqis surrendered and all opposition ceased.

The following waves of helicopters began stockpiling ammunition, food, water and the most important; fuel, for the planned attack. The dust kicked up from the helicopters was so thick that pilots were grounded for an hour until the dust had settled.

Less than 100 miles from Baghdad a split detachment from ODA 532, under the command of MSG Jeffrey Sims, had moved into position to observe the highway leading to Kuwait. The other half of the team, three more men, was located fifteen miles away near the town of Samawah. This detachment was discovered immediately and had to evade for three days until they could be extracted by helicopter.

Sims' team was able to move five miles to their hide site, beside the village of Oawam al Hamzah, and dig into position before the sun came up. Their first message sent by SATCOM radio was to send information about a fifty-car train that had rolled past their position. With Sims were SFC Ronald Torbett and SSG Roy Tabron.[149] The Team thought their position was in an unused farmer's field, since it was not the growing season, and traffic would be kept to a minimum. That turned out to not be the case and their hide site soon had a

---

[149] Tabron had been one of my evaluators during Robin Sage.

steady flow of villagers walking by. For eight hours the Team tried to remain concealed from both villagers and dogs, but at 1400 the Team's luck ran out.

No matter where I have been in the world, it always seems like children could find me. Maybe it is because children see the world differently and anything that is not "normal" stands out, but if there is a kid anywhere near where we trained, he would always find us. ODA 532 had the same problem. A little girl noticed the small observation hole that the Team was using and went to get her grandfather. The Team then had a dilemma that had been proposed to us several times in training. What do you do if you are discovered? Do you shoot the child? Do you take her hostage? If you do take her hostage more villagers would find her. Then what do you do? Shoot all the villagers? Take them all hostage? The cycle would never end. The Team decided to let the girl go.

When the grandfather showed up, he didn't seem to believe the little girl, but she ran to the hide site, pointing at the Special Forces soldiers inside. The three men inside the hole all had drawn their suppressed 9mm pistols and were aiming at the little girl's head. None would pull the trigger though. It was something they could not live with. The grandfather moved to within inches of the hole, squatted down and looked at it curiously. The Americans threw back the covering and SSG Tabron grabbed the old man, opening up his coat to show the man that he was unarmed. Sims spoke to him quickly, saying "We are your friends" in Arabic. The old man told Sims that there was an Iraqi garrison nearby. Sims told the other men to let them go, and the old man backed off, screaming "The American's are here! The American's are here!" and running to other shepherds nearby. Sims and his team now had to move quickly!

The Team quickly sent off a satellite message saying that they had been discovered. Sims had the team pack up all their equipment and move 500 yards away, along a drainage ditch by the field, to ground that looked more defensible. Thirty minutes after the old man had found them the team heard rifle fire from the village of Oawam al Hamzah. A bus arrived and unloaded Iraqi soldiers. Sims saw one officer, obviously in charge. He took careful aim and killed him with a single shot from his M16 rifle. The Special Forces soldiers began to snipe at the Iraqis from long distance with their rifles on semi-automatic. SFC Torbett saw an Iraqi girl driving a tractor, pulling a flatbed trailer loaded with Iraqi soldiers. Using the scope on his M16 Torbett fired and hit an Iraqi on the trailer a half a mile away.

Sims ordered his men to fire on semi-automatic to conserve their ammunition. Each soldier only had 300 rounds apiece. Soon two more busloads of Iraqi soldiers arrived, and began to move against Sims and his men. To further compound the situation Iraqi farmers had also armed themselves and moved against the Americans. Sims had Torbett move 200 yards to his right, while he moved to the left to try to stop the Iraqis from flanking them.

Some of the villagers stood on an old masonry wall, directing the Iraqi soldiers to where the Americans were. The Special Forces soldiers directed their fire against them, and one fell off the wall, making the rest scatter. The soldiers and villagers would creep near Sims and his men, only to be gunned down. The surviving Iraqis would retreat. After reorganizing the Iraqis would approach again. More busses arrived, and it began to look hopeless to the men of ODA 532.

After an hour and a half of holding off the Iraqis, hope arrived attached to an F-16 fighter. Sims fired fourteen flares, but finally was able to show the F-16 where he was at. The lone jet dropped cluster bombs and one thousand pounders on the ditches near the Americans.

One hundred and seventy miles away in Rafha the frantic distress call of the Special Forces Team was heard by CWO Jim Crisafulli. He had been resting after inserting Special Forces teams behind enemy lines the night before, with his Blackhawk helicopter. However, within minutes of hearing the call he took off, carrying two other Special Forces soldiers to man the door guns. Not even bothering with an evasive flight pattern, Crisafulli flew straight towards Oawam al Hamzah, skimming over the sand with only ten feet between the helicopter and the desert floor. As he got closer to Sims and his team Crisafulli had to fly under high power lines, and then over another pair, making the helicopter rise and fall like a roller coaster.

Sims fired his last pen flare and Crisafulli at first thought it was a surface to air missile, until his crew chief yelled it was a flare. The Blackhawk helicopter slammed down fifty yards from Sims and his team. The Special Forces soldiers in the Blackhawk jumped out and provided covering fire with their M16s as Sims and his men ran for the door. Three Iraqis chased right behind them but were cut down by the door gunner. As soon as the helicopter flew out of range all the men inside began cheering, glad to be alive! [150]

Seventy miles to the east, eight Special Forces soldiers from ODA 525 landed near the small village of Swayjghazi. CWO "Bulldog" Balwanz had his team dig two hide sites, 300 yards west of Highway 7. Unfortunately,

---

[150] A year after this I worked with CWO Crisafulli to try to figure out how to mount a Mark 19 automatic grenade launcher in his Blackhawk helicopter. We never did find an acceptable method to do this, but he was definitely a pilot who thought outside the box.

Balwanz's team had the same problem as Sims' Team. They did not think that villagers would be wandering around their position, located near a drainage ditch. However as soon as the sun came up shepherds led their goats near the Special Forces team, while women and children gathered firewood. Just like Sims' Team, the Iraqi children discovered the hide site. Balwanz knew that they could not shoot children, so after they were discovered, the Team moved 400 yards away and into a muddy ditch. The children returned with a young man, who spotted the Americans. "As-Salaam Aleykum" Balwanz called to him, but the young man turned and hurried back to the village.

Thirty villagers returned, armed with rifles. As they fanned out across the field four trucks arrived, along with a bus and a Land Rover. The vehicles unloaded 150 Iraqi soldiers. Balwanz called XVIII Corps headquarters and told them that he needed immediate extraction and air support. Balwanz ordered his men to pile all their rucksacks and equipment outside the ditch, and then primed it with a block of C-4 plastic explosive. The Americans only kept their weapons, ammo, and a single satellite radio. The C-4 exploded just as a group of Iraqis reached the pile.

Bullets tore through the dirt around them, as the Americans returned fire with M16s and M203 grenade launchers. The Iraqis resorted to human wave attacks, but they never quite understood the accuracy of American soldiers.[151] In just the first ten minutes of the fight Balwanz and his men killed about forty Iraqi soldiers. The fighting continued, with no end in sight. Though the field was littered with Iraqi bodies, the men did not see any rescue in their future. Balwanz saw his men wave farewell to each other across the sides of the ditch.

Just as the Iraqis were close enough to rush the Team, several F-16s arrived and dropped cluster bombs on the highway. Balwanz used his survival radio to direct bombs to within 200 yards of his position. A group of Iraqis charged down the ditch, trying to stay clear of the F-16s, but Balwanz and one of his sergeants stopped their attack.

Balwanz moved his men 300 yards away in all the chaos, without the Iraqis being aware of it. After an hour and a half, two rescue helicopters were able to land right on top of the team, and rescue Balwanz and his men. With 150 Iraqis dead in the field, it probably seemed to the Iraqis that they had been rescued from the demons that had been unleashed upon their small village.[152]

---

[151] Throughout the world the American soldier is the goal of training for the military. Special Operations takes that marksmanship to a higher degree.

[152] The Teams from Balwanz and Sims killed an estimated 250–300 Iraqis.

The next day we didn't do much of anything, except hang around near the radio, trying to learn anything about the war. There was a 24-hour news blackout, so we had no idea how the war was going. All of us were depressed, and wanted to get back into the fight. Throughout the day all the units around KKMC began packing up and moving towards the war, while we all stayed behind.

The news from CNN, ABC and NBC was almost always wrong, but we had a second source for what was going on, the Air Force CAS headquarters. They told us reports from their CAS Teams that were in the fight. We learned that there wasn't much resistance to the units moving into Kuwait, and they had taken 14,000 prisoners so far. At that time, we only had twelve dead and 24 wounded, which was amazing considering how it could have gone down. We also knew that if we didn't get into the fight, it would all be over in a few days.

The main force used in Schwarzkopf's "left hook" strategy was the VII Corps based out of Germany. The Corps took three months to move all their men and equipment into the war zone, but by the middle of February they were ready to strike. The 1st Cavalry Division would move up Wadi el Batin as a feint, making the Iraqis think the main attack would be using this massive river bed to advance into Kuwait.

While this was going on the 1st Infantry Division would breach the berm west of the Wadi, and head east towards Iraqi positions, while the British 1st Armoured Division followed. Much farther west the 2nd Armored Cavalry would breach the berm and head into the Iraqi desert, leading the 1st and 3rd Armored Divisions across 100 miles to envelope the Republican Guard. Those two tank divisions would only be spread out on tight 20-mile front as they advanced. General Fred Franks commented that the tight formation would be attacking the Iraqis with a "fist" and not just five disjointed fingers. The entire coalition stretched from the Persian Gulf, across a 400-mile front, ending with the VII Corps on the Iraqi border.[153]

Due to the success of the Marines advances through the minefields, Schwarzkopf could not let the Marines be flanked by the Iraqis, so he moved up the main attack by fifteen hours. At 1430 on February 23rd VII Corps began a

---

[153] Atkinson, *Crusade* pg. 258.

Triple Canopy

massive artillery barrage of Iraqi defenses. Thirteen artillery battalions and ten MLRS launchers rained down more than 11,000 rounds in less than 30 minutes.[154] The number of rounds falling on the Iraqis actually exceeded the rate of fire that fell on the Somme in July, 1916. Following the bombardment, the 1st Infantry Division pushed through the minefields, mainly using tanks with plows, rakes and steel rollers. It was similar to what the Marines had done in the east, but the Army had trained with the defective MK154 mine clearing lines and decided not to use them.

Once the 1st Division pushed into the trench lines, they forego trying to clear the trenches, which would slow down the attack and cause unnecessary casualties. Instead, an M9 armored bulldozer rode down the trenches, collapsing the walls on the enemy defenders. The Iraqis had made it easier by piling up all the dirt from the trench on the south side, facing the Americans. Bradley vehicles drove beside the bulldozers, pouring fire into bunkers that offered any return fire.

Joe Queen, one of the soldiers operating a bulldozer, later told CNN "What we did is we just took the dirt that the Iraqi soldiers had dug out, we just pushed that dirt right back into the trench. You could just look at the man's eyes and see fear. You know, you see him scared. You know, you're looking at a man's…the whites of his eyes as you're going through in the trench with this bulldozer, covering in the trench. And they were firing at the bulldozer and the first bullet that hit the blade, that made me know then, "Hey, look. This is for real. There's no game. Those are real bullets and a bullet would kill you…You don't think about, "Hey, what about this guy? What about that guy?" He had a chance to get out. He had every opportunity to get out and he took the way to die for his country, just like any American would." [155]

The defective MK154 mine clearing charges were also used against the Iraqis, by firing them down the length of the Iraqi 26th Division trenches. The Iraqis had screwed up and dug them in straight lines. When the MK154 lines detonated it killed the Iraqis and buried them at the same time. It only took the 1st Division thirty minutes to clear through the "Saddam Line" while only suffering two casualties. The Iraqis had 150 soldiers "plowed under" and 500 quickly surrendered. After the war the concept of burying enemy soldiers was

---

[154] A battalion of artillery would normally be three Batteries. A battery would normally consist of six artillery pieces. The MRLS launchers rockets dropped more than 600,000 bomblets onto the Iraqis.
[155] Extracted from CNN Frontline at :
http://www.pbs.org/wgbh/pages/frontline/gulf/script_b.html on January 6th, 2012.

questioned, but it was quickly dismissed since the tactic was consistent with the rules of war.[156]

That night we took the Kuwaitis out to MOUT city to train, but a SCUD missile attack stopped the training. We were averaging three SCUD attacks a day and we were getting pretty used to it. We were told our Kuwaitis would deploy to the front lines as soon as they got vehicles. Our Kuwaitis still had no uniforms and no weapons except their M16s. We learned that the Emir didn't want to have a fully equipped civilian army after the war and was limiting what would be issued to the newest battalions. We figured the Emir was afraid the Kuwaitis would overthrow his rule. My count of what I would be taking into the war was 201 men and 15 officers. My battalion actually had 330 men and 35 officers, but only those 201 and 15 "effectives" would be going to war.

Inside Kuwait there were four resistance movements. One was loyal to the Emir and the Sabah family, and the rest were against them. One resistance group was Palestinians and another was for a democratic style of government and not a monarchy. One of the groups was Shiite, which was a different religious sect than the Emir. None of us wanted to be with the Kuwaitis but all of us, except for Crowder, wanted to get into the fight. Crowder was the only one who wanted to stay back so he could spend more time with his new girlfriend Melanie.

On Sunday night General Franks decided to pause the attack of the VII Corps to "halt, refuel, ensure formation postured to achieve mass when the attack commences at first light."[157] Franks was worried that his right wing would outrun his left and leave it exposed to the Iraqi Republican Guard, waiting somewhere on the border of Kuwait. Schwarzkopf had also cancelled the Marine assault on Falayka Island, though Marine helicopters continued to dart towards the Kuwaiti coast. This was to keep the Iraqi's attention focused on the possibility of a beach assault.

On Monday morning the 3rd Brigade of the 101st (Air Assault) Division flew 150 miles to LZ Sand, located 25 miles south of Highway 8 in Iraq. The first sixty Chinook helicopters landed, each carrying two TOW mounted HUMVs. Due to the rain the road leading to Highway 8 was covered in mud and the

---

[156]Atkinson, *Crusade* pp 396-397.
[157] Atkinson, *Crusade* pg. 403.

Triple Canopy

infantrymen had to push their artillery and HUMVs through the mud, like soldiers had done in a time before motorized warfare.

Sixty-six other Blackhawks sped north with a thousand soldiers from the 187th Infantry Regiment (Rakkassans), landing at three landing zones near Al Khidr, south of Highway 8. The men sank up to their knees in the mud, but pushed on. Soon Colonel Robert Clark was able to send the satellite message, "The Screaming Eagles have landed on the Euphrates." Unfortunately, due to the mud and rain Clark cancelled the airlift of another 1,000 soldiers. Clark couldn't afford to have his men trapped in the mud against Iraqi artillery and tanks.

Men of the 3rd Battalion erected a sign on the Highway that stated "US Military Operations – Keep Out." Iraqi soldiers nearby investigated the signs but were driven off by 60mm mortar fire. Though civilian vehicles were allowed to pass through, all truck convoys were fired upon. The 101st soldiers envisioned that they were stopping SCUD missiles or some other military weaponry, but the first convoy that was destroyed was only carrying onions. When the mayor of Al Khidr approached the Americans to ask if they could go through the destroyed vehicles, the soldiers agreed. Hungry Iraqis swarmed over the battered vehicles, leaving with onions, flour sacks and loaves of bread.[158]

In the eastern part of Kuwait, the Marine assault had ground down due to the rainy weather and smoke covered skies. As Colonel Richard Hodory spread a map across his HUMV hood, an Iraqi T-55 and two other enemy vehicles appeared out of the smoke, fifty yards away! The surprised Marines opened fire on the tanks with machineguns, but soon white flags appeared out of the tank's hatches. Iraqi Major Adai surrendered to Colonel Hodory and explained to him that he was the commander of the 22nd Brigade of the 5th Mechanized Division. Adai handed over his map, stamped "Secret" and told Hodory that the Iraqi army was about to counterattack. He also told Hodory that he was the father of three children and had been wounded three times in the Iran-Iraq war.

Armed with Adai's map, the Marines began an artillery barrage at 0800. Three hundred rounds were fired at two Iraqi targets, and then the artillery barrels were raised and more rounds were fired at targets a thousand yards further out. Iraqi tanks appeared out of the fog and began firing their guns at the Marine vehicles. Colonel Hodory and his staff took cover on the ground as machinegun bullets zipped over their heads. The Iraqi rounds all flew too high,

---

[158]Atkinson, *Crusade* pp 409-411.

but the Marines fired back with precision, killing Iraqi tanks with DRAGON missiles and AT-4 rockets. When the Marine M-60 tanks were able to move to a ridgeline, the fog lifted showing the Iraqi troops below dashing about in confusion and ready for slaughter. The killing continued for three hours in what Marines would later call the "largest tank battle in Marine Corps history."[159]

Major Randy Hammond later told CNN "We'd got the word that there was an Iraqi counterattack. As we arrived out there, we saw, horizon to horizon, Iraqi tanks, armored vehicles, and as we started hitting these tanks things started happening in our favor, the attack basically just ground to a halt. All the tanks stopped. The armored vehicles stopped. The hatches flew open. The Iraqis started bailing out of the tanks and, you know, scrambling around in the desert, trying to figure out what was going on."[160]

After fighting through an evergreen grove, that the Marines nicknamed the "Emir's Ranch", the Marines pushed the Iraqis back out into the smoke covered desert. Hammond said "It was like...something out of a... spaghetti Western, with a trip wire and the horses falling. ...as you engaged these troops with 20 millimeter ... they'd just kind of pitch over in the desert."

The Marines destroyed over 100 vehicles and captured 300 prisoners but the Iraqi counter-attack had stopped the Marine forward momentum. The Marines would not move forward again for 24 hours.

By D+3 we were pretty pissed off at our Kuwaitis. They could have gotten trucks on the first day of the Ground War to take them into Kuwait, but instead they hesitated. We decided if they weren't going to sign for the trucks, we would, and then deliver them ourselves to the Kuwaiti compound. It was still raining, and had been since the Ground War began. I figured back home nobody would understand that our "desert" war was one of continual rain and cold weather. I didn't go back to the 5th MASH anymore, since the last farewell had been pretty emotional. Instead, Crowder and I would go over to the 44th EVAC each night after work. While we were there, we learned that a SCUD missile had gotten through and slammed into a postal unit down in Dhahran. The first

---

[159] Atkinson, *Crusade* pp 413-415.
[160]Extracted from http://www.pbs.org/wgbh/pages/frontline/gulf/script_b.html, *Frontline: The Gulf War* on January 6th, 2012

reports had twenty killed in the blast and ninety more wounded, so we knew it was bad.[161]

We had become complacent about the SCUD attacks, but this drove home that they were a danger if they hit the target. That one SCUD killed more Americans than the whole Iraqi army had killed so far. Each night after returning from the girl's barracks we would lay around in our beds, listening to the war on AFN, and become more and more frustrated.

By the third day of the war the reporters were saying that Saddam Hussein would pull out of Kuwait, but would not surrender. President Bush stated that we would continue to fight, but if the Iraqis laid down their arms they wouldn't be attacked. The reporters continued telling of a major firefight at the Kuwait City International airport between the Marines and the Iraqi Republican Guard.[162] While we listened to the war we had a hamburger cookout, and then we watched *The Longest Day*, about the Normandy invasion. Crowder's girlfriend Melinda came over to spend time with him, but none of the other girls came with her. We wouldn't have been very good company anyway.

Out in the Iraqi western desert the British SAS and the United States SFOD-D (Delta Force) continued searching for SCUDs. A reinforced company from the 1st Ranger Battalion was attached to the Delta Force soldiers to provide security during their hunts. On February 26th Delta soldiers attacked a radio relay site deep in Iraq using AH-6 "Little Bird" helicopters to spray the targets with mini guns and rocket fire. The attached Rangers then secured the compound, set explosive charges on the 100-meter-tall tower and destroyed it.[163]

These missions were not without loss. Five days earlier, on February 21st, four pilots and crew from Task Force 160th were killed when their MH60 helicopter slammed into a sand dune during their low-level flight. Three Delta soldiers were also killed in the crash. One of these men was my old first sergeant

---

[161] Due to a technical error, no Patriot missile was fired at the SCUD that exploded in the suburb of Khobar. Twenty-eight Americans were killed and 98 were wounded. Of the dead, thirteen were from the 14th Quartermaster Detachment out of Greensburg, Pennsylvania.

[162] This was the Battle at the Emir's Ranch also known as the Battle of Kuwait Airport.

[163] Extracted from *US Special Operations Forces in Desert Storm 1990-1991*, on January 3rd, 2012, located at:
http://www.specialoperations.com/Operations/Desert_Storm/Operations1.htm

from the 3rd Ranger Battalion, Patrick Hurley. The reason CSM Hurley was on the helicopter was that he had been injured when he fell from a cliff and was being medevac'd back to Saudi Arabia.

In a more famous case, told in the book *Bravo Two-Zero*, eight SAS troopers were compromised during a SCUD hunt mission and had to evade the Iraqis. Four of the British soldiers were killed during the pursuit and three were captured. Only one of the troopers escaped.

After the Battle of Kuwait International Airport, the Iraqis decided to abandon Kuwait City. Kuwaiti resistance, working with the CIA, reported that the Iraqis were loading vehicles and getting ready to withdraw. A US Air Force JSTAR aircraft was ordered to fly over the city and determine what was happening.[164] As the JSTAR flew north of the city it saw hundreds of vehicles on the Al-Jahra highway, heading north.

Colonel David Baker told CNN that when he briefed his pilots, "I told them of the importance of the mission. It's not just a retreating army. These guys are rapists, killers, murderers. And coincidentally, a Scud had just hit Dhahran airport and killed 60 Americans. And I convinced them, and I'm sure that I did, that they needed to put some hate in their heart and go out and stop the son of a bitches from getting out of Kuwait."[165]

Right before dawn a dozen F-15E Strike Eagles dropped bombs on the lead vehicles moving through the Al-Mutla pass. It blocked the six-lane highway, effectively putting a cork in the bottleneck. Captain Derrick Krause, one of the F-15 pilots, told CNN, "As we dove out of the clouds, the picture was absolutely astounding. There were thousands of headlights heading on every road that led north out of Kuwait City... We had twelve 500-pound bombs and we elected to drop them three at a time."

---

[164] JSTAR was a modified Boeing 707 that was equipped with radar that could track movement on the ground.
[165] Extracted from:
  http://www.pbs.org/wgbh/pages/frontline/gulf/script_b.html, *Frontline: The Gulf War* on January 6th, 2012

THE KILLING FIELDS   AL-MUTLA

Krause's weapon's officer, Major Joe Seidl, told of what happened next, "The bombs impacted in a string right across the highway, with the center bomb impacting … in between two trucks … causing both of the trucks to burst into flame." [166]

On the ground an Iraqi soldier named Sardar told CNN, "There were many wounded people on the road, some of them without arms or legs. They were just stranded there half dead. When they saw our car, they started to crawl towards us. We didn't have space for them. With all the strength they could muster, they were throwing themselves at the side of the car. The windows were smeared with blood. We had no space. We had to drive on." [167]

---

[166] Ibid.
[167]   Ibid.

The F-15s then struck the rear of the column, stopping all movement. For the next 48 hours Air Force, Navy and Marine aircraft attacked the Iraqis trapped in the killing field that would become known as the Highway of Death.

When we returned to the compound, we were shown a map of Kuwait City and told which part of the city our battalion would be responsible for. My Kuwaitis would be in charge of clearing the northwest part of the city that consisted of Sulaibikhat, Doha and Ashish Ad-Doha. From captured documents we learned that the Iraqis were told to inflict as many casualties and destroy as much property as they could on the way out of the city.

We were given large orange panels to duct tape to the hood of our DMVs so that the Air Force would recognize us as friendlies. Each of our vehicles flew a flag to identify ourselves. My vehicle had a Confederate flag, while others had Kuwaiti, British, American, Saudi and a pirate flag.

Our convoy consisted of five DMVs, one HUMV, one Air Force HUMV and a 2½ ton truck. The deuce and a half truck carried all our supplies and our captured Iraqi mortar from the Air War. Gator rode in the Air Force HUMV and it had HF, UHF and VHF capabilities, so we would be able to talk to anyone in our area. Pappa Doc's vehicle didn't have a .50 caliber anymore, since it was signed over to ODA 311, so he mounted our captured RPK on the turret.

The VII Corps had begun moving forward again on Monday morning, but by Tuesday the attack had slowed due to a sandstorm that grounded all aircraft. The 7th Cavalry was reduced to just a one-kilometer front, with Bradley and tank commanders standing up in the hatches trying to peer through the orange wall of dust surrounding them. As the lead platoon crested a ridge the Bradleys came under fire. Even though it was only 3:30 in the afternoon the Americans had to use thermal sights to see what was in front of them. They soon discovered they were in an Iraqi kill zone. By looking through the thermal scopes the seven Bradley vehicles could see six T-72 tanks and eighteen BMPs of the Tawalkana Division of the Iraqi Republican Guard.

Six more Bradley's quickly deployed to the left and right of the lead platoon and then signaled to attack using a star cluster.[168] One of the Bradleys was hit by a tank round, tearing through the turret and killing or wounding three soldiers. The shot had not come from the Iraqis, but from an M1A1 Abrams tank a half mile behind the Bradleys.

CSM Ronald Sneed ordered his driver to remove the burning Bradley, as his gunner fired 25mm rounds into the Iraqi trenches. The sergeant major jumped off his Bradley, dragging crewmen free as Iraqi bullets clanged off the sides of both Bradleys. All thirteen Bradleys returned fire with TOW missiles and cannon fire. One of the Bradleys was hit in the transmission by an Iraqi RPG and was immobilized. However, it continued to fire though it could no longer move. When another Bradley pulled up behind the wounded vehicle, it was struck a second time with a SAGGER missile, spraying shrapnel all over the crew. The crew, bloodied but not beaten, scrambled to the rescuing vehicle. As the Bradley drove away it was hit twice by M1A1 sabot rounds. The lieutenant commanding the track had been blinded, but he crawled down the front of the vehicle and yelled down to the wounded driver to stop the vehicle. Miraculously none of the eleven men in the Bradley were killed.

One more M1A1 Abrams tank fired, though it was never determined which battalion it came from. The round smashed into one of the undamaged Bradleys, killing the gunner, SGT Edwin Kutz, and wounding two others. What was left of Kutz was wedged in the turret so tightly that the body could not be retrieved until the next morning. All fourteen Bradleys involved in the fight had been hit by RPGs, shrapnel from mortar rounds and small arms fire. Three had been destroyed by US Army sabot rounds from the Abrams tanks. Two men were dead and a dozen were wounded by the friendly fire. Three men would receive Silver Stars for their actions during the battle. Though the Americans had been bloodied, the Iraqis had been mauled. Almost all of their BMPs and tanks had been destroyed by Bradley scouts. The M1A1 tanks had never moved up. Not wanting to risk anymore friendly fire casualties in the fading light, General Funk ordered a halt until morning.[169]

At 0600 on February 26th the VII Corps continued their blitzkrieg across the Iraqi desert. The 2nd Cavalry led the way, using grid lines on the map to mark their progress. The grid lines were known as "eastings" since the cavalry was

---

[168] A starcluster is a flare that doesn't come down on a parachute but has five brightly lit flares that fall through the sky, and usually burn out before they hit the ground.
[169] Atkinson, *Crusade* pp 430-433.

heading to the east. The men watched the desert pass under their vehicles; it was flat and featureless, like a white pool table. The Corps commander had ordered the 2nd Cavalry to the 70 Easting line, where the Tawalkana Division was suspected to be waiting. He also advised Colonel Don Holder to only lead with his scouts to minimize any damage if they bumped into the Republican Guard. Holder moved all of his vehicles forward, to include one hundred and twenty-five M1A1 tanks, announcing "We're all Scouts now!"

By 0715 the 2nd Cavalry began engaging the outer screen of the Tawalkana Division, killing two BMPs and capturing an Iraqi captain that gave them information on what they could expect. Due to the sandstorm the 2nd Cavalry crept forward slowly, only able to see as far as their thermal sights would let them. The grid lines slowly crept by as the Cav Scouts inched forward. Around 1600 the Americans came under machinegun fire from a group of buildings behind a sand berm. This turned out to be an Iraqi armor training center. Since this was the only target they had, Captain McMaster, commander of Eagle Troop, ordered his soldiers to open fire. The building was hit by M1A1 tank rounds, TOW missiles and 25mm chain gun rounds. The building exploded in a gray cloud of dust, collapsing the walls while flames shot out of the rubble.

McMaster took the lead and continued forward, until he crested a ridgeline near 70 Easting. His gunner yelled "CONTACT!" and McMasters looked through his thermal sight, spotting at least eight T-72 tanks dug into the desert in a "reverse slope" defense. McMaster ordered "Fire! FIRE Sabot!" The first Iraqi tank exploded in a fireball, the sabot round striking right above the turret ring. The loader slammed another round into the cannon and yelled "UP!" in just a matter of seconds. A second round flew downrange, smashing into another T-72 and sending the turret spinning off into the desert. Iraqi rounds flew towards the M1A1s, but slammed into the desert floor to the left and right of McMaster's tank. Machinegun bullets slammed into the steel harmlessly, but made a racket inside the tank. A third round flew from McMaster's Abrams, destroying an Iraqi tank 400 meters away.

McMaster later told CNN "As we crested the rise, my gunner identified tanks to our direct front and he said, "Identify tanks!" I yelled, "Fire!" As I was yelling "Fire" on the intercom, the gun erupted. The round impacted on the frontal slope of the tank and the tank commander was ejected out of the hatch and he himself was in flames."

"As the other tanks crested the rise, they began to select their targets and engaged in almost simultaneous manner. So, fifteen seconds before, there was a cohesive or coherent Iraqi defense. Fifteen seconds later, that defense was completely in flames. When we looked back over the armored vehicles, you

459

think- you think, "My God!" You know, "I"... this is what an armored cavalry troop in the assault can accomplish in that short amount of time." [170]

All this happened in a matter of seconds. The Iraqi gunners, using Soviet technology, had automatic loaders requiring ten seconds to reload each round. During this "eternity" the Iraqi guns aimed upwards, letting the Americans know that they were reloading. The Americans relied on their crewmen to reload, and they were quicker than ten seconds. The Iraqis tried to see through the sandstorm, but could not spot the Americans. They sat helplessly while tanks to their left and right were erupting in fireballs, flinging turrets, men and exploding ammunition around their vehicles. In just four minutes every single Iraqi tank and BMP in the first line of defense had been destroyed. Bradleys followed the tanks, shooting at Iraqi infantry that tried to engage the M1A1s from the rear with RPG rockets.

One anonymous Iraqi T-72 tanker of the 52nd Armoured Division wrote of the experience of being taken out by the Americans. "I was a T-72 driver. I actually fought against American tanks and survived to tell you this story today. My whole crew survived, but our tank's mobility was destroyed by one of yours. We were halted in a defense behind the ridge. The ground was flat in front of us. Perfect ground for tanks to drive and shoot. Also, perfect to get shot. It was a one-sided fight but I firmly believe we Iraqi's fought valiantly and bravely. I have no knowledge of if we attained any strikes on American tanks. My commander was 3 years my senior, and he ordered us to halt with the rest of our group, and fire on the Americans. Through our armor and headsets I could hear low 'pwoom. pwoom. pwoom.' noises of tank guns firing and striking each other. It was scary, but very exciting! As untested tank soldiers, it was difficult to understand what we were doing was real. An American tank struck ours. I heard a sudden loud noise like a metal bar hitting a container. I realized I could not move our tank. We were useless. We could not move. Our commander ordered the rest of us outside. My memory of the battle is like observing a movie through a thin tube. I saw very little." [171]

One mile behind the Iraqi first line lay seventeen more T-72s across a three-mile front. McMaster's executive officer tried to remind him that the 70 Easting

---

[170] Extracted from:
http://www.pbs.org/wgbh/pages/frontline/gulf/script_b.html, *Frontline: The Gulf War* on January 6th, 2012
[171] Facebook post on "Battles and Beers" on October 5, 2021,
https://www.facebook.com/BattlesAndBeersMilitaryHistory/photos/a.24175454107145 1/362924098954494/

line was the limit of advance for the brigade, but McMaster told him that he couldn't stop, he was still in contact.

By 1640 McMaster was able to stop his Troop at 74 Easting. Behind him vehicles lay burning in the desert, the smell of burning diesel and human flesh combining to create a one-of-a-kind stench. McMaster formed a circle with his vehicles, and then had his mortar crews set up their tubes and fire upon fleeing Iraqi infantrymen that headed to the east. The mortar rounds were set to explode above the ground, raining bits of steel down into the helpless Iraqis. McMaster's troop would be credited with killing twenty-eight tanks, sixteen BMPs and thirty-nine trucks, in a fight that only lasted twenty-three minutes. He had not lost a single man.[172]

To the east of Eagle Troop Captain Daniel Miller's Iron Troop moved by the buildings at 68 Easting that McMaster had supposedly destroyed. Amazingly the Iraqis inside fired upon Miller's troop, only to have more cannon rounds slam into the building. Miller's scouts spotted rectangular shapes two miles away in their thermal sights. Miller ordered "Action front, follow me" and led the way in a flying wedge. The M1A1s were in the center and two Bradley platoons covered the flanks of the wedge.

The Iron Troop quickly destroyed the Iraqi first line at 70 Easting and spotted a second line of tanks, moving to counterattack the Eagle Troop. The Iron Troop was able to hit the Iraqis from the front and flanks in a crossfire, TOW missiles and Abram's rounds turning all into blazing hulks. This fight also only lasted a matter of minutes. The only casualty to Iron Troop was when a friendly TOW missile slammed into a Bradley by mistake, wounding three soldiers, who were able to crawl away from the burning vehicle.

Captain Joseph Sartiano had his Ghost Troop move forward slowly past 70 Easting. He had seen the fighting going on to his left, where McMaster's Eagle Troop had rushed forward. Sartiano's gunner saw a shape in the distance, but it wasn't giving off a heat signature. When he looked through the regular scope, he still couldn't make out what was in front of his tank. When Sartiano dropped down the hatch to get a better look through the scope, his gunner saw a squad of Iraqis moving near a BMP. The gunner fired his machinegun, killing the

---

[172] Atkinson, *Crusade* pp 441-444.

surprised Iraqis as they ran to the ramp of their BMP. The Iraqis were torn apart by the machinegun, and the killing finally ended when the gunner fired a tank round into the Iraqi vehicle.

For fifteen minutes the Iraqi tanks, sitting still without their engines on, were methodically destroyed by the Ghost Troop. The Iraqis had dug all their tanks in behind dirt berms that did not allow them to see what was going on to the front and restricted their ability to fire. The American's nicknamed them "kill me berms." Ghost Troop did not run through the Iraqis, like Iron and Eagle Troop had done, but stayed at 73 Easting, firing, moving, firing and moving again. The soldiers of Sartiano's company said the slaughter was similar to being on a training range back home. Gunners on the flanks fired at targets from the outside, in, and those tanks in the middle fired from the inside to the outside edges. Bradley vehicles fired tracers at tanks, identifying the targets to the M1A1 crews.

The only casualties inflicted by the Iraqis happened towards the beginning of the fight, when a Bradley vehicle was struck by a 73mm round from a BMP, startling the crew, then getting hit a second time, that penetrated the hull, killing one crewman and wounding another. Soldiers from nearby vehicles rushed over and pried the ramp open with crowbars, while machinegun fire clanged off the sides of the Bradley.

Though the main fight was over in minutes, Ghost Troop remained at 73 Easting, destroying tanks, BMPS and trucks that moved through its kill sack trying to flee the slaughter. The killing continued for three more hours. At 5:40 the Iraqis attempted to launch a counterattack. Ghost Troop killed four BMPs with TOW missiles fired from 1,200 meters away. The Iraqis stopped and dismounted their infantrymen, only to have them cut down by machinegun and 25mm cannon fire. Sartiano called in artillery, which rained down upon the Iraqis, killing eleven more tanks and BMPs. Mortar rounds landed among the Iraqis and bombs dropped from aircraft that had finally been given clearance to attack. A second Iraqi company launched a counterattack, but met a similar fate. A third attack was launched at 20:00.

The Americans had the advantage with their thermal sights easily spotting the Iraqi tanks and BMPs in the cool night air. Ghost Troop fired at extreme ranges, destroying the vehicles, while 2,000 howitzer rounds and 130,000 bomblets from MLRS rockets rained down upon the enemy. F-16s and A-10s fired on the Iraqi supply trains, and artillery, trying to bring reinforcements to the battle. By 2200 the Battle of 73 Easting was over. Over 200 Iraqi vehicles were destroyed and burning, illuminating the battlefield with an orange glow. One battalion of the Tawalkana Division and elements of the Iraqi 12th Armored

Division were destroyed. Due to American training and superior technology the Iraqis never had a chance.[173]

While the Battle of 73 Easting was being fought the Army's Tiger Brigade of the 2nd Armored Division led the 2nd Marine Division to Al-Mutlaa Ridge, north of Kuwait City. The brigade captured the Al Jahra police station, killing fifty Iraqis that were trying to escape the city. The capital of Kuwait was now encircled, leaving the Iraqis no way to escape except down the Highway of Death. Now both tanks and aircraft continued the methodical killing of anything that moved on the highway.

Just about everything that was flying attacked the Iraqis fleeing the city. The Washington Post wrote that the aircraft "swarmed over the Iraqi armor and truck columns, slaughtering the scattering vehicles by the score in a combat frenzy variously described as a 'turkey shoot' and 'shooting fish in a barrel." After the war over 1,500 vehicles were found destroyed on the highway. Air droppable mines stopped any vehicles from moving north of Al Jahra. Vehicles traveled across the desert, only to be hit with cluster bombs, or traveled down the coastal road leading to Bubiyan Island. Those vehicles didn't fare any better and after the war over 400 vehicles were found burned on that highway.

Only a small percentage of vehicles were military, due to the Iraqis confiscating anything that moved and loading it with loot from Kuwait City. Since the Iraqis were not clearly surrendering, they were legitimate targets and would be torn to shreds by cluster bombs or burned alive in their vehicles. General Colin Powell began to question whether or not this slaughter was necessary. He was worried about a public relations backlash that could tarnish an otherwise impressive victory. Powell was able to convince President Bush that the slaughter needed to stop, and an order was given from the White House to stop the bombing of the Iraqis retreating from Kuwait, unless they fired upon Coalition forces.

At 0100 on February 27th the 24th Division seized the high ground overlooking Highway 8, only seventy-five miles to the west of Basra. Throughout the night the soldiers blocked the highway and captured any Iraqi soldiers that unsuspectingly came into their perimeter. By morning the

---

[173] Atkinson, *Crusade* pp 444-448.

Americans had captured 1,200 prisoners and destroyed a hundred vehicles with tank rounds and TOW missiles. At 0630 the division artillery began pounding Jalibah airfield. At first the Iraqis thought they were under attack from the sky and began firing their eighty anti-aircraft guns up into the air. As the tracers floated up to the sky two battalions of M1A1 tanks and Bradleys crashed through the perimeter fence and raced down the runway, shooting at hangers, helicopters, parked fighters and a tank battalion waiting to get refueled. By 10:00 the airfield had fallen and was in American hands. The 24th Division had originally been given four days to move across the Iraqi desert and take airfield. The Division had accomplished the mission in just sixty-seven hours.

Feeling lucky the Division commander, Barry McCaffrey, received authorization to attack Talil airfield forty-five miles to the west. McCaffrey thought that the mission was essential to ensure that the million-gallon fuel depot, that he was about to install near Highway 8, would not come under any counterattacks. One tank battalion raced up the Euphrates and rode down the runway, destroying MIG fighters, helicopters, and infantry in their bunkers. The raiders quickly left, leaving only death and smoking ruin in their wake, and returned to Jalibah. Though no Americans were hurt on the raid, the battalion did lose two M1A1s and two Bradleys when they had become stuck in the mud in the irrigation ditches throughout the area. The 24th Division then turned their eyes on Basra. McCaffrey believed that the city would fall in twelve hours.[174]

On February 27th we went over to the Kuwaiti compound to see what they had been issued to go to war. They were given RPG rocket launchers and PRC-77 radios for communication. While we were there the Kuwaiti battalion commander told me that they would be leaving at 1100 that day and headed to the front. This was news to us, so all of us immediately jumped in our DMV and went back to pack up our gear. We arrived back to the Kuwaiti compound just before 1100, and then waited, and waited and waited some more. I dozed in the DMV for most the afternoon, while Crowder went over to Melinda's unit and said goodbye to her.

I was assigned a Kuwaiti translator that would be with me for the rest of the war, Mohammed Al-Hambra. We just called him "Ham". I took him over to see the girls at the 44th EVAC, and to get Crowder back to the war. The girls gave me a hospital scrub with all their signatures on it, and they gave us a picture of all of them. I mounted it on the front windshield so we could look at it as we

---

[174] Atkinson, *Crusade* pp 454-456.

drove to war. Later that day we learned that the resistance in Kuwait City said that the city had been liberated. I didn't trust that and decided I would have to make that determination myself.

At 0900 on February 27th the first Kuwaiti troops, with their Special Forces advisors, drove past the 2nd Marine Division and began entering Kuwait City. Schwarzkopf had decided that the Kuwait military should liberate Kuwait City. Joint Force Command- North (JFC-N) consisted of the Liberation (Tahrir), Immortality (Khulud) and Martyr (Shaheed) Brigades. JFC-East consisted of Truth, Full Moon and Victory Brigades. Marine General Walter Boomer later told CNN, "We had made arrangements for the Arab coalition forces to go into the city. And I was sitting there, getting a little restless and wondering, "Okay, what are we going to do? I'm tired of sitting here." So, we drove into the city and the outpouring was something I'll never forget. I don't know where all the people came from. They came down to the side of the road by the thousands and they had Kuwaiti flags and some had American flags. Vehicles that the Iraqis hadn't stolen or destroyed, they had acquired some of those, so they were driving around us in this mad circle and I thought sure we were going to crush a vehicle. What they were saying was, "God bless you, America. God bless you." You know, "We love you." Very emotional moment for us, after all of this." [175]

However not all was celebration, bands of Kuwaitis roamed the streets in pickup trucks with .50 caliber machineguns mounted on them looking for any collaborators. Dozens of Palestinians, always the bastard step-child of the Middle East, were beaten or executed. By mid-March 6,000 Palestinians would be sitting in Kuwaiti jails throughout the city. Special Forces soldiers would be ordered to occupy the police stations in the city and, in the words of Colonel Jesse Johnson, tell the Kuwaitis to "knock that shit off." [176]

On the afternoon of February 27th JSTARS detected a mechanized infantry unit from the Republican Guard Adnan Division moving south towards the XVIII Airborne Corps sector. The Adnan Division had wounded twenty-three American soldiers with artillery fire the night before, but was devastated by a

---

[175] Extracted from http://www.pbs.org/wgbh/pages/frontline/gulf/script_b.html, *Frontline: The Gulf War* on January 6th, 2012
[176] Atkinson, *Crusade* pp 459-460.

counter attack of MLRS rockets and Apache gunships. Not taking any chances XVIII Corps quickly annihilated the Iraqis moving towards them.

The Iraqi Madinah Division knew that there was an American threat out to the west and deployed into a battle line seven miles long to counter any attack. Coming fast at the Iraqis were the 2nd Brigade of the 1st Armored Division. Nine tank companies … 166 M1A1s … were bringing death with them. Shortly after noon the American gunners fired on the Iraqis, over two miles away. Within just a few minutes the Iraqi line was filled with burning tanks, and exploding ammunition.

The Madinah Division had been caught unaware of the threat. The crews were outside their tanks, eating a lunch of tomatoes and rice. When the Iraqis returned fire their rounds fell harmlessly into the desert sand, short of their targets. The American tanks slowed down, ordered by their commander "Don't get bushwhacked!" After firing for forty minutes at the panicked Iraqis, the battle finally ended. The 1st Armored Division had destroyed sixty T-72s, while A-10s and F-16s annihilated any fleeing Iraqis or supply convoys. When some resistance was met, it was broken quickly after being bombarded by MLRS rockets. Over 300 vehicles had been destroyed, with only one American soldier being killed. [177]

The British 7th Armoured Brigade also moved towards Basra, from Kuwait City, and were ordered to secure the road between the two cities before the "UN deadline for the cessation of hostilities." This was news to the British, who had not even heard of a deadline. The Queen's Royal Hussars Museum described what would be known as "The Charge of the Heavy Brigade":

"After a few hours of hoisting out and in, reconnecting and testing, Tiffy declared our vehicle fit to fight again and we headed back towards our position in the line. Such was the thickness of the smog that we had to ask our Troop to listen out for our engine noise as we burbled along back to the west behind the Squadron line at no speed. Eventually we found our way back and eased into the position we had vacated a few hours before. Time for the usual combination of shut eye and stag." [178]

"One minute we were fast asleep the next it was dawn and everyone was shouting. The Immediate Action Drill for this eventuality is to get your kit packed and yourself back into your crew position as quickly as possible and then

---

[177] Atkinson, *Crusade* pp 466-467.
[178] Facebook post on The Queen's Royal Hussars Museum Facebook page, https://www.facebook.com/page/1087446214620464/search/?q=desert on 17 July 2021

find out what the hell was going on. It was madness. Tanks were firing up their engines all along the line. Columns of blue-black exhaust smoke shooting diagonally up into the air from the tanks' twin exhausts. Turbos screaming and turret systems coming on line. Brew got the tank started."[179]

"Pete flicked the radio systems into action. Gus got the Gun Control Equipment warmed up. I reached into the turret and pulled out my helmet with its integral radio fit so that I could start to get some situational awareness. Were we under attack, were we moving, where to and why? In minutes the whole Squadron was all on board in their crew positions." [180]

Either side of me Brad and Urby gave me a thumbs up. We were ready to go and fight. I looked left and right down the Squadron line. It was a similar madness of activity. Tank commanders were gesticulating to each other as if to say 'I haven't got a fu****g clue what just happened but now we're ready to go, what is it all about?' [181]

"Toby gave a quick set of battle orders. The Armoured Regiments of 7th Armoured Brigade are to advance in line to secure the road from Kuwait City to Basra and we were to do it before then UN deadline for the cessation of hostilities. What deadline? What time for the cessation of hostilities? This was news to us frontline mushrooms – kept in the dark and fed on shit. Apparently, some bright spark in the White House had come up with the catchy '100 Hours War' phrase."[182]

"Now we had no time whatsoever – an hour – to cover the 40km from our current position to the Basra Road before the expiry of the deadline. 'Four Zero, you're point navigator,' ordered Toby. Outstanding. On the point for the Squadron for a 40km cavalry charge. I plugged the eastward waypoints into the Trimble. 'Go, Brew, go'. We accelerated out of the line, Brad and Urby either side of me. 'Go f**king where, boss?' 'Turn right and go straight. I'll keep you straight'. Brew pulled a right stick and we came round to face the east. I glanced rearwards through my episcopes. The rest of the Squadron was falling in either side of me and slightly to the rear in a flat arrowhead with Squadron Headquarters tucking in behind me. We accelerated away to the east." [183]

---

[179] Ibid.
[180] Ibid.
[181] Ibid.
[182] Ibid.
[183] Ibid.

Triple Canopy

"We kept going east at full pelt. 40km/h. But this was no Valley of Death. It was a straight forward, old fashioned charge. It was hugely exhilarating. None of us had ever envisaged when we started our military careers in Munster that we would take part in the longest and fastest modern day cavalry charge. More than a hundred tanks in almost line abreast, charging to the east. If there had been any Iraqis in front of us, it would have been a truly awesome and frightening experience. The ground shook like an earthquake. Clouds of desert dust and exhaust dust billowed up behind the tanks. At the base of the clouds, small black dots, gradually becoming larger and then suddenly, they were tanks. We would have swept straight over any resistance like a steel tsunami, without hesitation, without break." [184]

"As the distance between us and the Basra Road counted down, so did the minutes to the deadline for the cessation of hostilities. The pollution clouds billowed in again, reducing visibility to only a few hundred metres. We slowed but maintained our movement eastwards. We were nearly on top of the highway now. A Gazelle recce helicopter materialized out of the smog on our left flank, flying at no higher than the turret of a tank. A couple of Iraqi soldiers appeared out of the murk. They were old men in greatcoats waving their hands above their heads and walking around in a dazed manner. Perhaps they had been victim of one too many air strikes or maybe they had been abandoned when their unit had pulled out. We by-passed them." [185]

"Over the radio came the shocking news that a US A-10 Warthog tank buster had taken out a Warrior killing a number of British troops. A Warrior would not have stood a chance against an A-10. It was armed with air-to-ground armoured-vehicle-killing Maverick missiles and a 30mm gatling gun firing depleted uranium tipped, milk bottle sized rounds which would have ripped open an Armoured Personnel Carrier. The frustration and rage was huge. So far the only battle casualties that we knew about had been ones that we had from being shot up by our Allies." [186]

"At a brief halt, and despite the still present clag all around us, soldiers could be seen out on their turrets adjusting the large Day-Glo air recognition panels fixed across the rear bins of their tanks. Others broke out the Union Flag and

---

[184] Ibid.
[185] Ibid.
[186] Ibid.

468

the Irish Tricolour and attached them to their radio antennas. Anything to make ourselves visible to the death to our sides and to the death above us." [187]

"Instead of Lancer pennants pointing at the enemy as they had at the Charge of the Light Brigade in 1854, our pennants now pointed skywards as a warning to a pilot at 8,000ft. It was almost a mediaeval display of pageantry. The wind changed again and the pollution blew away to somewhere else leaving us to drive up the highway embankment and cross the Basra Road in clear sunshine. Disappointingly there was nothing on the road. No convoys of retreating Iraqi troops to shoot up. Just a few burnt out vehicles on either side of the highway. But we had made it by the deadline. Objective secured." [188]

"It's now half past eight on the morning of the 28th and we have moved about forty Ks eastwards into Kuwait to intercept the main MSR. The whole Division appears to be moving into quite a small area. On the ground there are quite a lot of shell scrapes. Now we're waiting in a big circle waiting for the other call signs to catch up with us." [189]

Though we were ready to go, and waited in the vehicles all day to deploy, we ended up not going after all. The Kuwaitis stalled, because there still was fighting going on around Kuwait City. All of the advisors returned to KKMC extremely pissed off. Our morale was as low as possible. None of us could believe we ended up with such a group of cowards. We were told that 42 of Saddam's divisions were gone, either destroyed or surrendered, and that over 3,000 tanks had been destroyed. To make it even worse, as the Battle of Basra began Saddam Hussein said he would give in to the United Nations demands. All the Coalition forces were told to stop their offensive movements and hold in place. A cease fire was now in place as Washington and Baghdad worked out how to end the war. Bush told the American public, in a televised speech, "Kuwait is liberated. Iraq's army is defeated. Our military objectives are met."

Historian Bryan Perrett wrote "Just as the desert is incapable of compromise, battles fought therein result in total victory or total defeat." Marcus Licinius Crassus led 40,000 Roman legionnaires into the Parthian desert at the battle of Carrhae, with only 5,000 getting out alive. The British annihilated ten Italian divisions in North Africa, taking 130,000 prisoners. At El Alamein the British

---

[187] Ibid.. What he described, using different flags to try to make aircraft realize that they were Coalition vehicles, is the reason we flew various flags off or our DMVs.
[188] Ibid.
[189] Ibid.

defeated Rommel, leaving the Germans with 55,000 dead, wounded or captured. This was our desert war, and like the ones in the past it was extremely one sided.[190] With the news of the war being all but over, our Kuwaitis "heroes" finally decided that they would move out the next day and "liberate" their city.

At noon on February 28th, D+4, we arrived at the Kuwaiti compound and waited for them to go to war. Throughout the afternoon they attempted to cram as many of their men as they could onto large cattle trucks. Our Kuwaitis still didn't have a uniform, so they put on their camouflaged chemical suits so they would look like an army. The chemical suits had a green "splinter" pattern, similar to the British camouflage. Most of the soldiers forgot to leave their suits out, so they had to unload all of their suitcases off of one of the flatbed trucks. On top of all this chaos, there was a sandstorm that continued through the day. As the Kuwaiti chain of command tried to undo all of the mess around them, truckloads of Kuwaitis from the other brigades drove past us on the way to the border. They had their flags flying and the men raised their rifles in the air, cheering as they drove past, like they had already liberated their country. This made us even angrier because the Kuwaitis had done nothing so far except sit this war out, while others were killed and wounded freeing their homeland.

We listened to the radio traffic as we waited for our Kuwaitis to finally move out. One report stated that the 5th Group commander had taken 450 prisoners, while another report told us of how some of our guys were stuck in the sand with their vehicle, and an Iraqi T-72 drove up, pulled them out, and then surrendered to them. Supposedly seven engineers from the 82nd Airborne attacked the French by mistake and were killed in a friendly fire situation. Rumors also told of a United States airborne unit jumping in front of the Iraqis to cut off their escape to Basra.[191] We were told the Iraqis then took 800 Kuwaitis hostage and put them in their vehicles so that they could leave Kuwait without being bombed. Most of these rumors turned out to be false, but we didn't know that till later. Right then we were stuck, sitting in a DMV, in the middle of a sandstorm, waiting for our Kuwaitis to finally move out. It was extremely frustrating.[192]

---

[190] Atkinson, *Crusade* pp 250-251
[191] This was actually a mission that the 82nd Airborne was ready to execute, until Schwarzkopf called it off.
[192] In an account from a Marine in the 1st Marine Division, he told what happened to many of these hostages. "In Iraq, 2004 in the Al Anbar Province. Our platoon was tasked with providing security for a group of Kuwaiti forensic scientists who were

Chapter 3: SPECIAL FORCES – A Storm in the Desert

Our convoy finally began moving at 1600, but the Kuwaiti baggage truck was stacked so high with suitcases and packs that it toppled over, spreading the Kuwaiti's belongings all around the highway. After an hour of reloading the truck, the convoy began to move again. We had about sixty vehicles heading to Kuwait that included cattle cars, flatbed trucks, Mercedes Benz's, Ford F150 pickups and vans. All were flying Kuwaiti flags from their antennas. As we drove through Ruqi, staying on the road, we could see the damage done since we had left it a month ago. The Kuwaiti border was our first checkpoint and as each of the advisor's vehicles crossed the border, they would say something on the radio.

"Finally!"

"Kuwait City, here we come!"

"Humdullallah" (Thank God in Arabic)

"Let's get this shit over with!"

"Go tell the Spartans that we obey"[193]

Kuwait looked a lot like Saudi Arabia; the major difference was the burning skyline from all the oil well fires and the miles of destruction. All along the road to Kuwait City there were destroyed tanks that were so covered in holes they looked like Swiss cheese. When the Kuwaitis saw the first destroyed tank, they stopped the convoy and got out to pray. The entire battalion went down on their knees and while murmuring the call to prayer our turret gunners pulled security. The rest of us topped off our gas tanks with the five-gallon cans of

---

recovering the remains of Kuwaitis. The remains were from 1991 during the Gulf War. Saddam Hussein used captured Kuwaitis as human shields on his bases so U.S. and coalition forces would not bomb them.
After the war instead of allowing them to go back to Kuwait, he has them taken out to different areas in the Iraqi desert. From there, a big trench was dug and they were shot in the head execution style and buried. Some were still alive when buried because their arms and hands (skeletal remains) were covering their faces as dirt was being thrown on them.
I stood and watched as the Kuwaiti forensic scientists pulled the remains out of the ground. The skulls were men, women and children and they all head bullet holes in them. The Iraqi that led us to one of the sites was the same man who drove the bus in 1991 that took them to their death. Seeing this had me thinking that we were there for a reason and it was worth it." Facebook post on "Battles and Beers: Military History" on October 13, 2021.
[193]   From tape recordings owned by Patrick O'Kelley.

diesel. I started the diesel flowing by sucking through a garden hose, and then gulped down what felt like a gallon of diesel. While the Kuwaitis were on their knees praying, I was on my knees coughing up fuel.

As we continued down the road, we came to a minefield with handwritten signs in English that said it had been cleared. We didn't trust it and we slowly worked our way through the mines. This was tricky since part of the road had been blown away by bombs and there were huge craters that we had to maneuver around. We passed by a truck that had been blown onto its side by a mine, and beside it was mines stacked in piles, either waiting to be laid or had been picked up for destruction. At the end of the minefield there was an abandoned 120mm mortar that looked like it had not been scratched. I hooked it to our DMV and pulled it down the road. I was going to use it as a roadblock later on, but as we moved forward at a fast rate the mortar unhooked itself from the DMV, and began to pass us, then swerved off the road and tumbled into the desert.

There were anti-tank ditches spanning across the highways. These were giant trenches that were too wide for a tank to drive across. Some were filled with fuel to be set on fire when the tanks crossed over. However, our tanks didn't attack from that angle, and instead had gone around way to the west. There were destroyed buildings along the road, blown up by either aircraft or tank fire. Craters covered the desert as far as the eye could see.

Kuwait City

There were numerous delays on the road to Kuwait, mainly from either flat tires, or stopping to let everyone relieve themselves on the side of the road. We passed through the Egyptian armor, and then the Saudi army, and Kuwaiti army, and finally we slipped through the Marines.

Kuwait City was surrounded by concentric rings, like a bullseye, and each ring was given a number. The ring in the center was the 1st Ring Road, and then working out towards the desert was the 2nd Ring Road, and then the 3rd Ring Road, and so on. No one was allowed past 6th Ring Road, except the Kuwaitis and their advisers. The 6th Ring Road was not small and was as wide as an interstate in the United States.

At 0330 we finally pulled into a warehouse parking lot by the road and parked alongside another Kuwaiti division. I could hear automatic gunfire in the distance, and a series of explosions. A lone armored vehicle cruised the streets in the distance, but I wasn't able to tell which side it was on. There was a large fire in the city north of us, which told me the city had not been cleared yet. I was still able to see the Marines south of 6th Ring Road. There were two captured T-55s parked beside the Marine's M60 tanks. Blocking the west bound lane of the "interstate" was an Iraqi convoy that had been destroyed. There were huge holes in the armor of the tanks, caused by the Marine armor or our aircraft. A few of the Iraqi vehicles still burned, casting light upon the city. CH-46 and MI-8 helicopters flew over our position every few minutes.

We sat in our DMVs, in the western part of Kuwait City known as As-Sulaibkhat, listening to our Kuwaiti PRC-77 radios. They spoke rapid Arabic, sounding like they were arguing with each other, but the Kuwaiti Division had not made a decision on what to do yet, so we could only stand by and wait. Only the 16th Battalion of the Khulud Brigade was in Kuwait City, and the rest of the Brigade would have to be shuttled into the city, using whatever vehicles they could find.

I slept for about an hour until a nearby shot startled me awake. It was a Kuwaiti firing off his weapon by accident. Since we had issued them ammo they had a dozen accidental discharges, but luckily no one had been wounded. Everyone else on the Team was awake also. Though they could have slept in their vehicles, we all wanted to stay awake and see what would happen next. Eventually everyone was able to doze off a few hours until we moved.

Around 15:00 on March 1st we finally moved out of the warehouse parking lot and out to the center of our sectors. The base of operations for our Battalion was going to be a school that looked like any average High School. The Iraqis had used it as an interrogation center before we took the city back. As we drove to the school we passed by Major Star____. I think we expected a "Hello, how are you" or a "nice to see you're alive" or something, but instead all he did was get angry, and then complain because we hadn't stopped the convoy when we passed by him. We didn't think much of it, since we were attached to B Company and no longer under his command.

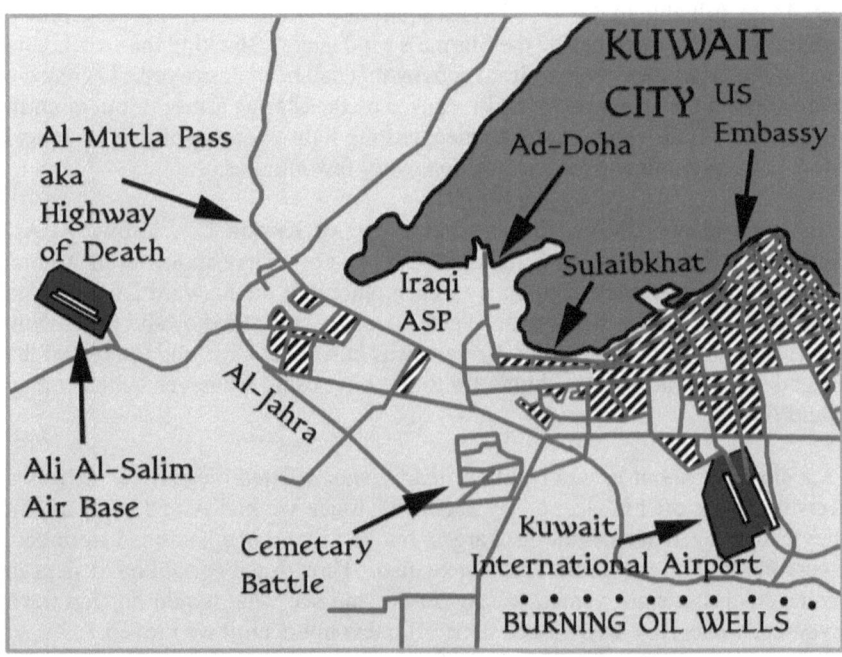

As we drove deeper into the city, we passed by the armor of the Kuwaiti Shahita Brigade. They had Soviet T-72s and BMPs and American M113 APCs, all flying the Kuwaiti flag. When we arrived at the school the four American DMVs covered the four approaches to the school, while a company of our Kuwaitis went inside the school to clear it and secure it. While the Kuwaitis went from room to room, I overwatched the school with my sniper rifle. After the school was cleared the Kuwaitis raised their flag on the school flag pole. When we questioned the people who lived near the school, we learned that the Iraqis had taken 5,000 civilians as they retreated from the city. Many of our Kuwaitis had family members who were taken. We also learned that in the last few hours twenty-two Iraqis had been killed, when they decided to not surrender and decided they were going to fight to the last man.

Throughout the day we could hear bombs going off around the city and continuous shooting. We weren't sure if this was a firefight or the Kuwaitis firing happily in the air because their city was being liberated. Our Kuwaitis moved into the school and set up operations while we parked the DMVs on the basketball court outside, beside the school. Our quarters were the floor of the gymnasium, beside the basketball court. There was still running water in the gym, unlike many parts of the city.

474

Chapter 3: SPECIAL FORCES – A Storm in the Desert

Later that night the rest of the Khulud Brigade showed up at the school, arriving in buses. When they got out of the vehicles our company cheered and then began firing in the air. We didn't know about this and had been sleeping in the gym on folding cots. When the gunfire erupted, I jumped off the cot and hit the floor. My pistol was in my hand, but I didn't remember pulling it out of the holster. It took a few minutes for my heart to slow down and get back to sleep. I was lulled asleep by the sound of the Marine Cobra gunships patrolling the beach.

The next day, March 2nd, we were told by the B Company commander, Major Phelan, that all that was left of the Iraqi army were five brigades, waiting for a bridge to be built in Basra so that they could retreat back into the perceived safety of their country. Earlier, when we had drove into Kuwait, we were told the military had taken 300,000 enemy prisoners of war and that the Iraqis had lost 150,000 killed and wounded.[194] He also told us that the United States had lost seventy-two killed so far, and about 150 wounded.

The question that many wanted to know after the war was exactly how large of an army did Iraq have? We had been told we were fighting the fifth largest army in the world with a one-million-man army. Even the day before the Ground War began Schwarzkopf had told Dick Cheney that if he wasn't allowed to delay the Ground War by a few more weeks there would be 5,000 coalition casualties in just the first few days. However, the estimates were from Iraqi reports after the Iran-Iraq war in the 1980s. No intelligence estimates had been updated since that time. Schwarzkopf did not want to underestimate his enemy, which every commander had done in almost every war, so he went with the worst-case scenario and used the totals from the Iran-Iraq war as his base.

So, how many Iraqis manned the trenches when the Ground War commenced? Post-war estimates figured that the Iraqi army had as many as 362,000 in the trenches, and as low as 183,000. This number faced a coalition force of 700,000 moving across the desert. Iraqi losses were estimated at 153,000 deserters and 26,000 killed during the Air War. Total casualties for the Unites States in Desert Storm were 293 killed (148 in combat, 145 killed by

---

[194] According to "Desert Storm, a forgotten war" there were 85,000 Iraqi prisoners of war, with less than 2,000 of these being wounded. The total killed and wounded is contested but it would have been between 30,000 and 150,000. The authors also wrote that the best estimate for Iraqi casualties was 50,000. Extracted from http://books.google.com/books on December 12th, 2010, *Desert Storm: a forgotten war*, Alberto Bin, Richard Hill, Archer Jones, Praeger Publishing 1998.

Triple Canopy

accidents) and 467 wounded. After the war critics downplayed the victory, as if a sea of blood was needed to achieve glory.

Though the cease fire had gone into effect at 0800 on February 28th, there was still intermittent artillery and mortar rounds fired by Iraqis who did not get the word, or just didn't care. Counter battery fire from the Americans was swift and massive, destroying whatever pockets of resistance were left. In one instance, two busloads of Iraqi soldiers stopped at a roadblock on Highway 8. The Americans at the roadblock thought that they were surrendering, and originally that might have been the plan for most of the Iraqis on board the buses, but when the buses slowed to a stop, two holdouts whipped out AK-47s and began firing at the Americans. The Americans returned fire, bullets tore through the buses, and the surrendering Iraqis, killing seven and wounding six.[195]

The American 24th Division was ordered to just observe the Iraqis fleeing north through Basra, and "if they're not shooting at you, leave them alone." The Iraqis bulldozed a ramp around several craters on the causeway across Lake Hawr al Hammar so that the Republican Guard Hammurabi Division could get past the Americans observing them. Around 0800 one of the Iraqis fired at the Americans with an RPG rocket. The soldiers returned fire with their 25mm chain gun. Other Iraqi BMPs and tanks turned their turrets towards the American and began moving against them. The commander of the 7th Infantry Regiment begged to return fire, General McCaffrey ordered him, "Take 'em apart!"

As if they were on a training range back home, the commander ordered "Ready on the right? FIRE! FIRE!" and a volley of M1A1 tank rounds and Bradley TOW missiles flew across the lake, slamming into the Iraqi T-72s. At the same time five American artillery battalions opened up on causeway, sending hundreds of artillery rounds and MLRS rockets raining death upon the outmatched Iraqis. Two Apache helicopter companies flew overhead, raining Hellfire missiles down upon anything that appeared to be hostile and not already on fire. The helicopters fired 107 Hellfire missiles, with only five missing their targets.

The American 64th Armored Regiment went around the flank of the 7th Infantry and into the Rumaylah oilfield, using the higher road there to fire upon Iraqi ammunition convoys heading into the battle. The regiment continued

---

[195]   Atkinson, *Crusade* pg. 482

476

traveling up the road, destroying all they could see for seven miles. Another company veered off the road and attacked an Iraqi garrison that had dared to fire at the attacking Americans.

General McCaffrey decided to get into this final action, ordered his helicopter to land, and then jumped into a Bradley fighting vehicle, charging into the fight. The Division commander captured a dozen panicked Iraqis waving a dirty white flag. The shooting finally ended around 3:00 in the afternoon, when there were no more targets to shoot at.

Many Iraqi soldiers had stripped off their boots and equipment so that they could run faster into the salt marshes. In an area that covered eight by twelve miles there were nothing but burning vehicles, and rotting corpses. Thirty Iraqi tanks, 147 BMPs and over 400 trucks were burning in the desert. Several hundred Iraqi soldiers had been killed. Tank ammunition continued to cook off in spectacular fireworks displays. The only American losses was one M1A1 tank damaged and another vehicle destroyed by the exploding ammunition from the burning Iraqi vehicles. One of the most destructive battles of the war happened two days after it had ended.

On March 3rd our senior advisors left to scout out a new headquarters, while we remained with our company. We weren't told what the mission was; only an ambiguous order to clear our sectors. There was no real plan, so we made it up as we went along. Our first mission was to take our Kuwaiti company and go clear a telephone center. When we arrived, each platoon was tasked to clear a part of the building complex and one of our advisors would go with them. I would be in a supporting role, as a sniper, watching for any possible enemy activity.

Through my scope I saw an American lieutenant colonel walking near the parking lot with a Kuwaiti civilian. I walked over and asked him what he was doing there. He said his name was LTC Richards, an advisor to the Minister of Public Works and that the building had already been cleared. Not totally trusting this unknown colonel, I told him my company was going to do it again anyway. I was suspicious of what he was doing there because no Americans were supposed to be inside the 6th Ring Road.

Trying to sound casual, I asked him if he had told the head American, SFC "Q" Balot, about the building already being cleared. He said he had, so I figured there was not much I could do, since I was more concerned with what was

477

happening with my Kuwaitis. I told him he needed to leave, since I couldn't ensure his safety around the Kuwaitis. The Kuwaitis didn't know who he was and they might shoot first and question later. He agreed and began to walk away. After taking a few steps he turned to say something to me, just as a Kuwaiti soldier rounded the corner and saw him. The Kuwaiti fired from the hip, from a distance of about two feet, and the LTC dropped like a rock.

I ran to the Kuwaiti, yelling in Arabic that he was an American and to "Cease Fire!" The Kuwaiti turned and aimed his M-16 at me, but I slapped it away, grabbed him by the collar and yelled in Arabic to clear his weapon... "NOW!"

I started yelling "American down! American down!" towards the direction of our medics. The LTC was face down in the parking lot, and I didn't see any blood. I rolled him over and asked him was he OK, and he said back "Fuck NO! I'm not OK!" I figured this was a good sign and he might live.

I asked him where he was hit and he said it was his chest. I ripped open his shirt, but saw no blood or bullet wounds. There was blood spattered around his face, from a small hole in his neck. He was holding his chest, so I moved his hand out of the way so I could get his bandage out of his first aid case. When Baby Doc arrived, I was holding the bandage over the neck wound. Baby Doc took over and I went over to the DMV, to make sure a MEDEVAC was coming. Crowder had already called in the 9-line request and K.C., back at the school, relayed it to the choppers.

While Be__ set up a landing zone, I stopped traffic on the road leading to the PZ. Baby Doc saw the Blackhawk helicopter in the distance and signaled it using his survival mirror. As the chopper got closer Be__ popped a green smoke grenade in the road. The Blackhawk came in, with four COBRAs in support flying nearby. The medics on the helicopter put the LTC on the stretcher and then carried him to the bird. When it flew away, I was surprised at how long the MEDEVAC took. From the time the LTC was shot, until the helicopter flew away was less than half an hour. Baby Doc said that the wound was an "in and out", just missing the artery, and it would not be serious.

After the dust from the MEDEVAC had settled the Kuwaitis continued clearing the phone center. When they finished, they gathered together and shouted "Allah Akhbar!", thanking God that they had finished their first combat mission without any casualties. Charlie Blake pulled me aside and told me what LTC Richards had really been doing there.

THE OIL FIELDS OF AL-MAWQA

When the LTC had been on the ground he told Charlie to take the pistol out of his pocket. He had a nickel plated .45 caliber pistol. Charlie didn't think anything of it, and told the LTC that he would put in the trunk of the LTC's vehicle that he had parked by the phone station. After the MEDEVAC chopper left Charlie popped open the trunk on LTC Richards' civilian vehicle, and found a large amount of jewelry, coins and other riches, taken from people's homes. Charlie immediately told CPT Mastrovito about what he had found, and Mastrovito told 5th Group. After the war I learned that LTC Richards had been approached by a representative from CENTCOM, and told that he would be going home… immediately. He was not punished, but he also would never see a general's star.[196]

In the afternoon we received our new mission, which was the security of a sector of Kuwait City. My battalion was ordered to clear Dohah and Sulaibkhat and restore order to that part of the city. We were also told that with the war

---

[196] Unfortunately, this is the reality of the military, and officers rarely get punished for an offense that if it had been done by an enlisted soldier, they would go to prison or be dishonorably discharged. I was told that LTC Richards ended his career teaching at the Command and General Staff College, telling the students of how he was wounded by enemy fire when he operated behind enemy lines with 5th Special Forces Group.

officially over, 5th Group would begin shipping teams home. Our company, C Company, would be returning home first, but our A-Team would not, since we were attached to B-580 (B Company).

Our Kuwaiti battalion commander decided to do a leader's recon and put his company commanders in his personal car and then took off like he was being chased. We tried to keep up with the speeding car in our DMVs, but we lost him several times. We finally found him and his officers while driving through Dohah. Our sector was right on the beach and had Iraqi tanks, trucks and vehicles all over the road. Most were destroyed by cluster bombs. The vehicles had been rummaged through by souvenir hunters and their contents had been thrown out on the highway. We stopped at a few supply trucks ourselves to see what we could find. I threw a case of Iraqi radios into my DMV and also got a periscope from a BTR-60.

Along the road were dead Iraqis, either laying in the dirt, or sitting up in their vehicles. We stopped at one crowd of Kuwaitis to see what was going on and found a bunch of women kicking a corpse and spitting on it. The dead Iraqi had been hit in the head by something and was missing the top of his skull. We covered him up with a blanket lying on the side of the road, and then drove on through our sector.

In our sector was "Entertainment City" which was the Kuwaiti version of Disneyland. However, after the Iraqis left it was full of mines, ammunition and explosives that we planned to clear out as soon as possible. Driving was a becoming hazardous because the road was crowded with civilian cars flying the Kuwaiti flag and honking their horns, happy over the liberation of their city. All of them wanted us to stop so they could shake our hands and take our pictures, while they thanked us for freeing their city. One Kuwaiti in a BMW sports car hopped out and gave me a box of Cuban cigars. Another, driving an old Ford sedan handed me two bottles of Johnny Walker Red, and, looking around, told me to not tell any other Kuwaitis where I got the liquor.

We passed by convoys of Egyptians, Syrians and Saudis roaming through the city, looking for loot or souvenirs. We also saw packs of Marines, hunting for souvenirs, which we pursued and told them to get back to the 6th Ring Road unless they wanted to be reported. The coalition forces had set up road blocks to check for Iraqi soldiers and around the road blocks normalcy was returning. A few gas stations had opened up, but the lines were incredibly long. Reporters from various news agencies flitted about, filming everything and looking for some big story that they could lead the news with that night. All were hoping there were scenes of atrocities that they could film and asked us if we had seen anything like that. We just smiled, didn't say anything, and drove away, leaving them with a questioned look on their face... wondering who we were. Like in every war, the media are the vultures, searching for someone else's pain and horror, so they can get it on film and look all concerned for their viewers.

The destruction caused by the Iraqis was everywhere. When the Iraqis retreated from Kuwait, they set all the oil wells on fire. The oil wells shot black smoke into the sky, and covered everything with a layer of oil as if it was coming out of a WD-40 spray can. The heat from the oil well fires was incredible, though we did back our vehicles up to one just to get a picture. There were downed power lines covering the roads from towers that had been knocked over. Luckily there was no power, or else we would not have been able to drive anywhere across the live wires. Bunkers were everywhere and covered the landscape. Each room in the city had been fortified for "the mother of all battles", with cinderblock bricks covering every window, except for small firing ports. Inside each room were cases of ammunition, grenades and RPG rockets. If the Iraqis had stayed and fought, we would have lost thousands of our guys trying to clear out this city.

Our leader's recon returned to the school and loaded up the whole battalion into any vehicle that they could commandeer. Some of the soldiers had their family members bring their car to the school so they would be able to move to the new location. We finally moved out at 8:00 that night, and drove to our new headquarters, another school.

The battalion commander didn't want to clear the new school, and told us that he had walked through it during the day. I told him that the Americans would clear it if he didn't want his men doing it. He admitted to us that when we had cleared the phone center that morning the Kuwaitis had six accidental discharges and he was afraid the men might shoot someone again. He finally agreed though and sent his men in, with us following them.

We discovered three Iraqis in the school, looking pretty worn out and wearing civilian clothes. They all had weak excuses while they were there, and told us, through our interpreter, that they were there on a work permit from Iraq. The Kuwaitis didn't care if they were Iraqi, but they told us the prisoners were Palestinians. The Kuwaitis hated the Palestinians and wanted to kill all of them. It seemed like everyone in the Middle East hated the Palestinians. We had the prisoners taken to the Kuwaiti brigade and then we moved into our new base of operations.

We parked our DMVs in a courtyard beside the school. Like most structures in the Middle East, the school had a large wall around the school and the courtyard to keep sand and trash out when the sandstorms came. We set up cots beside the DMVs and some of us put them up against the wall of the school. The Kuwaitis moved their men into rooms in the school, and had men on the roof providing security for the whole battalion. After the sun went down, I walked the perimeter and up on the roof, making sure that all was OK and letting the Kuwaitis see me, and then returned to the courtyard, laying back onto the cot to get some decent sleep.

Sleep was not meant to be. There was sporadic firing all around the neighborhood until 0100. The Kuwaitis were shooting at shadows, and no matter how many times we tried to get them to stop, the next relief would begin firing again. When the Kuwaitis began firing into our courtyard, I sent Nickolai up to the roof to tell them to stop. Nickolai saw them firing at abandoned Iraqi vehicles parked in the same courtyard. They told Nickolai that someone had been running between the vehicles and he wouldn't stop when they yelled at them.

The shooting must have waked up the rest of the building, because when I lay down again it seemed like the whole building opened fire on the Iraqi vehicles. All of us crawled over to the DMVs and got out our weapons, wondering what the firing was about. The sporadic shooting must have drawn the interest of the only Iraqi, or Palestinians, in the area, because a burst of AK fire raked the rooftop. I could see green tracers zipping over our heads and hitting the building. One of our Kuwaitis was hiding behind a rooftop air conditioner and a bullet ripped through the machine and tore into his thigh. A second AK-47 opened up on the other side of the courtyard wall, the tracers racing over the roof of the DMVs. All of our team ran behind the DMV for cover, trying to see where the enemy was. I grabbed two concussion grenades from my LCE that was hanging on my DMV seat and raced over to the wall. Charlie Blake ran beside me, both of us breathing hard with our backs to the wall. He had a frag grenade out, but I told him to let me throw the concussion grenade instead. I didn't want to harm any civilians who might be in the blast radius.

I threw the first concussion grenade over the wall, but all I heard was a pop. The concussion grenade's body was made up of something resembling cardboard, and not metal, so it wouldn't throw off any lethal fragments. When the grenade hit the street on the other side of the wall it broke in half and the grenade fuze flew out, exploding, but not detonating the grenade. I threw the second grenade over the wall, and it exploded, bouncing me off the wall and making my ears ring.

Chapter 3: SPECIAL FORCES – A Storm in the Desert

Gentz wanted to run into the school building, twenty-five yards away, but the Kuwaitis were still shooting into our courtyard at anything they thought was an Iraqi. I told Gentz to stay where he was or the Kuwaitis might shoot him. After 15 minutes of continuous firing, the shooting slowly died away. The Kuwaiti commanders finally realized that there was no more return fire and had finally gotten control of the situation. I could still hear single AK-47 shots firing in the distance, but I don't know if they were directed at us or not.

I took Nickolai with me to the roof of the building to figure out what had happened. He brought his M203 grenade launcher with him. We passed by Kuwaitis bringing the wounded man down the stairs. I hollered back to the courtyard for Baby Doc to take a look at him. When we got to the roof the company commander of 1st Company pointed to where the enemy fire had come from. He pointed to a warehouse, next to the school that was surrounded by a high wall. There were Iraqi vehicles parked beside the warehouse and next to it were three apartment buildings. I told the captain that the firing most likely came from the apartments or the warehouse roof. I had Nickolai fire two flares over the warehouse to see if we could spot any movement, but we saw nothing.

Baby Doc met us at the bottom of the stairs and told me that the wounded Kuwaiti was being taken to the hospital since the bullet was still lodged in his leg. I found the Kuwaiti battalion commander and told him that he needed to send his men to go clear that warehouse next door. He was not sure his men could do it at night and said he would wait until morning. His officers crowded into the room and began arguing with him about what to do. They wanted him to attack the warehouse or call in an airstrike. I told them that air strikes were out of the question in the city. Too many citizens would be killed by stray bombs. While we argued single AK-47 rounds kept snapping by the roof of the building.

I was getting frustrated by his lack of action, and told him that we could support him with flares and smoke grenades while his men rushed the warehouse. He still told us that he would wait until morning when it was safer. I finally suggested that we could use C.S. gas on the warehouse, while his men go in with their gas masks on. He told us that he wanted us to fire into the buildings with our machineguns. I asked him if there were any civilians in the buildings, and he said, "Yes, but only Palestinians." I told him we wouldn't fire upon the buildings if there were civilians there. His next solution was to line the roof with his company and fire into the buildings with M16s. I knew that would also be a disaster, so I told him that we would fire M203 rounds into the building for him. I did this so we would reduce civilian casualties, and maybe shut up the sniper for the night.

I took four Americans up to the roof, with our PVS-7 night vision goggles. I also took Faisal, our interpreter, with a loudspeaker. I told Faisal to point the loudspeaker in the direction of the warehouse and tell them that they had to give up, or we would blow up their buildings. While Faisal was yelling this into the loudspeaker, I had Nickolai fire two flares to see if we could detect any movement. There was still no sign of the enemy. I pointed my PAQ-4 laser to a spot on the ground and told Nickolai to fire a grenade there. The 40mm grenade detonated in the middle of the warehouse parking lot, 200 meters away from the nearest building. We waited, and still heard and saw nothing. One benefit that came from this was that the firing from the AK-47 had stopped. I aimed my PAQ-4 at a wall beside the warehouse and told Nickolai to fire. The round exploded against the wall, the sound echoing around the now quiet city. The explosions had made everyone take notice and the nervous rifle fire from both sides finally stopped. We left the roof and decided to get some sleep. It was 0300.

Right after we left the roof a small white flag began waving out of the window of one of the apartments. The Kuwaitis argued about what to do, but they ended up staying on the roof and not checking it out. Unfortunately, they didn't tell us about seeing the white flag until the next day.

Earlier that day B-580 had their own firefight with an unseen enemy. They nicknamed this fight "The Battle of Cemetery Ridge". Dave Langer, who was on the B Team, wrote "We were looking for a place to set up after we got there. We spent the first night in the old Iraqi HQ (loved the torture room), and then were moving to a more open area. We stopped in an open field, which is where they wanted to put us, but the ground was saturated with oil. While we were waiting for someone to make a decision, we started taking green tracers. One of the Kuwaitis had climbed over a wall into a cemetery to take a piss, and got captured by an Iraqi officer. From what I heard, when the officer turned to call to his men, the Kuwaiti jumped back over the wall and the bad guys started shooting at him."[197]

Langer continued, "It was 583 and 581 at the cemetery. As I recall, the Kuwaitis were doing some celebrating shooting in the air. I noticed some of the Kuwaitis leveling their rifles and shooting into the distance. I then noticed rounds impacting near our location. As the M2 .50 gunner, I climbed into the turret and returned fire where I saw the rounds coming from. After that,

---

[197] Yahoo Groups post from Dave Langer on 02 March 2015.

everyone with a rifle or RPG was returning fire. I must have fired about 100 rounds before the 581 team sergeant came running up to our position and said that we were engaging a friendly unit. It was never confirmed if it was a friendly unit or not."[198]

Kerry Barry on ODA 581 wrote "elements of 585 and 582, the B Team commander Mark Phelan and SGM Dink Jordan, were fired upon as we stopped outside the cemetery and all the Kuwaitis hauled ass. Myself, Kenny Chapman and… Joe Garrido provided 60mm mortar support to 581 as they moved forward on the cemetery. We fired about 30 rounds of 60mm mortar, both HE and illumination. 581 engaged with 40mm and machine guns but I don't believe we ever got anybody."[199]

Tony Rudeen, with ODA 583 wrote "Some Kuwaiti elements did jump over the wall to look over the cemetery, or more likely take a piss outta site of anyone, and jumped back over saying that he saw some Iraqi elements in the cemetery. He came back over the wall and things got very excitable. I can't remember who shot first but it started kind of slow and then picked up steam. Mostly what I remember was some small arms fire incoming plinking off the vehicles that made everyone react and take cover. Some .50 cal started going out and we received incoming, but because of our position it was all over head and not effective. I think that fire that went overhead hit the apartments to our rear and that caused Rob Glass and Buck Davis to turn around and fire up the buildings a little to our rear. Dink Jordan was yelling something, to the effect of "WTF were we shooting at" and we pointed up over head and told him to take a look. That's when he saw the green tracers and basically figured out what was going on. Soon after people started taking charge of the Kuwaiti forces and got them spread out and ready for whatever. The K's got very excitable and started randomly shooting. Mostly sending out RPG rounds that, unbeknownst to me until that moment, would eventually air burst down range if they didn't impact something. All of this probably took no more than a few minutes and when it died down, we moved out and sent back in a couple of elements that did some recon by fire back into the cemetery." [200]

Bobby Wholf of ODA 581 wrote, "81 did the recon by fire. I did see the Major [Phelan] run between a stream of .50 tracers. I remember thinking, with

---

[198] Ibid.
[199] Yahoo Groups post from Kerry Barry on 02 March 2015.
[200] Yahoo Groups post from Tony Rudeen on 02 March 2015.

all the Kuwaitis firing, which side of the cover should I be on? It was amazing no one got hit." [201]

Grant Winkler also remembered Major Phelan's run across the open ground. He wrote that he saw the major "running across that parking area trying to get to a vehicle with tracers going all directions and someone air-bursting an RPG, then a long, slow walk through the cemetery being lit up like Six Flags!" [202]

J.T. Wilson, also on the B Team remembered that himself and "Jack Bunting conducted a two platoon Kuwaiti sweep of the cemetery road that led into the mausoleum during night fall. We drew no fire. Prior to the sweep I asked either 81 or 82 to conduct a recon by fire with their Mark 19s. They had two HUMVs with Mark 19s. They both misfired several times prior to kicking out a belt of about 30 rounds. Several went high and I was worried of civilian damage, though none reported. There was one abandoned Chevy van with its parking lights on, I asked John Schuller prior to the sweep if I could shoot a 203 round into it, he denied it. However, all was good and one of our LT Kuwaitis took it over."[203]

"Prior to occupying the cemetery during the initial firefight, the funniest thing happened. As the brigade convoy was taking fire and Major Phellan and Dink were doing their thing, Jim Magnuson and his BN were doing the infamous Kuwaiti peel. As they passed us going the other way I yelled out to Jim and asked what the heck was going on, he replied "They're Running Away!" Just then we had green tracers coming onto us and an RPG exploded beside Troy Cornwell and his HUMV. I looked back from my pickup truck and saw a Kuwaiti fall face first off the bed of another pick up. Joe (Doc) Garrido bolts over to him, as I yell to Joe "where's he hit?" I watched Joe rear back and smack the Kuwaiti right in his face. Well hell, I'm stunned and thinking I never saw that medical procedure and then scream out to Joe what's going on? Joe looks up and yells back "He passed out from fright".[204]

Sergeant Major "Dink" Jordan wrote "After we got the Kuwaitis under control from shooting up the world, Major Phelan organized a search party with a team or two. A search was conducted and remnants of Iraqis were found, but

---

[201] Yahoo Groups post from Bobby Wholf on 02 March 2015.
[202] Yahoo Groups post from Grant Winkler on 02 March 2015.
[203] Yahoo Groups post from J.T. Wilson on 02 March 2015.
[204] Ibid.

no bodies. Somebody saw people fleeing further into the cemetery, but pursuit was not done due to oncoming darkness." [205]

"At the beginning of firefight, bullets and RPGs were all over the place, Major Phelan was moving toward the wall with tracer rounds going between his legs. I ran towards a truck where Walt Garren was and then moved towards Phelan, while all of this shoot-em-ups was happening and hollering and screaming, swallowing Redman chewing tobacco and having a good ole time. I had more fun trying to get our counterparts to stop all of that rampant indiscriminate shooting. They were shooting RPGs like a mortar. Some team members were crouched down at the wall and could not get over it because the Kuwaitis were shooting over it. After all of it was said and done, me and Major Phelan was leaning over the hummer, passing the chaw around and trying to get the heart rate back down to normal. I told him I was proud of him, (he was hell on wheels in his first firefight). Darkness fell upon us; we slept while drive-bys were happening all over town. This was known as the "Battle of Cemetery Ridge", a battle where an entire SF company was engaged by the enemy."[206]

Over in another part of the city ODA 575 had drive by shootings at the Police station where they stayed. Richard Toellner in 575 wrote that these drive by shooting continued to happen "until all the ordnance the Kuwaitis were stockpiling in our sleeping area exploded. Luckily we were out and about that day." [207]

The next morning, I sent one of the DMVs out to pick up a PSYOPs team that would be assigned to our sector. The PSYOPs vehicle had loudspeakers instead of machineguns mounted on their turrets.[208] I stayed back because I thought our Kuwaitis would be clearing the apartments where the shooting came from the night before. The Kuwaitis decided to set up a road block instead, and proceeded to pick up ten civilians who appeared suspicious when they discovered they were carrying stripper clips of AK-47 ammunition.

Captain Mastrovito and Charlie arrived to investigate what happened during the night. Unfortunately, Gentz had told Captain McDonald that we had been firing into a heavy civilian area. Captain McDonald became angry and drove

---

[205] Yahoo Groups post from "Dink" Jordan on 02 March 2015.
[206] Ibid.
[207] Yahoo Groups post from Richard Toellner on 03 March 2015/
[208] The PSYOPs unit was from the Texas National Guard. Their company was part PSYOPs and part Pathfinder. They called themselves the PSYCOPATHs.

back with the PSYOPs vehicle to find out what happened. I walked Captain McDonald around the area, and showed him the impact marks of the M203 grenades and the impacts of the enemy rounds into our building. He agreed that we had done a proper course of action, though we were limited by our Kuwaiti soldier's experience.

While we were conducting an AAR of the fight, our Kuwaitis began bringing in crates of ammunition and grenades found in the nearby buildings. They also brought in two more prisoners. The Kuwaiti soldiers were pretty rough on the prisoners and I told them they had to treat them in accordance to Geneva Convention standards.

Later that day Kuwaiti civilians began bringing in ammunition and weapons that were stored in their homes. When they brought in two cases of 60mm mortar illumination rounds Nickolai remembered the location of a 60mm mortar on the side of the road. He took a team out and brought back the mortar. After a long night and even longer morning, we moved out to our new location, tired but still alert after the previous night's firefight.

Our new home was a police station in the Ad-Doha area, near Sulaibikhat. The Kuwaiti resistance had been gathering up all the ammunition and putting it into the police station for later use, so the building was filled with Iraqi weapons and ammunition. We had to walk across the crates of ammunition just to get to the stairs leading to the second floor. There were cases of ammunition for rifles, grenades, RPG rockets, mortar rounds, gas masks, mines and web gear. Scattered among the crates were two flamethrowers, three PKM machineguns, one Soviet tank co-axial machinegun, thirty RPG rocket launchers, a 60mm mortar, and an SA-7 GRAIL anti-aircraft missile. Most of the ammunition we found was Soviet or Jordanian. I gave one of the PKM machineguns to the PSYOPs vehicle, along with some cases of ammunition so they could defend themselves if we were attacked again.

I sent all of our company commanders out to their new posts, which were a series of checkpoints on the road leading into Ad-Doha. Staying at the police station would be just our team and the Kuwaiti battalion staff. Our team now consisted of three DMVs, to include the PSYOPs DMV, six Special Forces soldiers, three PSYOPs soldiers, their Saudi lieutenant interpreter, and my Kuwaiti interpreter, Private Faisal.[209] Initially we were given three offices on the second floor of the police station as our barracks.

---

[209] The six SF were Charlie Blake, Nickolai, Baby Doc Westover, Darren Crowder, Brent Gentz, and myself. One of the PSYOPs was a Marine.

Chapter 3: SPECIAL FORCES – A Storm in the Desert

We cleared the building and surrounding offices and collected anything that looked like good souvenirs. Beside the police station we found an Iraqi supply truck full of RPG and mortar ammunition. We were not worried about anything being booby-trapped since there was ammunition all over the city and we had not heard of any of it being rigged to explode. While we were clearing out the police station reporters from NBC continued to drive by attempting to get our attention. They wanted to know if we were Special Forces, but we just ignored them.

Later that afternoon I heard cries of pain coming from inside the police station. When I went inside, I found the Kuwaitis beating the prisoners with sticks. The prisoners were bleeding from the head and were bruised pretty badly. I told the Kuwaitis to stop and that they could not do that. They looked confused and didn't understand why they couldn't beat them. They told me that this was how they got the prisoners to confess. I told them that if they continued beating the prisoners all of the Special Forces soldiers would have to leave, and I would have to report it to my government. They stopped and promised me they would no longer beat any more prisoners.

I immediately radioed Captain Mastrovito and told him of the incident, but I also told him that we had it under control. We had hooked up our little PRC-77 radios to a 70-foot radio tower beside the police station. I didn't think it would work, but Crowder had it running in no time. As soon as I got off the radio with Mastrovito a Marine intelligence officer drove up and talked to us. He told us that he was leaving soon, but he would assign his interpreter, Captain Habiib, to our team. Habiib had been raised in this part of Kuwait City, so he would be a great asset.

We conducted our first street patrol at sunset, taking Faisal with us. So far, we had been up since the firefight the night before with no sleep. We drove down the streets looking for our different companies and their checkpoints. The 2nd Company was living in a warehouse that was empty, but we discovered that beside the warehouse was an Iraqi intelligence jeep, full of communication gear. The jeep door was locked, but I was able to open the door, picking the lock by using my Swiss army knife awl. I was able to hotwire the vehicle and we started it up, driving it back to the police station.[210] Crowder drove it back, but it was difficult since the jeep had no clutch.

As soon as we returned to the police station one of the Kuwaiti civilians asked if we could come over to see something that he had in his home. There

---

[210] Learning to hotwire a car and picking a lock were skills I had learned while in the 3rd Ranger Battalion.

was no electricity yet, so the inside of his house was extremely dark. He told us that there was an explosive on the floor of his bedroom that he thought was a mine. I went into room, with a fading flashlight, and found what appeared to be a mine. Right when I found the mine the batteries died in my flashlight. I was worried about the mine being booby-trapped so I had to feel all around the mine for trip wires or for any other devices that may be attached to it. It was pitch black, and hot. I sweated in the windowless room as I felt for any type of anti-tampering device, but I found none. I soon discovered it wasn't a mine at all, but was a 75-round drum magazine for a Soviet machinegun. In the dark it was hard to tell the difference. I kept the drum, and used it with my captured AKM.

That night we slept beside our DMVs but not in the building, because we felt safer out in the open. It was also definitely cooler. Our interpreters slept in the police station offices, since the heat didn't bother them as much. We were able to get a few hours sleep before the Kuwaiti battalion commander woke us up and told us that he wanted to go to 1st Company. He told us that the company was under fire and they wanted to use the flares from our mortar to light up the area. We drove to the school where we had stayed the night before and found 1st Company. They were surprised to see us and told us that they were not under fire from anyone. Since we were there anyway, I had Nickolai test fire the mortar towards the ocean. He had been a mortar man before going into Special Forces and he knew more about the weapon than any of us. Rat fired two rounds and both failed to have their parachutes open. The Kuwaiti-made rounds splashed harmlessly into the sea.

When we drove back to the police station the Kuwaiti S-3 told us that he had made a mistake... it wasn't the 1st Company who was under fire, but it was the 3rd Company. Throughout the city there was intermittent shooting and tracers flying up in the air, but we didn't know if this was in anger or celebration. I had the S-3 give me all the details this time. He told me that the 3rd Company had surrounded a building with five to ten Iraqis inside. They wanted to use our loudspeakers and the mortar flares. We crawled back into the DMVs and headed over to their area.

We arrived at their police barracks and found the 3rd Company out in the street, all looking towards the buildings down the road. I found the company commander and asked him what he would like us to do for him. He told me that they would wait until morning to clear out the buildings (these guys did not like fighting at night), but right now they wanted us to use the loudspeakers. He told me that he could take seven of his men in one of his vehicles and go over to the target buildings with us. After all the running around that night I was getting angry and I told him that we were not going to be used like this, on an unknown objective, with no support. As we started to leave, he told me that he had two

490

of his squads around the enemy's building and there were two of his men still inside the building. I didn't want to leave any Kuwaitis trapped in a dangerous situation, so we drove over to the building with their civilian vehicles following us.

When we arrived at the building the seven Kuwaitis jumped out of their cars and laid down. The commander then rushed up to me and said "We are here, what do you want us to do?" There were no "two squads" surrounding the building, there were only his seven guys he brought with him. He also didn't have two men trapped in the building. There also wasn't any enemy or gunfire. I was pissed now and I chewed him out for wasting our time. When I returned to the S-3, I pulled him aside and instructed him on the proper way to use the Special Forces soldiers. I also told him that he needed to make sure his company commanders tell us the truth and not a bunch of bullshit stories. The S-3 looked embarrassed and told me that from now on the Battalion commander would be the only one who would send out the Special Forces soldiers to assist the Kuwaitis. Our day finally ended at 0300, when we collapsed in our cots, leaving one person awake to guard.

The next day, March 4th, we woke up around 0900 and had planned to move out on patrol around 10:00. I washed my clothes in a rain barrel, and cleaned up the best I could. Our clothes were black from all the oil in the air from the burning wells. The Battalion commander asked me if I would go back to the 3rd Company police post and investigate some possible booby traps. I agreed and decided to use the PSYOPs vehicle on the way so they could do their announcements over the loudspeakers, telling the civilians to turn in all their military weapons. Unfortunately for the PSYOPs soldiers, the loudspeakers broke right after we left our police station.

When we arrived at 3rd Company, a platoon leader took us to the suspected booby trap, an old suitcase. Gentz checked it out and after he made sure it was safe, we all went into the building where the suitcase had been. Inside I found an AGS-17, the Soviet version of our Mark 19 automatic grenade launcher. We also found an RPG launcher and put both in our vehicle. As I entered the last room there was a sudden blast and the ground shook throwing dust up into my eyes. I knelt on the floor, pistol now in my hand, with my ears ringing. There had been an explosion outside the window. I quickly recovered, and ran to each room, searching for any enemy and making sure no one had been hurt. We backed out of the building and I told the 3rd Battalion commander to clear the block. As he sent his company out on perimeter, I went back to investigate what happened. Someone had thrown a hand grenade into the walled area at the rear of the building. There was a small, smoking crater, and shrapnel holes dotting the wall outside the room where I had been standing.

491

While we were investigating the attack, we received word that all the Americans had to return to Brigade headquarters with "bag and baggage". We didn't know why, but our spirits were lifted with the thought that we would be going home. We returned to the police post and picked up our 60mm mortar, the PKM, and tried to take the Iraqi jeep with us, but it wouldn't go into gear. We left it and headed out to Brigade.

When we arrived at Brigade, we learned that Colonel Shaw had heard of atrocities and torture being done by the Kuwaitis, so he had recalled all the Special Forces teams. We had not heard of any atrocities and the only torture we knew of was when Charlie Blake and Darren Crowder saw the Kuwaitis beating on the prisoners. Our Battalion commander had been there, and we told him at that time that all of the beatings against prisoners had to stop or we would have to leave. At first, he didn't understand what all the fuss was about, but he soon agreed that he would make sure his men no longer torture or beat any prisoners. Since that time none of us had seen any more prisoners, so I suspected that they were keeping them in a different location. Colonel Shaw decided he wasn't going to have his Special Forces tainted with any hint of atrocities, but SOCOM over-ruled Colonel Shaw and told him to send us back to our battalions.

While we were with the Brigade, we learned that the advance parties (ADVON) were going back to the United States and that two men from ODA-584 were going back with them. While we waited for SOCOM to make a decision, I was sent out with my DMV team to escort an EOD team that was trying to find some Iraqi trucks that were loaded with white phosphorous rounds.[211] I was able to locate one of the Iraqi trucks beside "Mr. Green Jeans", an Iraqi corpse inside a vehicle. The Iraqi had been shot in the head while he waited at a stop sign. The car and the body remained on the side of the road and he had decomposed to a green color, which gave him the nickname. Kuwaitis had spray painted graffiti on the side of the car that said "Dog" and pointed an arrow towards him.

As we loaded up the rockets some Kuwaiti teenage girls came over and wanted our autographs. I thought this was a surreal scene. A bunch of soldiers, covered in a thin coat of oil from the burning wells, loading up rockets on the side of the road, while a dead Iraqi decomposed a few feet away, and teenage girls wanted us to sign autographs like we were rock stars.

---

[211] WP... nicknamed "Willy Pete".

Chapter 3: SPECIAL FORCES – A Storm in the Desert

When we got back to Brigade, we were told that we would be going back to our battalions in the morning. The Brigade didn't want us moving out at night because they had received small arms fire for about thirty minutes the night before. They wanted us to stay as reinforcements in case they were attacked again.

When we got up the next day, March 5th, it was still dark and it looked like nighttime for the rest of the day. Between the oil well smoke and rain clouds, all light had been shut off. I sent the PSYOPs DMV back to the SFOB to get their speakers repaired. Their commander gave us some leaflets and posters telling the Kuwaitis to "leave the ammo and explosives alone, do not pick it up."

I also learned about the atrocity that Colonel Shaw had spoken of. We were told that five women had been found hanging upside down in a house, with their skin cut off. The Kuwaitis who found them then took their Iraqi prisoners outside and began shooting them. As they gunned down the prisoners some of the bullets tore through the Iraqis and hit some Kuwaiti guards nearby. We were also told that twenty Kuwaiti women had been mutilated and their bodies were thrown in a cemetery. We drove back to our police post and I picked up Faisal, and then headed over to the 2nd Company to find out if these stories were true.

The 2nd Company was staying in an electric company administrative building. The Iraqis had used the building as a command post for the Mokhabbarat (secret police). The company commander of 2nd Company told me that he had not seen any bodies and had not heard the story about the atrocities. I noticed that there was Iraqi ammunition all over the 2nd Company area, so I told him to consolidate it into one area so they could keep an eye on it. The 2nd Company commander was pretty squared away, and told me that he had cleared his whole sector but found one object that might be a booby trap. We drove over to it and Gentz checked it out. It wasn't a booby trap, though it looked like one. It was in the middle of a mortar battery position and must have been part of their mortar ballistics computer. While we were there, we found another bunker full of ammunition.

Continuing to search for any evidence of an atrocity we drove over to 1st Company. They were staying in a mosque, which was also full of ammunition. I told the company commander of 1st Company that he needed to begin consolidating his ammo, like the 2nd Company was. When I asked him what he had cleared in his sector, the commander told me that they had cleared nothing so far. He told me that they had a firefight that lasted from 0100 until sunrise when a lone AK-47 gunman ran around on the beach and fired rounds in their direction. The commander had put his company on line, facing the beach, and ordered his men to fire in the direction of the sniper. After the initial volley his men charged across the beach and into some bunkers located there. They did not find any enemy gunman.

I dropped off the PSYOPs posters and then went down the beach to see the area he had attacked. The bunkers had been constructed very well and were made of concrete. Most of the bunkers had beds, and electric lighting. The bunker that was the headquarters briefing room had a long table and chairs. It reminded me of pictures from WWII of Hitler's bunker. The FO bunker had maps of all the TRPs (target reference points). I had my men gather up some of the Iraqi SOPs so we could turn them over to our intelligence. The Iraqis had contingency plans in the SOP in case they were attacked by Marines from the ocean, one for paratroopers from the rear and another for artillery and chemical attacks.

The 1st Company commander had not heard of any atrocities, so he took us to his 1st Platoon, located two kilometers north of his location to find out if they had heard anything. I told the 1st Platoon leader about consolidating ammunition, but he stopped me in mid-sentence and told me to come with him. Behind his CP was an Iraqi ammunition supply point. There were two kilometers of bunkers, as far as we could see, filled with every type of ammunition imaginable. On the edges of the supply point were ZSU-23-4 anti-aircraft guns. All around the anti-aircraft guns were piles of spent brass. Interestingly taped to the inside of a bunker by the guns was a poster of Samantha Fox. I thought that at least one Iraqi had good taste.

I asked the platoon leader about his perimeter to the north, another six more kilometers of ground. He told me that no one was there. We left the Kuwaitis behind and decided to check out those six kilometers. The first place we came upon was an amusement park known as "Entertainment City", but we referred to it as Kuwaiti Disneyland. We drove the DMVs right into the park and checked it out. There was the typical ammunition and RPGs stacked around the buildings, but there was so much ammunition that there was not much for us to do except note it was there. We also found some fighting positions and RPG rounds thrown in the water at the "Jungle Queen" ride.

Leaflets we handed out explaining about mines and grenades.

All of us posed around various rides and attractions and took pictures. We took a handful of plastic skulls from a shooting gallery and put them on the DMVs. When Charlie linked back up with us, I told him that that the skulls were real and he thought we had lost our minds.

Triple Canopy

We left Kuwaiti Disneyland and went north to the small fishing village of Ashish Ad-Doha that was on a peninsula. There were anti-aircraft guns and some fake mortars scattered amongst the beach defenses and a lot of bomb damage around some bunkers by the village, but there were no people. I told the other DMVs to clear this area and see what they could find, while I went on to the Brigade headquarters.

Gator kept a map at the headquarters so that we could fill in the areas we had cleared. I highlighted the area we had just gone through and turned in the Iraqi SOPs I had found in the FO bunker. I learned that we were now temporarily going to be assigned to 3rd Special Forces Group, since 5th Group would be going home. I also learned of events happening in Iraq. There was an uprising in Basra when the people there had overthrown the mayor and taken over the town. There was also fighting in Baghdad against Saddam's troops. Things were heating up inside Iraq, and we expected we would become part of an insurgency. I met with the Battalion commander and talked to him about consolidating ammunition and getting it off the streets. He agreed that it needed to be done, but he had no trucks.

The next day, Wednesday, March 6th I checked in on our Iraqi prisoners. We had twenty, ten of which were former Iraqi soldiers. One of the soldiers told me that he saw an Iraqi officer pull Kuwaitis out of their cars and shoot them with a pistol. He told me he would write a statement against this officer. I told the Battalion commander that we needed to get that Iraqi to Brigade, so he could be interrogated. After an MRE breakfast and a cup of instant coffee we started our rounds. The first stop was the 3rd Company, where I found the company commander moving all the ammunition to a Division storage area. He gave me two Italian mines he had found and told me of a cemetery that had been used as an Iraqi ASP (ammo supply point). We called these plastic mines "Gucci mines" due to their fashionable appearance. He also had found a prisoner who had been a MIG pilot and a major in the Iraqi Air Force. Between the two I was pretty interested in the ASP, so we headed to the cemetery.

As we drove down the roads our PSYOPs vehicle blasted out a recorded message, telling the people of Kuwait to leave any ammunition they found alone because it was deadly. We also passed out leaflets or threw them out the window as we drove. We stopped when I discovered an American convoy in my area and I asked them what they were doing there. They told us that they were waiting for one of their vehicles and then they would be heading into Iraq. We drove on to Mister Green Jeans to let the PSYOPs guys take a look at him. Mister Green Jeans was decaying a bit more each day and soon would be gone. As we drove into the cemetery, I found a Kuwaiti who was from the Ministry of Health. I told him about Mister Green Jeans, since I was worried that the

496

decaying body would start spreading some sort of disease. The cemetery turned out to be an ASP and we found bunkers of tank ammunition and some Iraqi trucks. If we could get these trucks working our Kuwaiti companies could use them to move all the ammo from the streets.

At the cemetery entrance there was an "infectious disease" clinic that Baby Doc wanted to check out. While we were there, we ate a lunch of MREs wondering how infectious the diseases were. The clinic was abandoned, but Baby Doc was able to find some medical items he could use. After lunch we went out on a PSYOPs patrol, warning Kuwaitis over the loudspeaker to turn in weapons or to tell any soldier where large caches of weapons were. Usually when we did this Kuwaitis would come out and tell us about weapons, bombs and many other destructive devices, but when we checked it out it usually turned out to be nothing. One old Kuwaiti called us in his home so that we could see the weapons, but it was only a pile of web gear left behind by the Iraqis.[212]

There had been no garbage pickup for awhile and the Kuwaitis would pile the garbage in the streets. One street was covered from end to end with garbage bags. The street looked like a shortcut to where we wanted to go, so I had Crowder drive over the bags. That was a big mistake. Many of the bags contained shit and as they burst it sprayed all over our DMV. After we drove through the garbage road we found a large puddle, and drove back and forth through it quickly, washing off as much shit as we could. The DMV still smelled bad though.

While we were driving through the puddle, we received word over the radio to check out two more places. One was another ASP located at a school and the other was a police station in Dohah. I decided to check out the police station first. When we arrived, we cleared it by moving from room to room with our M9 pistols, but we didn't find anything except cases of mortar ammo. Behind the police station was a trench leading to the beach and all along the trench were bunkers filled with ammunition. I had the team follow the trenches until they came out at the Ashish Ad-Dohah peninsula, where we had been at the day before. We found more trenches, bunkers, pillboxes and barbed wire. Every two feet in the wire were Gucci mines.

A ZSU-23-4 was located in the middle of the beach, aiming out across the ocean. Baby Doc crawled up onto it to see if he could disarm it and he ended up firing it off by accident. While we were trying to disarm the anti-aircraft gun Gentz dug up a mine and asked me if he could set it off. I saw no good coming

---

[212] Web Gear is the name of the pistol belts and suspenders that would hold ammo pouches and canteens.

of that and I told him no. About that time Rat started throwing rocks at the mines to set them off and I told him to stop. As all this chaos was going on a car approached us with a civilian and a Kuwaiti soldier inside. They stopped and the soldier asked me for food and cigarettes. I asked him why he needed food if he was a soldier. While I was talking to him, I slowly drew out my 9mm pistol, not trusting either one. I told him to show me his ID card and he did. He appeared to be legitimate, but I told both of them to get out of the area.

Since all we were doing at Ashish Ad-Dohah were being tourists, I had the team mount up and move on to the other location, the suspected ASP in a school. It turned out to be the school that Rat had fired his M203 at, on the night we received fire. A single T-55 tank sat in the courtyard, along with several Iraqi trucks and jeeps. There were sixty tank rounds stacked beside a wall.

After clearing the school to see if there were any Iraqi holdouts, we drove back to Brigade to update the map. While I colored in the map, we drank Pepsis and read our mail that came in that day. I got my pay stub (called an LES) that said I had made $2,400 that month, which was a huge amount since we weren't paying taxes while in the war zone. I also learned that we were now officially attached to 3rd Group. Some of the 5th Group had already deployed back to the states, but 3rd Group was going to stay behind and train the Kuwaiti army and clean up the city. We didn't know when we would be going home, but the word was that we might be there three more months.

I left the PSYOPs vehicle behind because they had to return their Marine officer, since he was going back home. When I got back to Battalion, I briefed the Battalion Commander about what we had discovered that day. He told me that the 2nd Company had received fire from a building near their position and they surrounded it without firing a shot. They discovered some Americans there firing off an Iraqi pistol. He was angry because they almost shot the Americans. I was pretty pissed off too with the wandering souvenir hunters going into my area. The Americans were just passing through on their way back to Saudi Arabia. They did not know how lucky they were, since my Kuwaitis almost killed them.

My Kuwaitis had found a stack of Saddam Hussein posters and had lined the sidewalk leading to the front door of the Battalion HQ with the posters. That night, for the first time since we arrived in Kuwait, I heard no shots at all, it was totally quiet and peaceful. As if it was a sign from God, the electricity came on, lighting up the building. That night we passed around a bottle of Johnny Walker that a Kuwaiti had given us to celebrate my 30th birthday. Since there were four of us, no one got drunk, but it was a nice night. While we were drinking, we decided that on my birthday, the next day, we would all go to Al Mutla.

498

Chapter 3: SPECIAL FORCES – A Storm in the Desert

Al Mutla was known by several names, "Highway to Hell", "Road to Nowhere", "The Killing Fields" but I think the official name was the Al Mutla Pass. If you wanted to leave Kuwait City and head to Iraq, you had to go through the pass in the Al Mutla ridgeline. As the Iraqis fled Kuwait City our Air Force bombed the lead elements, jamming up the pass. The only way around the pass was to try and go over the ridgeline, but it was too steep for most vehicles and they would get bogged down in the sand. After the initial bombings the Air Force continued with a two-day killing spree. Coalition aircraft, returning from a mission, would unload whatever unexploded ordnance they had on the "Highway of Death". Over a thousand vehicles were destroyed and hundreds, if not thousands, of Iraqis were killed.

Unfortunately, our quiet sleep that night was short lived. Around 2300 there was a message that the police post in 3rd Company's area was under fire and needed support, so off we went. By the time we arrived, the 3rd Company had already taken care of the problem. A lone shooter had fired at them with an AK-47 from a building across the street, and the Kuwaitis fired the building up. All we could give them, for support, was illumination rounds from our mortar. Instead of us doing it I decided to give them a mortar, and had Nikolai show them how to use it. They fired a few rounds, to see how the mortar worked, then when they discovered it was my birthday they sang "Happy Birthday" to me. We drank tea in celebration and then left them the mortar with two cases of illumination rounds.

The next day Gator and the Captain wanted to go with us to Al Mutla, but they still hadn't showed up by 11:00. They radioed us and said they were delayed because they had fifty Iraqi POWs and most likely would not be able to go, so we headed north to Al Mutla without them.

Though it was probably in my head, I swore I could smell the rotting flesh about five miles before we saw it. Once we arrived on the four-lane highway heading north, the view was incredible. For 2,000 meters there were destroyed vehicles spreading out from the highway. The Iraqis tried to run into the desert night, but they were destroyed out in the open. The Air Force had dropped cluster bombs and anti-tank Rockeyes all over their column of vehicles. There were unexploded Rockeyes all over the battlefield. There were so many vehicles around the kill zone that it was impossible to count them. In some places they were four vehicles deep, piled on top of each other, either blown up that way or pushed there by the tanks trying to get through. The fleeing tanks didn't survive much better. We found tanks that were upside down, or shredded like a burst Coke can from their ammunition exploding. The heat from the flames had been so intense that some of the tanks had just melted. Every type of vehicle was there, not just military. There were civilian cars, fire trucks, buses and pickup trucks. The buses were full of melted AKs, RPGs, and mortars. I didn't know if they had been full of Iraqis carrying those weapons or not. All had been incinerated and only the charred and melted metal parts were left.

A one lane path had been cleared through the destruction by our bulldozers so that refugees could travel back to Kuwait City. There was a carnival atmosphere in all this destruction, with representatives from all the armies going through the carnage, looking for souvenirs. French, British, American, Saudi, Syrian, Kuwaiti, Egyptian and Nigerian soldiers roamed freely. After our initial shock we fanned out to do our own exploring.

Going through the destruction I saw evidence of mass looting by the Iraqis. They had stolen everything from Kuwait City and then tried to run when our aircraft caught them in the open. I passed by a group of British soldiers that had cracked one of the safes found in a vehicle and were dividing up $6,000 in Iraqi Dinar. A French soldier had painted the word "France" on the side of an MTLB and was driving it south towards the city. A group of American soldiers had found a ZSU-23-4 that was undamaged and were towing it back to their unit. Many of these trophies would end up decorating the front yards of many Battalion headquarters back home.[213] Another group of Americans had found some Olympic gold medals and were cheering, thinking they were real gold. There was civilian gold and silver jewelry everywhere. I didn't take any of the civilian items, but I did focus on military equipment that we could use. I picked up two mortars that had not been harmed so I could take it back to my battalion. I found an RPG-18, still in its plastic wrapper, two ammo drums for the AGS-17 grenade launcher, and some surveying equipment that our team could use. I also found a case of SVD sniper rifles, still in the packing grease that had never been used.[214]

While I was digging through the chaos, I met up with some Americans who were the first on the scene right after the massacre. They told me that there had been about 2,000 bodies in there and they had pulled them out, most in pieces. While I was going through the wreckage, I found a few bodies that they had missed. Both the corpses looked like they didn't have a mark on them, but they had turned black from being in the sun for days. While I was moving through

---

[213] 5th Group grabbed one of these anti-aircraft guns and put it in their "front yard". A plaque was put on the gun saying it had been captured by one of the ODAs. I guess if we had thought of it, we could have "captured" dozens that day.

[214] In past wars soldiers were allowed to bring home trophies of war, such as pistols or semi-automatic rifles. However since the Vietnam War soldiers were not allowed to do this anymore. I wish I had been able to keep one of those sniper rifles though.

the vehicles there were occasional shots being fired and a few explosions due to some of the vehicles that were still on fire. The exploding ammunition went off like popcorn. The French, not knowing what was happening, became spooked and fired upon the vehicle which made it burn even faster.

We left before it became any more dangerous. On the way back to our sector we resumed the duties of "sheriff" and when we came across a lost American convoy. I gave them directions and sent them on their way. I also found an American warrant officer and SGM in my area, using some lame excuse on why they were there. I knew they were looking for souvenirs and I told them to leave immediately.

We stopped by our 2nd Company to have some tea and to learn if anything new had happened in their area. They told me that two Americans in BDUs and driving a green HUMV had put a picture of Saddam on an abandoned Iraqi supply truck and started firing at it with an AK-47. The idiots were quickly surrounded by my Kuwaitis and told to go back to their own area. I guess they wanted a picture of Saddam with "war damage" to show off back home.

By the end of the day, we drove over to the brigade and filled in the Intel map with the newest information. Be__'s team had found two rifles, an SVD sniper rifle and an old WWII Enfield. We learned that we were now authorized to blow up unsafe demolition, since the EOD techs were too busy to handle it all. On a darker side, we learned that the Kuwaitis were roughing up their own civilians at checkpoints. I had not seen this in my sector, but the Brigade Commander put out an order to stop all such attacks.

I wanted to see Gator's fifty POWs, so we checked in on the Iraqi holding area. There were still the original twenty POWs that had been there for the last two days but no new ones. Ten of those in the holding area were Iraqi soldiers, one was a SCUD driver and two were tied up. One of the prisoners had been beaten on the arms and back with a stick and was bruised heavily. I did try to get the Kuwaitis to untie them, but to no avail.

Near the holding area an old man approached us. He had his eye full of dried blood from where an Egyptian soldier had butt stroked him. Baby Doc examined him and told the Kuwaitis that he needed to go to a hospital. The Kuwaitis agreed and let him go, since they didn't consider him to be a threat. I approached the Battalion commander again, and told him he needed to get the prisoners out of the Battalion area. He said he would try, but he didn't know where to take them.

Triple Canopy

Back in our small shack in the parking lot the guys gave me a chocolate nut cake for my birthday with two candles on it. The number "30" had been glued to the front of the candles. They sang "Happy Birthday" to me and Faisal gave me a Kuwaiti flag with his name written on it. His sister had hidden the flag when the Iraqis came and only recently brought it back out and given it to Faisal.

Though we had a subdued birthday celebration, the war never took a break. Right after they lit the candles on my "cake" a woman came by our post and told the Battalion commander that her house was full of Iraqi soldiers. He told us to attack the house with our team. Since this was getting to be monotonous, I told him no. Every night we were running around chasing shadows and ghosts without ever seeing the enemy. I told him that he had to take control of the situation now, and if he wanted the house to be attacked, he needed to order one of his companies to do it. Right after that, around 2200, we got a message over the radio that the 1st Company ASP was under attack and they wanted us to come and shoot flares. I told them no. Each company had mortars and knew how to use them. If he needed flares he needed to go through his chain of command. I wanted to see if they would solve their own problems without us micro-managing the fight. That night they were able to do it, which became a turning point in how we interacted.

The day after my birthday we decided to head to the AT&T phone center. We learned that the Kuwaiti government was paying for the calls and they would all be free. When we arrived, there were huge lines to use the phones, so we decided to come back after curfew one night when there would be less people. On our return trip we drove down the entire Kuwaiti beachfront and for once we became the tourists. As we drove, I played the pirated tape of Chris Rea singing "On the Beach". We saw miles and miles of beach defenses, anti-aircraft guns and blown-up buildings. All the American fast-food places like Kentucky Fried Chicken and Hardees had been blown up. There was nothing left but burned-out buildings and the anti-aircraft guns in the parking lots. As we drove by the American Embassy, we were stopped by some American M.P.s. They told us that EOD was defusing a possible chemical mine and we had to be ready to put on our chemical gear. After EOD made the mine safe, we continued on to our sector.

When we returned, we took some RPG rounds that did not have the safety caps out to the beach to destroy them with C-4.[215] As we were doing this Captains McDonald and Mastrovito showed up and then told us to detonate the

---

[215] The tin caps were covers on the ends of the rockets and were a safety device. If the caps were removed the rockets could drop, crushing the warhead and setting off the piezoelectric charge that would detonate the warhead.

RPG rounds in a different area. They didn't trust us to detonate them, which really angered my team. We finally destroyed them and drove away from the officers, hoping they would not follow.

Every day since we had been there Kuwaitis would come into our post and tell us they had guns, or bombs, or knew of a minefield. Each time we went out it would turn out to be nothing. We called this exercise the "Wild Kuwaiti Hunt".[216] Every Kuwaiti citizen we met would lie through his teeth and not even blink an eye. We learned not to trust anything that any of them said. We never found the suspected minefield that was supposed to be extremely large. As we did our patrols, I continually bumped into souvenir hunting Americans and would tell them to leave. We once found a group of Saudi officers firing into the ocean, and our Kuwaitis radioed us and asked if they could attack them. The Kuwaitis did not trust any other soldiers. An example of why they did not trust them was when one of my companies had their gear stolen by some Americans who thought it was Iraqi gear. We had to chase down the American soldiers and threaten them with charges to get the gear back.

There were various EOD teams from the different nations trying to defuse all the thousands of explosives in the city. A Saudi EOD team had been working in our area but they ran out of explosives and asked if we could destroy a Rockeye they had found. Gentz rigged the Rockeye to blow, but when we called it in, the officers not trusting us again, called it off. This angered all of us, so we drove to Brigade to see what the heck was going on. Either we were authorized to destroy munitions, or we weren't, but give us guidance that doesn't counter previous orders. We learned that it was Gator who kept stopping us, for no apparent reason, except he didn't trust the team to do the job. He also ordered us to turn in all of our Iraqi weapons and equipment. We finally got permission to continue with the demolition by going over Gator's head, but by the time we returned it was dark. While we sat in our hut, three of the guys swore that when the war was over, they were going to get out of Special Forces. If our Team Leader and Sergeant didn't trust us to do our job, why were we here?

On the morning of March 9th, we returned to the Rockeye, re-rigged the demolitions, and then blew the charge in place. It was in a residential area, so we had to use minimum explosives. Gentz used a one-pound block of TNT and the M113 remote detonator. Since the previous day had been so frustrating, we decided to take the day off and do some maintenance. We cleaned the .50 caliber and the Mark 19 grenade launcher. Afterward we wrote letters home

---

[216] Like a wild goose chase.

and Charlie went to check on the prisoners. He discovered an Iraqi-American, who had an American passport and told him he was from Fort Wayne, Indiana. He had been carrying a fake Iraqi driver's license when he was stopped by our Kuwaitis and picked up. He told Charlie that his parents were Kuwaiti, and had lived there since 1968. When the Iraqis invaded, he decided to stay in Kuwait City to protect his parents. A Kuwaiti friend gave him the fake license so the Iraqis wouldn't bother him. When Charlie told me about the situation, I radioed CPT McDonald then tried to get the guy's passport back. The Kuwaitis wouldn't let me see the passport. CPT McDonald arrived and talked to the prisoner and then the Battalion Commander. McDonald wanted to take the prisoner to the embassy, but the Battalion Commander wouldn't let him go unless the Brigade Commander ordered it. I called CPT Mastrovito and asked if he could talk to the Brigade Commander. After several hours CPT McDonald was finally able to go with the prisoner to the American Embassy.

That afternoon we went to a fig orchard that was reported to have some tank ammunition stored somewhere in there. We found the ammunition and set it off using the fettuccine-looking powder charges stored with the ammunitions. Afterwards CPT McDonald radioed us to go clear the water towers behind the Brigade CP since a suspected sniper was there. When we arrived, we found ODA 584 had already cleared the towers so we ate lunch with them, got our mail and updated the Intel map. In the mail I got a newspaper from Cupertino, California, that had printed a letter I wrote them, and I also received a care package full of food from my parents. While we were at Brigade, we turned in any ammo that we didn't think was essential. All this ammo would go back to C Company. CPT Mastrovito approached me and wanted me to write up some phrases for awards he was submitting. He said I had a way with words and thought it would sound better. He was trying to put in our Team for Bronze Stars for what we did during the Air War. Since it started getting dark, we decided to stay at the Brigade until 2200 so we could go to the phone center. The regular Army troops were only allowed there until 2130, when the curfew took effect. However, the curfew was not for us. Unfortunately, when we got to the phone center, we learned that they also began to close when curfew took effect. The workers at the phone center told us of a new center that would be opening up in Al Jahra, which was only a few kilometers from our post.

The next morning the Iraqi-American that had been taken to the Embassy returned and thanked us for getting him out of the prison cell. He had been cleared by the Embassy and allowed to go home. Most of March 10th was spent trying to find the elusive phone center in Al Jahra so we could call home. We didn't find it but we found a medical bunker complex instead. We also found thirty British Engineers who had occupied an oil rig in our area.

504

Chapter 3: SPECIAL FORCES – A Storm in the Desert

Since we were in the area, I had the Team clear the Ashish Ad-Doha area.[217] There were numerous concrete pillboxes, concrete trenches and bunkers filled with ammunition. The Team found a lot of good souvenirs and every one of us got a treasured AK-47 bayonet. This was one of the few things that we were allowed to bring back home legally. Baby Doc even found two night vision sights that he was going to try to mail back to the States. When we visited the 2nd Company Crowder kept trying to trade what we had found for a couple of MP5K submachine guns, and some bottles of whiskey. They gave us the whiskey and some carved wooden camels that they had done in their spare time while sitting at the guard post. That afternoon our PSYOPs team said their goodbyes and told us they were returning home, however they returned to us after dark. They were as confused as we were.

We updated the Intel map at Brigade and learned that we might also be leaving around March 20th, on the date of the signing of the peace treaty with Iraq. CPT McDonald told us that no bayonets or gas masks could be returned back to the United States in the mail, which made our morale drop even lower. Our spirits did raise a bit when AFN came online in Kuwait City, but it lowered again as we heard the announcements of all the other troops going home.

I figured we needed to do something "real" instead of just wandering around the city keeping busy. I decided to do a PSYOPs mission. Before we left on this mission Charlie did his daily check on the prisoners and found a guy who had been there for four days, but was not seen by us until now. He had been beaten severely, which angered all of us. Charlie, Gentz and Baby Doc wanted to open the prison up NOW! I told them to quit making threats against the Kuwaitis and I simply asked Faisal to get me the key. Faisal returned with the key and we examined the beaten prisoner. He had been an Iraqi intelligence officer who admitted to killing ten Kuwaitis and raping three women. He confessed all this as he was being tortured. His whole body was bruised and he had cigarette burns everywhere. His penis had five cigarette burns on it and his left ear had been ripped in half. His fingernails had also been ripped out.

The Kuwaiti guards told us that he had to die for his crimes. I told the guards that they were not the judge and jury. Faisal told them that the Battalion Commander was the judge there, not them. I also told the guards that they had to take him to a hospital. The guards told me that they wouldn't do it, because he would escape. I told Crowder to call CPT Mastrovito and tell him what was going down at our post. I had Faisal ask the prisoner who beat him, but he was too afraid and said he didn't know. I asked him was it the military or was it the

---

[217] This was where the Army put Camp Doha that was used during the war in Iraq and Afghanistan.

police. He lied and said he didn't know that either. I then asked him when he had been beaten and he told Faisal it was ten days ago, but he had also been beaten at the police post and "somewhere else".

Captains McDonald and Mastrovito arrived and when they saw the prisoner they became as angry as we were. They both marched off to confront the Battalion Commander who said that he knew of the prisoner and that the prisoner had to die. McDonald angrily told the Battalion Commander that he was not a judge, and because of this he was going to have to pull all the American advisors out of his Battalion.

When I went through Phase III and the Robin Sage exercise, I never knew that I would be living the "moral dilemma" part of that training mission just a few years later. We quickly returned to our shack and began packing up all our gear, but ten minutes later CPT McDonald walked over and told us we weren't leaving after all. He had called the Battalion Commander's bluff and we would be taking the prisoner back to the Brigade under guard.

I sent one of our DMVs to escort the prisoner, while the rest of us conducted the PSYOPs mission in our sector. We drove around the different neighborhoods, broadcasting a recorded message and giving out leaflets that told them to tune into a radio station that would give them more information. We also threw out small American flags to the kids along the route, that they excitedly ran out in the road to pick up. By the end of the day, I told the guys we needed to go over to the American Embassy to pick up some more American flags. The PSYOPs guys with us said they had 30,000 flags and we wanted to get a thousand to hand out. All that the Embassy gave us was twenty flags, but while we were there, we were able to witness what the Marine guards called the "4:00 Parade".

Kuwaiti women dressed in skirts began strutting in front of the Embassy. Others cruised by slowly in cars, smiling at the American soldiers posted there. Reporters appeared out of nowhere and began interviewing people on the streets. The Marines told us it happened every day at 4:00. While we were there, we got a radio message that told us our PSYOPs vehicle were ordered back to the Brigade again because they might be getting sent back home. We drove back to our sector by the way of the oceanfront and stopped when we saw Charlie's DMV on the beach. They had stopped to clear a loaded anti-aircraft gun and decided the easiest way to do it, since they didn't know how to clear it, was to fire it out over the water. While they were doing that, we detonated a Soviet RPG-18 that couldn't be collapsed. On the return to our shack, we saw a gas station and a grocery store that were open, both with lines of people waiting to buy things.

Chapter 3: SPECIAL FORCES – A Storm in the Desert

That night I made the decision that the next day we would head over to the ARCENT PX at Camp Freedom and wash our clothes so we would have a better appearance.[218] They were black from all the burning oil well soot drifting through the air. God only knew what it was doing to our lungs.[219]

At ARCENT there were free hamburgers and sodas, a PX, a post office, a finance office and a laundry. We dropped the laundry off for one day service, cashed some checks at finance for cash and then mailed off all sorts of Iraqi souvenirs through the Post Office. We claimed all we sent back on the forms since no restrictions had been placed on the items yet. There were showers, but they only ran from 0730 to 0930. We decided we would do showers when we came back for our laundry the next day. Faisal picked up seven cases of water for his sister since they had no drinking water in her house.

Camp Freedom sat smack in the middle of the burning oil wells. Flames were shooting hundreds of feet in the air from dozens of wells. There was black smoke everywhere, so much that it looked like night most of the time. After only spending a day there I was coughing up soot and I could not believe that other soldiers were there all the time. Even though it seemed dangerous, on the way out we did back up our vehicles to a burning oil well fire to get a team picture because we thought it looked pretty cool. On the return trip we drove through the burning oil wells. It was hot, hard to breathe and as dark as night. We called it the road through Hell.

That day we learned from another Team how the U.S. Embassy had been cleared. Initially the Embassy was occupied by 5th Group after the Marines had cleared it. The 3rd Group then showed up and told 5th Group that they had to leave, because they had the mission to clear the Embassy. Even though 3rd Group was told that it had already been cleared, 5th Group was ordered to leave anyway. The CSM of 5th Group took the US flag down off the flag pole and withdrew his Teams. With the reporters in place 3rd Group then rode into the

---

[218] Army Central Command Post Exchange.

[219] In 2011 several of us learned that the Army study on the oil well fires stated that we only had low exposure to the oil well smoke. We all laughed at this. Bobby Wholf wrote "My hair was jet black every day we were in the city (My hair is light brown). I also remember one day when we started patrolling early one morning before the sun was up a long day, and then as the sun was going down or so I thought, I looked at my watch and it was only 2 in the afternoon. It was getting dark because of the oil smoke, a very common occurrence." Mike Jacquard wrote "How many people who worked on this "study" got up every morning and coughed up strangely colored phlegm, poured gasoline on their windshield to clean the oil off enough to see or still have the grey colored chocolate chip uniforms we wore?"

Embassy on helicopters and fast roped onto the building. Their Group commander was along for this televised mission. They detonated several charges to clear rooms to make it sound more menacing than it was. It was pretty hilarious to watch. There were rumors after we got back to the United States that their Group commander was put in for a Silver Star for that, but I don't know if it was true.

That night I stopped off at the home of Faisal's sister's and dropped off the cases of water for her. Her husband invited us in and we had tea, bread and some sort of syrup to dip it in. I stopped by Brigade and picked up a new PSYOPs HUMV from the 296th PSYOPS Battalion (Reserve) from Texas. Our original PSYOPs HUMV was returned to us, minus one of their soldiers who had to return home because his wife was having a baby in the next ten days. Gator and K.C. drove to KKMC so that they could return the Air Force HUMV and begin packing up our team gear. By the time I returned back to our shack I was now in command of two Special Forces DMVs, two PSYOPS HUMVs and a total of nine Americans.

Sleep didn't last long. Faisal came into our shack at 2300 and told me that 3rd Company was under attack. We listened for awhile but all we could hear was a lone MP-5 firing into the night. I told Faisal that we weren't going to be going on any rescue missions that night and rolled over back to sleep, my right hand always on my pistol.

The next morning, we got up at 0700 and headed to ARCENT to pick up our laundry and take showers. When we got to the front gate the MPs wouldn't let us in because we didn't have any helmets or LCEs on. The rule for the camp was that everyone had to have a helmet and an LCE on at all times. We argued with the MP that this rule wasn't in effect when we were here the day before. I had the MPs escort us to the post commander, Colonel Fields. He wasn't thrilled about Special Forces and told us that we were all in the Army and had to abide by his rules. Unfortunately, this was a tricky problem because we had not worn helmets since we crossed into Kuwait, and some of our guys had them buried in the gear in the vehicles and couldn't find them. So, to solve this problem we rounded up some Iraqi helmets that were lying about the town and Crowder had a firefighter helmet that he found in an abandoned fire station. They were helmets, and we looked incredibly stupid, but they let us in.

The showers were extremely hot which was great. We had been washing up in cold water out of buckets for a few weeks and it felt like heaven. I sent one of our PSYOPs HUMV over to the FOB along with Baby Doc's DMV to get the front end aligned. It had been wobbling for days after hitting the bomb craters all over the road. We had banged them up pretty bad driving as fast as we could around town. Speed equaled security for us.

When we returned to Brigade, we learned that the US Army CID was trying to convict Special Forces soldiers for torturing POWs and committing atrocities. Colonel Shaw had been told there would be a court of inquiry and everything happening in Kuwait City was our responsibility. We thought this was a load of bullshit and Colonel Shaw agreed with us. We were ordered to stop any CID agents we saw in the city trying to talk to Kuwaitis and apprehend them. I wasn't sure if that was legal or not, but what the CID was trying to do was pin the atrocities on us.

I took my vehicle to the FOB to get it "verified" for shipment home, and then we returned to ARCENT and ate lunch at the "Wolf Burger" shack. Nothing like oil spattered burgers for lunch. While we were there, we learned that B-580 came under attack and the fight was still going on. We quickly drove over to their location in the eastern part of our sector of Sulaibakhat. When we arrived, everyone was taking cover from a sniper firing from a high-rise office building. The Kuwaitis were all lined up behind a concrete barrier in the median of the road and aiming at the high-rise building, that I later learned was the Water Works. The top of the building had been burned out during the fight. The Kuwaitis had deployed a platoon on the building after they were attacked and then took fire from different locations. When we arrived, they were sending out two companies, one going to each location where they thought the snipers were. There was constant firing from the Kuwaitis behind the wall, firing into the high rise. Tracers bounced off the walls and set rooms on fire.

William Jeffreys was there at the start, and he wrote "I, Bruce Menard, Eric Meyers and Dave Be__ were across the building when the firing began. We were checking out the motor work shop across the street that were dug underground.... We were there when they opened up with the .50 and when the Kuwaiti's went in. As I remember it, the entire thing took well over two hours from start to finish."[220]

Richard Van Winkle on ODA 581 wrote, "Those of us from ODA-581, Dave Lange, Grant Winkler, Rip VanWinkle, James Williams, Bob Arnzen, and one other, were getting ready to go to the phone center. When the ODA was notified to respond, we went since we were ready. I was on the .50 cal with Dave and Grant spotting. James was on the Mark 19. Bob and Odell were another of the sniper teams. Mike was also another sniper out there looking for the

---

[220]Email received from William Jeffreys on 6 August 2011.

shooter. We never saw the shooter but did see the clearing teams up on the roof and the whole thing was called off. There was a Major Mohammed that was telling us to shoot everything, for they would rebuild. I do remember having to pee at least three times while on the gun. I had just drank two big bottles of water before we left the ODA site."[221]

Grant Winkler added to the story "Bob Odell stayed behind with Dave Brace for commo to FOB. I was the FNG on 581 so I was just following everyone else.[222] I do know that the Kuwaitis wanted us to fire into the windows to chase the guy out, but cooler heads prevailed."[223] Mike Jacquard summed up the day "What a fun day! Never did catch that asshole."[224]

We didn't have a job in this fight, and I noticed some Kuwaitis that were aiming RPGs at the high-rise building. However, they were looking through the sights while their RPG rockets were still touching the concrete median wall in front of them. It was an accident waiting to happen. I told one of the other Special Forces guys there about the Kuwaitis aiming the RPGs, and then we drove back to Battalion to get out of there.

We were surprised by our newest bit of eye candy, a ZU-23-4 parked in front of our police post. The Kuwaitis thought it would impress the civilians and had pulled it there. I figured they had seen and heard enough of those guns that it wouldn't be that big of a deal to them. Gator returned from KKMC and told us to come to Brigade ASAP. Once there we learned we would be losing Baby Doc and one of our DMVs to Brigade. Other men from our unit would be returning to KKMC to start palletizing all our gear for the return trip home. The peace treaty between Iraq and the United States was supposed to be signed in seven days on March 20th, but if it wasn't the war would begin again and we would start the drive to Baghdad.

On March 14th we were supposed to pick up an Air Force EOD team to help us blow up our minefield on the Doha peninsula, but they thought they were going to blow up a munitions dump. They said they couldn't handle taking out a minefield with only their six men. We still picked them up and drove them out to see the minefield. It was spread across the whole tip of the peninsula and most of the mines were uncovered due to the shifting sands. While we were there CPT McDonald investigated the hospital/bunker complex nearby that we had discovered earlier. The bunker complex had been a signal center and had

---

[221] Email received from Rip Van Winkle on 5 August 2011.
[222] FNG = Fucking New Guy.
[223] Email received from Grant Winkler on 6 August 2011.
[224] Email from Mike Jacquard on 2 August 2011.

taken a direct hit by a 500-pound bomb during the air war. The signal tower had been toppled over and the building it was connected to had collapsed. The EOD Team decided to stay with us because they said they hadn't been able to find any good Iraqi souvenirs and figured if they were with us, they would have a better chance.

I had the team split into two groups and search for any additional mine fields. The EOD team followed us, looking into all the trenches for Iraqi gear. I came across one ZU-23-4 and cleared it by firing over the Gulf. I heard Charlie clear another one with his group. When Charlie fired the anti-aircraft gun, his team found an Iraqi soldier in the bunkers. I gathered my guys and quickly got over there to see the prisoner. The Iraqi had been in a bunker with an SPG-9 recoilless rifle, aimed at our vehicles. He had on civilian clothes and told us that he wasn't a soldier. He told us, in Arabic, that he was the only survivor of a group that "came from the sea". I had him tied up, blindfolded and then taken to Brigade. I figure he could skip two or three days of torture at the Battalion if we sent him right to Brigade. This was a good call because when he got to Brigade, they gave him an MRE, and then turned him over to our S-2.

On our return we saw a green HUMV cruising through an orchard. I stopped the HUMV and had to tell a regular Army SFC to leave our area. He told me that his captain had found two bottles of booze and hid it behind a tree in the orchard and he was looking for it. I chased him away, but then we all went back and looked for the liquor. We didn't find it, but we did find a flamethrower and fourteen gas canister backpacks. Nick strapped one on and then fired it down the street. It scared the hell out of me and I yelled at him to cut it out and don't fire anymore gas. We didn't know if these things were defective or not. We took the canisters to the Battalion and put it in their storage room.

While there the Kuwaiti Battalion commander invited me over to his S-2's house to drink some whiskey they had stashed there and invited me to have sex with some Iraqi women they had found. I lied and told him that I had to go to Brigade and couldn't make it. I figured all we needed now was to have a rape charge over our heads to go with everything else they were accusing us of.

On Friday, March 15th, we started the day by having Charlie check on the prisoners. All of the old prisoners had been moved out, but three new ones had come in. Afterwards I broke the team down into two elements. My group would go to the west and do a PSYOPs mission, while Charlie would go east with the other PSYOPs vehicle. Unfortunately, as they tried to get started their HUMV battery died. Crowder gave them a jump from our vehicle and then they drove over to POFID (the PSYOPs headquarters) to get it fixed.

Triple Canopy

I scrapped the PSYOPs mission and instead returned to the trenches where we had found the Iraqi prisoner the day before. I took some Kuwaitis with me, with their vehicle, so they could pick up the SPG-9 and get it off the beach. While searching the trenches I found some Iraqi uniforms in the bunker of the Marine Navy Defense force. This may have been what unit the Iraqi prisoner said had "come from the sea". We continued looking through the trenches, when I turned a corner and found two MT-12 howitzers! One of them had a fifty-pound night sight, which I took off and threw in my vehicle. As we continued to the abandoned Ad-Doha police post, we turned the corner and found a British APC. I stopped them and asked them what they were doing in my sector. They said they were "patrolling". I told them that they were full of crap and to get out of my area.

That afternoon I finally took our PSYOPs team out to do the mission I had planned for earlier that morning. We drove through the streets shouting out the day's message on the loudspeaker and handing out leaflets and some paper American flags. We always had about fifty kids chasing us everywhere we went, so I threw out packs of gum to them as we drove by.

While we were doing the PSYOPs mission our loudspeaker was drowned out when a low flying Chinook helicopter buzzed over us. It hovered over the beach and then landed. This was the beach that might have mines on it, so we raced over to warn them. I asked them what they were doing there and the crew chief told me that they had spotted an anti-aircraft gun on the beach and wanted to take one back to the States. I told him that he may have just landed in a minefield and he needed to leave…NOW! As they took off, I had my guys hook up the anti-aircraft gun the pilot had been interested in to our DMV, and towed it to the front of the Battalion's police post with the other gun, so it wouldn't draw any more attention of roving helicopters or other souvenir hunters.

Every few days one of us would rotate back to KKMC for "R+R." It was mainly because CSM Sims wanted us to get haircuts. None of us wanted to go because there was nothing going on back there. Rumors would return with each of our guys that came back. We were told that everyone on 3rd Group's embassy "clearing" mission was put in for a Silver Star. We were told that ARCENT headquarters was already awarding Bronze and Silver Stars to each officer as they deployed back home. I don't know if it was true or not, but it was amazing to us… you know, the men who had actually been behind the lines for weeks. When Gentz returned with the broken PSYOPs HUMV he tried to get my MT-12 night sight to work by hooking it up to a car battery. It worked for a few seconds, and then died again. Gentz and one of the Reserve soldiers decided to dissect it and figure out how to make it work. We called Gentz "Inspector Gadget" because he liked technological toys and tinkering with them.

512

Chapter 3: SPECIAL FORCES – A Storm in the Desert

The stress on whether or not we would go home was getting to everyone. We didn't know if we were going to return to war or return home. Whenever we went to Brigade Gator would chew us out for one reason or another. He chewed me out when he saw the night sight and asked me if all we did all day was loot trenches. On that very day we had walked five kilometers, on foot, looking for evidence of minefields and he thought we were just goofing off. It pissed us all off.

The next day, March 16th, I requested trucks from Brigade so that we could start loading up all the ammunition and explosives sitting on the side of 5th Ring Road. The trucks never showed up, so we went back out to our beach to continue marking minefields and unexploded munitions. We found ten RPG rounds and blew them in place. It was a huge explosion! Afterwards I had the team split up, one half to go get gas in their vehicles and I would go and pick up K.C. who had just returned from KKMC. Gator met me halfway and dropped off K.C. then we continued over to a Kuwaiti gas station to fill up the vehicles.

There was a free butane gas line for people to fill up their tanks so they could cook food. It was supposed to have been one butane can per family, but each family loaded the lines with family members, and they ran out of butane gas before every family got a bottle. We learned about this when K.C. and I helped an old woman carry the empty bottles back to her house.

Near our police station a new clinic had just opened up in the neighborhood. K.C. talked to the doctors and translated what they said for me. They told him that they needed generator power to cool the drugs and they also needed water. I reported to the Kuwaiti Battalion commander all that I had seen in this neighborhood, but I doubted if anything was going to be fixed.

That afternoon I went back to Brigade and dropped off Nickolai because it was his turn to go to R&R in KKMC. While we were there, Gator did another odd thing that just had us shaking our heads. He had us take all of our gear out of our DMV and move it into his DMV. He then switched our DMV for his. We didn't know why and he wouldn't let us know the reason. It took us most of the afternoon to switch all the gear, radios, ammunition and food, so we didn't get much else done that day.

While there we learned that some of the Kuwaitis of the 16th Battalion had kidnapped a woman the night before. This was the same battalion that executed an Iraqi prisoner by shooting him in the back five times, and they had also raped a girl that they said was an Iraqi. I was getting pretty disgusted with the whole thing, but at least we tried to keep our battalion doing what was right. To add

513

more insult to injury I had been saving a bottle of Johnny Walker Red that we found for St. Patrick's Day, but someone stole the bottle out of the DMV where I hid it.

On the morning of St. Patrick's Day, we drove over to ARCENT to take showers and I mailed my night sight home. It was 65 pounds and cost a lot to get home! Afterwards I went over to the phone center and called my parents. They wondered when I would be coming home and I told them I had no idea. The rumor was that we would be there for an additional four to six weeks, but no one knew for sure. After the calls we drove over to the Embassy so I could get our Kuwaiti Battalion commander a picture of George Bush. The Kuwaitis loved pictures of their leaders and most of them wanted a picture of Bush to hang up. The Embassy told us that they weren't giving out any more photos because it was becoming too expensive to produce them. The Marines were no longer guarding the Embassy anymore and instead it was guys from 10th Special Forces Group out of Germany. While we were there, kids ran around us screaming "Give me some Candy!"

St. Patrick's Day was also the first day of Ramadan. We were dreading this day, because we knew little to nothing would get done. During Ramadan the Muslims would not eat all day, but they would also do little to no work. They could only eat from sunset, or around 9:00 until sunrise or around 0400. Due to this the nights turned into giant food orgies.

About 7:00 that night we heard a party of a different sort coming from the second floor of the police post. We all dropped what we were doing and ran up there, seeing LT Josem, our Intel officer, beating a prisoner. LT Josem looked like the stereotype for a shifty, sleazy person, sort of like a greased rat with bad teeth. I told Josem to stop and he told us that it was none of our business and we needed to leave. I told him that he needed to stop and both of us needed to go see the Battalion Commander. He shook his head, told me "No" and then told me that he would go with me in ten minutes…after he finished beating the prisoner. I grabbed him and told him that in ten minutes all the Special Forces would be gone and we would not return. I had Gentz call Brigade, while I found the Battalion commander. The Battalion Commander was surprised that this was going on in his building and he told me to bring him LT Josem. I heard the Battalion commander chew him out, but afterwards he told me that there had been a directive from the American commander of Coalition forces that said they could beat prisoners, but only on the bottoms of their feet. I found that hard to believe, but I went along with the lie. We both agreed that the prisoner needed to go to Brigade now, before anything else happened. As he was talking Charlie burst into the room and told us to come with him, quickly!

514

Chapter 3: SPECIAL FORCES – A Storm in the Desert

LT Josem was dragging the prisoner by his neck, with a bag over his head, and was trying to sneak him out of the building. The Battalion commander took the prisoner into his office and told Josem that he would be the interrogator, not Josem. The prisoner had a gash in his head that had blood flowing from it. There were whip marks on his back, a bleeding hole in his shin, and his wrists were tied together and bleeding from the rough cord. We learned that the prisoner was a Palestinian who had an ID that had expired. This was his only offense, but I had not met any Muslims who liked the Palestinians, so there was no love lost. The Battalion commander told Josem that the prisoner would go to the hospital. I had Charlie and Baby Doc escort the prisoner to make sure no "accidents" happened on the way there.

After that long night we hoped the next day would be a bit less chaotic. I checked in on the clinic to see if they were able to get their generator fixed. They told me that the generator had been taken away to be fixed, but there was still no water. Our police post was supplied with water from a truck, so I took the doctor over to talk to the driver of the water truck. The driver told me that the clinic could not get any water unless the Battalion commander signed off on it. I figured it was no problem, so I went to find the Battalion commander, but I discovered the whole post was empty. Damn! It was Ramadan and everyone took off for the day. This amazed me that they would do this during a crisis situation and I had to send the doctor back with no water. I knew that if this was America people would be bringing the clinic water from their own sources if they needed it. However, this definitely was not America. I tried to get Faisal to go with us to find water, but he said he was just going to sleep all day since he was fasting for Ramadan.

The whole city had fallen apart and now all the Kuwaitis wanted to do was sleep during the day and then eat and drink tea all night. I called CPT McDonald and he gave us a new directive. Since no infrastructure was being built, we would now start focusing on the civil affairs side of the mission. There were too many reports of non-Kuwaitis, mainly the Palestinians, being screwed over so we had to make sure that the food and water would get dispersed equally. This would have some exciting consequences.

Since no one was awake or working, and we could not get anything done, we decided to drive out to Al-Mutla...the Killing Fields. We had reports that some more dead bodies had been sighted in the wreckage and we wanted to track this down. The fear of disease spreading through Kuwait City due to rotting bodies and the garbage in the streets motivated us to locate these corpses for burial. Though Al Mutla was just out of our sector I figured we would follow up on the report. The battlefield had changed a bit since we were there ten days earlier. A lot of the vehicles had been towed away and the highway had been

515

cleared a bit more. We knew there were still corpses in the wreckage because we could smell them a mile before we got to where they lay. We found the "new" bodies beside the road, hidden on the other side of a depression. All of the bodies were around a large crater and they were in pieces. They had all turned dark green from decay. They most likely died the night of the attack and had been there for over three weeks. The bodies were so badly mangled that I couldn't tell which side was the chest and which side was the back. The day was dark again from the oil wells and the smoke covered the sight of the slaughter. Ash, looking like snow, fell among the rotting corpses. We left the bodies after taking grid coordinates of where they lay.

When I did my circuit of the different company guard posts I discovered that they either weren't manned, or they only had one guard on duty due to Ramadan. As we zoomed down the highway, doing 60 mph or more, some of the Kuwaiti kids would jump in the road hoping that we would stop and give them candy. It was a bit unnerving trying to dodge little kids jumping in front of our speeding vehicle. Frustrated with the whole situation due to Ramadan I decided to take my guys to ARCENT and get a Wolf Burger for lunch. When we got there the place was covered with reporters and about thirty Congressmen. We tried to avoid them as much as possible. A few of us mailed some more boxes back to the States. The mail clerk must have seen quite a few Special Forces in the last few days because when we walked in, she said "Oh no, more SF renegades!" However, everything we sent back we wrote on the customs forms, figuring that if it was illegal, someone would stop it along the way.

After our "civilized" lunch I drove over to 1ˢᵗ Company to see what was happening there and only found three Kuwaiti soldiers on duty. They told me that their captain was not going to show up until around 8:00 or so due to Ramadan. They had their RPG cocked and the safety cap off. Annoyed at this lack of safety I picked it up and fired it off into the ocean. I told them to not take the safety cap off until it is time to fire. They told me that LT Josem was bragging about how he was going to make us "go away". We no longer trusted being near that slime ball, so we decided to sleep in the Brigade building from now on. I also did this because we lost our PSYOPs vehicles and we were now down to one DMV and six guys. There wasn't enough to pull security each night and get any decent sleep. We drove back to Battalion to tell the commander we wouldn't be sleeping there anymore, but, like every other place we had been to, there was no one there due to Ramadan. It was a ghost house.

When I arrived at Brigade, we were told that we would be going out to the different food points the next day and see that it was being distributed equally. The building that Brigade was housed in was a major difference than where we had been. There was electricity, which we didn't have in our old sector. I was ordered to go back to KKMC for R&R, but when they tried to start up the HUMV it wouldn't work. I felt relieved. I really didn't want to go back to KKMC. There was nothing there for me and I wanted to stay with my team.

The next morning, I took Baby Doc and K.C. out with me to check on the food distribution centers. K.C. was there to interpret because Faisal would not go out during the daytime due to Ramadan. The other DMV on the team went out to gather up ammunition and launchers that were scattered throughout the city. Gator decided to let our team train on all these different weapons we were stockpiling at Brigade. It was also a way to get rid of the ammunition in the area.

Our first stop for our patrol was the food and bottled gas point in Sulaibakhat. The Kuwaiti that was running the point told K.C. that they get 315 bottles every other day and to get a bottle you needed to have an old empty one. The only problem with this system was the Beduns, who never had any bottles to begin with. The Beduns were the desert people who had no country and just roamed the desert. The word "Bedouin", commonly used in movies, came from these nomadic people.

When K.C. talked to the manager who ran the food place, Mr. Salaam, we got the same story about the Beduns. He told us that he only gave food to Kuwaitis, and all non-Kuwaitis had to go to another place, where he was also the manager. He explained that each family got a monthly ration card. One neighborhood of Doha and Sulaibakhat were notified each day to come and get their ration of food and gas.

As the manager talked to K.C. one of the women pulled me aside and told me, in English, that the manager was holding back food and selling it to the highest bidder. I asked her where she had learned English and she told me at the University of North Carolina. I told K.C. to start interviewing the women in the line to find out the truth of what was going on. If you want to know what is really happening in an Islamic country, talk to the women.[225] The women, covered in red henna in their hair and feet, and tattoos on their faces, excitedly told K.C. that if you were a Bedun you never got anything, and the same was true if you were a non-Kuwaiti, such as a Palestinian. They were amazed that a

---

[225] In 2011, during the war in Afghanistan, the Army and Marines realized that the women in an Afghan village had a lot of information, but they were not allowed to talk to the male soldiers. Female Engagement Teams were created to go on patrols with the Special Forces soldiers so that they could interview the women in the villages.

man would ask their opinion. The only way these women, without official papers, could get food was by begging off of others.

I was disgusted with the corruption at the food point and decided to check on the water point that was nearby. At the water point trucks would drive up to a large hose and fill up their tanks and then take the water out to the different areas of the city. I asked the men at the water point where the other food point was located, and they told me it was in Sulayba, which was the 16th Battalion's sector. We drove over there to see if the Beduns were getting water or food any better there. When we arrived, we found long lines everywhere, with women sitting by the entrance. K.C. talked to the women and they surrounded him and said that they had sat there all day but the gas point had not been open in five days. They also told him that they sometimes got food, but not on a regular basis. They told him that they had never received any gas to cook their food with, so they had to eat it raw "like animals."

When we entered the food point, we found a Kuwaiti LT and two SGTs picking up food for his platoon. I told him what he was doing was wrong, and this food was for the civilians. He told me that he didn't care and his men needed more food for Ramadan. The manager of the food point told K.C. that each family was supposed to get a ration card, but the people who compiled the list had put a whole family name under one family. There may have been as many as four families under one family name, so they couldn't get a ration card. I had K.C. ask the manager if he knew that Mr. Salaam, from the other point, was sending him the non-Kuwaitis. He told us that he had heard of that, but there was nothing he could do about it.

We drove back to talk to Mr. Salaam, but when we returned, we discovered that he had left for the day after we interrogated him, and he had put his assistant in charge. The assistant manager told us that the non-Kuwaiti store was not in Sulayba, but only 1,000 meters away. He told us we had been sent to the wrong place on purpose to get rid of us. He then took us to the non-Kuwaiti food point where K.C. asked the women if there had been any problems. They told us no, but we weren't sure they were telling the truth and didn't know if they had been threatened to not say anything bad about Mr. Salaam. One Bedun man told K.C. that he hadn't been given any food at all, but the assistant interrupted him and told us that the Beduns were too stupid to know which day they had gotten food and the man was just confused. The assistant then told us that the government had just sent down a list of Beduns, so they would all get food the next day. I told him we would be back to make sure it happened, but I knew he wasn't going to do it.

Chapter 3: SPECIAL FORCES – A Storm in the Desert

As we drove over to Doha to check out the food points located there, we came across a HUMV with a US Army colonel driving by himself. I asked him what he was doing in my area. He told me that he was on official business from N.A.C. I told him he was full of shit and he needed to get out of my area now. He looked around at the group of annoyed and dirty looking SF guys and quickly drove away. I was pretty pissed off about the Beduns and Palestinians being screwed by the Kuwaitis and wasn't going to tolerate any bullshit from a looting colonel.

We drove through Doha slowly, randomly passing out American flags to kids and talking to people about the conditions in the neighborhood. Most of them told us that they didn't have any problems getting food, but they also told us that they thought Mr. Salaam wasn't fair because he gave some neighborhoods food and withheld it from others. Whenever he was confronted by hungry people, he would close his store. While we were in Doha, we did find the first store open for business. It was a small store, just selling candy bars and cigarettes, but it was something. The owner had brought up all of the goods in his truck from Saudi Arabia.

On the way back to the 16th Battalion's food point I stopped a group of Marines whose excuse was that they were lost and looking for Camp Freedom. I told them to get out of my neighborhood. They asked who I was and I told them "I'm the sheriff here". When we finally got to the 16th Battalion's food point, I was going to find the owner, Mr. Rashiidi, and tell him that I had been wrong about Mr. Salaam and that he did have a non-Kuwaiti food point over in Sulaibakhat. When I tried to go in a Kuwaiti LT stopped me and told us we couldn't talk to Rashiidi, and instead we had to talk to him. I was already pissed off due to the corruption and the looters and I told him to get out of my way and let me talk to Rashiidi. He told me we had to talk to his major first. I was pretty pissed off now and I got inches from his face and told him that I would talk to anyone I wanted to and there wasn't a damn thing he could do about it! As I looked around, I saw that the store was full of armed Kuwaiti soldiers, taking all of the food that was meant for the civilians. The situation became pretty tense and all of the Americans dropped our hands to our pistols, ready to shoot our way out if we needed to. The LT realized that we were about to kill him and his men and we were not going to go away, and he quickly sent one of his soldiers to go get Mr. Rashiidi. A "Mr. Rashiidi" showed up, but it wasn't the same one we had talked to earlier. I asked him why the army was coming into his store and getting food that should be going to the civilians. He lied, telling me that the soldiers were actually getting it for their families. We weren't going to get anywhere so we left out the door, only to bump into a crowd of Kuwaiti women.

Triple Canopy

The women had heard that an American soldier would actually listen to them and they knew we were the people to go to. They also saw that we were not afraid of the Kuwaiti LT and his soldiers, so they began to barrage us with stories of how their husbands were still in Iraq as prisoners and their children couldn't get any papers without their husband being there. Without those papers they couldn't get any food at all.

After listening to them we drove back to Brigade and I wrote up several incident reports that we sent to Civil Affairs and to 5th Group. While I wrote the reports, I listened to stories on AFN that told of Kuwaitis beating prisoners and the American army was accused of covering it up. The main culprits were soldiers from "the Green Berets at the police posts to the ARCENT commander". We weren't covering up anything and we kept copies of all pictures, interviews and reports we had done for our own records in case they tried to come after us. I would not let my team go down for something we never did. We all realized we needed to get out of Kuwait City soon or the wave of corruption would overtake us and wash us away.

On the morning of March 20th, we went over to ARCENT to take hot showers. The manager of the PX recognized us and talked with us for awhile. He told us that all redeployments back to the States had been cancelled because the Iraqis had failed to show up to sign the peace treaty that morning. He also told us that our Air Force shot down an Iraqi MIG trying to fly against the Kurdish rebels in the northern part of Iraq. We left ARCENT and drove over to the non-Kuwaiti food point, to make sure that the Beduns were getting food, like Mr. Salaam told us they would. The food was being passed out and all the people in line thought that "their Americans" were the ones who had made it happen. One of the Palestinians gave me a black checked Kafiyah and told me to wear it, and it would show other Palestinians that I was not against them. It would protect me if something happened.

Afterwards we drove out to the small fishing village of Ashish Ad-Doha to get rid of some of the ammunition we had collected over the last week. When we arrived at the village, we found a Scottish mechanized unit living in one of the warehouses. I was surprised because we knew nothing about it and they had moved in undetected. The officer in charge told me that General Jobber of the Kuwaitis had placed them there to guard the refinery area. In the warehouse they showed me a giant sand table that spread out all over the floor. The Iraqis had used that warehouse sand table map to plan their defenses of the beach.

That afternoon we fired some RPGs, SPG-9s and the AGS-17 automatic grenade launcher. I gave classes to the other guys on how to fire these things. When I taught them how the SPG-9 worked I aimed at a car on the beach 900

520

meters away and fired. It hit the car right in the hood on the first shot, and it burst into flames. Gator brought a safe he had found lying in the middle of the road in town, and we blew it open using Chinese det cord. Inside the safe were a handful of "Saddam dollars" but otherwise it was empty. There were cases and cases of grenades and we pulled the pin and tossed them into trenches and bunkers that were empty.

In one of the scarier moments of the day we almost killed ourselves by throwing grenades at the same time. There were dozens of cases of grenades in the ASP, so we all gathered in a mortar pit, would all grab a grenade out of a case sitting in the bottom of the pit, then we would pull the pin and all throw at the same time. After we tossed the grenades, we would all duck down until the grenades went off. On one of these shots, we all pulled the pins, threw the grenades, and then ducked down into the hole. To our surprise, spinning in the middle of the floor was a grenade that someone had dropped and it was seconds away from exploding. We had to make a quick decision... do we stay in the hole with the grenade, or do we dive out and take our chances with the half dozen outside that hadn't gone off yet. We all dove out of the hole and lay as low as possible. Grenades went off in front and behind us, the shock waves slamming together above us, but no one was hurt. We never did find out who dropped the live grenade, since no one would admit it, but we were lucky.

We finished the day by firing MAG-58 machine guns, FN-FAL rifles, SVD sniper rifles, and finally by firing RPGs and SPG rounds at shipwrecks out in the water. The only bad point in the day was when I got back to Brigade, I learned that Crowder had wrecked my DMV, number C-11, when a Kuwaiti teenager ran into Crowder with his Mercedes. We were down one more DMV. I ended up with my original vehicle, C-10, that Gator had taken the week before.

The next day I finally got Faisal to start going out with us again. K.C. was a great translator, but he didn't always understand the local dialect as well as he could. We drove over to the clinic in Sulaibakhat to see if electricity had turned back on yet, but the place was still dark. The medicine was beginning to spoil due to the heat. Baby Doc got up with the clinic's doctor and made a list of all the medicines that they needed. While he was doing that an incredibly beautiful Kuwaiti girl came up to me and started flirting. She didn't speak a lot of English but we were able to understand each other.

When Baby Doc came out of the darkened clinic we went over to the gas line, but we found out it was closed and hadn't been open for six days. I drove over to the water point, where the trucks filled up with water each day, and drove into total chaos.

Triple Canopy

There was no one in charge at the water point and it had denigrated to "first come, first served." Each truck inched forward into the packed mob trying to get there first. There were three water spouts that the trucks would pull up to and fill up with water. Civilians were gathered around the trucks while they filled up, catching whatever ran off the truck in any container they had. I found some Army Civil Affairs guys who looked dazed by all the chaos, and they told me what had happened. The reason it had degraded into chaos was because Kuwaiti soldiers had been cutting the line, saying that they had priority over everyone else. When I asked where the Kuwaiti soldiers were the crowd around me all pointed over to a group, but none of those Kuwaitis were from my Battalion. I walked over and asked one of the soldiers what unit he was from. He told me that he was from the 10th Battalion. I asked him if they had a water point over in the 10th Battalion area and he told me no. Faisal was pretty pissed at the soldiers and began to write down their names.

While we were trying to talk to the first group of soldiers, the civilians at one of the other points began yelling and screaming. One of the civilians screamed at Faisal that another group of soldiers were cutting the line again, for the third time. I figured this might become dangerous, so I called Charlie over for some "muscle".

The newest group of soldiers had cut the line and was holding up the civilian trucks by aiming their weapons at them. Their own truck began cutting in front of everyone to get water. I found the soldier in charge and told him that he can't be doing that; he was supposed to be protecting the civilians, not terrorizing them! I told Gentz to have the military truck move back and then told the civilian truck that had been in line to move up to the water point. One soldier came up to me and began yelling in my face. When I ignored him, he took a step back, yanked on the charging handle and loaded a round into his M-16. All of us there heard the bolt slam forward and it got really quiet.

The soldier aimed the rifle at my face, which was a real stupid move. In the movies the bad guys always point a gun at the hero, from just a few feet, and the hero always backs down or drops his gun. That is the movies, not real life. When I saw the gun aimed at me, I didn't even think, I just reacted. I grabbed the rifle barrel and then punched him in the throat several times until he finally let it go. Another soldier came up to me, with a G3 rifle and put it to my head. I grabbed that barrel with my free hand, while Charlie grabbed him around the neck and put his pistol to his head. I began yelling at Charlie, "Don't shoot! Don't shoot!" because there were people all around these guys, grabbing at them. I was afraid that if Charlie fired, the bullet would go through the soldier's head and hit one of the angry civilians. Gentz dived into the fight and both of them wrestled the Kuwaiti soldier down to the ground. I still had a rifle in each hand, and I didn't want to let the crowd have a loaded weapon, so I quickly disassembled the rifles, throwing each piece as far as I could out over the heads of the crowd. I took the G3 and smashed it on the ground, over and over, until it split in two. I was pretty pissed off so I backed away, to cool off. If I didn't, I figured someone would die. I told Gentz to move the civilian truck forward to the water and I told Faisal to tell everyone that if anyone tried to stop that truck, we would shoot them. The civilians were all cheering us and kicking the soldiers who were on the ground. A few took off their sandals and were smacking them in the face with them, the ultimate insult in the Muslim world.

An Army Civil Affairs captain showed up, holding the soldier who had initially pointed the rifle at me, and showed me that soldier's ID card. The soldier was beat up pretty badly and looked terrified. It turns out that they were from the Al Fatah Brigade, not even from our own brigade. The other soldier, who had the G3, was able to escape with their water truck. The soldier with the CA captain had pieces of his M16 and asked me if he would be able to get a new one. This pissed me off even more because that was all he was concerned about. I blew up at him, yelling at him in Arabic, "You are less than a dog! You are only a child! You are the shit on the bottom of my boots! You do not deserve a rifle! Leave or I will kill you!" I was amazed that I could get all that out without an interpreter, but it shook him up, and he left quickly.

I radioed CPT McDonald and told him of the water riot. CPT McDonald arrived a short time later and took statements from Charlie, Gentz, the CA CPT and me. The CA CPT told him that we were totally justified in our actions and we had averted a pretty dangerous situation. He said that he had been worried for his men at the water point until we showed up. CPT McDonald took Faisal and drove off to the Al Fatah Brigade to see if they could find the soldiers. While this was going on Gentz, "Inspector Gadget", fixed one of the water hoses and got another point started up. The civilians looked on us as if we were rock stars. We handed out American flags, many of them with our autographs and then left to go to the Scottish unit to eat lunch.

During lunch we traded some badges with the Scots and I learned that their unit were the "Queen's Own Highlanders". We promised to bring them back some Iraqi stuff the next time we came by. They had not been able to find any good souvenirs, and being a disciplined unit, had not gone off on their own to do some souvenir hunting. After lunch we drove over to the gas point in our sector and found that it was finally open. Our actions at the water point were quickly passed around on the street and many of those who were corrupt decided to change their ways.

Triple Canopy

CPT Hasty, my old LRSU company commander, SGM "Dink" Jordan and a Kuwaiti division MAJ met us at the gas point to hear what happened during the water riot. The Kuwaiti MAJ told us that the soldier who put the rifle in my face would be punished, and he might be executed for embarrassing the Emir. He also told us that from now on there would be a military guard on the water point to make sure there was no more incidents like that one. Using the notoriety to my advantage I let CPT Hasty know that our clinic had no generator or water and he said it would be fixed. "Dink" Jordan said we had done good and then asked me why I hadn't just shot the soldier when he pulled his gun on me. I told him that I didn't have time to pull out a gun and I just took him down with my hands. While we were there another vehicle drove in our area that had the Dutch Ambassador inside. He wanted to see me, "The Sheriff of Sulaibakhat" so he could get a pass to go see the Al Mutla highway. I told him how to get there and wrote him a "pass" that would let him get through the road blocks.

As if the day was not chaotic enough, while we were at the gas point, shooting erupted all over the neighborhood! Bullets cracked off the sides of buildings and clanged off of rooftops. We learned that the power had come on in the city for a few seconds and all the Kuwaitis began firing in celebration. It scared the hell out of us!

When I returned to the brigade COL Shaw wanted to see me. He told me that we did well at the water point, and that the Kuwaiti soldier would most likely be executed. The Emir wanted to know if I would like to be there when it happened, and I politely declined. COL Shaw told me that my report went to both the Emir of Kuwait and to General Fritz, ARCENT commander.[226]

On March 23rd I was finally ordered to go and take R&R at KKMC. Gentz would be with me, but I really didn't want to go. Once we left Kuwait City it became hot again because the sun was not covered by oil well smoke. While we were living in the city it was never really hot, because of all the oil clouds covering the sky and blotting out the sun. As we drove back, we noticed that most of the destroyed Iraqi vehicles we had seen when we drove into Kuwait had been removed from the sides of the road. I told Gentz take a detour to the Iraqi OP that had been across from Gator's OP 3 during the Air War. Gator called in an airstrike on the OP when the Iraqis had attacked Ruqi and I wanted to do some bomb damage assessment. There was a large bomb crater right in the middle of the OP building and their radio tower had been knocked down. I picked up a piece of shrapnel from the OP as a souvenir. I had stared at this OP

---

[226] Before we left we received ceremonial Wilkinson swords from the Emir for what we had done for his country.

for four months on the other side of the border and wanted something to remember it.

We stopped at the gas station where we had foraged for our diesel during the Air War. The owner, a Pakistani, recognized us and greeted us like long lost brothers. Two American MPs were there, amazed at this reunion between an old Pakistani and some dirty looking Americans. They asked if any action had happened around Ruqi. We laughed and told them some of what we had done during the Air War. They were amazed and said they hadn't thought anything happened until the Ground War. On the way back we noticed MPs everywhere! Every few miles there was a checkpoint to make sure that we had our helmets. After the ARCENT incident we had dug up our helmets and kept them in the back of the vehicle. The MPs told us the new motto was "Not one more life!" Evidently the regular soldiers had been having some serious problems with accidental discharges and soldiers being killed by accident. Even in 5th Group, our only casualty had been Leonard Russ, killed by accident. Listening to AFN we learned that 40 Americans had been killed and 120 wounded by duds lying around the desert.

Whenever we approached an MP checkpoint, we put our helmets on, and when we passed, we took them off. There were signs, in English, that said "no passing" so we just took the DMV into the desert and followed the highway. We were not used to waiting for anyone and the idea of getting behind a slow vehicle, where you cannot maneuver in a firefight, was alien to us. When we finally got to KKMC we were allowed take off all the combat gear and walk around without it on. I didn't do much in KKMC. I got my second hair cut in six months and mailed some more things from the post office. The postal clerks now looked inside any box to make sure that we were not mailing back anything forbidden. I never was able to find out what was forbidden until after I returned. To kill time, I wrote letters and tried to wash the oil out of my clothes. MAJ Star____ came by and told us that all of our NBC gear had to be turned in because there was no more chemical threat. I was only in KKMC for two days, and I was glad to get back to Kuwait.[227]

On the return trip to Kuwait City, we stopped in Ruqi and I called my dad and Sue Cawood. My dad told me that Schwarzkopf said he wanted two Special Forces battalions at his disposal, which is why we had not returned yet. Rumor

---

[227] After the war 5th Special Forces Group published a Desert Storm yearbook. Each team submitted pictures. The only picture we had with the whole team was one taken with MSG Cr___ in them. None of us wanted to have that picture in the yearbook, so they picked another picture, taken of the team when both Gentz and I were at KKMC. So we were not in the yearbook.

was that we may not get home until June. As we drove out of Ruqi we saw the remains of a Bradley Fighting Vehicle that had the front end melted off from the heat of a fire. We also saw several M1 Abrams tanks, on flatbeds, that had been destroyed. The only thing that can really destroy an M1 Abrams is another M1 Abrams, so they must have been killed by friendly fire.[228]

On March 25th the Kuwaiti police began to take over from the Kuwait Army, but there was a ten-day transition period for this to happen. I was still riding the circuit, checking on the food and water points and checking on my companies. On one of our rides a Kuwaiti gave Nickolai three beers and then told us of a house that an Iraqi general had lived in. We drove over to check it out and it turned out that the house had been the quarters of General Ali Hussein Majid, the commander of the Iraqi forces in Kuwait City. He later became known as "Chemical Ali" because he was the one who was in charge when the Kurds were gassed.[229]

We didn't find any papers in the house, but it wasn't trashed either. The owner was still living there and told us that his nephew had read most of Ali's papers and had given them to the resistance during the liberation of the city. I radioed Gator and told him to bring K.C. and CPT Mastrovito to question the owner. The owner told us of the assassination of Kuwaitis during the occupation and the resistance poisoning of 300 Iraqis. He also told us we were the first Americans to be inside the house. I took pictures of the place, not for any intelligence value, but just to show the luxury of the home.

When I went back to the gas and water point Kuwaiti guards told me that the new rule was that no soldiers could be in the line at all. While we were there some soldiers did try, but the Kuwaitis turned them away. When they heard that the Emir was going to execute the one soldier, they decided to behave. To make it fair there was a separate line just for soldiers at the water point.

I learned from one of the kids at the water point that a Rockeye was laying in someone's driveway. The kid took us there and we decided to detonate it. We were told the Rockeye bombs were extremely unstable and if you moved it from horizontal to vertical, it would explode. The Rockeyes we found were sitting on some cement blocks in the kid's driveway. I had to move them to detonate the bomb, or else we would destroy the kid's home. I wasn't really thrilled about this, but I had to do it. I dug up the Kevlar flak jacket and put my

---

[228] Due to all the friendly fire in this war, the Army went out of its way to invent different ways to tell friendly from foe on the battlefield which saved many lives during the Global War on Terror.

[229] Chemical Ali would be killed in the second Gulf War in April 2003.

Chapter 3: SPECIAL FORCES – A Storm in the Desert

Kevlar helmet in front of my face. Not exactly a bomb disposal suit, but it was all I had. I picked up the Rockeye by one of the tailfins, keeping it horizontal, but it scared the shit out of me while I was doing it. I carefully placed it down on the ground, waiting for a loud explosion and the metal fragments tearing through my body. We had learned a more efficient way to get rid of the bomblets, by shooting them, so I shot at it with my M16 until it exploded. They didn't make that big of an explosion after all, though I didn't know that when I picked it up.

Later I discovered a bank in my sector that was working. I walked in and asked them about their money situation. They told me that all Kuwaitis would be getting 1500 Dinar each month from the government. The food points would begin selling food at the end of the month instead of giving it away. The government was printing new Kuwaiti Dinar and the old Dinars could be traded in. I asked about the Beduns and they told me that if you were a Bedun or a non-Kuwaiti you would get no money. The bank manager told me that even though the Beduns say they do not have any money, they had plenty that they had stolen from Kuwaitis. He said they were just like Jews and had plenty of money hidden somewhere.

I checked on other banks, to see if any were open, but found none that were. On top of one bank was an anti-aircraft gun. Nickolai and I decided to break into the bank, to see if we could find any cash deposits. If we did, we were going to use them to feed the Beduns and Palestinians. We tried to pull open the door with the DMV wench, but it didn't work. We then tried to rip out the bars from the window, but almost broke the wench. As we sat there, contemplating using C-4 to blow the doors open, I spied a ladder. The Iraqis had broken the second-floor window, so we climbed up through it, cutting ourselves on the glass. I checked out the AA gun while Nickolai looked for any money or safes. The AA gun would never work again, but I wanted it off the roof. I figured the easiest way was over the side, so I pushed it and it fell three stories to the ground below with a mighty crash. Nickolai didn't find any money in the bank, so we returned to Brigade.

After a T-Rat chicken dinner CPT Mastrovito and I went out to check on the transition of the checkpoints from the military to the police. My police post where we had lived for several weeks was now all run by the police. Each of our companies had moved out and gone elsewhere, but we didn't know where.

One of the big problems we were having during this transition was Kuwaiti civilians shooting AK-47s into the air each night in celebration. The rounds would bounce off buildings and wound citizens. I decided that this would be my next personal mission and we would hunt them down. As we were driving

527

around the people would point at our DMV and tell their children about us. I had picked up a reputation in the neighborhood as the American who took on the corruption in the Kuwaiti Army and made things run again. People in the bank, in the food and gas lines and at the police stations would come up and thank me and give me gifts of Kuwaiti flags, prayer beads and such. A few even offered to their daughters to me for marriage. That was me, Marshall Pat... the Sheriff of Sulaibkhat.

For several weeks the Team had been gathering various bottles of liquor on our patrols. We didn't drink it, because it was too dangerous to be running around drunk in a war zone. However, on the night of March 25th we had a big party at Brigade. There were two bottles of VAT 69 from Iraq, three beers, four bottles of homemade wine and a bottle of gin. A few of the B-Team stood guard while the rest of us drank up the booze, listened to "Q" Balot play the harmonica, and I drew pictures of naked girls on the chalk board. It was a lot better R& R than taking a trip to KKMC. The next morning though I paid for it with a pounding head.

Though I felt like crap, work had to be done. I went to the new home of our 3rd Company and tried to get the commander to wake up. It was noon and he was still asleep. The day before the Company commander had asked me to come over and pick up a bunch of AKs in the houses. When I tried to wake him up a second time he finally got up, got in his personal car and drove away. The Kuwaitis did not want to work at all due to Ramadan and avoided it if at all possible. I figured if he is going to play that way, I would also, and left. Gentz and I drove over to a building that would be our new home for awhile. Along the way we found a group of Marines trying to jump start a T-55 so they could drive it away. I chased them out of the area. Near the beach was a sports stadium, and when the power came back on in that sector the lights stayed on. We decided to drive over there to turn the lights off. Along the way we found a bunch of Rockeyes strewn across the main four lane highway. Gentz rigged them to blow with C-4 while I stopped traffic. The French EOD stopped and watched the two of us work. When the bombs exploded, they gave a cheer and came over praising us for doing what would normally take them a squad of men. Before we returned to Brigade we discovered another ASP, but this one was mainly full of tank, artillery and mortar ammunition.

Gator had found a hotel restaurant in downtown Kuwait City that would cost each of us $25 for a dinner meal.[230] Most of the city had electricity and running water, but our sector still didn't have running water. Papa Doc fabricated our

---

[230] This would be about a $50 dinner in 2011. In 2011 it cost about $200 a night to stay at the Safir International Hotel.

own "sit down" toilet by ripping up sidewalk slabs, putting a wall around it and putting a chair there that had the seat missing. We had planned to go eat at that hotel, but CPT McDonald told us to wait until he returned from a meeting. When he finally returned, he told us that our Team would be going back to KKMC tomorrow, for good. We didn't go to the hotel restaurant, but instead spent the night unloading our ammunition and grenades and then packing up all our gear. I drove out to the Battalion to tell our Kuwaiti counterparts that we would be leaving, but I couldn't find anyone there. Crowder and I drove out to the 2nd Company to tell them we would be leaving the next day, and as we drove, I popped flares and threw smoke grenades out the back of the DMV. Unfortunately, it was all premature because when we got back to Brigade, we were told we wouldn't be leaving after all. We wouldn't leave until 3rd Group drove up from Saudi and we did a face-to-face hand off of duties. During our packing we found two bottles of gin and two bottles of wine, but no one wanted to drink. I didn't either since I still had my hangover from the night before.

The next day, March 27th, was not a good day. Everyone was pissed off at each other due to us being teased with going back. I spent the morning gathering up Iraqi weapons that we were going to have shipped back to the States to use for training. To do this legally we were going to have the Army Test Team ship it back for us. I drove out to our sector, trying to find the Kuwaiti Battalion commander, and let him know we would be going soon, but I never found him or any of his staff. When I returned, we learned that one of our Kuwaiti soldiers had shot and killed a civilian while he was in a car. Nickolai and I were sent out to investigate.

When we drove out to the 3rd Company no one knew anything about it. We stopped at another checkpoint, where there were some tents set up, but no one was there. It was abandoned due to Ramadan. We drove over to 1st Company, but no one knew about anyone being killed there. We drove to the police post and they knew nothing about a civilian being shot. I finally found the Battalion commander, but he avoided me and didn't want to talk. This really pissed me off and I didn't know if I was being stonewalled or if nothing had happened at all. When I reported to CPT McDonald, he also became angry and told me that I no longer had anything to do with that Battalion.

I felt relieved and we went to ARCENT to get some Wolf Burgers and mail some packages. We were ordered to go wash the DMV down to get it ready for transportation back to the States. A wash point had been set up on the Doha peninsula and Bravo Company had created a pump that would wash the vehicles with sea water. Unfortunately, it only worked when the tide came in. When we saw the situation, we decided that we didn't want to wash our DMVs with sea water, so we drove back.

Triple Canopy

Doha had become the 5th Group hang out. Guys would roam the trenches, scrounging for souvenirs, get a tan beside the mined beaches or blow-up munitions. We would blow up all sorts of explosives each day. One of the larger explosions we pulled off was when we packed a concrete pillbox full of tank ammo and then detonated it with some Chinese C-4 we had also found. It rocked the peninsula. We also fired Mark 19 ammo, shot 9mm and 5.56mm ammunition and shot RPGs at the bunkers. That night we finally went to the Kuwaiti International Hotel and had our $25 dinner. The meal was a buffet of Arabic and Mediterranean food. While we ate, we were still in our uniforms, carrying our pistols, but the waiters didn't blink an eye. Reporters staying in the hotel kept trying to ask us questions but we told them to fuck off, we were trying to have a peaceful dinner. Those same reporters would broadcast their stories from the roof of the hotel, never venturing too far from their rooms because they thought it was dangerous out there.

The next day 3rd Group arrived and we briefed them on what was in our sectors and what we had done so far. I drove one of them out to the Battalion sector, but they decided they would not attach anyone with the 57th Battalion due to all the corruption. I learned from the 3rd Group guys that the Silver Stars for the Embassy mission had been downgraded to Bronze Stars. At least someone was using their head. While I was doing this the rest of the Team washed the DMVs in regular, unsalted water, to prepare them for movement.

We had noticed that the Syrian troops would come out every few days and jump start an Iraqi tank, then drive it off. We really didn't trust the Syrians and figured we would eventually have to go to war with them, so we came up with a plan to destroy the remaining tanks in our area. We also wanted to get rid of some of the munitions that we had left, so we figured we would kill two birds with one stone. We picked an Iraqi T-55, parked far enough away from the city where it could safely detonated, and packed 500 pounds of Chinese C-4 inside. The C4 was placed on all the tank rounds inside the tank on the ammunition rack, and throughout the vehicle. It was so full of plastic explosives that we almost couldn't close the commander's hatch. For safety we decided that the tank would be detonated using the M122 remote detonator. We drove back a safe distance, about 500 meters, and parked the DMV behind another abandoned T-55 to shelter it from the blast. Some of our guys crawled underneath this "safe" T-55, while Nickolai went inside the tank and would watch through the tank's periscope. I wanted to get a picture, so I stood up behind the tank, ready to duck as soon as explosion went off to escape any flying shrapnel. I braced myself against the "safe" tank, knowing that it was about to be a hell of an explosion. Gentz counted down, 3... 2...1... NOW!

530

Chapter 3: SPECIAL FORCES – A Storm in the Desert

Telling you that it was a "huge explosion" just doesn't do it justice. The tank disintegrated in a giant flaming ball in a fraction of a second. I didn't have time to react as the shock wave hit me and blew me backwards, landing on my butt. I was stunned for a second, but I crawled as fast as I could underneath the "safe" tank. Everyone under the tank was laughing out loud and screaming at each other to shut up! The parts and pieces of the tank began raining down all around us, hitting the "safe" tank and rocking it back and forth. One giant metal rod landed only inches from the DMV and drove itself into the ground several feet. Inside the tank Nickolai realized that the hatches were open and he ran to the sides, hugging the walls, hoping nothing would fly in through the opening. After the red-hot steel quit raining around us, we crawled out to see what we had done. There was no more tank, only a smoking hole in the ground. The turret had been thrown 100 feet away from that smoking hole. There were only ball bearings where the tank had been parked and a few road wheels. We all gathered up the ball bearings for souvenirs. The explosion was heard throughout the city and everyone wondered if Iraq had gone back on the surrender treaty and had fired a missile at Kuwait City. We innocently denied any knowledge of what happened.

That night we had two "goat grab" dinners, one with our Brigade and one with the Kuwaiti Division. We could care less; we just wanted to go home. On the morning of March 29th, we took off in a convoy with twenty DMVs and headed back to Saudi Arabia. All along the way there were signs by the road with "Not one more life!" and "Wear your Kevlars" printed on them. The MPs were everywhere but we didn't stop for them, and instead just pulled off in the desert when we saw any checkpoints and went around them. They were pissed but there was nothing they could do about it. The road was full of other convoys so it took us seven hours to finally get back to KKMC.

When we finally got to KKMC two MPs would not let us in. They told us that KKMC was off limits to US Army personnel. SFC Girrard was in the lead truck and he was pissed. He told us to ignore the MPs and drive through anyway. When we started to move forward one of the MPs pulled his .45 pistol out of the holster and aimed it at our convoy. This was not a smart move. Instantly every weapon in the convoy was brought out and aimed at the lone MP. The MP looked like he was ready to piss on himself and realizing how incredibly stupid that move had been, slowly put the pistol back in his holster. We drove on to our barracks, ignoring the angry looks from the humiliated MPs.

As we were cleaning the DMVs and weapons, I learned I would be going back to the States on the "Love Boat". The Love Boat was the nickname for the cruise ship, *Cunard Princess*, which was parked in the Gulf and had been used for R&R. None of us ever saw it, though many soldiers in the Gulf had taken

531

R&R on it. When word came down that there were five slots, and only one for our company, I volunteered. I was to drive down to Dhahran with our DMVs then I would be staying on the Love Boat for the return trip.

On Easter Sunday we steam cleaned the DMVs for shipping. We disassembled most of the vehicles, removing all the padding and boxes so we could get at the hidden dirt. I packed all my gear away and turned in my weapons since I would not need them for the cruise home. I only packed one uniform, since the Love Boat didn't require anyone to be in uniform. For Easter dinner I made everyone a big Mexican dinner of chili, refried beans and tacos, sort of my specialty. I then tried to get some sleep because at 0100 we would be driving down to King Fahd International Airport to turn in the DMVs. At midnight I went down to Support Company with all the other drivers and got our briefing. I would be the commander of the convoy, but no one had a map and no one in the convoy had ever driven to KFIA before. I was given a hand drawn, not to scale, strip map of how to get there. To add more confusion none of the vehicles had any radios, since they were all packed away.

As I was about to leave Gator came up to my vehicle and told me that the cruise ship story was all bullshit. He hadn't known about it and he was pissed. It turned out that they needed one volunteer to escort the DMVs back on a regular ship and they used the Love Boat lie to get the volunteer. I was going to be on a naval transport for three weeks. There was nothing I could do about it now, so off the convoy went.

It took us 9½ hours to get to Dammam. I started with twenty vehicles, but I lost five for about two hours before I finally got them back on track. Amazingly all of us got there in one piece. The good bit of news I learned when we arrived was that I was not going to go back by naval transport because Gator raised a big stink about how they tricked me into volunteering. Someone else, who really volunteered, would be going in my place.

At KFIA we stayed in the "Bat Cave", which was the nickname for the tunnel underneath the airport leading to the parking garage. The SFOB had lived there during the war, but now there was only a skeleton crew left. The main group living down there was the 101st Division. Crowder, K.C., Be__ and Gentz had driven some of the other DMVs to KFIA with me and we went out that night getting burgers from Hardees in Dammam. We had brought Katzia and another puppy, Magua, with us to try and get them back to the States.

In the morning I put on my civilian clothes because it was the only clean ones I had left, and we took the puppies to the Saudi veterinary and agriculture

Ministry to get them the shots needed to go the US. Katzia would be going to the home of CPT Mastrovito since he was footing the bill. He told K.C. to try to get him out of country, but he also told him to make sure it wasn't too expensive. We were able to pick up a cage for cheap and then we went and then started negotiating with Saudi Cargo, Air France and KLM to get Katzia out for the cheapest price. K.C. decided to go with KLM because it was the cheapest, $518.[231] Gentz put the amount on his credit card, since the Captain only gave us $200. Initially Darren was going to send Katzia to his home in Tennessee, but he found out how expensive it was. We couldn't afford to send the other dog, Magua, to the United States. Magua was a puppy that Bravo Company had carried around with them during the war.

While we were figuring out what to do with Magua we met Bill Valbracht, an American who worked for ARAMCO (Arab-American Oil Company). He told us that he could put Magua in the ARAMCO kennel and someone would probably adopt the dog. He also invited us to his house for sandwiches and drinks. It was almost 4:00 in the afternoon and we hadn't had anything to eat all day due to Ramadan and due to all the stores being closed.

When we drove into the ARAMCO complex it was like driving into a suburb in Texas. There were no Arabs anywhere and all the houses were American style homes on half acre lots, all covered in grass! There were American cars on the streets and joggers running down the sidewalks. It was a total culture shock. We dropped off Magua at the kennel and I wrote up an elaborate and heroic war time history of what the dog did, then pinned it to the bulletin board.

Bill's wife was surprised when we drove the DMV into their driveway. Inside the home there were three kids playing Nintendo on the television. We stayed outside, in the backyard, watching Katzia and their dog. The whole scene was surreal and we figured we fit in better outside and not in a suburban kitchen. Katzia didn't know what grass was, so whenever he stepped on it, he looked confused. He also didn't know how to play with other dogs. It was a culture shock for him too.

Bill cooked up chicken on a grill while his wife made us fries, corn and rolls. No goat and rice at all! We washed it down with sweet iced tea! We were trying our hardest to not be barbarians but we must have looked out of place. We told them stories of what we did, as the C.L.O.W.N.S. during the Air War. They both sat there, fascinated by all we said. When we left, we gave them some Iraqi Dinars (Saddam dollars), some leaflets and I gave them an Iraqi beret badge.

---

[231] To do this in 2013 would cost over $2,000.

Triple Canopy

We left ARAMCO drove to the KLM hanger and turned over Katzia. By the time we finished it was 9:00 and we drove into Dammam to get some burgers from Hardees. There were American soldiers everywhere, with vendors speaking broken English trying to sell them tourist junk. When the vendors spoke to us, we told them, in Arabic, that we weren't tourists. It amazed them that there were Americans who spoke Arabic.

We returned to the Bat Cave and noticed that all the vehicles were gone. Be__ began freaking out and screaming that we had missed movement! He panicked and said we had to get to the port NOW! We rushed around until we realized that Be__ had overreacted again, like he did every time he thought there was enemy around during the Air War. We didn't need to go anywhere because only half of the vehicles had left. After a few hours sleep we took the remaining vehicles to the docks and washed them off one more time. Some of the Customs Inspectors were assholes and some were OK. We cleaned the vehicles again and again with fire hoses to please them, becoming soaked by the end of the day. My vehicle was finally inspected at 3:00 and passed on the first try. The female sergeant who inspected us had the longest beard I had ever seen on a soldier. She had three long hairs coming from her chin, which were at least six inches long. As she briefed us all I could do was stare at her chin hairs waving around like Indiana Jones's bullwhip.

After the inspection we drove all the vehicles to the dockside then waited a couple of hours for a bus to take us back to the Bat Cave. CPT Davis found us and initially was angry, wondering where we had been, why was I in civilian clothes and why did we leave the airfield. I told him about the whole Love Boat bullshit and that we were off the airfield trying to get CPT Mastrovito's puppy back to the States. He loosened up after that.

The next day we didn't do anything but sleep, get some sun on the roof of the airport and read books. It was a nice change from the rushed day we had yesterday. On April 5th we flew out of Dhahran on a C-130 and landed back at KKMC. Be__ had left the day before, so when we arrived both CPT Mastrovito and Gator were pissed at us from what he had told them. Be__ had lied and said that we had missed movement and that we were joy riding all day, when we were actually trying to get Katzia out of the country. I got angry and told them what really happened. Gator sided with me, but CPT Mastrovito was still pissed, mainly because he did not want to spend $500 on getting a dog back to the States.

Now we had to wait again, but this time it was for a plane back to the States. CPT Mastrovito told me that he had got a message that my dad might be there when I return, but I shouldn't rely on that. The wives back at FT Campbell had

534

been told many times that we were coming home, but each time it was just rumor. MAJ Star____ gave us a briefing and told us that 5th Group would only be getting two days off when we returned to the States. This angered us because the 101st and the 82nd Airborne were getting 30 days off. Our 3rd Battalion commander was going to try and get us 30 days off also, but there were no promises. We had been in the desert for seven months and he figured we deserved more than just two days off.

We finally left Saudi Arabia on Sunday, April 7th. We flew out of KKMC on a Hawaiian Airlines L-1011 and landed in Athens to refuel. Unfortunately, we weren't allowed to get off. We continued on to Ireland, where we were able to get off for 45 minutes. Almost everyone got an Irish beer and some souvenirs. A plane load of Russian civilians landed and they all gawked at us in our Desert camouflage. K.C. talked to a few of them, in Russian, and found one of the older guys had fought the Nazis in WWII.[232]

We continued on to Bangor, Maine, amazed that there was still snow on the ground there. As I walked down the concourse hallway into the terminal there were lines of people cheering, while a band played behind them. We were a bit shocked by this and didn't expect a crowd like this to welcome us home. We weren't the first to come home... heck, we weren't second or third, we were some of the last. We figured all the welcome committees would have done it so many times that they would be bored by now. After the initial welcome died down, we settled back into the lobby to wait for the plane to refuel. When we learned that our airliner had broken somehow, we started drinking beers. Some of the guys got really drunk, since it was the first beer they had in seven months. I went outside and had a snowball fight with a few other guys. The thing that was the most shocking was the green trees, plants and grass. The darker greens actually hurt my eyes, since it had been used to white sand and blinding sunlight for all that time.

The plane didn't look like it was going to be fixed anytime soon, so they loaded us all on a bus and took us to a hotel. We hung out in the lounge, ate a free dinner and talked to the barflies. We never spent the night at the hotel because Hawaiian airlines brought in a second plane and we loaded up around 9:00.

We finally landed at Fort Campbell around midnight and marched to a lit-up hanger. Inside wives and girlfriends lined the sides of the building. A band began to play patriotic music as we entered. As if on cue the women all rushed

---

[232] After I left 5th Group I lost track of K.C. He eventually got out of the Army, got married and is now a chef in Kansas City.

forward, hugging and kissing their soldiers. I moved off to the side, avoiding the crowd. I didn't have anyone to greet me and felt a bit awkward. I just wanted to get back to my apartment and see what still remained of it.

We took trucks to the barracks and then turned in our weapons. Once each Team had their items secure, we were allowed to go home. I picked up my pickup truck from the long-term storage and drove to my dark apartment. The electricity and phone had both been turned off, like I knew it would. I sat in the dark, contemplating what I had just gone through. We were given two days off. I used it to start my life again.

Parades

After Desert Storm I led a wilder life than before. I started seeing all the different women who wrote to me while I had been in the desert. It seemed I had a new "serious" girlfriend each week. For some reason I ended up attracting married women more than others. Though they were married, they were younger than me. I usually found out they were married after the third or fourth date. I became really serious with one girl I met at a reenactment, named Kim, who lived in Milwaukee. At one time she was set to leave her doctor husband and move in with me in Fort Campbell, but she finally realized this was not a smart move.

There were parties put on by my friends every month welcoming me home and there were several parades that we marched in. The biggest were the large parades in Washington, D.C. and New York City. There were fifty of us from 5th Group who went, mainly consisting of the single soldiers. We rode a Greyhound bus to D.C., and only took rubber weapons so no one would have to pull guard and miss out on any fun. We stayed in the dorm rooms at Georgetown University for several days. When we went into the bars, we didn't have to buy any beers once they found out who we were. On the first night we took over a bar in Georgetown called "The Tombs" and it became the Special Forces hang out.[233] The Group commander sat with us one night and bought us all beers.

When it came time to march in the parade, we were the first unit in the column, directly behind General Schwarzkopf. While we waited, we mocked the 3rd Group support guys and gave no one any mercy. One CSM came over and told us to tone it down and we all laughed at him. Whenever any reporter asked us what we thought of the war, we all replied with the same exact saying,

---

[233] The Tombs restaurant was still operating when this book was written and its most interesting feature is a long steel staircase on the outside of the bar that leads to the Potomac River. The staircase was used in the final scenes of the move *The Exorcist.*

Chapter 3: SPECIAL FORCES – A Storm in the Desert

"Desert Storm was a complete success!" One of our guys, Russell, tried to do a serious interview, but we all crowded behind him and hummed "The Ballad of the Green Berets". The reporter thought we were all assholes.

During the parade most units were being somber and dignified, but we were having fun with it and would do the "wave" every few minutes. There were millions of people lining the streets as we marched down the avenue. Giant yellow ribbons adorned all the trees on the parade route. At one-point Schwarzkopf stopped the parade and strode over to the main viewing booth, where he shook hands with President Bush, got up into the stand with him, and then ordered the parade to continue. When we continued the march, it seemed like a hundred aircraft did flyovers, to include the stealth fighter. It was the first time I had seen one. As the jets screamed overhead there were fireworks being shot up at them, looking like anti-aircraft explosions. As we watched this, we lost our smiles and had to remember it was only a parade. We ended the parade in Arlington cemetery, where they gave us free cokes. After all the cheering and shouting on the parade route, Arlington sat silent, letting us know the reality of war.

When we got bussed back to D.C., we were let out beside a group of war protesters. It wasn't many, only about a hundred or so, but we thought it was funny as hell. A lot of the soldiers avoided the protesters, but we didn't. We went over to them, put our arms around them and all started singing "Give Peace a Chance" as loud as we could. It pissed off the protesters and they left to find someone who was more argumentative.

We had lunch behind the White House, in the Ellipse. There was great food, but no beer, so I left to buy some beer, and then stretch out beside the Washington monument. I took off my shirt, rolled up my desert camouflage pants and laid there getting some sun, drinking my slowly warming beer. I dozed off and woke up when a shadow passed over me. My right hand had been on my pistol, or where my pistol would have been during the war. Two beautiful girls were sitting on either side of me and smiling. Their names were Chris and Deb. I offered them my warm beer, and then offered to take them to the big concert at the bottom of the hill that was for servicemen only. I told the "bouncer" at the entrance that these two girls were my sisters and they let us in. The rest of the night with them was a blur but it was a great time. I finally got back to the dorm the next morning, right before the buses left for New York.

For the New York parade we stayed at Fort Dix, New Jersey for a couple of days. On the day of the parade, we took the Staten Island Ferry over to NYC. Blackhawk helicopters escorted us the entire way, hovering a few feet above the water. We lined up at Battery Park and waited for the march to begin. As we waited all sorts of celebrities came down and talked to us. Brooke Shields and Kevin Dobson showed up, but we wanted to hang around with Christina Applegate the most. She played up being the dumb blonde character she portrayed in the TV series "Married with Children" and kept calling the war "Dessert" Storm.

As we marched down Broadway Schwarzkopf and Colin Powell rode in a convertible in front of us. Ticker tape came down so thick that I couldn't see the guy a few feet in front of me. It was like walking in the snow. Millions of people lined the roads and cheered. While we marched, we did the "wave" and tried to look like steely-eyed killers. People cheered us on, yelling out "USA! USA! USA!" and "GREEN BERETS! GREEN BERETS!"

When the parade ended, we got on buses and were taken to the USS Intrepid aircraft carrier and museum. The parade organizers had filled the aircraft carrier with food, bands, free beer and a heck of a lot of pretty girls. They all told us they were models or actresses. The bad part was that there were around 30,000 soldiers and about 200 women. The odds didn't look good, so I grabbed a couple of guys and we headed downtown.

We stopped in the middle of Times Square, not knowing where to go. We were looking for a bar, so when I saw a gorgeous looking girl walking across the street, I stopped her and asked her if she knew where a good bar was. She initially thought we were the typical nutty people in the streets, but I told her that we were part of the parade that morning. Her name was Donna Cipallone and she decided she would show us the town. She had long curly blonde hair, a short mini-skirt and lots of cleavage! My kind of girl!

For the rest of the day, and night, Donna and I went everywhere. I lost the other two guys somewhere along the way after they found their own girls. Donna and I went to bars down by the Brooklyn Bridge that was full of bagpipe bands and cheering crowds. She had lived in New York City all her life and thought all men were either psychos or gay. When she heard me talk, she loved how Southern it sounded.[234] She told me I seemed like a "real" person and not a fake.

Sometime during that night fireworks went off by the Brooklyn Bridge. We both kissed while thousands of people around us sang "God Bless the USA". Unfortunately, the night ended too soon, with me almost not making it back to the *Intrepid.* I put Donna in a cab and told her goodbye. I said I would try to

---

[234] To this day I don't think I have a Southern accent unless I drink a lot.

make it back to New York City, and I meant it, but I never saw her again. I felt like I had just lived the famous picture of the sailor kissing the nurse in Times Square at the end of WWII.

Once we got back to Fort Campbell my Team was all awarded Bronze Stars for what we had done during the war. The only one on our Team who did not get a Bronze Star was Nickolai. MAJ Star____ downgraded his medal to an ARCOM due to Nick getting a DUI before the war ever began.[235] We were all angry when we learned about it, and the Team refused to accept the medals. Finally, CPT Mastrovito convinced MAJ Star____ to resubmit the medal as a Bronze Star.[236]

Weeks after we got back Papa Doc was told to keep an eye on all of us, to see if any of us had post traumatic stress disorder. I thought it was bull, and I felt great, but there were some indications that other guys were taking it hard.

Towards the end of June, the Team went down to Texas to attend the Special Forces Association Convention. We were there to display a table of equipment and answer questions about what we had done in the war. I ran the Desert Storm Memorabilia table. Each night we didn't have to buy drinks at all due to all the old Special Forces Vietnam vets attending the convention.

During one of my breaks, I met one of the girls manning a souvenir table, a dark-haired beauty named Lynn. I asked her out and she ended up bringing along several of her friends, who spent the weekend with us.[237] Unfortunately Gator ended up getting in a fight with Charlie after he had been drinking. Charlie was a big mountain of a man and he ended up punching Gator out.[238] We all blamed the beer and figured it would be ignored. Long before Las Vegas came up with their famous slogan, the Special Forces had their own slogan... what happens on TDY, stays on TDY.[239]

---

[235] The Bronze Star certificate on the next page is not the actual wording on the award. The actual award had a pretty generic description. The certificate on the next page was the description that was used to justify the award of the Bronze Star for our A-Team, the "award recommendation".
[236] Shortly after this Nickolai left the army and started work as a private investigator. He then went to a stock market advising company for day traders
[237] Lynn and her girlfriends had all been extras on the set of Oliver Stone's movie "JFK" and were hired for the convention by the old Special Forces vets due to their looks.
[238] Charlie ended up being a 3rd Special Forces Group with me in 1999. He stayed in after 9-11 and deployed with the first units in Afghanistan. He finally retired as a CSM in 2009.
[239] TDY = Temporary Duty.

About a week after we got back to Fort Campbell Gator had a few beers and then went to the home of Papa Doc. Gator ended up fighting him in front of Doc's wife and kids. This was one fight too many and Gator was relieved as the Team Sergeant and moved to the Battalion S-3 shop.[240]

THE BRONZE STAR
TO SFC PATRICK J. O'KELLEY

FOR EXCEPTIONALLY MERITORIOUS SERVICE FROM 17 JANUARY 1991 TO 20 MARCH 1991 DURING OPERATION DESERT STORM AS A MEMBER OF ODA 592, 3RD BATTALION, 5TH SPECIAL FORCES GROUP (AIRBORNE) SFC O'KELLEY SERVED AS AN ASSISTANT TEAM LEADER FOR A SPECIAL RECONNAISANCE MISSION ON THE SAUDI- IRAQI BORDER. DURING THIS TIME HE REPORTED CRITICAL INFORMATION ON ENEMY ACTIVITIES, INCLUDING 11 CONFIRMED SCUD- B MISSILE LAUNCHES, MULTIPLE MLRS AND ARTILLERY BARRAGES AGAINST COALITION FORCES, AND ASSISTED IN THE INTELLIGENCE PROCESSING OF 5 ENEMY PRISONERS OF WAR. SFC O'KELLEY ALSO CONDUCTED CLOSE AIR SUPPORT AGAINST ENEMY POSITIONS WHILE OFTEN UNDER ENEMY DIRECT FIRES. HE ALSO PARTICIPATED IN AN EIGHT HOUR RESCUE/ RECON MISSION IN AN ADJOINING SECTOR AFTER A SPECIAL FORCES TEAM WAS REPORTED MISSING AND THEIR OBSERVATION POST OCCUPIED BY ENEMY FORCES. HIS CALM BEHAVIOR AND SOUND ADVICE LED TO THE SUCCESSFUL COMPLETION OF A DIFFICULT NIGHT MISSION IN A HOSTILE AND UNFAMILIAR AREA. SFC O'KELLEY FURTHER DISTINGUISHED HIMSELF WHILE SERVING AS THE SENIOR ADVISOR TO A KUWAITI MOTORIZED INFANTRY BATTALION. HIS ADVISORY ROLE EXTENDED FROM INITIAL ORGANIZATION AND TRAINING TO COMBAT OPERATIONS INSTRUMENTAL IN THE LIBERATION AND CLEARING OF KUWAIT CITY. SFC O'KELLEY'S TACTICAL EXPERTISE AND TIRELESS DEDICATION TO MISSION ACCOMPLISHMENT ENSURED THE KUWAITI BATTALION SUCCESSFULLY PERFORMED THEIR MISSION. SFC O'KELLEY'S EXEMPLARY ACTIONS REFLECT GREAT CREDIT UPON HIMSELF, THE 5TH SPECIAL FORCES GROUP, AND THE UNITED STATES ARMY.

Company B, 3rd Bn, 5th SFG (Abn)

Shortly after the convention in Texas the 3rd Battalion XO pretty much begged the Battalion commander to put an 18B Weapons Sergeant in the Arms Room. The officers in Battalion thought for sure they were all going to jail because the Arms Room was messed up, no one knew where anything was and there was no accountability. I was chosen to do the job for six months, with the promise that I could go to any team or school when I finished. While I ran the arms room, I got everything back into working order, but I also took home different weapons, such as one of the silenced pistols. If I had been caught, I would be looking at a court martial, just like Gator was, but I didn't care.

I was over 30 and had never been able to keep a relationship with a woman longer than a few weeks. When I got too lonely, I would drink a bottle of liquor

---

[240] Gator ended up being court martialed and discharged from the Army. After he got out he ended up doing contract work with one other 5th Group soldier.

and go out shooting at streetlights or mailboxes with the suppressed pistol. I was never caught, but it was pretty stupid. I think my way of dealing with the war, with being all alone, was not very good and this may have been a form of PTSD. I continued to pursue women, but I was getting nowhere. They were all temporary and I really didn't want any long-term commitments due to past failures.

Like most things in life, you don't really know when something amazing is going to happen in your life and you are never ready for it. In October there was a huge reenactment in Ninety-Six, South Carolina. I initially wasn't going to go to the reenactment, but at the last second, I was able to get a pass. When I got to Ninety-Six, I had a great time, linking up with Bert Puckett and other friends, but I also met the sister of one of the Kentucky guys, Jeff Fraley.[241] We really didn't hit it off, but I continued to hang around her. At the camp fire that night I told her that I thought all women were evil. She told me she thought all men are jerks.

Her name was Alice.

Life is full of coincidences. If I had not accepted the offer of a friend, Brian Jeznach, to go to a nearby reenactment back in 1980, I would have not gotten into this odd hobby of mine. If I had not had a series of bad relationships with several girlfriends, I would not have gone to a reenactment in the small town of Ninety-Six, South Carolina in 1991. If I had not decided to make a movie, I would not have had to have a second person to hold the camera, that person was Jeff Fraley's sister.[242] She just happened to go to her first reenactment that weekend, with her brother. I was stuck in Fort Campbell during Christmas, and I decided to go to a party put on by the Kentucky reenactors that I knew. Alice just happened to be there.

We were married two years later and are still married to this day.[243] She gave me meaning to my chaotic life, and she also gave me three beautiful

---

[241] Bert Puckett ended up in the 82nd Airborne Division, doing one tour in Iraq and three tours in Afghanistan before he finally retired as a CSM in 2010.

[242] The movies was called *"Do you Sleep in those Tents"* and was for and about reenactors.

[243] For our 20th wedding anniversary, I took Alice to France for a week, and then to Grenada for the 30th Anniversary of the invasion.

daughters, Cailin, Katriane and Adelise. I literally thank God every night that she is in my life.[244]

The first Gulf War ended not with a crushing defeat of an enemy, but with a cease fire, leaving a large army in place that continued to defy the UN treaty and waste millions of US taxpayer's dollars every year. Untold thousands of Iraqis lost their lives due to the conditions under the emboldened Saddam Hussein. After twelve years of this the United States went to war with Iraq again. That second war became known as the Iraq War, though some call it the Second Gulf War. Over 4,000 American soldiers were killed and over 32,000 were wounded. The war seemed to finally end for the United States in 2011 when the last major US forces returned home, but the coalition forces pulled out too quickly, under President Obama, and ISIS rose up in the vacuum. Finally, under President Trump ISIS was destroyed in Iraq and Syria and Iraq finally has some sort of peace. [245]

When I finished writing this book the first time, in 2014, the United States was still involved in Iraq and the "Forever War" of Afghanistan. I was too old to go to war anymore, but I sent my Junior ROTC cadets off each year. The "Forever War" went on, with no end in sight, but amazingly President Trump put a plan in place to get us out of Afghanistan. When President Biden took over the job in the Oval Office, he decided to do his own plan, which turned out to be a disastrous withdrawal, that was compared to how the United States left

---

[244] My father, James O'Kelley, passed away when Hurricane Florence was shutting off all transportation to South Carolina. He had been living his last years in his life, battling dementia, in a Veteran's Home. Due to this I was not able to be there when he died, though I was able to see him one final time a few weeks before he passed away. He was cremated and his ashes given to me for a later memorial. In the Spring of 2019, I had an official memorial, but I wanted it to be epic and not just the usual. So, for this I bought a mannequin, and wrapped it up in kerosene-soaked rags. I put a pair of highly spit-shined jump boots on the mannequin. I then built a giant funeral pyre on my 15-acre property, where I was going to place his ashes and headstone. On the day of the memorial 7th Special Forces Group sent out an honor guard, and there was a bugler, bagpiper and representatives from the military. After doing the "normal" service, which included the 21-gun salute, and folding of the flag, Bert Puckett and myself both lit torches, and then set the funeral pyre on fire, as the bagpiper played traditional dirges in the dark. It was epic and the ceremony can be seen on YouTube by searching for "*James O'Kelley Warrior Pyre*".
[245] ISIS stands for Islamic State in Iraq and Ash-Sham, which is Syria, Cyprus, Turkey, Iraq, Syria, Lebanon, Jordan, Palestine and Israel, also known as ISIL, which stands for Islamic State of Iraq and the Levant (the Eastern Mediterranean states).

Vietnam.   But now, as of the re-release of this book in 2021, we are no longer involved in any wars in the Middle East.

Maybe, just maybe, our children and grandchildren will not have to fight there again, but as Plato said… "Only the dead have seen the end of war".

Patrick O'Kelley
Barbecue Township, North Carolina
October 2021

Triple Canopy